U0417761

Linux

系统架构与运维实战

明哲 著

電子工業出版社
Publishing House of Electronics Industry
北京•BEIJING

内 容 简 介

本书第 1 章主要讲解 Linux 的基础运维，可以使读者快速了解和掌握日常运维的技巧；第 2 章～第 7 章讲解 Web 网站的架构模式和在大型网站架构中实现集群的高可用与负载均衡的方法，线上的项目发生故障时可以借助监控快速定位、排错和解决问题；第 8 章讲解 NoSQL 非关系型数据库，它具有更高的写入负载，可以采集数据进行大量的写入，如果数据查询具有更高的读取速度且有确定位置，则会查得更快；第 9 章～第 13 章讲解 Jenkins 持续化集成、Docker 的安装和应用，以及使用 Kubernetes 容器编排工具进行容器的管理和调度。

无论你是运维人员，还是开发人员，学习本书都会有所收获。

图书在版编目（CIP）数据

Linux 系统架构与运维实战 / 明哲著. —北京：电子工业出版社，2019.1
ISBN 978-7-121-32533-5

Ⅰ. ①L… Ⅱ. ①明… Ⅲ. ①Linux 操作系统 Ⅳ.①TP316.85

中国版本图书馆 CIP 数据核字（2017）第 202717 号

责任编辑：黄爱萍
印　　刷：三河市双峰印刷装订有限公司
装　　订：三河市双峰印刷装订有限公司
出版发行：电子工业出版社
　　　　　北京市海淀区万寿路 173 信箱　邮编：100036
开　　本：787×1092　1/16　印张：26　字数：540 千字
版　　次：2019 年 1 月第 1 版
印　　次：2019 年 1 月第 1 次印刷
定　　价：99.00 元

凡所购买电子工业出版社图书有缺损问题，请向购买书店调换。若书店售缺，请与本社发行部联系，联系及邮购电话：（010）88254888，88258888。

质量投诉请发邮件至 zlts@phei.com.cn，盗版侵权举报请发邮件至 dbqq@phei.com.cn。

本书咨询联系方式：（010）51260888-819，faq@phei.com.cn。

前　　言

20 世纪 90 年代初，Linux 操作系统诞生，随着虚拟化、云计算、大数据、容器技术的出现和人工智能时代的来临，Linux 以迅雷不及掩耳之势飞速发展，占据着整个服务器行业的半壁江山，但同时也面临着巨大的挑战。当今互联网企业的需求多种多样、业务复杂且难度大，这都需要使用合理的管理模式来保证 Linux 服务器的安全、稳定和高可用性。

虚拟化一般分为硬件级虚拟化（Hardware-Level-Virtualization）和操作系统级虚拟化（OS-Level-Virtualization）。硬件级虚拟化是运行在硬件上的虚拟化技术，其管理软件是 Hypervisor 或 Virtual Machine Monitor，需要模拟一个完整的操作系统，也就是通常所说的基于 Hyper-V 的虚拟化技术，VMWare、Xen、VirtualBox、亚马逊 AWS 和阿里云用的都是这种技术。操作系统级虚拟化是运行在操作系统上的，模拟的是运行在操作系统上的多个不同的进程，并将其封装在一个密闭的容器里，也称为容器化技术。Docker 正是容器虚拟化中目前较流行的一种实现。

我们知道，销售传统的服务器或计算机主机基本上都是一锤子买卖，商家销售出去之后基本就很难再从消费者身上获得其他收入。随着云概念的出现，越来越多的商家意识到卖硬件是不可能获得长期利润的，只有服务才能持续盈利。因此，在 2010 年左右，出现了大批提供云服务的公司，大体可以归为下面几种类型。

- 基础设施即服务（Infrastructure as a Service，IaaS），通常指在云端为用户提供基础设施，如虚拟机、服务器、存储、负载均衡、网络等。亚马逊的 AWS 就是这个领域的佼佼者，在国内则以阿里云为首。
- 平台即服务（Platform as a Service，PaaS），通常指在云端为用户提供可执行环境、数据库、网站服务器、开发工具等。国外的 OpenShift、Red Hat、Cloudera Cloud Foundry、Google App Engine 都是这个领域的佼佼者，当然还有一个非常有名的公司，那就是 dotCloud。
- 软件即服务（Software as a Service，SaaS），通常指在云端为用户提供软件，如 CRM 系统、邮件系统、在线协作、在线办公等。国内的有道、麦客、Tower 都是这个领域的产品。

一般认为以上三种类型是最基本的云服务模式，其分层结构如图 1 所示。

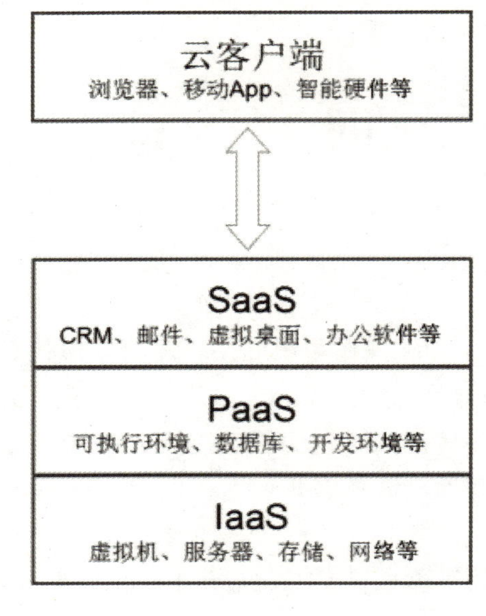

图 1

目前市场上 Linux 相关图书众多，但是普遍带有局限性，要么只有理论和基础知识，要么侧重于介绍软件的安装，大多偏离了企业生产场景。

本书以 RHEL/CentOS 7 为背景，从基础知识讲起，逐步深入，结合大量的实际应用案例，使读者迅速掌握 Linux 运维技术的经验和各种使用技巧，从而达到提升技术能力的效果。

轻松注册成为博文视点社区用户（www.broadview.com.cn），扫码直达本书页面。

- **提交勘误**：您对书中内容的修改意见可在 提交勘误 处提交，若被采纳，将获赠博文视点社区积分（在您购买电子书时，积分可用来抵扣相应金额）。
- **交流互动**：在页面下方 读者评论 处留下您的疑问或观点，与我们和其他读者一同学习交流。

页面入口：*http://www.broadview.com.cn/32533*

目　　录

1

第 1 章

Linux 日常运维管理

作为一名运维工程师或系统管理员，对系统不了解就不能进行问题排查，例如查看系统资源的耗费情况。对系统进行排查会用到各种各样的命令。

1.1 w命令

使用 w 命令可以监控系统的状态，运行 w 命令后会列出一些信息，代码如下，“14:51:32”表示当前时间；“up 11:17”表示系统启动 11 小时 17 分钟；“1 user”表示目前有一个用户登录，可以从第三行查看登录用户是谁，下面的代码中显示的用户是 root；FROM 表示从哪里登录，IP 地址是多少；TTY 表示登录的终端是哪一个（pts/0 表示用远程管理工具登录，在 VMware 终端窗口登录显示 tty1，有 tty1~tty6 共 6 个终端）；load average 表示系统负载，load average 后面有三个数字，分别表示 1 分钟、5 分钟和 15 分钟时间段内系统的负载值。具体来说，第一个数字表示 1 分钟内使用 CPU 的活动进程有多少个，该数值为平均值，数值可以是零点几，也可以是一百多；当前值为 0 表示系统没有负载，也就是系统中没有活动的进程，说明系统处于空闲状态，该状态对服务器来说比较浪费。数值的理想状态取决于系统的 CPU 数量（该数量针对的是逻辑 CPU，并非物理 CPU。例如，Intel、AMD 都会有几颗 CPU，每颗 CPU 又有很多逻辑 CPU），CPU 颗数可以在/proc/cpuinfo 中进行查看（processor 表示 CPU 颗数，3 表示该 CPU 为 4 颗）。1 分钟内数字为 4 是系统的最佳状态。

```
[root@centos7 ~]# w
 14:51:32 up 11:17,  1 user,  load average: 0.00, 0.01, 0.05
USER     TTY      FROM             LOGIN@   IDLE   JCPU   PCPU WHAT
root     pts/1    192.168.222.1    12:11    4.00s  0.07s  0.01s w
[root@centos7 ~]# date
```

```
Sun Jan 14 14:51:33 CST 2018
[root@centos7 ~]# cat /proc/cpuinfo
processor       : 3
vendor_id       : GenuineIntel
cpu family      : 6
model           : 45
model name      : Intel(R) Xeon(R) CPU E5-2670 0 @ 2.60GHz
stepping        : 7
microcode       : 0x710
cpu MHz         : 2593.057
cache size      : 20480 KB
```

运行 uptime 命令后显示的信息与运行 w 命令后显示的第一行信息相同，操作命令如下。

```
[root@centos7 ~]# uptime
 15:23:18 up 6 min,  2 users,  load average: 0.06, 0.09, 0.05
```

1.2 vmstat命令

使用 w 命令可查看系统的负载值，当负载值大于 CPU 核数时，说明 CPU 不够用。是什么原因导致 CPU 不够用？此时进程在做什么？都有哪些任务在使用 CPU？若想进一步查看系统的瓶颈在哪里，可运行 vmstat 命令。使用 vmstat 命令可查看 CPU、内存、虚拟磁盘、交换分区、I/O 磁盘和系统进程的信息，操作命令如下。

```
[root@centos7 ~]# vmstat
procs -----------memory---------- ---swap-- -----io---- -system--
------cpu-----
 r  b   swpd   free   buff  cache   si   so    bi    bo   in   cs us sy id wa st
 1  0      0 3545704   2076 169640    0    0    41     3   70   50  0  1 99  0  0
```

在 vmstat 命令后加数字 1 表示每秒动态显示一次，不停地进行显示，结束需按 Ctrl+C 组合键，操作命令如下。

```
[root@centos7 ~]# vmstat 1
procs -----------memory---------- ---swap-- -----io---- -system--
------cpu-----
 r  b   swpd   free   buff  cache   si   so    bi    bo   in   cs us sy id wa st
 1  0      0 3545440   2076 169672    0    0    38     3   67   48  0  1 99  0  0
 0  0      0 3545440   2076 169672    0    0     0     0  141  112  0  0 100  0  0
 0  0      0 3545440   2076 169672    0    0     0     0  104   87  0  0 100  0  0
 0  0      0 3545440   2076 169672    0    0     0     0  120  100  0  0 100  0  0
```

还可以用 vmstat 命令设置每秒显示一次、共显示 2 次，或者每 3 秒显示一次、共显示 3 次，操作命令如下。

```
    [root@centos7 ~]# vmstat 1 2
    procs -----------memory---------- ---swap-- -----io---- -system--
------cpu-----
     r  b   swpd   free   buff  cache   si   so    bi    bo   in   cs us sy id wa st
     1  0      0 3545404   2076 169868    0    0·    15     1   36   27  0  0 99  0  0
     0  0      0 3545264   2076 169868    0    0     0     0  101   73  0  0 100  0  0
    [root@centos7 ~]# vmstat 3 3
    procs -----------memory------- ---swap-- -----io---- -system--
------cpu-----
     r  b  swpd   free   buff  cache   si   so    bi    bo   in   cs us sy id wa st
     1  0      0 3545404   2076 169868    0    0    15     1   36   27  0  0 99  0  0
     0  0      0 3545388   2076 169868    0    0     0     0   54   45  0  0 100  0  0
     0  0      0 3545388   2076 169868    0    0     0     0   59   46  0  0 100  0  0
```

在运行 vmstat 命令后显示的信息中，需要关注 r、b、swpd、si、so、bi、bo、us、wa 这几列。其中，r（run）表示有多少个进程处于运行状态。

b（block）表示被 CPU 之外的资源阻断了（如网络、硬盘等），处于等待状态。

swpd 表示内存和交换分区，数字处于静止状态时没有任何问题，一旦数值不停变化就需要额外注意，说明交换分区和内存在频繁地交换数据，内存空间不够使用，需增加交换分区的内存。

si、so 与 swpd 是关联的，如果 swpd 的数值在不断变化，那么 si 和 so 的数值也会随之变化。si 表示有多少 Kb（单位为 Kb）的数据从交换分区进入到内存中。so 表示有多少 Kb 的数据从内存进入到交换分区中。

bi、bo 与磁盘是相关的。bi 表示磁盘读的数据量有多少，bo 表示磁盘写的数据量有多少。bi、bo 列的数值若在不停地变化，说明磁盘在不停地读写数据。

us 表示用户使用的资源占 CPU 的百分比，该数值不会超过 100%，如果该数值长时间大于 50%，则说明系统资源不够用。sy 表示系统本身的进程/服务占 CPU 的百分比。id 表示空闲 CPU 的百分比。us、sy、id 三者相加等于 100%。

wa 表示有多少个进程处于等待状态，类似于 b 列，单位是百分比，该列数值过大说明 CPU 不够用。

1.3 top命令

使用 top 命令可以查看具体哪个进程出现了问题，运行 top 命令后会动态显示出结果，该结果每 3 秒变化一次，如图 1-1 所示。

在图 1-1 中，执行 top 命令后的第一行显示的内容和运行 w 命令后第一行显示的内容一致。

第二行的 total 表示有多少个进程，running 表示有多少个正在运行的进程，sleeping 表示有多少个进程处于睡眠状态，stopped 表示有多个进程处于停止状态，zombie 表示有多少个进程处于僵尸状态（主进程被意外终止，留下一些子进程）。

第三行是 CPU 的百分比。us、sy、id 三者数值总和等于 100%。ni 表示改变过优先级的进程占用 CPU 的百分比，wa 表示 IO 等待占用 CPU 的百分比，hi 表示硬中断（Hardware IRQ）占用 CPU 的百分比，si 表示软中断（Software Interrupts）占用 CPU 的百分比。

第四行和第五行是内存使用情况，Mem 表示物理内存，Swap 表示交换分区。一般关注 total（总内存）、free（剩余内存）、used（使用内存）和 buff/cache（缓冲/缓存）。

```
[root@docker ~]# w
 17:52:48 up 5 days, 20:42,  1 user,  load average: 0.00, 0.01, 0.05
USER     TTY      FROM             LOGIN@   IDLE   JCPU   PCPU WHAT
root     pts/0    218.67.233.250   17:52    0.00s  0.01s  0.00s w
[root@docker ~]# top
top - 17:52:50 up 5 days, 20:42,  1 user,  load average: 0.00, 0.01, 0.05
Tasks:  93 total,   1 running,  92 sleeping,   0 stopped,   0 zombie
%Cpu(s):  1.6 us,  0.0 sy,  0.0 ni, 98.4 id,  0.0 wa,  0.0 hi,  0.0 si,  0.0 st
KiB Mem :  8009052 total,  1992888 free,   282316 used,  5733848 buff/cache
KiB Swap:        0 total,        0 free,        0 used.  7391156 avail Mem
 Unknown command - try 'h' for help
  PID USER      PR  NI    VIRT    RES    SHR S  %CPU %MEM     TIME+ COMMAND
    1 root      20   0  190992   3904   2588 S   0.0  0.0   0:02.84 systemd
    2 root      20   0       0      0      0 S   0.0  0.0   0:00.00 kthreadd
    3 root      20   0       0      0      0 S   0.0  0.0   0:00.00 ksoftirqd/0
    5 root       0 -20       0      0      0 S   0.0  0.0   0:00.00 kworker/0:0H
```

图 1-1

第五行下面的数据是需要时刻关注的，默认按照 CPU 使用率从高到低进行排序。若想按照内存大小进行排序，可按大写 M 键，使用内存最多的为 firewalld，如图 1-2 所示。若想返回继续按 CPU 使用率进行排序，可按大写 P 键。按数字 1 键可显示所有的 CPU 使用情况，如图 1-3 所示。按小写字母 q 键可退出。

PID：进程 ID。

USER：进程所有者。

PR：进程优先级。

NI：nice 值，负值表示高优先级，正值表示低优先级。

VIRT：进程使用的虚拟内存总量，单位为 Kb，VIRT=Swap+RES。

RES：进程使用的、未被换出的物理内存大小，单位为 Kb，RES=CODE+DATA。

SHR：共享内存大小，单位为 Kb。

S：进程状态，D 为不可中断的睡眠状态，R 为运行状态，S 为睡眠状态，T 为跟踪/停止状态，Z 为僵尸状态。

%CPU：上次更新到现在的 CPU 时间占用百分比。

%MEM：进程使用的物理内存百分比。

TIME+：进程使用的 CPU 时间总计，单位为 1/100 秒。

COMMAND：进程名称（命令名/命令行）。

```
top - 17:19:35 up  2:02,  2 users,  load average: 0.00, 0.01, 0.05
Tasks: 108 total,   1 running, 107 sleeping,   0 stopped,   0 zombie
%Cpu(s):  0.0 us,  0.1 sy,  0.0 ni, 99.9 id,  0.0 wa,  0.0 hi,  0.0 si,  0.0 st
KiB Mem :  3865304 total,  3543728 free,   148832 used,   172744 buff/cache
KiB Swap: 10239996 total, 10239996 free,        0 used.  3484864 avail Mem

  PID USER      PR  NI    VIRT    RES    SHR S  %CPU %MEM     TIME+ COMMAND
  606 root      20   0  334064  28688   6692 S   0.0  0.7   0:01.95 firewalld
  927 root      20   0  562392  16560   5884 S   0.0  0.4   0:01.73 tuned
  565 polkitd   20   0  534784  12828   4856 S   0.0  0.3   0:00.40 polkitd
  622 root      20   0  471816   8868   6588 S   0.0  0.2   0:00.93 NetworkManager
    1 root      20   0  128164   6812   4048 S   0.0  0.2   0:04.96 systemd
  566 root      20   0  216388   6356   3436 S   0.0  0.2   0:01.01 rsyslogd
  588 root      20   0  305080   6140   4772 S   0.0  0.2   0:08.12 vmtoolsd
  574 root      20   0   99608   6080   4472 S   0.0  0.2   0:00.11 VGAuthService
  405 root      20   0   47960   5960   2808 S   0.0  0.2   0:00.81 systemd-udevd
 1263 root      20   0  145700   5204   3932 S   0.0  0.1   0:00.81 sshd
  928 root      20   0  105996   4068   3088 S   0.0  0.1   0:00.04 sshd
 1168 postfix   20   0   89716   4012   3008 S   0.0  0.1   0:00.03 qmgr
 1656 postfix   20   0   89648   3992   2988 S   0.0  0.1   0:00.13 pickup
  380 root      20   0   36828   3200   2884 S   0.0  0.1   0:00.44 systemd-journal
 1267 root      20   0  116316   2988   1684 S   0.0  0.1   0:00.14 bash
 1241 root      20   0  116316   2916   1636 S   0.0  0.1   0:00.09 bash
  597 root      20   0   94384   2448   1788 S   0.0  0.1   0:00.44 login
 1166 root      20   0   89544   2176   1140 S   0.0  0.1   0:00.12 master
 1670 root      20   0  157712   2172   1504 R   0.0  0.1   0:00.05 top
  575 dbus      20   0   32776   1852   1400 S   0.0  0.0   0:00.34 dbus-daemon
  578 chrony    20   0  115640   1776   1320 S   0.0  0.0   0:01.79 chronyd
  572 root      20   0   24204   1704   1396 S   0.0  0.0   0:00.20 systemd-logind
```

图 1-2

```
top - 17:23:32 up  2:06,  2 users,  load average: 0.04, 0.03, 0.05
Tasks: 108 total,   1 running, 107 sleeping,   0 stopped,   0 zombie
%Cpu0  :  0.0 us,  0.0 sy,  0.0 ni,100.0 id,  0.0 wa,  0.0 hi,  0.0 si,  0.0 st
%Cpu1  :  0.0 us,  0.0 sy,  0.0 ni,100.0 id,  0.0 wa,  0.0 hi,  0.0 si,  0.0 st
%Cpu2  :  0.0 us,  0.0 sy,  0.0 ni,100.0 id,  0.0 wa,  0.0 hi,  0.0 si,  0.0 st
%Cpu3  :  0.0 us,  0.0 sy,  0.0 ni,100.0 id,  0.0 wa,  0.0 hi,  0.0 si,  0.0 st
KiB Mem :  3865304 total,  3542344 free,   150048 used,   172912 buff/cache
KiB Swap: 10239996 total, 10239996 free,        0 used.  3483492 avail Mem
```

图 1-3

使用-c 选项可查看具体的命令全局路径，如图 1-4 所示。

```
top - 17:26:49 up  2:09,  2 users,  load average: 0.00, 0.01, 0.05
Tasks: 108 total,   1 running, 107 sleeping,   0 stopped,   0 zombie
%Cpu(s):  1.4 us,  1.4 sy,  0.0 ni, 97.1 id,  0.0 wa,  0.0 hi,  0.0 si,  0.0 st
KiB Mem :  3865304 total,  3542856 free,   149556 used,   172892 buff/cache
KiB Swap: 10239996 total, 10239996 free,        0 used.  3484004 avail Mem

  PID USER      PR  NI    VIRT    RES    SHR S  %CPU %MEM     TIME+ COMMAND
 1724 root      20   0  157776   2212   1568 R   6.2  0.1   0:00.03 top -c
    1 root      20   0  128164   6812   4048 S   0.0  0.2   0:05.03 /usr/lib/systemd/systemd --switched-root --system --deserialize 21
    2 root      20   0       0      0      0 S   0.0  0.0   0:00.07 [kthreadd]
    3 root      20   0       0      0      0 S   0.0  0.0   0:00.06 [ksoftirqd/0]
    5 root       0 -20       0      0      0 S   0.0  0.0   0:00.00 [kworker/0:0H]
    6 root      20   0       0      0      0 S   0.0  0.0   0:00.31 [kworker/u256:0]
    7 root      rt   0       0      0      0 S   0.0  0.0   0:00.12 [migration/0]
```

图 1-4

使用-bn1 选项可静态、一次性地把所有进程信息显示出来，该用法适合在写脚本时采用，操作命令如下。

```
[root@centos7 ~]# top -bn1
```

运行 top 命令时通过某些按键可增加特效，按小写 b 键可打开/关闭加亮效果，如图 1-5 所示。

按小写 x 键可打开/关闭排序列的加亮效果，如图 1-6 所示。

通过按 Shift + >组合键或 Shift + <组合键可以向右或向左定位排序列，如定位

到 PID 列，如图 1-7 所示。

```
top - 17:47:03 up  2:29,  2 users,  load average: 0.00, 0.01, 0.05
Tasks: 108 total,   1 running, 107 sleeping,   0 stopped,   0 zombie
%Cpu(s):  0.0 us,  0.2 sy,  0.0 ni, 99.8 id,  0.0 wa,  0.0 hi,  0.0 si,  0.0 st
KiB Mem :  3865304 total,  3542848 free,   149552 used,   172904 buff/cache
KiB Swap: 10239996 total, 10239996 free,        0 used.  3483996 avail Mem

  PID USER      PR  NI    VIRT    RES    SHR S  %CPU %MEM     TIME+ COMMAND
 1730 root      20   0  157712   2168   1504 R   0.6  0.1   0:00.33 top
```

图 1-5

```
  PID USER      PR  NI    VIRT    RES    SHR S  %CPU %MEM     TIME+ COMMAND
  606 root      20   0  334064  28688   6692 S   0.0  0.7   0:01.97 firewalld
  927 root      20   0  562392  16560   5884 S   0.3  0.4   0:02.10 tuned
  565 polkitd   20   0  534784  12828   4856 S   0.0  0.3   0:00.40 polkitd
  622 root      20   0  471816   8868   6588 S   0.0  0.2   0:00.98 NetworkManager
    1 root      20   0  128164   6812   4048 S   0.0  0.2   0:05.04 systemd
  566 root      20   0  216388   6364   3444 S   0.0  0.2   0:01.13 rsyslogd
  588 root      20   0  305080   6140   4772 S   0.0  0.2   0:10.15 vmtoolsd
  574 root      20   0   99608   6080   4472 S   0.0  0.2   0:00.11 VGAuthService
  405 root      20   0   47960   5960   2808 S   0.0  0.2   0:00.81 systemd-udevd
 1263 root      20   0  145700   5204   3932 S   0.0  0.1   0:01.78 sshd
  928 root      20   0  105996   4068   3088 S   0.0  0.1   0:00.05 sshd
 1168 postfix   20   0   89716   4012   3008 S   0.0  0.1   0:00.04 qmgr
 1656 postfix   20   0   89648   3992   2988 S   0.0  0.1   0:00.14 pickup
  380 root      20   0   36828   3232   2916 S   0.0  0.1   0:00.48 systemd-journal
 1267 root      20   0  116316   2988   1684 S   0.0  0.1   0:00.17 bash
 1241 root      20   0  116316   2916   1636 S   0.0  0.1   0:00.09 bash
  597 root      20   0   94384   2448   1788 S   0.0  0.1   0:00.44 login
 1166 root      20   0   89544   2176   1140 S   0.0  0.1   0:00.14 master
 1734 root      20   0  157712   2172   1504 R   0.3  0.1   0:00.65 top
  575 dbus      20   0   32776   1852   1400 S   0.0  0.0   0:00.39 dbus-daemon
```

图 1-6

```
  PID USER      PR  NI    VIRT    RES    SHR S  %CPU %MEM     TIME+ COMMAND
 1734 root      20   0  157712   2172   1504 R   0.0  0.1   0:00.96 top
 1726 root      20   0       0      0      0 S   0.0  0.0   0:00.08 kworker/0:1
 1656 postfix   20   0   89648   3992   2988 S   0.0  0.1   0:00.14 pickup
 1651 root      20   0       0      0      0 S   0.0  0.0   0:01.64 kworker/0:0
 1650 root      20   0       0      0      0 S   0.0  0.0   0:00.00 kworker/u256:2
 1321 root      20   0       0      0      0 S   0.0  0.0   0:02.87 kworker/0:2
 1267 root      20   0  116316   2988   1684 S   0.0  0.1   0:00.17 bash
 1263 root      20   0  145700   5204   3932 S   0.0  0.1   0:01.81 sshd
 1241 root      20   0  116316   2916   1636 S   0.0  0.1   0:00.09 bash
 1168 postfix   20   0   89716   4012   3008 S   0.0  0.1   0:00.04 qmgr
 1166 root      20   0   89544   2176   1140 S   0.0  0.1   0:00.15 master
```

图 1-7

按小写 f 键可进入 top 命令的另一个视图，在这里可编排基本视图中的显示字段，如图 1-8 所示。按 Esc 键可返回基本视图。

```
Fields Management for window 1:Def, whose current sort field is %CPU
   Navigate with Up/Dn, Right selects for move then <Enter> or Left commits,
   'd' or <Space> toggles display, 's' sets sort.  Use 'q' or <Esc> to end!

* PID     = Process Id              ENVIRON = Environment vars
* USER    = Effective User Name     vMj     = Major Faults delta
* PR      = Priority                vMn     = Minor Faults delta
* NI      = Nice Value              USED    = Res+Swap Size (KiB)
* VIRT    = Virtual Image (KiB)     nsIPC   = IPC namespace Inode
* RES     = Resident Size (KiB)     nsMNT   = MNT namespace Inode
* SHR     = Shared Memory (KiB)     nsNET   = NET namespace Inode
* S       = Process Status          nsPID   = PID namespace Inode
* %CPU    = CPU Usage               nsUSER  = USER namespace Inode
* %MEM    = Memory Usage (RES)      nsUTS   = UTS namespace Inode
* TIME+   = CPU Time, hundredths
* COMMAND = Command Name/Line
  PPID    = Parent Process pid
  UID     = Effective User Id
  RUID    = Real User Id
  RUSER   = Real User Name
  SUID    = Saved User Id
  SUSER   = Saved User Name
  GID     = Group Id
  GROUP   = Group Name
  PGRP    = Process Group Id
```

图 1-8

1.4　sar命令

使用 sar 命令可以非常全面地分析系统状态，在日常工作中一般用来查看网卡流量的使用情况。sar 命令和 w 命令一样都可以用于查看 CPU、内存、磁盘的状况。sar 命令被运维工程师称为 Linux 操作系统中的“瑞士军刀”，功能非常强大。

系统中若没有 sar 命令，则需要使用 yum 命令安装，直接运行 sar 命令会提示“无法打开/var/log/sa/sa15：没有这样的文件或目录”。提示这个错误是因为在 sar 命令中没有加具体选项参数时，会默认调用 Linux 系统中 sar 命令保留的历史文件所在的目录。sar 命令有一个特性，每 10 分钟就会把系统的状况过滤一遍并保存在文件中，该文件存在/var/log/sa/目录下，刚安装完 sar 命令还没有生成历史文件，等待 10 分钟再去查看该目录，就会发现已生成文件，操作命令如下。

```
[root@centos7 ~]# yum install -y sysstat
[root@centos7 ~]# sar
Cannot open /var/log/sa/sa15: No such file or directory
[root@centos7 ~]# ls /var/log/sa/
[root@centos7 ~]#
```

使用-n DEV 选项可查看网卡流量的使用状况，后面可加数字参数，如每隔 1 秒显示一次，共显示 3 次，如图 1-9 所示，各列的含义如下。

第一列：显示当前时间。

第二列：显示网卡信息，输出的结果为两块网卡，即 lo 和 ens33。

第三列：接收到的数据包，单位为个。

第四列：发送出去的数据包，单位为个。

第五列：接收到的数据量，单位为 Kb。

第六列：发送出去的数据量，单位为 Kb。

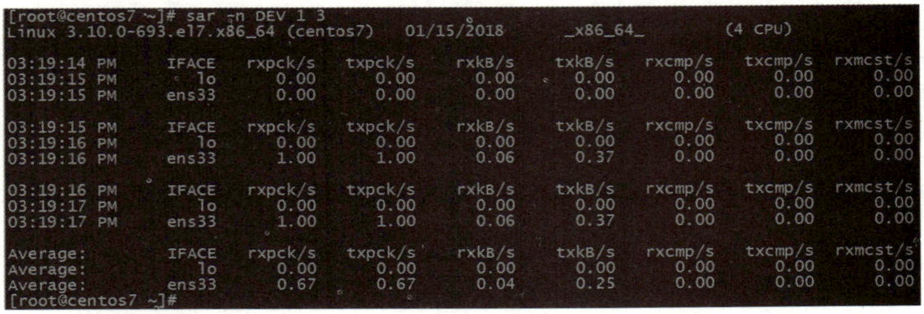

```
[root@centos7 ~]# sar -n DEV 1 3
Linux 3.10.0-693.el7.x86_64 (centos7)   01/15/2018      _x86_64_        (4 CPU)

03:19:14 PM     IFACE   rxpck/s   txpck/s    rxkB/s    txkB/s   rxcmp/s   txcmp/s  rxmcst/s
03:19:15 PM        lo      0.00      0.00      0.00      0.00      0.00      0.00      0.00
03:19:15 PM     ens33      0.00      0.00      0.00      0.00      0.00      0.00      0.00

03:19:15 PM     IFACE   rxpck/s   txpck/s    rxkB/s    txkB/s   rxcmp/s   txcmp/s  rxmcst/s
03:19:16 PM        lo      0.00      0.00      0.00      0.00      0.00      0.00      0.00
03:19:16 PM     ens33      1.00      1.00      0.06      0.37      0.00      0.00      0.00

03:19:16 PM     IFACE   rxpck/s   txpck/s    rxkB/s    txkB/s   rxcmp/s   txcmp/s  rxmcst/s
03:19:17 PM        lo      0.00      0.00      0.00      0.00      0.00      0.00      0.00
03:19:17 PM     ens33      1.00      1.00      0.06      0.37      0.00      0.00      0.00

Average:        IFACE   rxpck/s   txpck/s    rxkB/s    txkB/s   rxcmp/s   txcmp/s  rxmcst/s
Average:           lo      0.00      0.00      0.00      0.00      0.00      0.00      0.00
Average:        ens33      0.67      0.67      0.04      0.25      0.00      0.00      0.00
[root@centos7 ~]#
```

图 1-9

10 分钟后再次运行 sar 命令就不会报错了，因为在/var/log/sa 目录下生成了 sa15 文件，操作命令如下。

```
[root@centos7 ~]# sar
Linux 3.10.0-693.el1.x86_64 (centos7)  01/15/2018   _x86_64_    (4 CPU)

03:20:01 PM    CPU    %user    %nice    %system    %iowait    %steal    %idle
03:30:01 PM   all     0.01      0.00      0.05       0.01       0.00      99.93
Average:      all     0.01      0.00      0.05       0.01       0.00      99.93
[root@centos7 ~]# ls /var/log/sa/
sa15
```

在/var/log/sa/目录下不仅有 sa15 文件，还有 sar15 文件。sa15 文件是二进制文件，不能使用 cat 命令进行查看，只能用 sar －f 命令加载此类文件；sar15 文件可以用 cat 命令进行查看。

在-n DEV 选项后加上-f 选项可查看指定的历史数据文件，如查看当天的历史数据，如图 1-10 所示，/var/log/sa/目录下最多保留一个月的历史数据文件。

```
[root@centos7 ~]# date
Mon Jan 15 15:36:38 CST 2018
[root@centos7 ~]# sar -n DEV -f /var/log/sa/sa15
Linux 3.10.0-693.el7.x86_64 (centos7)   01/15/2018      _x86_64_        (4 CPU)

03:20:01 PM     IFACE   rxpck/s   txpck/s    rxkB/s    txkB/s   rxcmp/s   txcmp/s  rxmcst/s
03:30:01 PM        lo      0.00      0.00      0.00      0.00      0.00      0.00      0.00
03:30:01 PM     ens33      0.11      0.10      0.01      0.01      0.00      0.00      0.00
Average:           lo      0.00      0.00      0.00      0.00      0.00      0.00      0.00
Average:        ens33      0.11      0.10      0.01      0.01      0.00      0.00      0.00
```

图 1-10

使用-q 选项可以查看系统负载信息（该方法等同于运行 w 命令），用 sar －q 命令可查看系统负载的历史数据，例如，设置每秒显示 1 次、共显示 3 次，操作命令如下。

```
[root@centos7 ~]# sar -q 1 3
Linux 3.10.0-693.el1.x86_64 (centos7)   01/15/2018    _x86_64_   (4 CPU)

03:39:17 PM   runq-sz  plist-sz   ldavg-1   ldavg-5  ldavg-15   blocked
03:39:18 PM         0       130      0.00      0.01      0.05         0
03:39:19 PM         1       130      0.00      0.01      0.05         0
03:39:20 PM         0       130      0.00      0.01      0.05         0
Average:            0       130      0.00      0.01      0.05         0
```

使用-b 选项可以查看系统磁盘数据信息，如每秒显示 1 次、共显示 3 次，操作命令如下。

```
[root@mingzhe ~]# sar -b 1  3
Linux 3.10.0-693.2.2.el1.x86_64 (mingzhe)  01/15/2018 _x86_64_ (1 CPU)

03:43:19 PM       tps      rtps      wtps   bread/s   bwrtn/s
03:43:20 PM      0.00      0.00      0.00      0.00      0.00
03:43:21 PM      0.00      0.00      0.00      0.00      0.00
```

```
03:43:22 PM      0.00      0.00      0.00      0.00      0.00
Average:         0.00      0.00      0.00      0.00      0.00
```

1.5　nload命令

使用 nload 命令可以监控网卡流量，但是 Linux 操作系统中没有该命令，若想使用，需要用 yum 命令进行安装，在安装 nload 命令前要先安装 epel 扩展源，操作命令如下。

```
[root@centos7 ~]# nload
-bash: nload: command not found
[root@centos7 ~]# yum install -y epel-release
[root@centos7 ~]# yum install -y nload
```

安装完成后直接运行 nload 命令，按 Enter 键后会在终端窗口动态显示网卡实时信息，如图 1-11 所示。第一行所显示的信息为网卡信息和 IP 地址，按向右的方向键可切换网卡。Incoming 表示进入网卡的流量，Outgoing 表示从该网卡流出的流量。每部分都有 Curr（当前流量）、Avg（平均流量）、Min（最小流量）、Max（最大流量）和 Ttl（总流量）。按 Q 键可退出该窗口。

图 1-11

1.6　监控I/O性能

使用 iostat 和 iotop 命令可以查看磁盘的使用情况，在日常运维工作中，除 CPU 和内存外，磁盘 I/O 也是非常重要的指标。有时 CPU 和内存明明有剩余，但系统负载仍然很高，用 vmstat 命令查看时发现 b 列和 wa 列数值较大，说明系统磁盘有瓶颈。若想更详细地查看磁盘状态，就需要使用上述两个命令。

在安装 sysstat 包时会默认安装 iostat 命令，iostat 命令和 sar 命令同在一个软件包中。运行 iostat 命令后可查看磁盘状况，如图 1-12 所示。在 iostat 命令后加数字 1 运行的效果和运行 vmstat 1 的效果一致，都会每隔 1 秒显示一次。

```
[root@mingzhe ~]# iostat
Linux 3.10.0-693.2.2.el7.x86_64 (mingzhe)       01/15/2018      _x86_64_        (1 CPU)

avg-cpu:  %user   %nice %system %iowait  %steal   %idle
           0.09    0.00    0.06    0.01    0.00   99.84

Device:            tps    kB_read/s    kB_wrtn/s    kB_read    kB_wrtn
vda               0.40         2.81         3.14     287121     320152

[root@mingzhe ~]# iostat 1
Linux 3.10.0-693.2.2.el7.x86_64 (mingzhe)       01/15/2018      _x86_64_        (1 CPU)

avg-cpu:  %user   %nice %system %iowait  %steal   %idle
           0.09    0.00    0.06    0.01    0.00   99.84

Device:            tps    kB_read/s    kB_wrtn/s    kB_read    kB_wrtn
vda               0.40         2.81         3.13     287129     320176

avg-cpu:  %user   %nice %system %iowait  %steal   %idle
           0.00    0.00    0.00    0.00    0.00  100.00

Device:            tps    kB_read/s    kB_wrtn/s    kB_read    kB_wrtn
vda               0.00         0.00         0.00          0          0

avg-cpu:  %user   %nice %system %iowait  %steal   %idle
           0.00    0.00    0.99    0.00    0.00   99.01

Device:            tps    kB_read/s    kB_wrtn/s    kB_read    kB_wrtn
vda               0.00         0.00         0.00          0          0

^C
[root@mingzhe ~]#
```

图 1-12

使用-x 选项可以显示和 I/O 相关的扩展数据。有一个非常重要的指标%util，如图 1-13 所示。%util 表示在统计时间内所有处理 I/O 的时间除以总共统计时间的百分比。例如，统计时间为 1 秒，该设备有 0.8 秒在处理 I/O，有 0.2 秒闲置，那么该设备的%util = 0.8/1×100% = 80%。该参数表示设备的繁忙程度。一般情况下，如果该参数是 100%，则表示设备已经满负荷运行了（如果是多磁盘，即使%util 是 100%，也未必会出现瓶颈，因为磁盘有并发能力）。

```
[root@mingzhe ~]# iostat -x
Linux 3.10.0-693.2.2.el7.x86_64 (mingzhe)       01/15/2018      _x86_64_        (1 CPU)

avg-cpu:  %user   %nice %system %iowait  %steal   %idle
           0.09    0.00    0.06    0.01    0.00   99.84

Device:         rrqm/s   wrqm/s     r/s     w/s    rkB/s    wkB/s avgrq-sz avgqu-sz   await r_await w_await  svctm  %util
vda               0.00     0.09    0.12    0.27     2.80     3.12    29.86     0.00    5.76    1.32    7.79   0.34   0.01

[root@mingzhe ~]#
```

图 1-13

磁盘 I/O 频繁工作时，并不知道是哪个进程在频繁地读写，此时可用 iotop 命令进行查看。Linux 系统中默认没有 iotop 命令，需要使用 yum 命令安装，操作命令如下。

```
[root@mingzhe ~]# yum install -y iotop
```

iotop 命令和 top 命令相似，都是动态显示进程信息的。运行 iotop 命令后即可查看具体是哪个进程使用的 I/O 较多，如图 1-14 所示。

```
Total DISK READ :0.00 B/s | Total DISK WRITE :0.00 B/s      0.00 B
 1040 be/4 rootD:0.00 B/s | Actual DISK WRITE:0.00 B/s      0.00 B
  TID  PRIO  USER     DISK READ  DISK WRITE  SWAPIN     IO>    COMMAND 1 be/4 root
  256 be/3 root        0.00 B/s   35.62 K/s  0.00 %  0.17 % [jbd2/vda1-8]tched-root --system --deserialize 21
    2 be/4 root        0.00 B/s    0.00 B/s  0.00 %  0.00 % [kthreadd]
    3 be/4 root        0.00 B/s    0.00 B/s  0.00 %  0.00 % [ksoftirqd/0]
    5 be/0 root        0.00 B/s    0.00 B/s  0.00 %  0.00 % [kworker/0:0H]
    6 be/4 root        0.00 B/s    0.00 B/s  0.00 %  0.00 % [kworker/u2:0]
    7 rt/4 root        0.00 B/s    0.00 B/s  0.00 %  0.00 % [migration/0]
    8 be/4 root        0.00 B/s    0.00 B/s  0.00 %  0.00 % [rcu_bh]
    9 be/4 root        0.00 B/s    0.00 B/s  0.00 %  0.00 % [rcu_sched]
   10 rt/4 root        0.00 B/s    0.00 B/s  0.00 %  0.00 % [watchdog/0]
   12 be/4 root        0.00 B/s    0.00 B/s  0.00 %  0.00 % [kdevtmpfs]
   13 be/0 root        0.00 B/s    0.00 B/s  0.00 %  0.00 % [netns]
   14 be/4 root        0.00 B/s    0.00 B/s  0.00 %  0.00 % [khungtaskd]
   15 be/0 root        0.00 B/s    0.00 B/s  0.00 %  0.00 % [writeback]
   16 be/0 root        0.00 B/s    0.00 B/s  0.00 %  0.00 % [kintegrityd]
   17 be/0 root        0.00 B/s    0.00 B/s  0.00 %  0.00 % [bioset]
   18 be/0 root        0.00 B/s    0.00 B/s  0.00 %  0.00 % [kblockd]
   19 be/0 root        0.00 B/s    0.00 B/s  0.00 %  0.00 % [md]
   25 be/4 root        0.00 B/s    0.00 B/s  0.00 %  0.00 % [kswapd0]
   26 be/5 root        0.00 B/s    0.00 B/s  0.00 %  0.00 % [ksmd]
   27 be/7 root        0.00 B/s    0.00 B/s  0.00 %  0.00 % [khugepaged]
   28 be/0 root        0.00 B/s    0.00 B/s  0.00 %  0.00 % [crypto]
 1053 be/4 root        0.00 B/s    0.00 B/s  0.00 %  0.00 % python -Es /usr/sbin/tuned -l -P
  517 be/4 polkitd     0.00 B/s    0.00 B/s  0.00 %  0.00 % polkitd --no-debug [JS Sour~ Thread]
   36 be/0 root        0.00 B/s    0.00 B/s  0.00 %  0.00 % [kthrotld]
  518 be/4 polkitd     0.00 B/s    0.00 B/s  0.00 %  0.00 % polkitd --no-debug
   38 be/0 root        0.00 B/s    0.00 B/s  0.00 %  0.00 % [kmpath_rdacd]
   39 be/0 root        0.00 B/s    0.00 B/s  0.00 %  0.00 % [kpsmoused]
   40 be/0 root        0.00 B/s    0.00 B/s  0.00 %  0.00 % [ipv6_addrconf]
  513 be/4 polkitd     0.00 B/s    0.00 B/s  0.00 %  0.00 % polkitd --no-debug [JS GC Helper]
   59 be/0 root        0.00 B/s    0.00 B/s  0.00 %  0.00 % [deferwq]
 1086 be/4 root        0.00 B/s    0.00 B/s  0.00 %  0.00 % AliYunDun
 1087 be/4 root        0.00 B/s    0.00 B/s  0.00 %  0.00 % AliYunDun
 1088 be/4 root        0.00 B/s    0.00 B/s  0.00 %  0.00 % AliYunDun
 1089 be/4 root        0.00 B/s    0.00 B/s  0.00 %  0.00 % AliYunDun
 1090 be/4 root        0.00 B/s    0.00 B/s  0.00 %  0.00 % AliYunDun
21347 be/4 root        0.00 B/s    0.00 B/s  0.00 %  0.00 % AliYunDunUpdate
   91 be/4 root        0.00 B/s    0.00 B/s  0.00 %  0.00 % [kauditd]
   92 be/4 root        0.00 B/s    0.00 B/s  0.00 %  0.00 % [kworker/0:2]
 1117 be/4 root        0.00 B/s    0.00 B/s  0.00 %  0.00 % AliYunDun
                       0.00 B/s    0.00 B/s  0.00 %  0.00 % AliYunDun
```

图 1-14

1.7　free命令

使用 free 命令可以查看内存使用情况，运行 free 命令后显示的信息如图 1-15 所示。第一行是列标题，第二行是内存使用情况，第三行是 Swap 交换分区的使用情况。需要关注的是第二行的内存使用情况。

第一列（total）：内存总大小，默认单位是 Kb。

第二列（used）：内存使用大小，默认单位是 Kb。

第三列（free）：空闲内存大小，默认单位是 Kb。

第四列（shared）：共享内存大小，默认单位是 Kb。

第五列（buff/cache）：缓冲/缓存大小，默认单位是 Kb。

第六列（available）：可用内存大小，包含 free 和 buff/cache 的剩余部分，默认单位是 Kb。

```
[root@centos7 ~]# free
              total        used        free      shared  buff/cache   available
Mem:        3865304      161960     3343320        8828      360024     3422664
Swap:      10239996           0    10239996
[root@centos7 ~]#
```

图 1-15

使用-m 选项可以以 MB 为单位查看数据，如图 1-16 所示。

图 1-16

使用-h 选项可以在具体的数值后面加上单位，方便查看，如图 1-17 所示。一般认为 total = used + free，但是 155MB+3.2GB≠3.7GB，这是因为 Linux 操作系统会把内存分一部分出来给 buff 和 cache，所以正确公式为：total=used+free+buff/cache。

图 1-17

1.8　ps命令

ps 命令可用来查看进程，运行 ps 命令所显示的结果和运行 top 命令所显示的结果相似。日常运维工作中使用 ps 命令有两种方法，一种是用 ps aux 命令把系统中所有的进程列出来；另一种是使用 ps -elf 命令，该方法的作用和 ps aux 的作用类似，操作命令如下。

```
    [root@centos7 ~]# ps aux |grep mysql
    root      3162  0.0  0.0 112660   972 pts/0    S+   19:13   0:00 grep
--color=auto mysql
    [root@centos7 ~]# ps aux |grep nginx
    root      3164  0.0  0.0 112660   972 pts/0    S+   19:13   0:00 grep
--color=auto nginx
```

上述查看的是系统中全部的进程，若想查看指定的进程信息，可使用管道符进行过滤，如查看是否存在 mysqld 服务和 nginx 服务，操作命令如下。

```
    [root@centos7 ~]# ps aux |grep mysqld
    root      3169  0.0  0.0 112660   976 pts/0    S+   19:17   0:00 grep
--color=auto mysqld
    [root@centos7 ~]# ps aux |grep nginx
    root      3173  0.0  0.0 112660   976 pts/0    S+   19:17   0:00 grep
--color=auto nginx
```

使用 ps aux 命令查看系统进程时会显示很多列，每列的含义分别如下。

USER：用户名。

UID：用户 ID（User ID）。

PID：进程 ID（Process ID），“杀死”进程时经常用到（使用 kill 命令+进程 ID

可“杀死”该进程，pkill 命令用于强制“杀死”某进程）。比如上线一款游戏时，突然后台进程卡死不动，此时需要进入服务器并重启，但是这样做浪费时间，可以写一个 shell 脚本，获取该进程的 PID，再用 PHP 或其他编程语言调用 shell 脚本，调用后还需要写一个 Web 页面，页面上只需设置三个 button 按钮，即启动、停止、重启按钮。“杀死”进程脚本和用 PHP 调用 shell 脚本的代码如下。该方法类似 CMDB 资产管理平台。

```
#!/bin/bash
##根据进程名“杀死”指定进程
##__author__ is humingzhe
##email admin@humingzhe.com
#从终端接收第一个参数，系统本身默认当前 shell 为第 0 个参数$0

param=$1

#启动进程函数
start()
{
    fpms=`ps aux | grep -i "mysqld" | grep -v grep | awk '{print $2}'`

     #当前进程不为空，-n 用于判断变量的值是否为空
    if [ ! -n "$fpms" ]; then
     #启动进程
        /usr/local/mysql/bin/mysqld&
        echo "mysqld Start"
    else
        echo "mysqld Already Start"
    fi
}

#停止进程
stop()
{
    fpms=`ps aux | grep -i "mysqld" | grep -v grep | awk '{print $2}'`
    echo $fpms | xargs kill -9

    for pid in $fpms; do
        if echo $pid | egrep -q '^[0-9]+$'; then
            echo "MySQLD Pid $pid Kill"
        else
            echo "$pid IS Not A MySQLD Pid"
        fi
```

```
    done
}

#switch 调用
case $param in
    'start')
        start;;
    'stop')
        stop;;
    'restart')
        stop
        start;;
    *)
        echo "Usage: sh kill.sh start|stop|restart";;
esac
<?php
$output = shell_exec('bash /data/web/default/kill.sh');
echo "<pre>$output</pre>";
?>
```

PPID：父进程的进程 ID（Parent Process ID）。

SID：会话 ID（Session ID）。

%CPU：进程的 CPU 占用率。

%MEM：进程的内存占用率。

TTY：与进程关联的终端（TTY）。

STAT：进程的状态，使用字符表示（STAT 的状态码）。

R：正在运行或在运行队列中等待（Runnable 或 On Run Queue）。

S：睡眠（Sleeping）、休眠、受阻，在等待某个条件的形成或等待接收信号。

I：空闲（Idle）。

Z：僵死（Zombie），进程已终止，但进程描述符还存在，在父进程调用 wait4()后释放。

D：不能中断的进程（Uninterruptible sleep），收到信号不唤醒且不可运行，进程必须等待，直到有中断发生。

T：暂停的、终止的进程（Terminate），进程收到 SIGSTOP、SIGSTP、SIGTIN、SIGTOU 信号后停止运行。

<：高优先级或高优先序的进程。

N：低优先级、低优先序的进程。

L：页面被锁定在内存中（用于实时和自定义 IO）。

s：进程的领导者（主进程，其下有子进程）。

l：多线程的进程。

+：前台进程。

START：进程启动的时间和日期。

TIME：进程使用的总 CPU 时间。

COMMAND：正在执行的命令行命令。

NI：优先级（Nice）。

PRI：进程优先级编号（Priority）。

1.9　查看网络状态

netstat 是查看网络状态的命令。Linux 是服务器的操作系统，服务器上有很多服务，服务往往是和客户端相互通信的，这就意味着要有监听端口，要有对外的通信端口，使用 netstat 命令查看的就是 TCP/IP 的状态。比如，给服务器安装 Nginx，提供 Web 服务，或者给服务器安装 MySQL，提供数据库服务。这些服务是要监听端口的（正常情况下，一台服务器是没有任何监听端口的，没有监听端口意味着无法和其他的计算机通信，要想提供 Web 服务，让其他人访问你的网站就需要监听端口。可以把端口与一台代理服务器相连，连接成功后让该代理服务器与远端的设备连接，数据通过该代理服务器进入到服务器中通信）。

使用 netstat －lnp 命令可以查看系统中哪些服务处于监听状态，如图 1-18 所示。其中，tcp 表示监听 IPv4，tcp6 表示监听 IPv6。使用 netstat 命令也能查看系统中存在哪些 socket 文件处于监听状态。

```
[root@centos7 ~]# netstat -lnp
Active Internet connections (only servers)
Proto Recv-Q Send-Q Local Address           Foreign Address         State       PID/Program name
tcp        0      0 0.0.0.0:22              0.0.0.0:*               LISTEN      928/sshd
tcp        0      0 127.0.0.1:25            0.0.0.0:*               LISTEN      1166/master
tcp6       0      0 :::22                   :::*                    LISTEN      928/sshd
tcp6       0      0 ::1:25                  :::*                    LISTEN      1166/master
udp        0      0 127.0.0.1:323           0.0.0.0:*                           578/chronyd
udp6       0      0 ::1:323                 :::*                                578/chronyd
raw6       0      0 :::58                   :::*                    7           622/NetworkManager
Active UNIX domain sockets (only servers)
Proto RefCnt Flags       Type       State         I-Node   PID/Program name     Path
unix  2      [ ACC ]     STREAM     LISTENING     14136    1/systemd            /run/systemd/private
unix  2      [ ACC ]     STREAM     LISTENING     19216    1166/master          private/tlsmgr
unix  2      [ ACC ]     STREAM     LISTENING     19219    1166/master          private/rewrite
unix  2      [ ACC ]     STREAM     LISTENING     19222    1166/master          private/bounce
```

图 1-18

使用-an 选项可以查看系统中所有服务的 TCP/IP 连接状态，操作命令如下。

```
[root@centos7 ~]# netstat -an
```

使用 ss 命令也可以查看服务是否处于监听状态，与 rep 命令一起使用可以查看指定端口，如查看 22 端口是否处于监听状态，如图 1-19 所示。

```
[root@centos7 ~]# ss -tnl
State      Recv-Q Send-Q                    Local Address:Port                   Peer Address:Port
LISTEN     0      128                                   *:22                                *:*
LISTEN     0      100                           127.0.0.1:25                                *:*
LISTEN     0      128                                  :::22                               :::*
LISTEN     0      100                                 ::1:25                               :::*
[root@centos7 ~]# ss -tnl | grep 22
LISTEN     0      128          *:22                  *:*
LISTEN     0      128         :::22                 :::*
[root@centos7 ~]#
```

图 1-19

netstat 命令还可以与 awk 命令一起使用，查看所有状态的数字，操作命令如下。

```
[root@centos7 ~]# netstat -an | awk '/^tcp/ {++sta[$NF]} END {for(key in sta) print key,"\t",sta[key]}'
LISTEN    4
ESTABLISHED      1
```

1.10　Linux操作系统下的网络抓包

tcpdump 命令是用来抓包的，但有时会遇到攻击，这时网卡流量会产生异常，进入的包数量可能会超过 1 万个，想知道都有哪些包，可以使用 tcpdump 命令查看。若系统中没有 tcpdump 命令，需使用 yum 命令进行安装。指定查看某块网卡，要在 tcpdump 命令后加-i 选项，如指定 ens33 网卡；也可指定 port（端口），如指定 22 端口；还可以不要 22 端口而指定 host，如指定 192.168.222.129 的包，操作命令如下。

```
[root@centos7 ~]# yum install -y tcpdump
[root@centos7 ~]# tcpdump -nn -i ens33
[root@centos7 ~]# tcpdump -nn -i ens33 port 22
[root@centos7 ~]# tcpdump -nn -i ens33 not port 22 and host 192.168.222.129
```

使用-c 选项可以指定抓取数据包的数量，如指定抓取 3 个，如图 1-20 所示。

```
[root@docker ~]# tcpdump -nn -i eth0 -c 3
tcpdump: verbose output suppressed, use -v or -vv for full protocol decode
listening on eth0, link-type EN10MB (Ethernet), capture size 262144 bytes
18:04:09.927065 IP 172.16.0.62.22 > 218.67.233.250.33645: Flags [P.], seq 3573774823:3573775035, ack 2345551631, win 273, length 212
18:04:09.927170 IP 172.16.0.62.22 > 218.67.233.250.33645: Flags [P.], seq 212:408, ack 1, win 273, length 196
18:04:09.927207 IP 172.16.0.62.22 > 218.67.233.250.33645: Flags [P.], seq 408:572, ack 1, win 273, length 164
3 packets captured
3 packets received by filter
0 packets dropped by kernel
[root@docker ~]#
```

图 1-20

使用-w 选项可以把抓取到的数据包存储到某个文件中，如保存到/root/文件夹的 tcpdump.txt 文件中，操作命令如下。

```
[root@centos7 ~]# tcpdump -nn -i ens33 -c 10 -w /root/tcpdump.txt
tcpdump: listening on ens33, link-type EN10MB (Ethernet), capture size 262144 bytes
```

```
10 packets captured
10 packets received by filter
0 packets dropped by kernel
```

对于 tcpdump.txt 文件，用 file 命令可查看其文件类型，不能用 cat 命令查看该文件，会出现乱码，因为该文件是从网卡中捕获的数据包信息，也可以使用-r 选项进行查看，如图 1-21 所示。

```
[root@centos7 ~]# file /root/tcpdump.txt
/root/tcpdump.txt: tcpdump capture file (little-endian) - version 2.4 (Ethernet, capture length 262144)
[root@centos7 ~]# tcpdump -r /root/tcpdump.txt
reading from file /root/tcpdump.txt, link-type EN10MB (Ethernet)
17:22:43.609655 IP centos7.ssh > 192.168.222.1.62193: Flags [P.], seq 2143810259:2143810391, ack 3462195579, win 264, length 132
17:22:43.610131 IP 192.168.222.1.62193 > centos7.ssh: Flags [.], ack 132, win 253, length 0
17:22:45.719889 IP 192.168.222.1.53520 > 239.255.255.250.ssdp: UDP, length 133
17:22:48.720606 IP 192.168.222.1.53520 > 239.255.255.250.ssdp: UDP, length 133
17:22:51.720927 IP 192.168.222.1.53520 > 239.255.255.250.ssdp: UDP, length 133
17:22:54.171065 IP6 fe80::9808:b549:fd50:3ebc > ff02::16: HBH ICMP6, multicast listener report v2, 1 group record(s), length 28
17:22:54.171127 IP 192.168.222.1 > igmp.mcast.net: igmp v3 report, 1 group record(s)
17:22:54.250338 IP 192.168.222.1.54530 > 239.255.255.250.ws-discovery: UDP, length 992
17:22:54.288465 IP6 fe80::9808:b549:fd50:3ebc > ff02::16: HBH ICMP6, multicast listener report v2, 1 group record(s), length 28
17:22:54.288507 IP 192.168.222.1 > igmp.mcast.net: igmp v3 report, 1 group record(s)
[root@centos7 ~]#
```

图 1-21

1.11　Firewalld和Netfilter

在配置某些服务时需要关闭 selinux，临时关闭 selinux 可运行 setenforce 0 命令。永久关闭 selinux 需编辑配置文件/etc/selinux/config，把 SELINUX=enforcing 改为 SELINUX=disabled。修改 selinux 配置文件可以使用 vim 编辑器，也可以使用 sed 命令的替换功能，修改完毕后重启 Linux 操作系统 selinux 才会生效，操作命令如下。

```
[root@centos7 ~]# setenforce 0
[root@centos7 ~]# getenforce
Permissive
[root@centos7 ~]# vim /etc/selinux/config
[root@centos7 ~]# sed -i s#SELINUX=enforcing#SELINUX=disabled#
/etc/selinux/config
[root@centos7 ~]# cat /etc/selinux/config
SELINUX=disabled
[root@centos7 ~]# getenforce
Disabled
```

CentOS 7 之前的版本中防火墙是 Netfilter，而 CentOS 7 之后的版本中防火墙默认用 Firewalld。Netfilter 和 Firewalld 这两个防火墙的机制不一样，但内部工具 iptables 的用法是一致的。通过 iptables 工具可以添加一些规则。

CentOS 7 操作系统默认使用的是 Firewalld，Netfilter 在 CentOS 7 操作系统中处于关闭状态。可以通过一系列命令在 CentOS 7 操作系统中把 Firewalld 关闭，开启 Netfilter，也可以理解为在 CentOS 7 操作系统中使用 Netfilter 是没有任何问题的。

在开启 Netfilter 之前需安装 iptables-services 软件包，操作命令如下。

```
    [root@centos7 ~]# systemctl disable firewalld.service
    Removed symlink /etc/systemd/system/multi-user.target.wants/
firewalld.service.
    Removed symlink /etc/systemd/system/dbus-org.fedoraproject.
FirewallD1.service.
    [root@centos7 ~]# systemctl stop firewalld.service
    [root@centos7 ~]# yum install -y iptables-services
    [root@centos7 ~]# systemctl enable iptables.service
    Created symlink from /etc/systemd/system/basic.target.
wants/iptables.service to /usr/lib/systemd/system/iptables.service.
    [root@centos7 ~]# systemctl start iptables.service
```

使用 iptabless －nvL 命令可以查看 iptables 的默认规则，如图 1-22 所示。

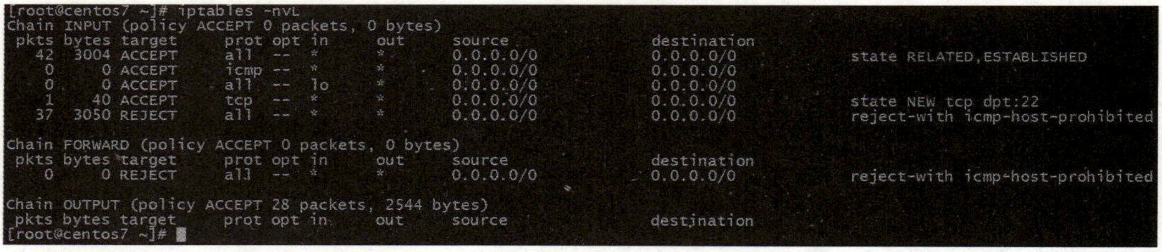

```
[root@centos7 ~]# iptables -nvL
Chain INPUT (policy ACCEPT 0 packets, 0 bytes)
 pkts bytes target     prot opt in     out     source               destination
   42  3004 ACCEPT     all  --  *      *       0.0.0.0/0            0.0.0.0/0            state RELATED,ESTABLISHED
    0     0 ACCEPT     icmp --  *      *       0.0.0.0/0            0.0.0.0/0
    0     0 ACCEPT     all  --  lo     *       0.0.0.0/0            0.0.0.0/0
    1    40 ACCEPT     tcp  --  *      *       0.0.0.0/0            0.0.0.0/0            state NEW tcp dpt:22
   37  3050 REJECT     all  --  *      *       0.0.0.0/0            0.0.0.0/0            reject-with icmp-host-prohibited

Chain FORWARD (policy ACCEPT 0 packets, 0 bytes)
 pkts bytes target     prot opt in     out     source               destination
    0     0 REJECT     all  --  *      *       0.0.0.0/0            0.0.0.0/0            reject-with icmp-host-prohibited

Chain OUTPUT (policy ACCEPT 28 packets, 2544 bytes)
 pkts bytes target     prot opt in     out     source               destination
[root@centos7 ~]#
```

图 1-22

1.12 Netfilter 5表5链简介

Netfilter 由 filter、nat、mangle、raw 和 security5 个表组成。

（1）filter 是默认的表，用于过滤包，filter 表包含 3 个内置链，即 INPUT、FORWARD 和 OUTPUT。

INPUT 链：数据包进来时，要经过该链。

FORWARD 链：数据包进入到计算机中，但并不会进入到内核中，该数据包不是给用户处理的，而是给另外一台计算机处理的，所以需要判断目标地址是不是本机，如果不是本机要经过 FORWARD 链。经过 FORWARD 链时也要做一些操作，比如把目标地址进行更改，或者进行转发。

OUTPUT 链：在本机上生成的一些包，在包发出去之前要做一些操作。这个包是发给某 IP 的，比如把发给 120.71.158.34 的包禁掉（加入黑名单，发到此 IP 的数据包都被禁掉）。

（2）nat 表多用于网络地址转换和共享。nat 表也有 3 个链，即 PREROUTING、OUTPUT 和 POSTROUTING。

PREROUTING 链：用来更改数据包在进来时的一些相关操作。

OUTPUT：在本机上生成的一些包，在包发出去之前要做一些操作。这个包是发给某 IP 的，比如把发给 120.71.158.34 的包禁掉（加入黑名单，发到此 IP 的数据包都被禁掉），并且要和 filter 表中 OUTPUT 链一致。

POSTROOUTING：用来更改数据包在发出去时的一些相关操作。

（3）mangle 表用于给数据包做标记。

（4）raw 表可以实现不追踪某些数据包。

（5）security 表用于设置强制访问（MAC）的网络规则。

1.13　iptables语法

使用-nvL 选项可以查看 iptables 的默认规则，规则默认保存在/etc/sysconfig/iptables 文件中，操作命令如下。

```
[root@centos7 ~]# iptables -nvL
[root@centos7 ~]# cat /etc/sysconfig/iptables
```

使用-F 选项可以清空 iptables 规则，删除后用-nvL 选项查看会发现不存在任何规则，如图 1-23 所示。使用-F 选项清空 iptables 规则后，/etc/sysconfig/iptables 文件中的规则不会被删除。

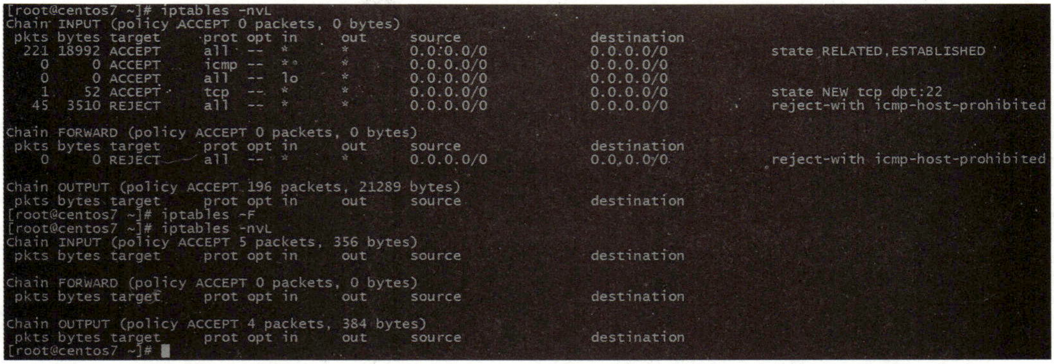

```
[root@centos7 ~]# iptables -nvL
Chain INPUT (policy ACCEPT 0 packets, 0 bytes)
 pkts bytes target     prot opt in     out     source               destination
  221 18992 ACCEPT     all  --  *      *       0.0.0.0/0            0.0.0.0/0            state RELATED,ESTABLISHED
    0     0 ACCEPT     icmp --  *      *       0.0.0.0/0            0.0.0.0/0
    0     0 ACCEPT     all  --  lo     *       0.0.0.0/0            0.0.0.0/0
    1    52 ACCEPT     tcp  --  *      *       0.0.0.0/0            0.0.0.0/0            state NEW tcp dpt:22
   45  3510 REJECT     all  --  *      *       0.0.0.0/0            0.0.0.0/0            reject-with icmp-host-prohibited

Chain FORWARD (policy ACCEPT 0 packets, 0 bytes)
 pkts bytes target     prot opt in     out     source               destination
    0     0 REJECT     all  --  *      *       0.0.0.0/0            0.0.0.0/0            reject-with icmp-host-prohibited

Chain OUTPUT (policy ACCEPT 196 packets, 21289 bytes)
 pkts bytes target     prot opt in     out     source               destination
[root@centos7 ~]# iptables -F
[root@centos7 ~]# iptables -nvL
Chain INPUT (policy ACCEPT 5 packets, 356 bytes)
 pkts bytes target     prot opt in     out     source               destination

Chain FORWARD (policy ACCEPT 0 packets, 0 bytes)
 pkts bytes target     prot opt in     out     source               destination

Chain OUTPUT (policy ACCEPT 4 packets, 384 bytes)
 pkts bytes target     prot opt in     out     source               destination
[root@centos7 ~]#
```

图 1-23

更改的规则在重启操作系统后默认不会生效，若想重启操作系统后依然生效，就需要保存 iptables 规则，操作命令如下。

```
[root@centos7 ~]# service iptables save
iptables: Saving firewall rules to /etc/sysconfig/iptables: [  OK  ]
```

被清空的规则也可以进行恢复，方法有两种，一种是重启 iptables 服务，另一种是重启操作系统。重启 iptables 服务的操作命令如下。

```
[root@centos7 ~]# service iptables restart
Redirecting to /bin/systemctl restart iptables.service
```

使用 iptables －nvL 命令默认查看的是 filter 表，增加-t 选项可以指定表的规则，操作命令如下。

```
[root@centos7 ~]# iptables -nvL
[root@centos7 ~]# iptables -t filter -nvL
[root@centos7 ~]# iptables -t nat -nvL
```

使用-Z 选项可以把计算器清零，在使用-nvL 选项查看时，第一列、第二列都会显示一些数据，操作命令如下。

```
[root@centos7 ~]# iptables -Z; iptables -nvL
```

使用-A 选项可以增加一条规则，规则默认增加到最后。除-A 选项外，还有-I 选项，使用-I 选项也可以插入规则，默认将规则插在最前面。-A 或-I 选项前不加-t 选项默认指定添加到 filter 表中。-A 和-I 选项的区别在于优先过滤的顺序，规则靠前的，会被优先过滤。

使用-s 选项可以指定来源 IP。

使用-p 选项可以指定网络协议，如 TCP、UDP、ICMP。

使用--sport 选项可以指定来源端口，如 1234 端口。使用--dport 选项可以指定目标端口，如指定目标端口为 80 端口。指定目标端口和来源端口时前面必须加-p 选项指定协议类型。

使用-d 选项可以指定目标 IP，如指定 192.168.122.128。

使用-j 选项可以指定处理的方式，-j 选项后可跟两个参数，分别为 DROP 和 REJECT。DROP 表示直接扔掉数据包，REJECT 表示拒绝。扔掉和拒绝最终实现的效果是一致的，都是让数据包进不来。DROP 和 REJECT 的区别在于，DROP 是数据包进来后看都不看直接扔掉；而 REJECT 是数据包进来后，看完告诉它“对不起，你不能进来”。

添加一条 INPUT 链，指定来源 IP 为 192.168.122.1，协议为 TCP 协议，来源端口为 1234，目标 IP 为 192.168.122.128，目标端口为 80，最后直接进行 DROP 处理，操作命令如下。

```
[root@centos7 ~]# iptables -A INPUT -s 192.168.122.1 -p tcp --sport 1234
-d 192.168.122.128 --dport 80 -j DROP
```

使用-D 选项可以删除某条规则，例如删除上述创建的规则，如图 1-24 所示。

```
[root@centos7 ~]# iptables -nvL
Chain INPUT (policy ACCEPT 0 packets, 0 bytes)
 pkts bytes target     prot opt in     out     source               destination
  214 15072 ACCEPT     all  --  *      *       0.0.0.0/0            0.0.0.0/0            state RELATED,ESTABLISHED
    0     0 ACCEPT     icmp --  *      *       0.0.0.0/0            0.0.0.0/0
    0     0 ACCEPT     all  --  lo     *       0.0.0.0/0            0.0.0.0/0
    0     0 ACCEPT     tcp  --  *      *       0.0.0.0/0            0.0.0.0/0            state NEW tcp dpt:22
  288 22917 REJECT     all  --  *      *       0.0.0.0/0            0.0.0.0/0            reject-with icmp-host-prohibited
    0     0 DROP       tcp  --  *      *       192.168.122.1        192.168.122.128      tcp spt:1234 dpt:80

Chain FORWARD (policy ACCEPT 0 packets, 0 bytes)
 pkts bytes target     prot opt in     out     source               destination
    0     0 REJECT     all  --  *      *       0.0.0.0/0            0.0.0.0/0            reject-with icmp-host-prohibited

Chain OUTPUT (policy ACCEPT 26 packets, 3688 bytes)
 pkts bytes target     prot opt in     out     source               destination

Chain VL (0 references)
 pkts bytes target     prot opt in     out     source               destination
[root@centos7 ~]# iptables -D INPUT -s 192.168.122.1 -p tcp --sport 1234 -d 192.168.122.128 --dport 80 -j DROP
[root@centos7 ~]# iptables -nvL
Chain INPUT (policy ACCEPT 0 packets, 0 bytes)
 pkts bytes target     prot opt in     out     source               destination
  318 23964 ACCEPT     all  --  *      *       0.0.0.0/0            0.0.0.0/0            state RELATED,ESTABLISHED
    0     0 ACCEPT     icmp --  *      *       0.0.0.0/0            0.0.0.0/0
    0     0 ACCEPT     all  --  lo     *       0.0.0.0/0            0.0.0.0/0
    0     0 ACCEPT     tcp  --  *      *       0.0.0.0/0            0.0.0.0/0            state NEW tcp dpt:22
  300 23853 REJECT     all  --  *      *       0.0.0.0/0            0.0.0.0/0            reject-with icmp-host-prohibited

Chain FORWARD (policy ACCEPT 0 packets, 0 bytes)
 pkts bytes target     prot opt in     out     source               destination
    0     0 REJECT     all  --  *      *       0.0.0.0/0            0.0.0.0/0            reject-with icmp-host-prohibited

Chain OUTPUT (policy ACCEPT 4 packets, 544 bytes)
 pkts bytes target     prot opt in     out     source               destination

Chain VL (0 references)
 pkts bytes target     prot opt in     out     source               destination
[root@centos7 ~]#
```

图 1-24

刚删除的规则因为是刚才创建的，所以知道规则的写法，如果需要删除的是一年前写的一条规则，可能非常复杂，现在想删除，但是不清楚当时写的什么规则了，那么这时可以使用--line-number 选项打印规则的序列编号，针对某一个编号进行删除，例如删除第 6 条规则，如图 1-25 所示。

```
[root@centos7 ~]# iptables -A INPUT -s 192.168.122.1 -p tcp --sport 1234 -d 192.168.122.128 --dport 80 -j DROP
[root@centos7 ~]# iptables -nvL --line-number
Chain INPUT (policy ACCEPT 0 packets, 0 bytes)
num   pkts bytes target     prot opt in     out     source               destination
1      473 34936 ACCEPT     all  --  *      *       0.0.0.0/0            0.0.0.0/0            state RELATED,ESTABLISHED
2        0     0 ACCEPT     icmp --  *      *       0.0.0.0/0            0.0.0.0/0
3        0     0 ACCEPT     all  --  lo     *       0.0.0.0/0            0.0.0.0/0
4        0     0 ACCEPT     tcp  --  *      *       0.0.0.0/0            0.0.0.0/0            state NEW tcp dpt:22
5      460 36484 REJECT     all  --  *      *       0.0.0.0/0            0.0.0.0/0            reject-with icmp-host-prohibited
6        0     0 DROP       tcp  --  *      *       192.168.122.1        192.168.122.128      tcp spt:1234 dpt:80

Chain FORWARD (policy ACCEPT 0 packets, 0 bytes)
num   pkts bytes target     prot opt in     out     source               destination
1        0     0 REJECT     all  --  *      *       0.0.0.0/0            0.0.0.0/0            reject-with icmp-host-prohibited

Chain OUTPUT (policy ACCEPT 6 packets, 776 bytes)
num   pkts bytes target     prot opt in     out     source               destination

Chain VL (0 references)
num   pkts bytes target     prot opt in     out     source               destination
[root@centos7 ~]# iptables -D INPUT 6
[root@centos7 ~]# iptables -nvL --line-number
Chain INPUT (policy ACCEPT 0 packets, 0 bytes)
num   pkts bytes target     prot opt in     out     source               destination
1      489 35992 ACCEPT     all  --  *      *       0.0.0.0/0            0.0.0.0/0            state RELATED,ESTABLISHED
2        0     0 ACCEPT     icmp --  *      *       0.0.0.0/0            0.0.0.0/0
3        0     0 ACCEPT     all  --  lo     *       0.0.0.0/0            0.0.0.0/0
4        0     0 ACCEPT     tcp  --  *      *       0.0.0.0/0            0.0.0.0/0            state NEW tcp dpt:22
5      460 36484 REJECT     all  --  *      *       0.0.0.0/0            0.0.0.0/0            reject-with icmp-host-prohibited

Chain FORWARD (policy ACCEPT 0 packets, 0 bytes)
num   pkts bytes target     prot opt in     out     source               destination
1        0     0 REJECT     all  --  *      *       0.0.0.0/0            0.0.0.0/0            reject-with icmp-host-prohibited

Chain OUTPUT (policy ACCEPT 4 packets, 432 bytes)
num   pkts bytes target     prot opt in     out     source               destination

Chain VL (0 references)
num   pkts bytes target     prot opt in     out     source               destination
[root@centos7 ~]#
```

图 1-25

1.14　iptables nat表应用

有两台服务器，A 服务器有两块网卡（CentOS 1.4（2）），一块网卡是外网，另一块网卡是内网。B 服务器（CentOS 1.4（2）复制）只有内网网卡，默认不能和外网通信，需求是让 B 服务器可以连接外网。解决方法是做一个路由器，只要路由器可以上网，计算机自然可以实现上网功能。

具体步骤是使用两台虚拟机，第一台虚拟机为“CentOS 1.4（2）”，第二台虚拟机为“CentOS 1.4（2）复制”。给虚拟机 CentOS 1.4（2）添加一块网卡，使该虚拟机拥有两块网卡。打开“虚拟机设置”对话框，“网络连接”选择“LAN 区段”（LAN 区段相当于将这块网卡连接到内网交换机），单击“LAN 区段”按钮，在打开的对话框中进行添加，名称自定义，如图 1-26 所示。添加完毕后返回“虚拟机设置”对话框，在“LAN 区段”下拉列表框中选择“内网交换机”，如图 1-27 所示。区段选择完毕后单击“确定”按钮保存。

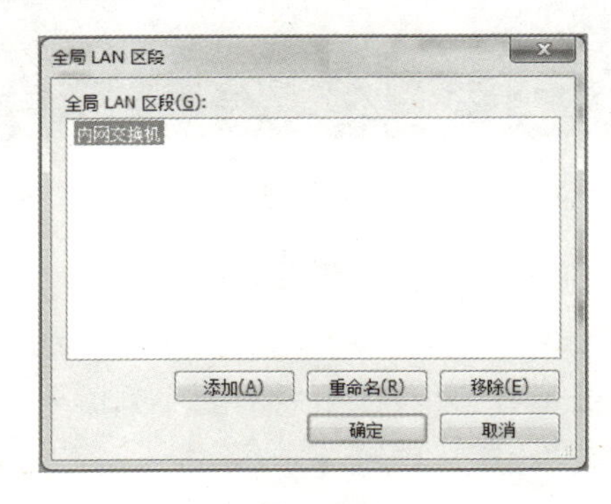

图 1-26

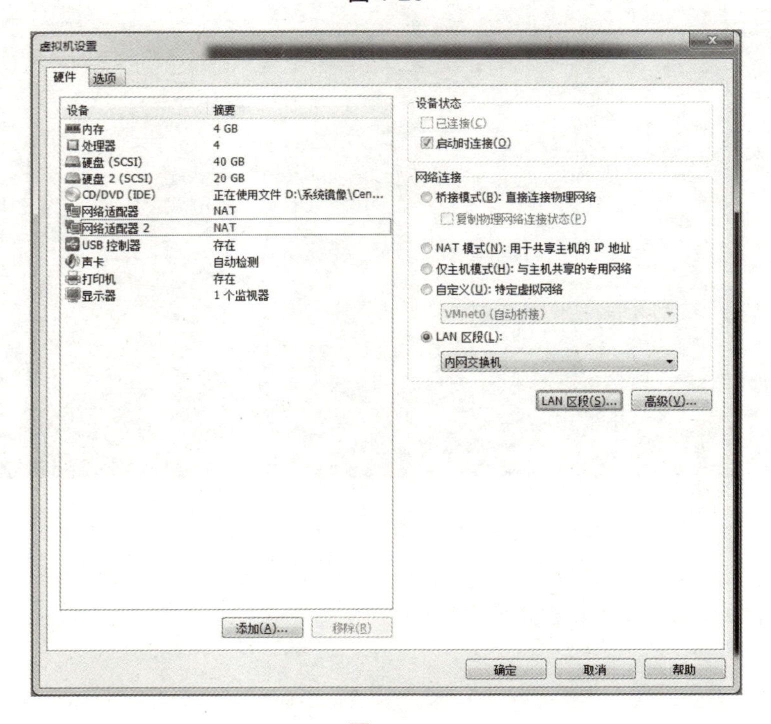

图 1-27

第二台虚拟机在复制时已经有了一块网卡，并且该网卡设置过 IP，需要禁掉该网卡。要禁掉网卡，可在“虚拟机设置”对话框中找到“设备状态”，取消选择“启动时连接”复选框即可。然后添加一块网卡，在“虚拟机设置”对话框右侧设置“LAN 区段”为“内网交换机”，保持和第一台虚拟机一致即可。

两台虚拟机全部设置完毕后，启动两台虚拟机。启动完毕后，第一台虚拟机可以在 Windows 操作系统中使用远程连接工具进行连接，但第二台虚拟机不能被连接，因为没有外网 IP。

使用 ifconfig 命令查看第一台虚拟机的网卡，如图 1-28 所示。图 1-28 中有前面设置的 ens33 网卡和刚添加的 ens37 网卡。现在给 ens37 网卡设置 IP，可以使用命令添加 IP，也可以设置配置文件。使用命令添加 IP 地址 192.168.100.1，网段为 24，如图 1-29 所示。此 IP 只是临时存在的，重启虚拟机会丢失。若想永久生效，需编辑 ens37 网卡配置文件，但 ens37 没有网卡配置文件，需复制 ens33 网卡配置文件，复制完成后，需更改 ens33 网卡为 ens37 网卡并修改网卡配置文件中的设置。

```
[root@centos7 ~]# ifconfig
ens33: flags=4163<UP,BROADCAST,RUNNING,MULTICAST>  mtu 1500
        inet 192.168.222.129  netmask 255.255.255.0  broadcast 192.168.222.255
        inet6 fe80::6be1:bf5e:d449:d3a5  prefixlen 64  scopeid 0x20<link>
        ether 00:0c:29:d6:dd:d9  txqueuelen 1000  (Ethernet)
        RX packets 120  bytes 12624 (12.3 KiB)
        RX errors 0  dropped 0  overruns 0  frame 0
        TX packets 98  bytes 11679 (11.4 KiB)
        TX errors 0  dropped 0 overruns 0  carrier 0  collisions 0

ens37: flags=4163<UP,BROADCAST,RUNNING,MULTICAST>  mtu 1500
        inet6 fe80::add4:11da:d655:b591  prefixlen 64  scopeid 0x20<link>
        ether 00:0c:29:d6:dd:e3  txqueuelen 1000  (Ethernet)
        RX packets 11  bytes 3762 (3.6 KiB)
        RX errors 0  dropped 0  overruns 0  frame 0
        TX packets 35  bytes 5914 (5.7 KiB)
        TX errors 0  dropped 0 overruns 0  carrier 0  collisions 0

lo: flags=73<UP,LOOPBACK,RUNNING>  mtu 65536
        inet 127.0.0.1  netmask 255.0.0.0
        inet6 ::1  prefixlen 128  scopeid 0x10<host>
        loop  txqueuelen 1  (Local Loopback)
        RX packets 72  bytes 5720 (5.5 KiB)
        RX errors 0  dropped 0  overruns 0  frame 0
        TX packets 72  bytes 5720 (5.5 KiB)
        TX errors 0  dropped 0 overruns 0  carrier 0  collisions 0

[root@centos7 ~]#
```

图 1-28

```
[root@centos7 ~]# ifconfig ens37 192.168.100.1/24
[root@centos7 ~]# ifconfig
ens33: flags=4163<UP,BROADCAST,RUNNING,MULTICAST>  mtu 1500
        inet 192.168.222.129  netmask 255.255.255.0  broadcast 192.168.222.255
        inet6 fe80::6be1:bf5e:d449:d3a5  prefixlen 64  scopeid 0x20<link>
        ether 00:0c:29:d6:dd:d9  txqueuelen 1000  (Ethernet)
        RX packets 322  bytes 30786 (30.0 KiB)
        RX errors 0  dropped 0  overruns 0  frame 0
        TX packets 167  bytes 19679 (19.2 KiB)
        TX errors 0  dropped 0 overruns 0  carrier 0  collisions 0

ens37: flags=4163<UP,BROADCAST,RUNNING,MULTICAST>  mtu 1500
        inet 192.168.100.1  netmask 255.255.255.0  broadcast 192.168.100.255
        inet6 fe80::20c:29ff:fed6:dde3  prefixlen 64  scopeid 0x20<link>
        ether 00:0c:29:d6:dd:e3  txqueuelen 1000  (Ethernet)
        RX packets 19  bytes 6498 (6.3 KiB)
        RX errors 0  dropped 0  overruns 0  frame 0
        TX packets 61  bytes 10074 (9.8 KiB)
        TX errors 0  dropped 0 overruns 0  carrier 0  collisions 0

lo: flags=73<UP,LOOPBACK,RUNNING>  mtu 65536
        inet 127.0.0.1  netmask 255.0.0.0
        inet6 ::1  prefixlen 128  scopeid 0x10<host>
        loop  txqueuelen 1  (Local Loopback)
        RX packets 72  bytes 5720 (5.5 KiB)
        RX errors 0  dropped 0  overruns 0  frame 0
        TX packets 72  bytes 5720 (5.5 KiB)
        TX errors 0  dropped 0 overruns 0  carrier 0  collisions 0

[root@centos7 ~]#
```

图 1-29

第二台虚拟机由于无法进行远程连接，所以需要在虚拟机中进行设置，在设置之前为避免发生意外，需要把 ens33 网卡禁掉，操作命令为 ifdown ens33。禁掉 ens33 网卡后，同样为 ens37 网卡编辑 IP 地址 192.168.100.100，网段为 24 段，如图 1-30 所示。设置完 IP 地址后，若能 ping 通 192.168.100.1，则说明没有问题，如图 1-31 所示。

```
[root@centos7 ~]# ifconfig ens37 192.168.100.100/24
[root@centos7 ~]# ifconfig
ens33: flags=4099<UP,BROADCAST,MULTICAST>  mtu 1500
        ether 00:0c:29:f7:65:5b  txqueuelen 1000  (Ethernet)
        RX packets 0  bytes 0 (0.0 B)
        RX errors 0  dropped 0  overruns 0  frame 0
        TX packets 0  bytes 0 (0.0 B)
        TX errors 0  dropped 0 overruns 0  carrier 0  collisions 0

ens37: flags=4163<UP,BROADCAST,RUNNING,MULTICAST>  mtu 1500
        inet 192.168.100.100  netmask 255.255.255.0  broadcast 192.168.100.255
        inet6 fe80::20c:29ff:fef7:6565  prefixlen 64  scopeid 0x20<link>
        ether 00:0c:29:f7:65:65  txqueuelen 1000  (Ethernet)
        RX packets 21  bytes 6618 (6.4 KiB)
        RX errors 0  dropped 0  overruns 0  frame 0
        TX packets 111  bytes 19532 (19.0 KiB)
        TX errors 0  dropped 0 overruns 0  carrier 0  collisions 0

lo: flags=73<UP,LOOPBACK,RUNNING>  mtu 65536
        inet 127.0.0.1  netmask 255.0.0.0
        inet6 ::1  prefixlen 128  scopeid 0x10<host>
        loop  txqueuelen 1  (Local Loopback)
        RX packets 270  bytes 23410 (22.8 KiB)
        RX errors 0  dropped 0  overruns 0  frame 0
        TX packets 270  bytes 23410 (22.8 KiB)
        TX errors 0  dropped 0 overruns 0  carrier 0  collisions 0

[root@centos7 ~]#
```

图 1-30

```
[root@centos7 ~]# ping 192.168.100.1
PING 192.168.100.1 (192.168.100.1) 56(84) bytes of data.
64 bytes from 192.168.100.1: icmp_seq=1 ttl=64 time=1.83 ms
64 bytes from 192.168.100.1: icmp_seq=2 ttl=64 time=1.14 ms
64 bytes from 192.168.100.1: icmp_seq=3 ttl=64 time=1.04 ms
^C
--- 192.168.100.1 ping statistics ---
3 packets transmitted, 3 received, 0% packet loss, time 2004ms
rtt min/avg/max/mdev = 1.041/1.342/1.838/0.354 ms
[root@centos7 ~]# _
```

图 1-31

若想让 B 服务器连网，则 A 服务器必须实现路由转发，nat 表进行网络转发必须要修改内核参数，/proc/sys/net/ipv4/ip_forward 默认参数是 0，0 表示没有开启内核转发，把该配置文件中的 0 改为 1 即可，操作命令如下。

```
[root@centos7 ~]# cat /proc/sys/net/ipv4/ip_forward
0
[root@centos7 ~]# echo "1" > !$
echo "1" > /proc/sys/net/ipv4/ip_forward
[root@centos7 ~]# !cat
cat /proc/sys/net/ipv4/ip_forward
1
```

增加一条规则，该规则可实现上网功能，表示让 192.168.100.0/24 这个网段可以上网，如图 1-32 所示。

```
[root@centos7 ~]# iptables -t nat -A POSTROUTING -s 192.168.100.0/24 -o ens33 -j MASQUERADE
[root@centos7 ~]# iptables -t nat -nvL
Chain PREROUTING (policy ACCEPT 0 packets, 0 bytes)
 pkts bytes target     prot opt in     out     source               destination

Chain INPUT (policy ACCEPT 0 packets, 0 bytes)
 pkts bytes target     prot opt in     out     source               destination

Chain OUTPUT (policy ACCEPT 0 packets, 0 bytes)
 pkts bytes target     prot opt in     out     source               destination

Chain POSTROUTING (policy ACCEPT 0 packets, 0 bytes)
 pkts bytes target     prot opt in     out     source               destination
    0     0 MASQUERADE  all  --  *      ens33   192.168.100.0/24     0.0.0.0/0
[root@centos7 ~]#
```

图 1-32

设置 B 服务器的网关为 192.168.100.1，因为数据从 A 服务器到 B 服务器，或者从 B 服务器到 A 服务器是需要一个网关的，所以要设置该网关，设置完毕后 ping A 服务器的外网 IP 地址 192.168.222.129，如图 1-33 所示。

```
[root@centos7 ~]# route -n
Kernel IP routing table
Destination     Gateway         Genmask         Flags Metric Ref    Use Iface
192.168.100.0   0.0.0.0         255.255.255.0   U     0      0        0 ens37
[root@centos7 ~]# route add default gw 192.168.100.1
[root@centos7 ~]# route -n
Kernel IP routing table
Destination     Gateway         Genmask         Flags Metric Ref    Use Iface
0.0.0.0         192.168.100.1   0.0.0.0         UG    0      0        0 ens37
192.168.100.0   0.0.0.0         255.255.255.0   U     0      0        0 ens37
[root@centos7 ~]# ping 192.168.222.129
PING 192.168.222.129 (192.168.222.129) 56(84) bytes of data.
64 bytes from 192.168.222.129: icmp_seq=1 ttl=64 time=0.877 ms
64 bytes from 192.168.222.129: icmp_seq=2 ttl=64 time=1.08 ms
^C
--- 192.168.222.129 ping statistics ---
2 packets transmitted, 2 received, 0% packet loss, time 1001ms
rtt min/avg/max/mdev = 0.877/0.979/1.082/0.107 ms
[root@centos7 ~]#
```

图 1-33

也可以配置 DNS 使之访问外网，DNS 配置文件为/etc/resolv.conf，添加一行“nameserver 114.114.114.114”，然后保存并退出即可。ping 114.114.114.114 和外网是可以 ping 通的，但是用 Windows 系统中的 cmd 窗口 ping192.168.100.100 仍然不通。现在每次操作都需要进入虚拟机中，很麻烦，使用远程操作会更加方便、快捷。

不能直接登录 192.168.100.100，但可以连接 192.168.222.129。通过 A 服务器跳转到 B 服务器叫作端口映射，把 192.168.100.100 的 22 端口映射出来，但 192.168.222.129 的 22 端口已经被占用，于是可以把 192.168.222.129 的 22 端口映射成其他端口，如映射成 1122 端口。可以理解为访问 192.168.222.129 的 1122 端口实际上是访问 192.168.100.100 的 22 端口。在添加 iptables 规则之前，需要把前面设置的规则删除，避免影响当前操作，如图 1-34 所示。设置完 A 服务器后需要在 B 服务器上设置网关，因为前面已经讲解过，故不再介绍设置操作。

全部设置完毕后，使用远程连接管理工具新建一个连接，设置主机名为 192.168.222.129，端口为 1122，用户名为 root，能够登录进去就表示设置成功了。

```
[root@centos7 ~]# echo "1" > /proc/sys/net/ipv4/ip_forward ^C
[root@centos7 ~]# iptables -t nat -D POSTROUTING -s 192.168.100.0/24 -o ens33 -j MASQUERADE
[root@centos7 ~]# iptables -t nat -A PREROUTING -d 192.168.222.129 -p tcp --dport 1122 -j DNAT --to 192.168.100.100:22
[root@centos7 ~]# iptables -t nat -A POSTROUTING -s 192.168.100.100 -j SNAT --to 192.168.222.129
[root@centos7 ~]# iptables -t nat -nvL
Chain PREROUTING (policy ACCEPT 0 packets, 0 bytes)
 pkts bytes target     prot opt in     out     source               destination
    0     0 DNAT       tcp  --  *      *       0.0.0.0/0            192.168.222.129      tcp dpt:1122 to:192.168.100.100:22

Chain INPUT (policy ACCEPT 0 packets, 0 bytes)
 pkts bytes target     prot opt in     out     source               destination

Chain OUTPUT (policy ACCEPT 1 packets, 76 bytes)
 pkts bytes target     prot opt in     out     source               destination

Chain POSTROUTING (policy ACCEPT 1 packets, 76 bytes)
 pkts bytes target     prot opt in     out     source               destination
    0     0 SNAT       all  --  *      *       192.168.100.100      0.0.0.0/0            to:192.168.222.129
[root@centos7 ~]#
```

图 1-34

1.15 iptables规则备份与恢复

使用 service iptables save 命令可以把规则保存在/etc/sysconfig/iptables 文件中，若想将规则保存到其他文件中，可以使用 iptables-save 命令重定向到/root 目录下的 ipt.txt 文件中，如图 1-35 所示。

```
[root@centos7 ~]# iptables-save > /root/ipt.txt
[root@centos7 ~]# cat !$
cat /root/ipt.txt
# Generated by iptables-save v1.4.21 on Tue Jan 23 17:01:44 2018
*nat
:PREROUTING ACCEPT [263:20375]
:INPUT ACCEPT [1:52]
:OUTPUT ACCEPT [9:692]
:POSTROUTING ACCEPT [9:692]
-A PREROUTING -d 192.168.222.129/32 -p tcp -m tcp --dport 1122 -j DNAT --to-destination 192.168.100.100:22
-A POSTROUTING -s 192.168.100.100/32 -j SNAT --to-source 192.168.222.129
COMMIT
# Completed on Tue Jan 23 17:01:44 2018
# Generated by iptables-save v1.4.21 on Tue Jan 23 17:01:44 2018
*filter
:INPUT ACCEPT [0:0]
:FORWARD ACCEPT [0:0]
:OUTPUT ACCEPT [1230:129647]
-A INPUT -m state --state RELATED,ESTABLISHED -j ACCEPT
-A INPUT -p icmp -j ACCEPT
-A INPUT -i lo -j ACCEPT
-A INPUT -p tcp -m state --state NEW -m tcp --dport 22 -j ACCEPT
-A INPUT -j REJECT --reject-with icmp-host-prohibited
-A FORWARD -j REJECT --reject-with icmp-host-prohibited
COMMIT
# Completed on Tue Jan 23 17:01:44 2018
[root@centos7 ~]#
```

图 1-35

使用 iptables -F 命令可以清空 iptables 规则，清空后可以使用 iptables-restore 命令恢复上述保存的 iptables 规则，如图 1-36 所示。

```
[root@centos7 ~]# iptables -F
[root@centos7 ~]# iptables -nvL
Chain INPUT (policy ACCEPT 33 packets, 2452 bytes)
 pkts bytes target     prot opt in     out     source               destination

Chain FORWARD (policy ACCEPT 0 packets, 0 bytes)
 pkts bytes target     prot opt in     out     source               destination

Chain OUTPUT (policy ACCEPT 14 packets, 1304 bytes)
 pkts bytes target     prot opt in     out     source               destination
[root@centos7 ~]# iptables-restore < /root/ipt.txt
[root@centos7 ~]# iptables -nvL
Chain INPUT (policy ACCEPT 0 packets, 0 bytes)
 pkts bytes target     prot opt in     out     source               destination
    5   356 ACCEPT     all  --  *      *       0.0.0.0/0            0.0.0.0/0            state RELATED,ESTABLISHED
    0     0 ACCEPT     icmp --  *      *       0.0.0.0/0            0.0.0.0/0
    0     0 ACCEPT     all  --  lo     *       0.0.0.0/0            0.0.0.0/0
    0     0 ACCEPT     tcp  --  *      *       0.0.0.0/0            0.0.0.0/0            state NEW tcp dpt:22
    0     0 REJECT     all  --  *      *       0.0.0.0/0            0.0.0.0/0            reject-with icmp-host-prohibited

Chain FORWARD (policy ACCEPT 0 packets, 0 bytes)
 pkts bytes target     prot opt in     out     source               destination
    0     0 REJECT     all  --  *      *       0.0.0.0/0            0.0.0.0/0            reject-with icmp-host-prohibited

Chain OUTPUT (policy ACCEPT 4 packets, 496 bytes)
 pkts bytes target     prot opt in     out     source               destination
[root@centos7 ~]#
```

图 1-36

1.16 Firewalld的9个zone

前面在操作 Netfilter 时把 Firewalld 关闭了，现在重新打开 Firewalld，操作命令如下。

```
    [root@centos7 ~]# systemctl disable iptables.service
    Removed symlink /etc/systemd/system/basic.target.wants/
iptables.service.
    [root@centos7 ~]# systemctl stop iptables.service
    [root@centos7 ~]# systemctl enable firewalld.service
    Created symlink from /etc/systemd/system/dbus-org.fedoraproject.
FirewallD1.service to /usr/lib/systemd/system/firewalld.service.
    Created symlink from /etc/systemd/system/multi-user.target.wants/
firewalld.service to /usr/lib/systemd/system/firewalld.service.
    [root@centos7 ~]# systemctl start firewalld.service
    [root@centos7 ~]# systemctl start firewalld.service
```

Firewalld 默认有 9 个 zone，zone 是 Firewalld 的单位，默认的 zone 为 public。每个 zone 相当于一个规则集，zone 里自带一些规则，比如规则里面放行 80 端口、22 端口，禁掉 8080 端口，这叫作规则集。

使用 firewall-cmd --get-zones 命令可以查看 Firewalld 中的所有 zone，操作命令如下。

```
    [root@centos7 ~]# firewall-cmd --get-zones
    block dmz drop external home internal public trusted work
```

9 个 zone 的含义分别如下。

- drop（丢弃）：接收的任何网络数据包都会被丢弃，没有任何回复，仅有发送出去的网络连接。
- block（限制）：接收的任何网络连接都被 IPv4 的 icmp-host-prohibited 信息和 IPv6 的 icmp6-adm-prohibited 信息所拒绝。
- public（公共）：在公共区域内使用，不相信网络内的其他计算机不会对自己的计算机造成危害，只能接收经过选择的连接。
- external（外部）：为路由器启用了伪装功能的外部网。不能信任来自网络的其他计算机，不相信其他计算机不会对自己的计算机造成危害，只能接收经过选择的连接。
- dmz（非军事区）：用于非军事区内的计算机，在此区域内可公开访问，可以有限地进入内部网络，仅接收经过选择的连接。
- work（工作）：用于工作区，基本相信网络内的其他计算机不会危害自己的计算机，仅接收经过选择的连接。

- home（家庭）：用于家庭网络，基本相信网络内的其他计算机不会危害自己的计算机，仅接收经过选择的连接。
- internal（内部）：用于内部网络，基本相信网络内的其他计算机不会危害自己的计算机，仅接收经过选择的连接。
- trusted（信任）：可接收所有的网络连接。

查看 Firewalld 默认的 zone，操作命令如下。

```
[root@centos7 ~]# firewall-cmd --get-default-zone
public
```

1.17 Firewalld关于zone的操作

设置 Firewalld 默认的 zone 为 work，操作命令如下。

```
[root@centos7 ~]# firewall-cmd --set-default-zone=work
success
[root@centos7 ~]# firewall-cmd --get-default-zone
work
```

zone 可以针对网卡进行操作，如查看 Firewalld 的 ens33、ens37 和 lo 网卡的 zone，操作命令如下。

```
[root@centos7 ~]# firewall-cmd --get-zone-of-interface=ens33
work
[root@centos7 ~]# firewall-cmd --get-zone-of-interface=ens37
work
[root@centos7 ~]# firewall-cmd --get-zone-of-interface=lo
no zone
```

如果 ens37 显示 no zone，就需要在/etc/sysconfig/network-scripts/目录下，把 ifcfg-ens33 的配置文件复制一份并重命名为 ifcfg-ens37，在 ens37 中写入一些配置信息，然后重启网络服务，再重新加载 Firewalld 服务，操作命令如下。

```
[root@centos7 ~]# cd /etc/sysconfig/network-scripts/
[root@centos7 network-scripts]# cp ifcfg-ens33 ifcfg-ens37
[root@centos7 network-scripts]# vim ifcfg-ens37
#删除 UUID 行
NAME=ens37
DEVICE=ens37
:wq
[root@centos7 network-scripts]# systemctl restart network.service
```

```
[root@centos7 network-scripts]# systemctl restart firewalld.service
```

给 ens37 网卡设置 zone 为 dmz，给 lo 网卡设置 zone 为 public，操作命令如下。

```
[root@centos7 ~]# firewall-cmd --zone=dmz --add-interface=ens37
The interface is under control of NetworkManager, setting zone to 'dmz'.
success
[root@centos7 ~]# firewall-cmd --get-zone-of-interface=ens37
dmz
[root@centos7 ~]# firewall-cmd --zone=public --add-interface=lo
success
[root@centos7 ~]# firewall-cmd --get-zone-of-interface=lo
public
```

更改 ens37 网卡的 zone 为 block，操作命令如下。

```
[root@centos7 ~]# firewall-cmd --zone=block --change-interface=ens37
The interface is under control of NetworkManager, setting zone to 'block'.
success
[root@centos7 ~]# firewall-cmd --get-zone-of-interface=ens37
block
```

针对网卡删除 zone，ens37 网卡已经被设置为 block 网卡的 zone，删除 block 中的 ens37，删除后变为默认的 zonework，操作命令如下。

```
[root@centos7 ~]# firewall-cmd --zone=block --remove-interface=ens37
The interface is under control of NetworkManager, setting zone to default.
success
[root@centos7 ~]# firewall-cmd --get-zone-of-interface=ens37
work
```

查看当前操作系统中所有网卡所在的 zone，操作命令如下。

```
[root@centos7 ~]# firewall-cmd --get-active-zones
work
  interfaces: ens33 ens37
public
  interfaces: lo
```

1.18　Firewalld关于services的操作

services 是 zone 下面的子单元，可以理解为 services 是一个指定的端口，防火墙无外乎就是针对某一个端口做一些限制，HTTP 操作的是 80 端口，HTTPS 操作的是 443 端口，SSH 操作的是 22 端口，查看当前操作系统中所有的 services，操作命令如下。

```
[root@centos7 ~]# firewall-cmd --get-services
RH-Satellite-6 amanda-client amanda-k5-client bacula bacula-client
bitcoin bitcoin-rpc bitcoin-testnet bitcoin-testnet-rpc ceph ceph-mon
cfengine condor-collector ctdb dhcp dhcpv6 dhcpv6-client dns docker-registry
dropbox-lansync elasticsearch freeipa-ldap freeipa-ldaps
freeipa-replication freeipa-trust ftp ganglia-client ganglia-master
high-availability http https imap imaps ipp ipp-client ipsec iscsi-target
kadmin kerberos kibana klogin kpasswd kshell ldap ldaps libvirt libvirt-tls
managesieve mdns mosh mountd ms-wbt mssql mysql nfs nrpe ntp openvpn
ovirt-imageio ovirt-storageconsole ovirt-vmconsole pmcd pmproxy pmwebapi
pmwebapis pop3 pop3s postgresql privoxy proxy-dhcp ptp pulseaudio
puppetmaster quassel radius rpc-bind rsh rsyncd samba samba-client sane sip
sips smtp smtp-submission smtps snmp snmptrap spideroak-lansync squid ssh
synergy syslog syslog-tls telnet tftp tftp-client tinc tor-socks
transmission-client vdsm vnc-server wbem-https xmpp-bosh xmpp-client
xmpp-local xmpp-server
```

当前的默认 zone 为 work，查看 work 下面有哪些 services，操作命令如下。

```
[root@centos7 ~]# firewall-cmd --list-services
ssh dhcpv6-client
```

查看指定 zone 下面有哪些 services，如查看 home 下面有哪些 services，操作命令如下。

```
[root@centos7 ~]# firewall-cmd --zone=home --list-services
ssh mdns samba-client dhcpv6-client
```

日常工作中经常用到的 zone 是 public 和 work，针对 public 进行一些相关操作，例如把 HTTP、FTP、NFS 服务增加到 public zone 中，操作命令如下。

```
[root@centos7 ~]# firewall-cmd --zone=public --add-service=http
success
[root@centos7 ~]# firewall-cmd --zone=public --list-services
ssh dhcpv6-client http nfs ftp
```

上述增加的服务只保存在内存中，若想永久保存，需要保存到配置文件中，保存到配置文件中的方法是在增加的服务后面加上-permanent，配置文件在/etc/firewalld/zones/目录下的 public.xml 解析文件中，如图 1-37 所示。

/usr/lib/firewalld/zones/目录下是所有 zone 的配置文件模板，操作命令如下。

```
[root@centos7 ~]# ls /usr/lib/firewalld/zones/
block.xml  dmz.xml  drop.xml  external.xml  home.xml  internal.xml
public.xml  trusted.xml  work.xml
```

```
[root@centos7 ~]# firewall-cmd --zone=public --add-service=ftp --permanent
success
[root@centos7 ~]# ls /etc/firewalld/zones/
public.xml  public.xml.old
[root@centos7 ~]# cat /etc/firewalld/zones/public.xml
<?xml version="1.0" encoding="utf-8"?>
<zone>
  <short>Public</short>
  <description>For use in public areas. You do not trust the other computers on networks to not harm your computer. Only selected incoming connect
ions are accepted.</description>
  <service name="ssh"/>
  <service name="dhcpv6-client"/>
  <service name="ftp"/>
</zone>
[root@centos7 ~]#
```

图 1-37

现在有一个需求，把 FTP 服务的默认端口更改为 1121 端口，在 work zone 下面放行 FTP 服务，操作命令如下。

```
[root@centos7 ~]# cp /usr/lib/firewalld/services/ftp.xml
/etc/firewalld/services/
[root@centos7 ~]# vim /etc/firewalld/services/ftp.xml
  <port protocol="tcp" port="1121"/>
:wq
[root@centos7 ~]# cp /usr/lib/firewalld/zones/work.xml
/etc/firewalld/zones/
[root@centos7 ~]# vim /etc/firewalld/zones/work.xml
  <service name="ssh"/>
  <service name="dhcpv6-client"/>
  <service name="ftp"/>
:wq
[root@centos7 ~]# firewall-cmd --reload #重新加载
success
[root@centos7 ~]# firewall-cmd --zone=work --list-services
ssh dhcpv6-client ftp
```

1.19　Linux任务计划

在 Linux 操作系统中，任务计划是必不可少的，有时需要在凌晨做一件事情，如备份数据、重启某项服务，操作的对象可能是 shell 脚本，也有可能是单独的某个命令，无论是什么，总要去执行它。

/etc/crontab 是任务计划的配置文件，如图 1-38 所示。该配置文件中定义了几个变量，例如 SHELL=/bin/bash、PATH=/sbin:/usr/sbin:/usr/bin/（命令路径）、MAILTO=root（发送邮件给一个用户）。

变量下面是任务计划的格式，从左到右为 5 个*，这 5 个*表示 5 个位，*表示全部。第一个位表示分钟，范围是 0~59；第二个位表示小时，范围是 0~23；第三个位表示日期，范围是 1~31；第四个位表示月份，范围是 1~12，可以用数字表示，也可以用英文月份简写方式表示，如 jan、feb、mar；第五个位表示星期几，范围是 0~6，

0 表示周日，也可以用英文简写方式表示，如 sun、mon、tue。

格式后面的 user-name 表示执行的用户，如不写则默认是 root 用户。user-name 后面是执行的命令。

```
[root@centos7 ~]# cat /etc/crontab
SHELL=/bin/bash
PATH=/sbin:/bin:/usr/sbin:/usr/bin
MAILTO=root

# For details see man 4 crontabs

# Example of job definition:
# .---------------- minute (0 - 59)
# |  .------------- hour (0 - 23)
# |  |  .---------- day of month (1 - 31)
# |  |  |  .------- month (1 - 12) OR jan,feb,mar,apr ...
# |  |  |  |  .---- day of week (0 - 6) (Sunday=0 or 7) OR sun,mon,tue,wed,thu,fri,sat
# |  |  |  |  |
# *  *  *  *  * user-name  command to be executed

[root@centos7 ~]#
```

图 1-38

使用 crontab －e 命令可以进入到 crontab 配置文件中，其用法和 vim 编辑器的用法一致，按 i 键可进入编辑模式。现在有一个任务计划需要在每天凌晨三点去执行/usr/local/sbin/lnmp.sh，还可把正确的输出重定向到/tmp/lnmp.log 文件中，错误的输出也重定向到/tmp/lnmp.log 文件中，操作命令如下。

```
[root@centos7 ~]# crontab -e
0 3 * * * /bin/bash /usr/local/sbin/lnmp.sh >>/tmp/lnmp.log
2>>/tmp/lnmp.log
```

也可以指定某个范围，如 1~10 日的每天凌晨 3 点执行该脚本，操作命令如下。

```
[root@centos7 ~]# crontab -e
0 3 1-10 * * /bin/bash /usr/local/sbin/lnmp.sh >>/tmp/lnmp.log
2>>/tmp/lnmp.lo
```

还可指定月份，如双月份的每个月 1~10 日，每周周二和周五凌晨 3 点执行该脚本，操作命令如下。

```
[root@centos7 ~]# crontab -e
0 3 1-10 */2 2,5 /bin/bash /usr/local/sbin/lnmp.sh >>/tmp/lnmp.log
2>>/tmp/lnmp.log
```

要想使 crontab 任务计划正常使用，就需要启动 crond 服务，启动完成后可以用 ps aux 或 systemctl status 命令查看是否启动成功，操作命令如下。

```
[root@centos7 ~]# systemctl start crond.service
[root@centos7 ~]# ps aux |grep crond
root        612  0.0  0.0 126264  1620 ?        Ss   03:31   0:00
/usr/sbin/crond -n
root       3396  0.0  0.0 112664   976 pts/0    S+   11:24   0:00 grep
```

```
--color=auto crond
    [root@centos7 ~]# systemctl status crond.service
```

使用-l 选项可以显示某个用户的 crontab 文件内容，如果不指定用户，则显示当前用户的 crontab 文件内容，操作命令如下。

```
    [root@centos7 ~]# crontab -l
    0 3 1-10 */2 2,5 /bin/bash /usr/local/sbin/lnmp.sh >>/tmp/lnmp.log
2>>/tmp/lnmp.log
```

使用-r 选项可以从/var/spool/cron 目录中删除某个用户的 crontab 文件，如果不指定用户，则默认删除当前用户的 crontab 文件，操作命令如下。

```
    [root@centos7 ~]# crontab -r
    [root@centos7 ~]# crontab -l
    no crontab for root
```

-u 选项用来设置某个用户的 crontab 服务，如-u ixdba 命令表示设置 ixdba 用户的 crontab 服务，此参数一般由 root 用户来运行，操作命令如下。

```
    [root@centos7 ~]# crontab -u ixdba -l
    no crontab for ixdba
```

1.20　Linux系统服务管理工具chkconfig

iptables、Firewalld、httpd、PHP、Nginx、MySQL 等都被称之为服务，这些服务是需要工具来管理的，通过工具可以控制服务的启动、开机自启动和在指定级别启动。CentOS 6 操作系统之前的版本中会用到 chkconfig，CentOS 7 操作系统已经不再使用 chkconfig 工具，但 CentOS 7 系统为了和之前的版本兼容，允许继续使用 chkconfig 工具。

使用 chkconfig--list 命令可以列出当前系统中有哪些服务使用 chkconfig 工具来管理，如图 1-39 所示。图 1-39 中只列出了两个服务，即 netconsole 服务和 network 服务，它们在/etc/init.d/目录下。

```
[root@centos7 ~]# chkconfig --list

Note: This output shows SysV services only and does not include native
      systemd services. SysV configuration data might be overridden by native
      systemd configuration.

      If you want to list systemd services use 'systemctl list-unit-files'.
      To see services enabled on particular target use
      'systemctl list-dependencies [target]'.

netconsole      0:off   1:off   2:off   3:off   4:off   5:off   6:off
network         0:off   1:off   2:on    3:on    4:on    5:on    6:off
[root@centos7 ~]# ls /etc/init.d/
functions  netconsole  network  README
[root@centos7 ~]#
```

图 1-39

使用 chkconfig 可以关闭 network 服务，off 表示关闭；既然可以关闭，那么也能打开，on 表示开启，操作命令如下。

```
[root@centos7 ~]# chkconfig network off
[root@centos7 ~]# chkconfig network on
```

使用 chkconfig 可以控制某个服务的某个级别的开启或关闭，如关闭 network 服务的 3 级别，也可指定关闭多个级别，如关闭 4 级别和 5 级别，操作命令如下。

```
[root@centos7 ~]# chkconfig --level 3 network off
[root@centos7 ~]# chkconfig --level 45 network off
```

使用--add 选项可以添加服务到列表中，使用--del 选项可以把某服务从列表中删除，如图 1-40 所示。

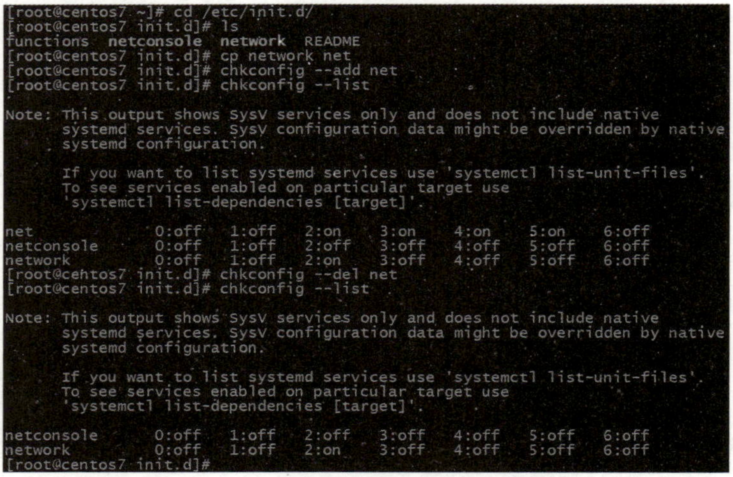

```
[root@centos7 ~]# cd /etc/init.d/
[root@centos7 init.d]# ls
functions  netconsole  network  README
[root@centos7 init.d]# cp network net
[root@centos7 init.d]# chkconfig --add net
[root@centos7 init.d]# chkconfig --list

Note: This output shows SysV services only and does not include native
      systemd services. SysV configuration data might be overridden by native
      systemd configuration.

      If you want to list systemd services use 'systemctl list-unit-files'.
      To see services enabled on particular target use
      'systemctl list-dependencies [target]'.

net             0:off   1:off   2:on    3:on    4:on    5:on    6:off
netconsole      0:off   1:off   2:off   3:off   4:off   5:off   6:off
network         0:off   1:off   2:on    3:off   4:off   5:off   6:off
[root@centos7 init.d]# chkconfig --del net
[root@centos7 init.d]# chkconfig --list

Note: This output shows SysV services only and does not include native
      systemd services. SysV configuration data might be overridden by native
      systemd configuration.

      If you want to list systemd services use 'systemctl list-unit-files'.
      To see services enabled on particular target use
      'systemctl list-dependencies [target]'.

netconsole      0:off   1:off   2:off   3:off   4:off   5:off   6:off
network         0:off   1:off   2:on    3:off   4:off   5:off   6:off
[root@centos7 init.d]#
```

图 1-40

1.21 systemd管理服务

使用 systemctl list-unit-files 命令可以罗列出 CentOS 7 操作系统的所有服务，不仅包含 service，还有 target 和 socket，如果只想查看 service 服务，可在命令后面加上类型，操作命令如下。

```
[root@centos7 ~]# systemctl list-unit-files
[root@centos7 ~]# systemctl list-units --all --type=service
```

设置某服务开机自启动，如设置 crond 服务开机自启动，操作命令如下。

```
[root@centos7 ~]# systemctl enable crond.service
Created symlink from
```

```
/etc/systemd/system/multi-user.target.wants/crond.service to
/usr/lib/systemd/system/crond.service.
```

设置某服务开机不自启动，如设置 crond 服务开机不自启动，操作命令如下。

```
    [root@centos7 ~]# systemctl disable crond.service
    Removed symlink
/etc/systemd/system/multi-user.target.wants/crond.service.
```

可以查看某服务的状态，如查看 crond 服务的状态，操作命令如下。

```
    [root@centos7 ~]# systemctl status crond.service
```

可以停止某服务，如停止 crond 服务，操作命令如下。

```
    [root@centos7 ~]# systemctl stop crond.service
```

可以开启某服务，如开启 crond 服务，操作命令如下。

```
    [root@centos7 ~]# systemctl start crond.service
```

可以重启某服务，如重启 crond 服务，操作命令如下。

```
    [root@centos7 ~]# systemctl restart crond.service
```

可以检查某服务是否开机自启动，如查看 crond 服务是否开机自启动，操作命令如下。

```
    [root@centos7 ~]# systemctl is-enabled crond.service
    enabled
```

1.22　unit和target简介

1.22.1　unit 简介

/usr/lib/systemd/system 目录下包含了系统中所有的 unit，分为以下几种类型。

- service：系统服务。
- target：多个 unit 组成的组。
- device：硬件设备。
- mount：文件系统挂载点。
- automount：自动挂载点。
- path：文件或路径。
- scope：不是由 systemd 启动的外部进程。

- slice：进程组。
- snapshot systemd：快照。
- socket：进程间通信套接字。
- swap：swap 文件。
- timer：定时器。

列出正在运行的 unit，操作命令如下。

```
[root@centos7 ~]# systemctl list-units
```

列出所有的 unit，包括失败的或闲置的，操作命令如下。

```
[root@centos7 ~]# systemctl list-units --all
```

列出闲置状态的 unit，操作命令如下。

```
[root@centos7 ~]# systemctl list-units --all --state=inactive
```

列出活跃状态的系统服务，操作命令如下。

```
[root@centos7 ~]# systemctl list-units --type=service
```

查看某个服务是否为活跃的，如查看 crond 服务是否为活跃的，操作命令如下。

```
[root@centos7 ~]# systemctl is-active crond.service
active
```

1.22.2 target 简介

为了方便管理，系统用 target 来管理 unit，如查看系统中所有的类型为 target 的服务，操作命令如下。

```
[root@centos7 ~]# systemctl list-unit-files --type=target
```

查看指定 target 下面有哪些 unit，如查看 mulit-user.target 下面有哪些 unit，操作命令如下。

```
[root@centos7 ~]# systemctl list-dependencies multi-user.target
```

查看系统默认的 target，操作命令如下。

```
[root@centos7 ~]# systemctl get-default
multi-user.target
```

设置系统默认的 target，如设置默认的 target 为 multi-user.target，操作命令如下。

```
[root@centos7 ~]# systemctl set-default multi-user.target
Removed symlink /etc/systemd/system/default.target.
```

```
    Created symlink from /etc/systemd/system/default.target to
/usr/lib/systemd/system/multi-user.target.
```

一个 service 属于一种类型的 unit，unit 包含多种类型，其中一种类型是 service。多个 unit 组成了一个 target，一个 target 中包含了多个 service。

查看 sshd.service，判断其属于哪个 target，这时查看[install]信息即可，它属于 multi-user.targe，如图 1-41 所示。

```
[root@centos7 ~]# cat /usr/lib/systemd/system/sshd.service
[Unit]
Description=OpenSSH server daemon
Documentation=man:sshd(8) man:sshd_config(5)
After=network.target sshd-keygen.service
Wants=sshd-keygen.service

[Service]
Type=notify
EnvironmentFile=/etc/sysconfig/sshd
ExecStart=/usr/sbin/sshd -D $OPTIONS
ExecReload=/bin/kill -HUP $MAINPID
KillMode=process
Restart=on-failure
RestartSec=42s

[Install]
WantedBy=multi-user.target
```

图 1-41

1.23　Linux系统日志

在日常运维工作中，会有很多的服务，这些服务涉及启动和运行，在启动时或运行过程中难免会遇到一些问题，如 MySQL 服务不能启动，解决方法除看启动过程中的报错外，还应该关注 MySQL 日志。很多错误会记录到日志中，不会出现在屏幕上，所以看日志非常重要。

/var/log/message 文件是 Linux 操作系统的总日志，很多信息都记录在这个日志文件中。有些服务会定义单独的日志文件，如果服务没有定义单独的日志文件，就会记录到系统日志中，操作命令如下。

```
    [root@centos7 ~]# ls /var/log/messages
    /var/log/messages
```

Linux 操作系统中有一个服务是 logrotate，它是用来切割日志的。为了防止日志无限制地在一个文件中增加，于是产生了这种机制。/var/log/message 文件后加上*会列出很多带有日期的日志文件，这些带有日期的日志文件就是用 logrotate 切割的，logrotate 服务在/etc/logrotate.conf 中，使用 cat 命令可以查看该配置文件中的具体配置信息，操作命令如下。

```
    [root@centos7 ~]# ls /var/log/messages*
    /var/log/messages  /var/log/messages-20180107
/var/log/messages-20180122
```

```
[root@centos7 ~]# ls /etc/logrotate.conf
/etc/logrotate.conf
```

1.23.1 dmesg 命令

使用 dmesg 命令可以把系统中与硬件相关的日志列出来，保存在内存中，注意它并不是某个文件。例如，硬盘损坏、网卡出现问题都会记录在 dmesg 命令中。使用-c 选项可清空日志，清空后遇到某些问题或重启操作系统还会生成日志，操作命令如下。

```
[root@centos7 ~]# dmesg
[root@centos7 ~]# dmesg -c
```

/var/log/dmesg 是日志文件，该日志文件和 dmesg 命令没有任何关联，日志是系统启动时记录的一些信息。

1.23.2 last 命令

使用 last 命令可以查看正确的历史登录信息，执行 last 命令会调用/var/log/wtmp 文件，操作命令如下。

```
[root@centos7 ~]# last
[root@centos7 ~]# ls /var/log/wtmp
/var/log/wtmp
```

1.23.3 lastb 命令

使用 lastb 命令可以查看登录失败的用户信息，执行 lastb 命令会调用/var/log/btmp 文件，操作命令如下。

```
[root@centos7 ~]# lastb
boo      pts/0                         Sat Jan  6 13:14 - 13:14  (00:00)
root     pts/0                         Fri Jan  5 17:27 - 17:27  (00:00)

btmp begins Fri Jan  5 17:27:57 2018
[root@centos7 ~]# ls /var/log/btmp
/var/log/btmp
```

/var/log/secure 文件中记录的是安全日志，登录操作系统时验证成功会在该日志文件中记录一条信息，登录不成功也会记录一条信息。

第 2 章

2

LAMP 服务架构

2.1　LAMP服务架构介绍

LAMP 即 Linux（操作系统）+Apache（httpd，Web 服务软件）+MySQL/MariaDB/MongoDB（数据库）+PHP/Python/Perl（编程语言）。

数据库可以单独使用一台服务器，但 Apache 和 PHP 必须在同一台服务器上使用。

2.2　MySQL和MariaDB数据库简介

MySQL 是关系型数据库，由 MySQL AB 公司开发，MySQL 在 2008 年被 Sun 公司以约 10 亿美元收购，2009 年 Sun 公司被 Oracle 公司以 74 亿美元收购。

MySQL 官网是 https://www.mysql.com/，最新版本是 MySQL5.7GA/2.0DMR。MySQL 在 5.6 版本时发生了较大的变化，在 5.7 版本时性能上有很大提升。

MariaDB 是 MySQL 数据库的一个分支，官网是 https://mariadb.com，最新版本是 MariaDB 10.2.12GA。

MariaDB 由 SkySQL 公司（现更名为 MariaDB 公司）维护，SkySQL 公司是 MySQL 原作者带领大部分原班人马创立的。

MariaDB 5.5 版本对应 MySQL 5.5 版本，MariaDB 10.0 版本对应 MySQL 5.6 版本。

Community 是社区版本；Enterprise 是企业版本；GA（Generally Available）是通用版本，在生产环境中使用；DMR（Development Milestone Release）是开发里程碑发布版本；RC（Release Candidate）是发行候选版本；Beta 是开发测试版本；Alpha 是内部测试版本。

2.3 MySQL和MariaDB数据库安装

2.3.1 MySQL 数据库安装

MySQL 数据的安装方法有三种：RPM 包安装、源码编译安装和二进制免编译安装。

进入/usr/local/src/目录，根据系统的版本（32 位或 64 位）下载 MySQL 5.6 二进制包，下载完毕后进行解压，如图 2-1 所示。

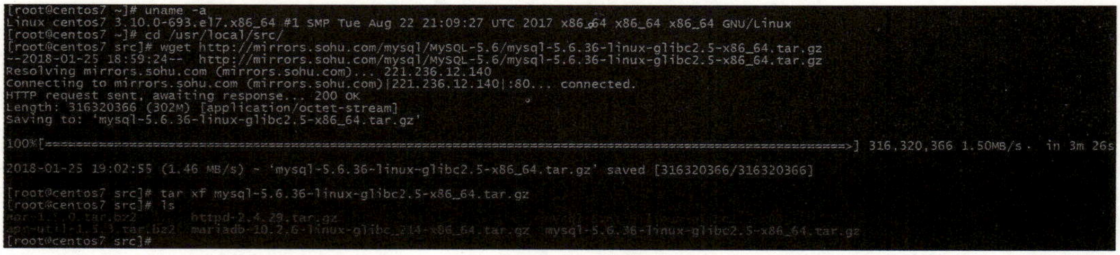

```
[root@centos7 ~]# uname -a
Linux centos7 3.10.0-693.el7.x86_64 #1 SMP Tue Aug 22 21:09:27 UTC 2017 x86_64 x86_64 x86_64 GNU/Linux
[root@centos7 ~]# cd /usr/local/src/
[root@centos7 src]# wget http://mirrors.sohu.com/mysql/MySQL-5.6/mysql-5.6.36-linux-glibc2.5-x86_64.tar.gz
--2018-01-25 18:59:24--  http://mirrors.sohu.com/mysql/MySQL-5.6/mysql-5.6.36-linux-glibc2.5-x86_64.tar.gz
Resolving mirrors.sohu.com (mirrors.sohu.com)... 221.236.12.140
Connecting to mirrors.sohu.com (mirrors.sohu.com)|221.236.12.140|:80... connected.
HTTP request sent, awaiting response... 200 OK
Length: 316320366 (302M) [application/octet-stream]
Saving to: 'mysql-5.6.36-linux-glibc2.5-x86_64.tar.gz'

100%[=====================================================================>] 316,320,366 1.50MB/s   in 3m 26s

2018-01-25 19:02:55 (1.46 MB/s) - 'mysql-5.6.36-linux-glibc2.5-x86_64.tar.gz' saved [316320366/316320366]

[root@centos7 src]# tar xf mysql-5.6.36-linux-glibc2.5-x86_64.tar.gz
[root@centos7 src]# ls
[illegible]          httpd-2.4.29.tar.gz
[illegible]          mariadb-10.2.6-linux-glibc_214-x86_64.tar.gz  mysql-5.6.36-linux-glibc2.5-x86_64.tar.gz
[root@centos7 src]#
```

图 2-1

将解压后的 mysql-5.6.36 目录移动至/usr/local/目录下，并更名为 mysql，运行命令如下。

```
    [root@centos7 src]# mv mysql-5.6.36-linux-glibc2.5-x86_64
/usr/local/mysql
    [root@centos7 src]# cd /usr/local/mysql/
    [root@centos7 mysql]# ls
    bin  COPYING  data  docs  include  lib  man  mysql-test  README  scripts
share  sql-bench  support-files
```

创建 MySQL 用户和/data/目录，/data/目录用来存储 MySQL 数据，创建完成后初始化 MySQL 数据库，运行命令如下。

```
    [root@centos7 mysql]# useradd mysql
    [root@centos7 mysql]# mkdir /data
    [root@centos7 mysql]# ./scripts/mysql_install_db --user=mysql
--datadir=/data/mysql
    [root@centos7 mysql]# echo $?
    0
```

复制配置文件和启动脚本，配置文件是 support-files/目录下的 my-default.cnf 文件，把该配置文件复制到/etc/目录下，并重命名为 my.cnf，运行命令如下。

```
    [root@centos7 mysql]# cp support-files/my-default.cnf /etc/my.cnf
```

启动脚本是 support-files/mysql.server，把该启动脚本复制到/etc/init.d/目录下，并重命名为 mysqld 文件，运行命令如下。

```
[root@centos7 mysql]# cp support-files/mysql.server /etc/init.d/mysqld
[root@centos7 mysql]# vim /etc/init.d/mysqld
basedir=/usr/local/mysql
datadir=/data/mysql
```

设置 MySQL 数据库开机自启动，并运行 MySQL 服务，启动成功后使用 ps 命令查看 mysqld 进程，使用 netstat 命令可查看 3306 端口是否被启动，运行命令如下。

```
[root@centos7 mysql]# chkconfig --add mysqld
[root@centos7 mysql]# /etc/init.d/mysqld start
Starting MySQL.Logging to '/data/mysql/centos7.err'.
. SUCCESS!
[root@centos7 mysql]# ps aux |grep mysqld
[root@centos7 mysql]# netstat -lntp
Active Internet connections (only servers)
Proto Recv-Q Send-Q Local Address           Foreign Address         State
PID/Program name
tcp6       0      0 :::3306                 :::*                    LISTEN
3648/mysqld
```

如果没有启动脚本模板去复制，那么还可以使用命令行的方法去启动，操作命令如下。

```
[root@centos7 mysql]# service mysqld stop
Shutting down MySQL.. SUCCESS!
[root@centos7 mysql]# !ps
ps aux |grep mysqld
root      3736  0.0  0.0 112660   976 pts/0    S+   18:05   0:00 grep
--color=auto mysqld
[root@centos7 mysql]# /usr/local/mysql/bin/mysqld_safe
--defaults-file=/etc/my.cnf --user=mysql --datadir=/data/mysql &
```

2.3.2　MariaDB 数据库安装

使用 wget 命令下载 MariaDB 需要半个小时左右的时间，为了节约时间，可以把 wget 命令后面的链接复制到迅雷中，用迅雷下载的时间在 1 分钟左右，下载完毕后用 FTP 上传即可。上传完毕后进行解压操作，运行命令如下。

```
[root@centos7 src]# wget https://downloads.mariadb.com/MariaDB/
```

```
mariadb-10.2.6/bintar-linux-glibc_214-x86_64/mariadb-10.2.6-linux-glibc_
214-x86_64
    .tar.gz
    [root@centos7 src]# ls
    apr-1.5.0.tar.bz2      httpd-2.4.29.tar.gz
mysql-5.6.36-linux-glibc2.5-x86_64.tar.gz
    apr-util-1.5.3.tar.bz2  mariadb-10.2.6-linux-glibc_214-x86_64.tar.gz
    [root@centos7 src]# tar xf mariadb-10.2.6-linux-glibc_214-x86_64.tar.gz
```

将解压后的 MariaDB 目录放至/usr/local/目录下并重命名为 mariadb，进入到 mariadb 目录中，初始化 MariaDB 数据库，用户还是 MySQL 用户，目录变更为/data/mariadb，方便和 MySQL 做区分，运行命令如下。

```
    [root@centos7 src]# mv mariadb-10.2.6-linux-glibc_214-x86_64
/usr/local/mariadb
    [root@centos7 src]# cd !$
    cd /usr/local/mariadb
    [root@centos7 mariadb]# ./scripts/mysql_install_db --user=mysql
    -basedir=/usr/local/mariadb--datadir=/data/mariadb
    [root@centos7 mariadb]# echo $?
    0
    [root@centos7 mariadb]# ls /data/mariadb/
    aria_log.00000001  aria_log_control  ib_buffer_pool  ibdata1
ib_logfile0  ib_logfile1  mysql  performance_schema  test
```

复制 MariaDB 配置文件，配置文件会根据系统内存情况选择 my-small.cnf 或 my.my-huge.cnf，内存大就选择 huge.cnf，内存小就选择 small.cnf，笔者计算机中是 support-files/my-small.cnf。复制该配置文件至/usr/local/mariadb/目录下，并重命名为 my.cnf（这个操作是为了防止和/etc/my.cnf 发生冲突），运行命令如下。

```
    [root@centos7 mariadb]# cp support-files/my-small.cnf
/usr/local/mariadb/my.cnf
```

复制启动脚本至/etc/init.d/目录下，并重命名为 mariadb，然后修改启动脚本，运行命令如下。

```
    [root@centos7 mariadb]# cp support-files/mysql.server
/etc/init.d/mariadb
    [root@centos7 mariadb]# vim /etc/init.d/mariadb
    basedir=/usr/local/mariadb
    datadir=/data/mariadb
    conf=$basedir/my.cnf
    $bindir/mysqld_safe --defaults-file="$conf"--datadir="$datadir" --pid-
file="$mysqld_pid_file_path" "$@" &  #--defaults-file="$conf"为后期添加
```

启动 MariaDB 数据库前要确保 mysqld 服务处于关闭状态，否则二者之间会产生冲突，关闭 mysqld 服务后开启 MariaDB 数据库，如图 2-2 所示。

```
[root@centos7 mariadb]# ps aux |grep mysqld
root      4615  0.0  0.0 112660   972 pts/0    S+   18:56   0:00 grep --color=auto mysqld
[root@centos7 mariadb]# /etc/init.d/mariadb start
Starting mariadb (via systemctl):  [  OK  ]
[root@centos7 mariadb]# ps aux |grep mariadb
root      4635  0.3  0.0 115388  1716 ?        S    18:56   0:00 /bin/sh /usr/local/mariadb/bin/mysqld_safe --defaults-file=/usr/local/mariadb/my
.cnf --datadir=/data/mariadb --pid-file=/data/mariadb/centos7.pid
mysql     4751  3.0  1.4 1649312 57864 ?       Sl   18:56   0:00 /usr/local/mariadb/bin/mysqld --defaults-file=/usr/local/mariadb/my.cnf --basedi
r=/usr/local/mariadb --datadir=/data/mariadb --plugin-dir=/usr/local/mariadb/lib/plugin --user=mysql --log-error=/data/mariadb/centos7.err --pid-f
ile=/data/mariadb/centos7.pid --socket=/tmp/mysql.sock --port=3306
root      4787  0.0  0.0 112660   976 pts/0    S+   18:57   0:00 grep --color=auto mariadb
[root@centos7 mariadb]# netstat -lntp
Active Internet connections (only servers)
Proto Recv-Q Send-Q Local Address           Foreign Address         State       PID/Program name
tcp        0      0 0.0.0.0:22              0.0.0.0:*               LISTEN      964/sshd
tcp        0      0 127.0.0.1:25            0.0.0.0:*               LISTEN      1186/master
tcp6       0      0 :::22                   :::*                    LISTEN      964/sshd
tcp6       0      0 ::1:25                  :::*                    LISTEN      1186/master
tcp6       0      0 :::3306                 :::*                    LISTEN      4751/mysqld
[root@centos7 mariadb]#
```

图 2-2

2.4　Apache服务安装

Apache 是软件基金会的名字，httpd 才是我们需要安装的软件包，早期的 httpd 叫做 Apache。

Apache 官网是 http://www.apache.org/

安装 Apache 2.4 版本需要依赖 apr 和 apr-util，apr 和 apr-util 是一个通用的函数库，可以让 httpd 不关心底层的操作系统平台，从而很方便地进行移植，如从 Linux 移植到 Windows。Apache、apr 和 apr-util 下载地址如下。

```
    [root@centos7 src]# wget
http://mirrors.cnnic.cn/apache/apr/apr-1.6.3.tar.gz
    [root@centos7 src]# wget
https://mirrors.cnnic.cn/apache/apr/apr-util-1.6.1.tar.gz
    [root@centos7 src]# wget
https://mirrors.cnnic.cn/apache/httpd/httpd-2.4.33.tar.gz
    [root@lanmp src]# ls
    apr-1.6.3.tar.gz  apr-util-1.6.1.tar.gz  httpd-2.4.33.tar.gz
```

下载完毕后解压 Apache、apr 和 apr-util，如图 2-3 所示。

```
[root@lanmp src]# tar xf apr-1.6.3.tar.gz
[root@lanmp src]# tar xf apr-util-1.6.1.tar.gz
[root@lanmp src]# tar xf httpd-2.4.33.tar.gz
[root@lanmp src]# ls
apr-1.6.3  apr-1.6.3.tar.gz  apr-util-1.6.1  apr-util-1.6.1.tar.gz  httpd-2.4.33  httpd-2.4.33.tar.gz
[root@lanmp src]#
```

图 2-3

进入 apr 软件包目录，设置 apr 安装路径，设置完毕后进行编译和编译安装，运行命令如下。

```
    [root@centos7 src]# cd apr-1.6.3/
    [root@centos7 apr-1.6.3]# ./configure --prefix=/usr/local/apr
```

```
[root@centos7 apr-1.6.3]# echo $?
0
[root@centos7 apr-1.6.3]# make && make install
[root@centos7 apr-1.6.3]# echo $?
0
[root@centos7 apr-1.6.3]# ls /usr/local/apr
bin  build-1  include  lib
```

进入 apr-util 目录，设置 apr-util 安装路径，设置完毕后进行编译和安装，运行命令如下。

```
[root@centos7 apr-1.6.3]# cd ../apr-util-1.6.1/
[root@centos7 apr-util-1.6.1]# ./configure
--prefix=/usr/local/apr-util --with-apr=/usr/local/apr
[root@centos7 apr-util-1.6.1]# echo $?
0
[root@centos7 apr-util-1.6.1]# make && make install
[root@centos7 apr-util-1.6.1]# echo $?
0
```

进入 httpd 目录，设置 httpd 安装路径和模块，设置完毕后进行编译和安装，运行命令如下。

```
[root@centos7 apr-util-1.6.1]# cd ../httpd-2.4.29/
[root@centos7 httpd-2.4.29]# yum groupinstall -y "Development Tools"
"Server Platform Development"
[root@centos7 httpd-2.4.29]# yum install -y pcre-devel
[root@centos7 httpd-2.4.29]# yum install -y openssl-devel
[root@centos7 httpd-2.4.29]# ./configure --prefix=/usr/local/apache2.4
--enable-so --enable-ssl --enable-cgi --enable-rewrite --with-zlib
--with-pcre --with-apr=/usr/local/apr --with-apr-util=/usr/local/apr-util
--enable-modules=most --enable-mpms-shared=all --with-mpm=prefork
[root@centos7 httpd-2.4.29]# echo $?
0
[root@centos7 httpd-2.4.29]# make && make install
[root@centos7 httpd-2.4.29]# echo $?
0
```

进入/usr/local/apache2.4 目录，bin/httpd 文件是启动 Apache 的文件，也是 httpd 的核心文件。conf 目录是配置文件所在的目录。htdocs 目录用于存放默认的网站页面。logs 目录是与日志（如错误日志、访问日志）相关的目录。modules 目录存放的是扩展模块，如图 2-4 所示。

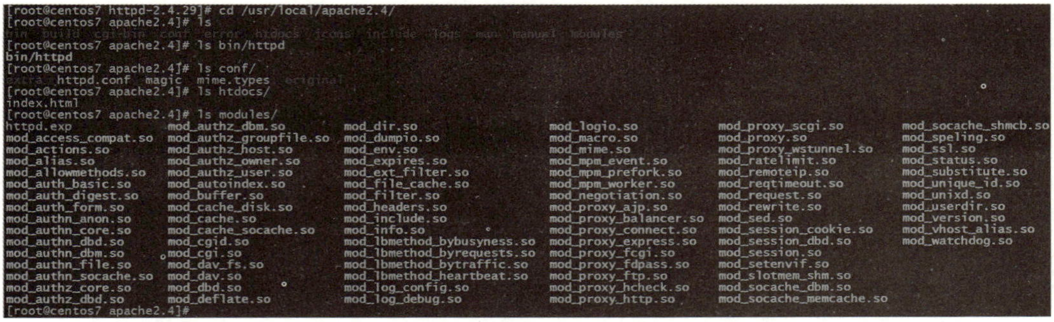

图 2-4

使用/usr/local/apache2.4/bin/apachectl -M 可以查看扩展模块，如果不想使用全路径，也可以编辑/etc/profile.d/httpd.sh 配置文件，添加全局变量，全局变量设置完毕后用 httpd -M 也可实现扩展模块的查看，运行命令如下。

```
[root@centos7 apache2.4]# vim /etc/profile.d/httpd.sh
export PATH=/usr/local/apache2.4/bin:$PATH
:wq
[root@centos7 apache2.4]# . /etc/profile.d/httpd.sh
[root@centos7 apache2.4]# echo $PATH
/usr/local/apache2.4/bin:/usr/local/bin:/usr/local/sbin:/usr/local/bin:/usr/sbin:/usr/bin:/root/bin
[root@centos7 apache2.4]# httpd -M
```

可以使用绝对路径启动 httpd 服务，也可使用更改全局变量后的相对路径启动 httpd 服务，如图 2-5 所示。

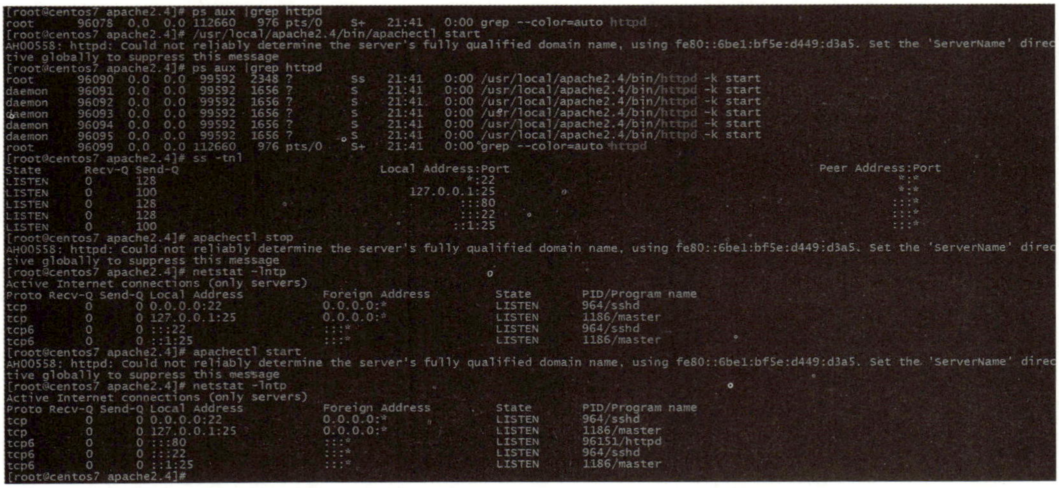

图 2-5

2.5 PHP源码编译安装

2.5.1 PHP 版本介绍

PHP 官网是 http://www.php.net。PHP 主流版本是 PHP 5.6 和 PHP 7.2，大多数企业还在使用 PHP 5.X 版本，部分新的项目采用 PHP 7.X 版本。PHP 7.X 版本是在 2015 年发布的，截至目前最新稳定版本为 7.2.1。PHP 7 版本自问世以来获得了很多开发者的喜爱和追捧，因为 PHP 7 版本的性能有很大的提升。

2.5.2 PHP 5.6 源码编译安装

进入/usr/local/src/目录，下载 PHP 5.6 版本的源码编译安装包，如图 2-6 所示。

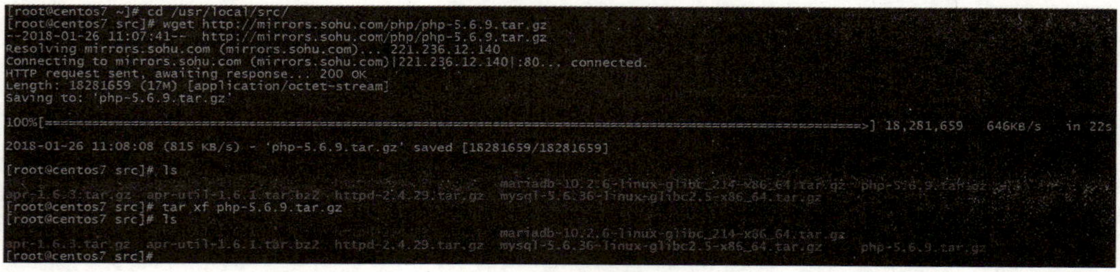

```
[root@centos7 ~]# cd /usr/local/src/
[root@centos7 src]# wget http://mirrors.sohu.com/php/php-5.6.9.tar.gz
--2018-01-26 11:07:41--  http://mirrors.sohu.com/php/php-5.6.9.tar.gz
Resolving mirrors.sohu.com (mirrors.sohu.com)... 221.236.12.140
Connecting to mirrors.sohu.com (mirrors.sohu.com)|221.236.12.140|:80... connected.
HTTP request sent, awaiting response... 200 OK
Length: 18281659 (17M) [application/octet-stream]
Saving to: 'php-5.6.9.tar.gz'

100%[==========================================================================>] 18,281,659   646KB/s   in 22s

2018-01-26 11:08:08 (815 KB/s) - 'php-5.6.9.tar.gz' saved [18281659/18281659]

[root@centos7 src]# ls
                                                      mariadb-10.2.6-linux-glibc_214-x86_64.tar.gz  php-5.6.9.tar.gz
apr-1.6.3.tar.gz  apr-util-1.6.1.tar.bz2  httpd-2.4.29.tar.gz  mysql-5.6.36-linux-glibc2.5-x86_64.tar.gz
[root@centos7 src]# tar xf php-5.6.9.tar.gz
[root@centos7 src]# ls
                                                      mariadb-10.2.6-linux-glibc_214-x86_64.tar.gz
apr-1.6.3.tar.gz  apr-util-1.6.1.tar.bz2  httpd-2.4.29.tar.gz  mysql-5.6.36-linux-glibc2.5-x86_64.tar.gz    php-5.6.9.tar.gz
[root@centos7 src]#
```

图 2-6

解压后进入 php5.6.9 目录，先设置 PHP 扩展库和扩展模块，运行命令如下。

```
[root@centos7 src]# cd php-5.6.9/
[root@centos7 php-5.6.9]# yum -y install bzip2-devel libmcrypt-devel
libxml2-devel libjpeg-devel libpng-devel freetype-devel
[root@centos7 php-5.6.9]# ./configure --prefix=/usr/local/php
--with-apxs2=/usr/local/apache2.4/bin/apxs --with-mysql=/usr/local/mysql
--with-pdo-mysql=/usr/local/mysql --with-mysqli=/usr/local/mysql/
bin/mysql_config --with-libxml-dir --with-gd --with-jpeg-dir --with-png-dir
--with-freetype-dir --with-icony-dir --with-zlib-dir --with-bz2
--with-openssl --with-mcrypt --enable-soap --enable-gd-native-ttf
--enable-mbstring --enable-sockets --enable-exif
[root@centos7 php-5.6.9]# echo $?
0
```

PHP 编译时间比较长，CPU 和内存的性能也影响安装的时间，安装时需要耐心等待一段时间，运行命令如下。

```
[root@centos7 php-5.6.9]# make && make install
[root@centos7 php-5.6.9]# echo $?
0
```

PHP 核心二进制文件在/usr/local/php/bin/目录下，使用 du－sh 命令可查看核心文件的大小，PHP 和 Apache 结合是通过/usr/local/apache2.4/modules/libpng5.so 文件实现的，如图 2-7 所示。

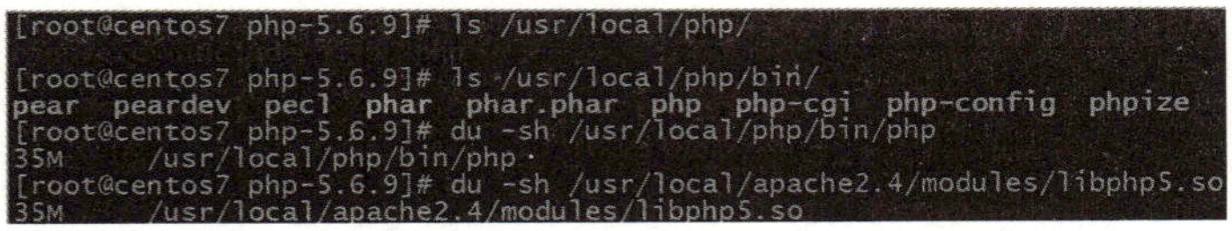

```
[root@centos7 php-5.6.9]# ls /usr/local/php/

[root@centos7 php-5.6.9]# ls /usr/local/php/bin/
pear  peardev  pecl  phar  phar.phar  php  php-cgi  php-config  phpize
[root@centos7 php-5.6.9]# du -sh /usr/local/php/bin/php
35M     /usr/local/php/bin/php
[root@centos7 php-5.6.9]# du -sh /usr/local/apache2.4/modules/libphp5.so
35M     /usr/local/apache2.4/modules/libphp5.so
```

图 2-7

使用/usr/local/php/bin/php－m 命令可查看 PHP 加载的模块有哪些，如果觉得路径长也可按照讲解 Apache 时的方法，编辑 profile.d 配置文件进行环境变量设置，全局变量设置完毕后使用 php－m 命令也可实现上述效果，运行命令如下。

```
[root@centos7 php-5.6.9]# vim /etc/profile.d/php.sh
export PATH=/usr/local/php/bin:$PATH
[root@centos7 php-5.6.9]# source /etc/profile.d/php.sh
[root@centos7 php-5.6.9]# echo $PATH
/usr/local/php/bin:/usr/local/apache2.4/bin:/usr/local/sbin:/usr/loca
l/bin:/usr/sbin:/usr/bin:/root/bin
[root@centos7 php-5.6.9]# php -m
```

Apache 和 MySQL 默认需要启动才能使用，PHP 不需要启动。使用 httpd－M 命令可查看 Aapche 安装的模块，在下面会看到 php5_module （shared），如图 2-8 所示。如果删除 php5_module 文件，Apache 就无法和 PHP 进行关联工作。

```
[root@centos7 php-5.6.9]# httpd -M
AH00558: httpd: Could not reliably determine the server's fully qualified domain name, using fe80::6be1:bf5e:d449:d3a5. Set the 'ServerName' direc
tive globally to suppress this message .
Loaded Modules:
 core_module (static)
 so_module (static)
 http_module (static)
 mpm_prefork_module (shared)
 authn_file_module (shared)
 authn_core_module (shared)
 authz_host_module (shared)
 authz_groupfile_module (shared)
 authz_user_module (shared)
 authz_core_module (shared)
 access_compat_module (shared)
 auth_basic_module (shared)
 reqtimeout_module (shared)
 filter_module (shared)
 mime_module (shared)
 log_config_module (shared)
 env_module (shared)
 headers_module (shared)
 setenvif_module (shared)
 version_module (shared)
 unixd_module (shared)
 status_module (shared)
 autoindex_module (shared)
 dir_module (shared)
 alias_module (shared)
 php5_module (shared)
[root@centos7 php-5.6.9]#
```

图 2-8

运行/usr/local/php/bin/php -i |less 时，发现 Loaded Configuration File => (none)并没有加载 php.ini 配置文件，需复制 php.ini-production 配置文件至/usr/local/php/etc/

目录下，并重命名为 php.ini 文件，再运行/usr/local/php/bin/php -i |less 时就有配置文件了，运行命令如下。

```
    [root@centos7 php-5.6.9]# /usr/local/php/bin/php -i |less
    [root@centos7 php-5.6.9]# cp php.ini-production
/usr/local/php/etc/php.ini
```

2.5.3 PHP 7.X 源码编译安装

源码编译安装 PHP 7.2 前需下载 7.2 版本的源码包至/usr/local/src/目录下，下载完毕后进行解压，如图 2-9 所示。

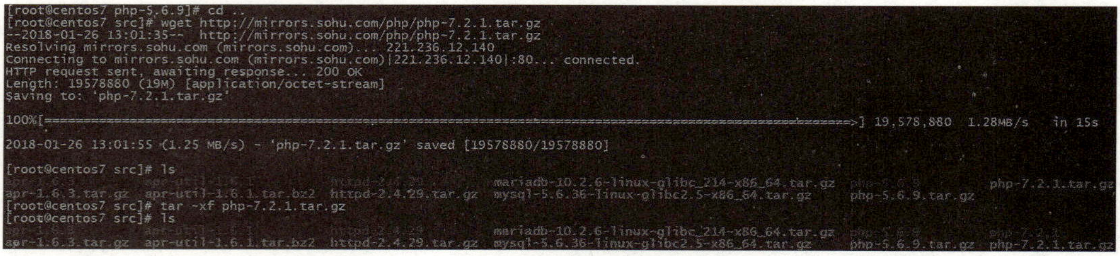

```
[root@centos7 php-5.6.9]# cd ..
[root@centos7 src]# wget http://mirrors.sohu.com/php/php-7.2.1.tar.gz
--2018-01-26 13:01:35--  http://mirrors.sohu.com/php/php-7.2.1.tar.gz
Resolving mirrors.sohu.com (mirrors.sohu.com)... 221.236.12.140
Connecting to mirrors.sohu.com (mirrors.sohu.com)|221.236.12.140|:80... connected.
HTTP request sent, awaiting response... 200 OK
Length: 19578880 (19M) [application/octet-stream]
Saving to: 'php-7.2.1.tar.gz'

100%[==================================================================>] 19,578,880  1.28MB/s   in 15s

2018-01-26 13:01:55 (1.25 MB/s) - 'php-7.2.1.tar.gz' saved [19578880/19578880]

[root@centos7 src]# ls
                                                 mariadb-10.2.6-linux-glibc_214-x86_64.tar.gz              php-7.2.1.tar.gz
apr-1.6.3.tar.gz  apr-util-1.6.1.tar.bz2  httpd-2.4.29.tar.gz  mysql-5.6.36-linux-glibc2.5-x86_64.tar.gz  php-5.6.9.tar.gz
[root@centos7 src]# tar -xf php-7.2.1.tar.gz
[root@centos7 src]# ls
                                                 mariadb-10.2.6-linux-glibc_214-x86_64.tar.gz
apr-1.6.3.tar.gz  apr-util-1.6.1.tar.bz2  httpd-2.4.29.tar.gz  mysql-5.6.36-linux-glibc2.5-x86_64.tar.gz  php-5.6.9.tar.gz  php-7.2.1.tar.gz
```

图 2-9

解压后进入 php-7.2 目录，进行参数配置和扩展模块设置，运行命令如下。

```
    [root@centos7 src]# cd php-7.2.1/
    [root@centos7 php-7.2.1]# ./configure --prefix=/usr/local/php7
--with-config-file-path=/usr/local/php7/etc --with-apxs2=/usr/
local/apache2.4/bin/apxs --with-pdo-mysql=/usr/local/mysql
--with-mysqli=/usr/local/mysql/bin/mysql_config --with-libxml-dir
--with-gd --with-jpeg-dir --with-png-dir --with-freetype-dir
--with-icony-dir --with-zlib-dir --with-bz2 --with-openssl --with-mcrypt
--enable-soap --enable-gd-native-ttf --enable-mbstring --enable-sockets
--enable-exif
    [root@centos7 php-7.2.1]# echo $?
    0
```

PHP 7.2 的编译过程与时间和 PHP 5.6 的编译过程与时间差不多，需要耐心等待，运行命令如下。

```
    [root@centos7 php-7.2.1]# make && make install
    [root@centos7 php-7.2.1]# echo $?
    0
```

PHP 7 版本也有和 PHP 5 版本一样的 libphp7.so 文件，如图 2-10 所示。

```
[root@centos7 php-7.2.1]# ls /usr/local/apache2.4/modules/libphp
libphp5.so  libphp7.so
[root@centos7 php-7.2.1]# ls /usr/local/apache2.4/modules/libphp7.so
/usr/local/apache2.4/modules/libphp7.so
[root@centos7 php-7.2.1]# du -sh !$
du -sh /usr/local/apache2.4/modules/libphp7.so
42M     /usr/local/apache2.4/modules/libphp7.so
[root@centos7 php-7.2.1]#
```

图 2-10

使用/usr/local/php7/bin/php－m 命令进行查看时，如果觉得命令长，可以设置全局环境变量，使用相对路径进行查看，运行命令如下。

```
[root@centos7 php-7.2.1]# /usr/local/php7/bin/php -m
[root@centos7 php-7.2.1]# vim /etc/profile.d/php7.sh
export PATH=/usr/local/php7/bin:$PATH
[root@centos7 php-7.2.1]# php -m
```

运行 httpd－M 命令，会看到不仅有 php 5 的模块，还有 php 7 的模块，如图 2-11 所示。

```
[root@centos7 php-7.2.1]# httpd -M
AH00558: httpd: Could not reliably determine the server's fully qualified domain name, using fe80::6be1:bf5e:d449:d3a5. Set the 'ServerName' direc
tive globally to suppress this message
Loaded Modules:
 core_module (static)
 so_module (static)
 http_module (static)
 mpm_prefork_module (shared)
 authn_file_module (shared)
 authn_core_module (shared)
 authz_host_module (shared)
 authz_groupfile_module (shared)
 authz_user_module (shared)
 authz_core_module (shared)
 access_compat_module (shared)
 auth_basic_module (shared)
 reqtimeout_module (shared)
 filter_module (shared)
 mime_module (shared)
 log_config_module (shared)
 env_module (shared)
 headers_module (shared)
 setenvif_module (shared)
 version_module (shared)
 unixd_module (shared)
 status_module (shared)
 autoindex_module (shared)
 dir_module (shared)
 alias_module (shared)
 php5_module (shared)
 php7_module (shared)
[root@centos7 php-7.2.1]#
```

图 2-11

这是因为同时安装了 PHP 5 和 PHP 7 两个版本，在生产环境中会产生一些冲突，若不想使用 PHP 5 或 PHP 7 版本，可在 httpd.conf 配置文件中找到该行进行注释，运行命令如下。

```
[root@centos7 php-7.2.1]# vim /usr/local/apache2.4/conf/httpd.conf
#LoadModule php5_module        modules/libphp5.so
LoadModule php7_module        modules/libphp7.so
:wq
```

2.6　Python源码编译安装

2.6.1　Python 3.6 编译安装

安装 Python 依赖包，运行命令如下。

```
[root@lanmp ~]# yum install -y openssl-devel readline-devel unzip gcc-c++ wget
```

下载最新版本的 Python 源码包，运行命令如下。

```
[root@lanmp ~]# wget https://www.python.org/ftp/python/3.6.5/Python-3.6.5.tgz
```

编译安装 Python 源码包，运行命令如下。

```
[root@lanmp ~]# ls
anaconda-ks.cfg  Python-3.6.5.tgz
[root@lanmp ~]# du -sh Python-3.6.5.tgz
22M Python-3.6.5.tgz
[root@lanmp ~]# tar xf Python-3.6.5.tgz
[root@lanmp ~]# cd Python-3.6.5
[root@lanmp Python-3.6.5]# ./configure --prefix=/usr/local/python36
[root@lanmp Python-3.6.5]# echo $?
0
[root@lanmp Python-3.6.5]# make && make install
[root@lanmp Python-3.6.5]# ls /usr/local/python36/bin/
2to3      easy_install-3.6  idle3.6  pip3.6  pydoc3.6  python3.6  python3.6m        python3-config  pyvenv-3.6
2to3-3.6  idle3             pip3     pydoc3  python3   python3.6-config  python3.6m-config  pyvenv
```

配置 pip 命令，pip 默认使用官方源，但在国内访问官方源速度较慢，需更改为国内源，如豆瓣源，运行命令如下。

```
[root@lanmp Python-3.6.5]# vim /etc/pip.conf
[gloabl]
index-url = http://pypi.douban.com/simple
trusted-host = pypi.douban.com
[list]
format=columns
```

安装 virtualenv，如图 2-12 所示。

```
[root@lanmp Python-3.6.5]# /usr/local/python36/bin/pip3 install virtualenv
Collecting virtualenv
  Downloading https://files.pythonhosted.org/packages/ed/ea/e20b5cbebf45d3096e
    100% |████████████████████████████████| 2.6MB 338kB/s
Installing collected packages: virtualenv
Successfully installed virtualenv-15.2.0
```

图 2-12

2.6.2 安装 Python 扩展 MySQL 数据库

Python 默认调用的是 SQLite 数据库，不支持 MySQL 数据库，因此需要手动安装 MySQL 数据库并进行配置，安装 MySQL 数据库的运行命令如下。

```
[root@lanmp ~]# wget
http://dev.mysql.com/get/mysql-community-release-el7-5.noarch.rpm
[root@lanmp ~]# ls
anaconda-ks.cfg  mysql-community-release-el7-5.noarch.rpm
Python-3.6.5  Python-3.6.5.tgz
[root@lanmp ~]# rpm -ivh mysql-community-release-el7-5.noarch.rpm
[root@lanmp ~]# yum install -y mysql mysql-server mysql-devel zlib-devel
[root@lanmp ~]# systemctl start mysqld.service
[root@lanmp ~]# netstat -lntp
tcp6       0      0 :::3306          :::*          LISTEN   24593/mysqld
```

调用 python/bin/virtualenv，在目录下生成 python36env 配置文件，运行命令如下。

```
[root@lanmp ~]# /usr/local/python36/bin/virtualenv ./python36env
Using base prefix '/usr/local/python36'
New python executable in /root/python36env/bin/python3.6
Also creating executable in /root/python36env/bin/python
Installing setuptools, pip, wheel...done.
```

进入 python36env 环境，python36env 环境拥有所有权限，在该环境下使用 pip list 命令可列出安装的包，运行命令如下。

```
[root@lanmp ~]# source python36env/bin/activate
(python36env) [root@lanmp ~]# pip list
Package    Version
---------- -------
pip        10.0.0
setuptools 39.0.1
wheel      0.31.0
```

使用 pip 命令安装 MySQL 数据库，安装后使用 pip list 命令查看是否安装成功，运行命令如下。

```
(python36env) [root@lanmp ~]# pip install pymysql
Collecting pymysql
   100% |████████████████████████████████| 92kB 730kB/s
Installing collected packages: pymysql
Successfully installed pymysql-0.2.0
(python36env) [root@lanmp ~]# pip list
Package    Version
---------- -------
pip        10.0.0
```

```
PyMySQL    0.2.0
setuptools 39.0.1
wheel      0.31.0
```

编辑 MySQL 主配置文件 my.cnf，配置时一定要注意，修改所有字符集编码为 UTF-8，否则后续操作会出现问题，运行命令如下。

```
[root@lanmp ~]# vim /etc/my.cnf
[mysqld]
default-storage-engine = innodb
innodb_file_per_table
collation-server = utf8_general_ci
init-connect =  'SET NAMES utf8'
character-set-server = utf8
```

MySQL 数据库设置完毕后，默认密码为空，需要给 MySQL 设置密码，可以使用 mysqladmin 命令进行设置，运行命令如下。

```
[root@lanmp ~]# mysqladmin -uroot password 'humingzhe'
[root@lanmp ~]# mysql -uroot -p
Enter password:
```

2.7 Apache结合PHP进行操作

编辑 httpd 主配置文件/usr/local/apache2.4/conf/httpd.conf，打开 ServerName，运行命令如下。

```
[root@centos7 php-7.2.1]# vim /usr/local/apache2.4/conf/httpd.conf
ServerName www.example.com:80
:wq
```

在浏览器的地址栏中输入虚拟机 IP 地址，笔者的 IP 地址是 192.162.222.129，但是并不能打开，显示无法访问，并提示检查网络连接。打开 cmd，ping 此 IP 可以 ping 通，但是用 Telnet 访问此 IP 的 80 端口显示访问被拒绝，如图 2-13 所示。

针对不能访问 80 端口的问题，可以在防火墙中添加 80 端口规则，运行命令如下。

```
[root@centos7 php-7.2.1]# iptables -I INPUT -p tcp --dport 80 -j ACCEPT
```

添加完 80 端口规则后，在浏览器的地址栏中输入 IP 地址 192.162.222.129 发现可以访问了，如图 2-14 所示。

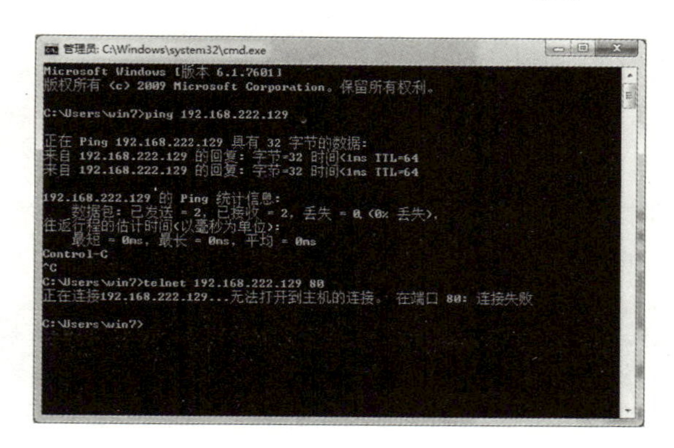

图 2-13

图 2-14

继续在 httpd.conf 配置文件中搜索 Require all denied，把 denied 改为 granted，再搜索 AddType，在 AddType application/x-gzip .gz .tgz 下面添加一行解析 PHP 的代码，完成后再添加一行解析.php 文件的代码，运行命令如下。

```
 <Directory />
   AllowOverride none
   Require all granted
</Directory>
AddType application/x-compress .Z
AddType application/x-gzip .gz .tgz
AddType application/x-httpd-php .php
<IfModule dir_module>
    DirectoryIndex index.html index.php
</IfModule>
:wq
```

保存配置文件后，可以使用-t 选项检查是否有语法错误，没有错误可重新加载 httpd.conf 配置文件，操作命令如下。

```
[root@centos7 php-7.2.1]# /usr/local/apache2.4/bin/apachectl -t
Syntax OK
[root@centos7 php-7.2.1]# /usr/local/apache2.4/bin/apachectl graceful
```

httpd.conf 配置文件更改完毕后还是无法得知是否能解析 PHP，可以在/usr/local/apache2.4/htdocs/目录下创建一个 test.php 文件，在文件中写入 phpinfo 函数，代码如下。

```
<?php
   phpinfo();
:wq
```

在浏览器的地址栏中输入 192.162.222.129/test.php，按 Enter 键后得到如图 2-15 所示的结果，输出该页面表示解析 PHP 成功。

PHP Version 7.2.1

System	Linux centos7 3.10.0-693.el7.x86_64 #1 SMP Tue Aug 22 21:09:27 UTC 2017 x86_64
Build Date	Jan 27 2018 20:23:57
Configure Command	'./configure' '--prefix=/usr/local/php7' '--with-config-file-path=/usr/local/php7/etc' '--with-apxs2=/usr/local/apache2.4/bin/apxs' '--with-pdo-mysql=/usr/local/mysql' '--with-mysqli=/usr/local/mysql/bin/mysql_config' '--with-libxml-dir' '--with-gd' '--with-jpeg-dir' '--with-png-dir' '--with-freetype-dir' '--with-icony-dir' '--with-zlib-dir' '--with-bz2' '--with-openssl' '--with-mcrypt' '--enable-soap' '--enable-gd-native-ttf' '--enable-mbstring' '--enable-sockets' '--enable-exif'
Server API	Apache 2.0 Handler
Virtual Directory Support	disabled
Configuration File (php.ini) Path	/usr/local/php7/etc
Loaded Configuration File	(none)
Scan this dir for additional .ini files	(none)
Additional .ini files parsed	(none)
PHP API	20170718
PHP Extension	20170718
Zend Extension	320170718
Zend Extension Build	API320170718,NTS
PHP Extension Build	API20170718,NTS
Debug Build	no
Thread Safety	disabled
Zend Signal Handling	enabled
Zend Memory Manager	enabled
Zend Multibyte Support	provided by mbstring
IPv6 Support	enabled
DTrace Support	disabled
Registered PHP Streams	https, ftps, compress.zlib, compress.bzip2, php, file, glob, data, http, ftp, phar
Registered Stream Socket Transports	tcp, udp, unix, udg, ssl, sslv3, tls, tlsv1.0, tlsv1.1, tlsv1.2
Registered Stream Filters	zlib.*, bzip2.*, convert.iconv.*, string.rot13, string.toupper, string.tolower, string.strip_tags, convert.*, consumed, dechunk

This program makes use of the Zend Scripting Language Engine:
Zend Engine v3.2.0, Copyright (c) 1998-2017 Zend Technologies

图 2-15

2.8 Apache默认虚拟主机

一台服务器既可以访问 humingzhe.com，又可以访问 humingzhe.cn，这是两个不同的网站，同时在一台服务器上运行，而这台服务器上只运行了一个 httpd 服务，这就是一台服务器有多个域名，每个域名对应一个虚拟主机。

它们默认都指向 httpd.conf 配置文件，若不想让它们都指向 httpd.conf 配置文件，可在 httpd.conf 配置文件中搜索 extra，找到 httpd-vhosts.conf 这行代码即可打开虚拟主机配置文件，操作命令如下。

```
[root@centos7 php-7.2.1]# vim /usr/local/apache2.4/conf/httpd.conf
Include conf/extra/httpd-vhosts.conf
:wq
```

htdocs 目录下只能定义一个虚拟主机，在日常工作环境中一台服务器要运行多个网站，若只运行一个网站会相当浪费，所以需要在虚拟主机配置文件中增加多个网站。

在/usr/local/apache2.4/conf/extra/httpd-vhosts.conf 文件中可以定义多个虚拟主机配置文件，打开 httpd-vhosts.conf 配置文件，会列出两个“VirtualHost *:80”，每个“VirtualHost *:80”代表一个主机（一个主机相当于一个网站）。定义 VirtualHost 后即可进行单独网站访问，操作命令如下。

```
<VirtualHost *:80>
    ServerAdmin admin@humingzhe.com     #定义管理员邮箱
    DocumentRoot "/data/www/humingzhe_blog"     #定义网站根目录路径
    ServerName humingzhe.com            #网站域名，如 humingzhe.com
    ServerAlias www.humingzhe.com       #网站别名，Alias 可以是多个域名，如
www.humingzhe.com、www.humingzhe.cn、www.humingzhe.net
    ErrorLog "/data/logs/humingzhe.com-error_log"   #网站错误日志路径
    CustomLog "/data/logs/humingzhe.com-access_log" common  #网站访问日
志路径
</VirtualHost>
```

虚拟主机配置文件修改完毕后，需要在对应的目录下创建相应的目录，操作命令如下。

```
[root@centos7 php-7.2.1]# mkdir /data/www/
[root@centos7 php-7.2.1]# mkdir /data/www/humingzhe_blog
[root@centos7 php-7.2.1]# mkdir /data/www/www.humingzhe.com
[root@centos7 php-7.2.1]# mkdir /data/logs/
```

在对应的站点下分别创建 index.php 文件，并在每个站点的 index.php 文件中定义站点的名字，如在 humingzhe_blog 站点中定义 humingzhe_blog，在 www.humingzhe.com 站点中定义 www.humingzhe.com，操作命令如下。

```
[root@centos7 php-7.2.1]# cd /data/www/
[root@centos7 www]# ls
humingzhe_blog www.humingzhe.com
[root@centos7 www]# vim humingzhe_blog/index.php
<?php
    echo "humingzhe_blog";
[root@centos7 www]# vim www.humingzhe.com/index.php
<?php
```

```
    echo "humingzhe.com";
```

定义完毕后检查语法是否有误，确认无误后重新加载此配置文件，操作命令如下。

```
[root@centos7 www]# /usr/local/apache2.4/bin/apachectl -t
Syntax OK
[root@centos7 www]# /usr/local/apache2.4/bin/apachectl graceful
```

直接访问 mingzhe.com 和 www.humingzhe.com 会解析到外网去，若想解析到本机 IP（192.162.222.129），需更改/etc/hosts 配置文件。若不想更改 hosts 配置文件，可使用 curl -x+IP+Port 进行访问，操作命令如下。虚拟主机配置文件明明指定的是 www.humingzhe.com，但是访问 mingzhe.com 时得出的结果是 humingzhe_blog，这是因为 humingzhe_blog 是虚拟主机配置文件中的第一个虚拟主机，也是默认虚拟主机，无论访问任何地址，只要解析到此 IP 上都会访问网站对应的虚拟主机配置。

```
[root@centos7 www]# curl -x192.162.222.129:80 mingzhe.com
humingzhe_blog
[root@centos7 www]# curl -x192.162.222.129:80 mingzhe.com
humingzhe_blog
```

2.9 Apache相关配置

2.9.1 Apache 用户认证

在登录某个博客或门户网站时，需要输入用户名和密码才有权限进行编辑和设置，而有些网站在访问该网站域名时就要求输入用户名和密码，输入正确即可访问网站，输入错误不能访问该网站。这样做虽然增强了网站的安全性，但是用户体验感差。

定义一个新的虚拟主机，在新的虚拟主机中设置以下配置信息，操作代码如下。

```
<VirtualHost *:80>
    ServerAdmin admin@humingzhe.cn
    DocumentRoot "/data/www/www.humingzhe.cn"
    ServerName humingzhe.cn
    ServerAlias www.humingzhe.cn
    <Directory /data/www/www.humingzhe.cn>       #指定认证的目录
        AllowOverride AuthConfig                #开启认证的开关
        AuthName "humingzhe.cn user auth"        #自定义认证名称，作用不大
        AuthType Basic                          #认证类型，一般为 Basic
        AuthUserFile /data/.htpasswd             #指定密码文件所在的位置
```

```
        require valid-user                  #指定需要认证的用户为全部可用用户
    </Directory>
    ErrorLog "/data/logs/humingzhe.cn-error_log"
    CustomLog "/data/logs/humingzhe.cn-access_log" common
</VirtualHost>
```

生成 htpasswd 密码文件，生成的密码文件的路径是/usr/local/apache2.4/bin/htpasswd。使用-c 选项可以指定生成密码文件，使用-m 选项可以指定 MD5 加密。生成密码文件目录和登录的用户名后，按 Enter 键提示输入密码和确认密码，如果是增加用户，就不需要再指定-c 选项，前面已经添加过密码文件，新增用户设置完毕后重载配置文件使之生效，操作命令如下。

```
    [root@centos7 www]# /usr/local/apache2.4/bin/htpasswd -c -m
/data/.htpasswd humingzhe
    New password:
    Re-type new password:
    Adding password for user humingzhe
    [root@centos7 www]# cat /data/.htpasswd
    humingzhe:$apr1$u.IuLFfs$oUqdRLVUcKu7po947A/2ql
    [root@centos7 www]# /usr/local/apache2.4/bin/apachectl -t
    Syntax OK
    [root@centos7 www]# /usr/local/apache2.4/bin/apachectl graceful
```

设置完毕后，在浏览器的地址栏中输入 www.humingzhe.cn，弹出如图 2-16 所示的界面，提示输入用户名和密码才可以访问。输入用户名和密码后，弹出如图 2-17 所示的界面，界面中的内容就是前面编辑的 index.php 文件中的内容。

图 2-16

图 2-17

在日常工作中，一般不要求对某个目录认证成功后才能访问，而是针对某个文件进行认证，验证成功后才可以访问。比如，一个网站的后台是 admin.php，每个人都可以在域名后输入 admin 访问后台登录界面，这是很不安全的行为，可以针对 admin 页面做访问认证。做页面认证也需要编辑虚拟主机配置文件，将用户认证代码更改为如下代码。

```
    [root@centos7 www]# vim
/usr/local/apache2.4/conf/extra/httpd-vhosts.conf
    <VirtualHost *:80>
        ServerAdmin admin@humingzhe.cn
        DocumentRoot "/data/www/www.humingzhe.cn"
        ServerName humingzhe.cn
        ServerAlias www.humingzhe.cn
        <FilesMatch admin.php>
            AllowOverride AuthConfig
            AuthName "humingzhe.cn user auth"
            AuthType Basic
            AuthUserFile /data/.htpasswd
            require valid-user
        </FilesMatch>
        ErrorLog "/data/logs/humingzhe.cn-error_log"
        CustomLog "/data/logs/humingzhe.cn-access_log" common
    </VirtualHost>
```

更改完虚拟主机配置文件后一定要记得检查语法是否有误，无误后重载虚拟主机配置文件，在/data/www/www.humingzhe.cn 目录下创建 admin.php 文件，在 admin.php 文件中输出 Hello HuMingZhe，运行命令如下。

```
[root@centos7 www]# /usr/local/apache2.4/bin/apachectl -t
Syntax OK
[root@centos7 www]# /usr/local/apache2.4/bin/apachectl graceful
[root@centos7 www]# vim /data/www/www.humingzhe.cn/admin.php
<?php
   echo "Hello HuMingZhe I Love You~";
```

在浏览器的地址栏中输入 192.162.222.129/admin.php，弹出登录框，提示输入用户名和密码才可以进行访问，如图 2-18 所示。输入刚才创建的用户名和密码，会显示如图 2-19 所示的界面。

图 2-18

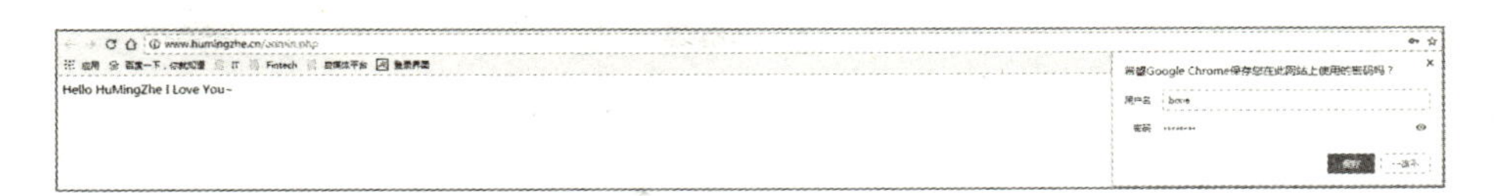

图 2-19

2.9.2　域名跳转机制

域名跳转是网站在更换域名时进行的一些操作，笔者以前博客用的域名是 humingzhe.cn，但是后期注册 humingzhe.com 时，会把网站解析到新的域名 humingzhe.com 上，有些用户可能不知道笔者更改了域名，此时可以做一个域名跳转，当用户访问 humingzhe.cn 时会自动跳转到 humingzhe.com，这样做的目的是防止用户丢失。

编辑虚拟主机配置文件，更改如下代码。

```
    <IfModule mod_rewrite.c>        #需要 mod_rewrite 模块支持
        RewriteEngine on            #打开 rewrite 功能
        RewriteCond %{HTTP_HOST} !^www.humingzhe.cn$    #定义 rewrite 条件，
主机名（域名）不是 www.humingzhe.com 就满足条件
        RewriteRule ^/(.*)$ http://www.humingzhe.com/$1 [R=301,L]    #
定义 rewrite 规则，当满足上面的条件时，这条规则才会执行
    </IfModule>
```

文件保存后仍需检测文件是否存在语法错误，无误后重载配置文件，并查询是否加载 rewrite 模块，若未加载，则需要在 httpd.conf 配置文件中找到 rewrite 模块，把前面的注释去掉，然后保存配置文件，操作命令如下。

```
[root@centos7 www]# /usr/local/apache2.4/bin/apachectl -t
Syntax OK
[root@centos7 www]# /usr/local/apache2.4/bin/apachectl graceful
[root@centos7 www]# /usr/local/apache2.4/bin/apachectl -M |grep rewrite
[root@centos7 www]# vim /usr/local/apache2.4/conf/httpd.conf
LoadModule rewrite_module modules/mod_rewrite.so
[root@centos7 www]# /usr/local/apache2.4/bin/apachectl -t
Syntax OK
[root@centos7 www]# /usr/local/apache2.4/bin/apachectl graceful
[root@centos7 www]# /usr/local/apache2.4/bin/apachectl -M |grep rewrite
rewrite_module (shared)
```

用 curl 命令可以查看跳转的状态码，操作命令如下。

```
[root@centos7 www]# curl -x192.162.222.129:80 humingzhe.cn -I
```

```
HTTP/1.1 301 Moved Permanently
Date: Fri, 26 Jan 2018 13:35:46 GMT
Server: Apache/2.4.29 (Unix) PHP/7.2.1
Location: http://www.humingzhe.com/
Content-Type: text/html; charset=iso-8859-1
```

在浏览器的地址栏中访问 humingzhe.cn，也会自动跳转到 humingzhe.com 界面，如图 2-20 所示。

图 2-20

2.9.3 Apache 访问日志

用 curl 命令访问网站或用浏览器访问网站，网站的后台都会产生一个访问日志信息，本书前面定义的访问日志路径在/data/logs/目录下，access_log 是记录访问日志信息的，error_log 是记录错误日志信息的，如查看 humingzhe.com-access_log，如图 2-21 所示。

```
[root@centos7 ~]# cd /data/logs/
[root@centos7 logs]# ls
humingzhe.cn-access_log  humingzhe.cn-error_log  humingzhe.com-access_log  humingzhe.com-error_log
[root@centos7 logs]#.cat humingzhe.com-access_log
192.168.222.129 - - [26/Jan/2018:16:24:51 +0800] "GET HTTP://mingzhe_blog/ HTTP/1.1" 400 347
192.168.222.129 - - [26/Jan/2018:16:24:56 +0800] "GET HTTP://mingzhe_blog/ HTTP/1.1" 400 347
192.168.222.129 - - [26/Jan/2018:16:25:01 +0800] "GET HTTP://mingzhe.com/ HTTP/1.1" 200 14
192.168.222.129 - - [26/Jan/2018:16:27:12 +0800] "GET HTTP://www.humingzhe.com/ HTTP/1.1" 200 14
192.168.222.129 - - [26/Jan/2018:16:27:41 +0800] "GET HTTP://mingzhe.com/ HTTP/1.1" 200 14
192.168.222.129 - - [26/Jan/2018:16:27:42 +0800] "GET HTTP://mingzhe.com/ HTTP/1.1" 200 14
192.168.222.129 - - [26/Jan/2018:16:57:32 +0800] "GET HTTP://www.wxample.com/ HTTP/1.1" 200 14
192.168.222.129 - - [26/Jan/2018:16:58:06 +0800] "GET HTTP://www.humingzhe.com/ HTTP/1.1" 200 14
192.168.222.129 - - [26/Jan/2018:16:58:10 +0800] "GET HTTP://mingzhe.com/ HTTP/1.1" 200 14
192.168.222.129 - - [26/Jan/2018:17:04:52 +0800] "GET HTTP://www.humingzhe.cn/ HTTP/1.1" 200 14
192.168.222.1 - - [26/Jan/2018:17:09:12 +0800] "GET / HTTP/1.1" 200 14
192.168.222.1 - - [26/Jan/2018:17:09:12 +0800] "GET /favicon.ico HTTP/1.1" 404 209
192.168.222.1 - bone [26/Jan/2018:21:28:17 +0800] "GET / HTTP/1.1" 200 14
192.168.222.129 - - [26/Jan/2018:21:29:44 +0800] "HEAD HTTP://humingzhe.com/ HTTP/1.1" 200 -
192.168.222.129 - - [26/Jan/2018:21:31:11 +0800] "HEAD HTTP://humingzhe.com/ HTTP/1.1" 200 -
192.168.222.129 - - [26/Jan/2018:21:31:55 +0800] "HEAD HTTP://humingzhe.cn/ HTTP/1.1" 200 -
192.168.222.1 - bone [26/Jan/2018:21:37:41 +0800] "GET / HTTP/1.1" 200 14
192.168.222.1 - bone [26/Jan/2018:21:37:42 +0800] "GET / HTTP/1.1" 200 14
192.168.222.1 - bone [26/Jan/2018:21:37:42 +0800] "GET / HTTP/1.1" 200 14
192.168.222.1 - bone [26/Jan/2018:21:37:43 +0800] "GET / HTTP/1.1" 200 14
192.168.222.1 - bone [26/Jan/2018:21:37:43 +0800] "GET / HTTP/1.1" 200 14
192.168.222.1 - bone [26/Jan/2018:21:37:43 +0800] "GET / HTTP/1.1" 200 14
192.168.222.1 - bone [26/Jan/2018:21:37:43 +0800] "GET / HTTP/1.1" 200 14
192.168.222.1 - bone [26/Jan/2018:21:37:44 +0800] "GET / HTTP/1.1" 200 14
192.168.222.1 - bone [26/Jan/2018:21:37:47 +0800] "GET / HTTP/1.1" 200 14
[root@centos7 logs]#
```

图 2-21

日志格式在 httpd.conf 主配置文件中进行定义，在配置文件中搜索"/LogFormat"，有 combined 和 common 两种格式，默认定义为 common 格式，操作命令如下。

```
[root@centos7 logs]# vim /usr/local/apache2.4/conf/httpd.conf
    LogFormat "%h %l %u %t \"%r\" %>s %b \"%{Referer}i\"
\"%{User-Agent}i\"" combined
```

```
LogFormat "%h %l %u %t \"%r\" %>s %b" common
```

日志访问格式在虚拟主机配置文件 httpd-vhosts.conf 中进行定义，把默认的 common 格式更改为 combined 格式，更改完毕后重载配置文件，操作命令如下。

```
ErrorLog "/data/logs/humingzhe.com-error_log"
    CustomLog "/data/logs/humingzhe.com-access_log" combined
[root@centos7 logs]# /usr/local/apache2.4/bin/apachectl graceful
```

重载后用 curl 命令执行几个操作，用浏览器刷新该域名，再找到 humingzhe.com-access_log 访问日志文件，用 curl 命令访问会记录 user_agent，如图 2-22 所示；而用浏览器访问会记录 Referer，如从“开源中国”跳转至 humingzhe.com。

```
[root@centos7 logs]# tail -f humingzhe.com-access_log
192.168.222.129 - - [27/Jan/2018:11:24:31 +0800] "HEAD HTTP://humingzhe.com/ HTTP/1.1" 200 - "-" "curl/7.29.0"
192.168.222.129 - - [27/Jan/2018:11:24:32 +0800] "HEAD HTTP://humingzhe.com/ HTTP/1.1" 200 - "-" "curl/7.29.0"
192.168.222.129 - - [27/Jan/2018:11:24:32 +0800] "HEAD HTTP://humingzhe.com/ HTTP/1.1" 200 - "-" "curl/7.29.0"
192.168.222.129 - - [27/Jan/2018:11:24:33 +0800] "HEAD HTTP://humingzhe.com/ HTTP/1.1" 200 - "-" "curl/7.29.0"
192.168.222.129 - - [27/Jan/2018:11:24:33 +0800] "HEAD HTTP://humingzhe.com/ HTTP/1.1" 200 - "-" "curl/7.29.0"
192.168.222.1 - - [27/Jan/2018:11:27:30 +0800] "GET / HTTP/1.1" 200 14 "-" "Mozilla/5.0 (Windows NT 6.1; WOW64) AppleWebKit/537.36 (KHTML, like Ge
cko) Chrome/64.0.3278.0 Safari/537.36"
192.168.222.1 - - [27/Jan/2018:11:27:30 +0800] "GET /favicon.ico HTTP/1.1" 404 209 "http://www.humingzhe.cn/" "Mozilla/5.0 (Windows NT 6.1; WOW64)
 AppleWebKit/537.36 (KHTML, like Gecko) Chrome/64.0.3278.0 Safari/537.36"
192.168.222.1 - - [27/Jan/2018:11:27:32 +0800] "GET / HTTP/1.1" 200 14 "-" "Mozilla/5.0 (Windows NT 6.1; WOW64) AppleWebKit/537.36 (KHTML, like Ge
cko) Chrome/64.0.3278.0 Safari/537.36"
192.168.222.1 - - [27/Jan/2018:11:27:32 +0800] "GET / HTTP/1.1" 200 14 "-" "Mozilla/5.0 (Windows NT 6.1; WOW64) AppleWebKit/537.36 (KHTML, like Ge
cko) Chrome/64.0.3278.0 Safari/537.36"
192.168.222.1 - - [27/Jan/2018:11:27:33 +0800] "GET / HTTP/1.1" 200 14 "-" "Mozilla/5.0 (Windows NT 6.1; WOW64) AppleWebKit/537.36 (KHTML, like Ge
cko) Chrome/64.0.3278.0 Safari/537.36"
192.168.222.1 - - [27/Jan/2018:11:28:29 +0800] "GET / HTTP/1.1" 200 14 "-" "Mozilla/5.0 (Windows NT 6.1; WOW64) AppleWebKit/537.36 (KHTML, like Ge
cko) Chrome/64.0.3278.0 Safari/537.36"
192.168.222.1 - - [27/Jan/2018:11:28:42 +0800] "GET / HTTP/1.1" 200 14 "-" "Mozilla/5.0 (Windows NT 6.1; WOW64) AppleWebKit/537.36 (KHTML, like Ge
cko) Chrome/64.0.3278.0 Safari/537.36"
192.168.222.1 - - [27/Jan/2018:11:28:46 +0800] "GET / HTTP/1.1" 200 14 "-" "Mozilla/5.0 (Windows NT 6.1; WOW64) AppleWebKit/537.36 (KHTML, like Ge
cko) Chrome/64.0.3278.0 Safari/537.36"
192.168.222.1 - - [27/Jan/2018:11:28:46 +0800] "GET / HTTP/1.1" 200 14 "-" "Mozilla/5.0 (Windows NT 6.1; WOW64) AppleWebKit/537.36 (KHTML, like Ge
cko) Chrome/64.0.3278.0 Safari/537.36"
192.168.222.1 - - [27/Jan/2018:11:28:46 +0800] "GET / HTTP/1.1" 200 14 "-" "Mozilla/5.0 (Windows NT 6.1; WOW64) AppleWebKit/537.36 (KHTML, like Ge
cko) Chrome/64.0.3278.0 Safari/537.36"
192.168.222.1 - - [27/Jan/2018:11:28:46 +0800] "GET / HTTP/1.1" 200 14 "-" "Mozilla/5.0 (Windows NT 6.1; WOW64) AppleWebKit/537.36 (KHTML, like Ge
cko) Chrome/64.0.3278.0 Safari/537.36"
192.168.222.1 - - [27/Jan/2018:11:28:46 +0800] "GET / HTTP/1.1" 200 14 "-" "Mozilla/5.0 (Windows NT 6.1; WOW64) AppleWebKit/537.36 (KHTML, like Ge
cko) Chrome/64.0.3278.0 Safari/537.36"
192.168.222.1 - - [27/Jan/2018:11:28:49 +0800] "GET / HTTP/1.1" 200 14 "-" "Mozilla/5.0 (Windows NT 6.1; WOW64) AppleWebKit/537.36 (KHTML, like Ge
cko) Chrome/64.0.3278.0 Safari/537.36"
```

图 2-22

2.9.4　静态文件不记录日期和时间

用浏览器访问 humingzhe.com 时，按 F12 键会弹出一个框，定位到 NetWork 处刷新会加载很多 CSS、JS 和图片文件，而这些请求和访问都会被记录到访问日志中。一个网站由数十个页面组成，每个页面都有几十个这样的图片文件或样式文件，每访问一次就记录一次，这样会使日志信息增大 *N* 倍，一天下来少则几 GB，多则数十 GB，浪费磁盘内存。

所以，没有必要去记录这些信息，清除这些文件需要在虚拟主机配置文件中进行定义，其定义在 CustomLog 中，操作命令如下。

```
[root@centos7 ~]# vim
/usr/local/apache2.4/conf/extra/httpd-vhosts.conf
    ErrorLog "/data/logs/humingzhe.cn-error_log"
        SetEnvIf Request_URI ".*\.gif$" mingzhe
        SetEnvIf Request_URI ".*\.jpg$" mingzhe
```

```
            SetEnvIf Request_URI ".*\.png$" mingzhe
            SetEnvIf Request_URI ".*\.bmp$" mingzhe
            SetEnvIf Request_URI ".*\.swf$" mingzhe
            SetEnvIf Request_URI ".*\.js$" mingzhe
            SetEnvIf Request_URI ".*\.css$" mingzhe
            SetEnvIf Request_URI ".*\.jpeg$" mingzhe
        CustomLog "/data/logs/humingzhe.cn-access_log" combined
env=!mingzhe
```

2.9.5 访问日志切割

日志写在文件中，总有一天会把磁盘写满，100GB 的硬盘，每天写 1GB 日志，100 天即可把磁盘写满，在 Linux 操作系统下这种事情随时都会发生。所以，有必要每天对日志文件进行切割，例如每天凌晨把前一天的日志归档，当天从凌晨开始写新的日志文件，超过 30 天后把旧日志删除或复制到其他地方，这样做可以保证磁盘永远不会被写满。

实现日志切割的方法是编辑虚拟主机配置文件 httpd-vhosts.conf，在 CustomLog 中进行定义，rotatelogs 是 Apache 中的日志切割工具，和操作系统中的不一样。使用-l 选项表示以当前系统的时间进行切割，默认为 CST，不使用-l 选项默认以 UTC 时间进行切割，时区不一样。每天进行切割，天换算成秒是 86400 秒，操作代码如下。

```
    [root@centos7 ~]# vim
/usr/local/apache2.4/conf/extra/httpd-vhosts.conf
    CustomLog "|/usr/local/apache2.4/bin/rotatelogs -l
/data/logs/humingzhe.cn-access_%Y%m%d.log 86400" combined env=!img
    [root@centos7 ~]# /usr/local/apache2.4/bin/apachectl -t
    Syntax OK
    [root@centos7 ~]# /usr/local/apache2.4/bin/apachectl graceful
```

进入日志所在的目录，用 ls 命令查看没有生成当天的日志文件，在浏览器中刷新几次就会生成当天的日志文件，如图 2-23 所示。

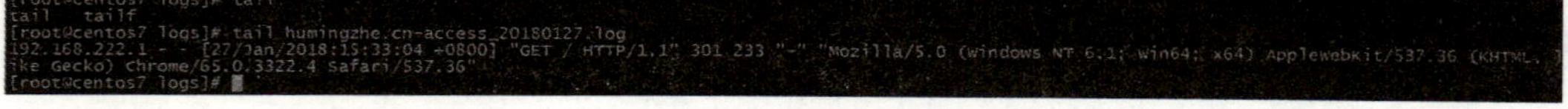

```
[root@centos7 data]# ls

[root@centos7 data]# cd logs/
[root@centos7 logs]# ls
humingzhe.cn-access_log  humingzhe.cn-error_log  humingzhe.com-access_log  humingzhe.com-error_log
[root@centos7 logs]# ls
humingzhe.cn-access_20180127.log  humingzhe.cn-access_log  humingzhe.cn-error_log  humingzhe.com-access_log  humingzhe.com-error_log
[root@centos7 logs]# tail
tail   tailf
[root@centos7 logs]# tail humingzhe.cn-access_20180127.log
192.168.222.1 - - [27/Jan/2018:15:33:04 +0800] "GET / HTTP/1.1" 301 233 "-" "Mozilla/5.0 (Windows NT 6.1; Win64; x64) AppleWebKit/537.36 (KHTML, l
ike Gecko) Chrome/65.0.3322.4 Safari/537.36"
[root@centos7 logs]#
```

图 2-23

最后写进任务计划中，把大于 30 天的日志文件删除，如果不删除，切割后和切割前是没有任何区别的，操作命令如下。

```
[root@centos7 logs]# vim /data/shell/del-30days-log.sh
#!/bin/bash
##删除大于 30 天的日志文件脚本
##__author__ is humingzhe
##email admin@humingzhe.com
find /data/logs/ -mtime +30 -name "*.log" -exec rm -rf {} \;
[root@centos7 logs]# chmod +x /data/shell/del-30days-log.sh
[root@centos7 logs]# crontab -e
   10 0 * * * /data/shell/del-30days-log.sh >/dev/null 2>&1
```

2.9.6　静态元素过期时间

用浏览器访问网站的图片时，会把静态文件缓存到本地计算机中，这样下次再访问时就不用远程加载了。缓存的具体时间是在服务器上定义的，如果不定义，浏览器就不会把这些文件清空，或者浏览器使用自带的机制清空缓存文件，或者在一段时间后计算机中的腾讯助手、360 管家就会帮助检测垃圾并提示清理。

定义失效时间需要编辑虚拟主机配置文件，在配置文件中增加几行代码，操作命令如下。

```
    [root@centos7 logs]# vim
/usr/local/apache2.4/conf/extra/httpd-vhosts.conf
       <IfModule mod_expires.c>
          ExpiresActive on #开启该功能
          ExpiresByType image/gif "access plus 1 days"
          ExpiresByType image/jpeg "access plus 24 hours"
          ExpiresByType image/jpg "access plus 24 hours"
          ExpiresByType image/png "access plus 24 hours"
          ExpiresByType image/css "now plus 2 hour"
          ExpiresByType image/application/x-javascript "now plus 2 hours"
          ExpiresByType image/application/javascript "now plus 2 hours"
          ExpiresByType image/application/x-shockwave-flash "now plus 2 
hours"
          ExpiresDefault "now plus 0 min"
       </IfModule>
```

查看 expires 模块是否被打开，如未被打开，需在配置文件中开启，开启后需重载配置文件，操作命令如下。

```
[root@centos7 logs]# vim /usr/local/apache2.4/conf/httpd.conf
LoadModule expires_module modules/mod_expires.so
[root@centos7 logs]# /usr/local/apache2.4/bin/apachectl -t
Syntax OK
[root@centos7 logs]# /usr/local/apache2.4/bin/apachectl graceful
[root@centos7 logs]# apachectl -M |grep expires
 expires_module (shared)
```

2.9.7 配置防盗链

通过限制 referer 可实现防盗链功能，编辑 httpd-vhosts.conf 配置文件，在该文件中增加如下代码。

```
    [root@centos7 logs]# vim
/usr/local/apache2.4/conf/extra/httpd-vhosts.conf
        <Directory /data/www/humingzhe.cn>  #定义目录
            SetEnvIfNoCase Referer "http://www.humingzhe.cn" local_ref  #白
名单
            SetEnvIfNoCase Referer "http://humingzhe.com" local_ref  #白名单
            SetEnvIfNoCase Referer "^$" local_ref          #白名单
            <FilesMatch "\.(txt|doc|mp3|zip|rar|jpg|gif|png|jpeg)">  #针对这
些文件类型做防盗链
                Order Allow,Deny
                Allow from env=local_ref
            </FilesMatch>
        </Directory>
```

2.9.8 访问控制 Directory

有时需要对某个目录进行访问控制，如访问 admin 目录。admin 目录只有公司内部人员才可以访问，对外不开放。如果是用户认证方式，用户名和密码泄露后外人就可以进行访问，所以要针对某个目录进行白名单 IP 限制，只有用白名单中的 IP 登录才可以访问。

公司公网 IP 是固定的，可以把公司公网 IP 放入白名单中，只有用公司的 IP 登录才可以正常访问，否则全部拒绝。编辑虚拟主机配置文件，增加如下代码。

```
    [root@centos7 logs]# vim
/usr/local/apache2.4/conf/extra/httpd-vhosts.conf
        <Directory /data/www/humingzhe.cn/admin/>
```

```
        Order deny,allow
        Deny from all
        Allow from 127.0.0.1
    </Directory>
```

设置完毕后在 www.humingzhe.cn 目录下创建 admin 目录，在 admin 目录下创建 index.php 文件并输出一些信息，操作命令如下。

```
[root@centos7 www]# cd humingzhe.cn/
[root@centos7 www]# mkdir admin
[root@centos7 www]# vim admin/index.php
<?php
    echo "我好喜欢你";
```

使用 curl 命令访问 127.0.0.1，返回 200 状态码显示正常，但用其他 IP 访问会显示“403 Forbidden”，如图 2-24 所示。用浏览器访问该地址时也显示 403 状态码，如图 2-25 所示。

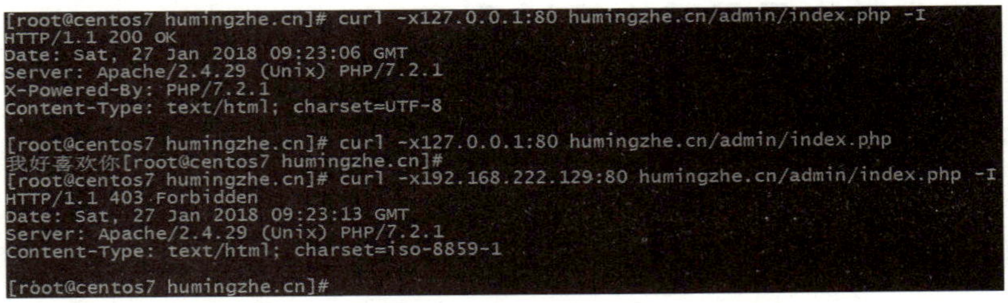

```
[root@centos7 humingzhe.cn]# curl -x127.0.0.1:80 humingzhe.cn/admin/index.php -I
HTTP/1.1 200 OK
Date: Sat, 27 Jan 2018 09:23:06 GMT
Server: Apache/2.4.29 (Unix) PHP/7.2.1
X-Powered-By: PHP/7.2.1
Content-Type: text/html; charset=UTF-8

[root@centos7 humingzhe.cn]# curl -x127.0.0.1:80 humingzhe.cn/admin/index.php
我好喜欢你[root@centos7 humingzhe.cn]#
[root@centos7 humingzhe.cn]# curl -x192.168.222.129:80 humingzhe.cn/admin/index.php -I
HTTP/1.1 403 Forbidden
Date: Sat, 27 Jan 2018 09:23:13 GMT
Server: Apache/2.4.29 (Unix) PHP/7.2.1
Content-Type: text/html; charset=iso-8859-1

[root@centos7 humingzhe.cn]#
```

图 2-24

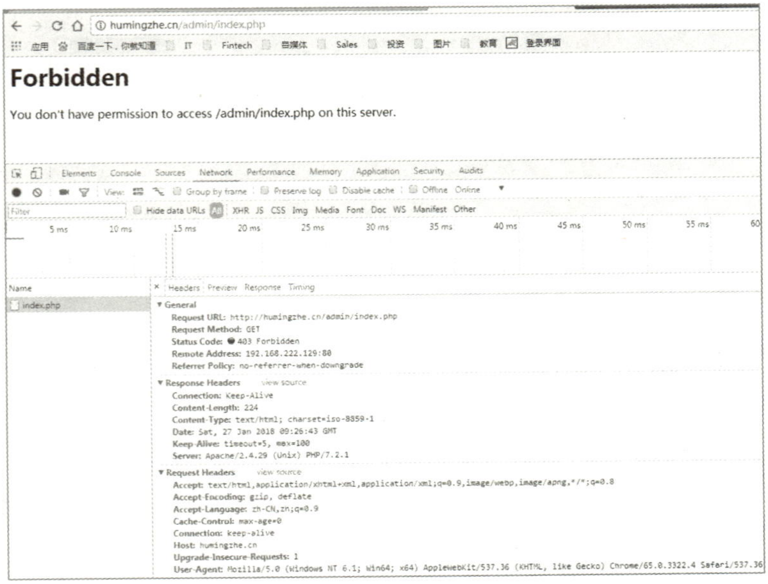

图 2-25

2.9.9 访问控制 FilesMatch

访问控制除限制目录外，还可以限制某个文件名，如后台登录界面是 admin.php，不想让外人访问，只想让公司内部人员访问，可以采取和限制目录一样的措施，限定 IP 地址。限定某个文件的做法是用 FilesMatch 进行匹配，在虚拟主机配置文件中增加如下代码。

```
<Directory /data/www/humingzhe.cn/admin/>
    <FilesMatch admin.php(.*)>
    Order deny,allow
    Deny from all
    Allow from 127.0.0.1
    </FilesMatch>
</Directory>
```

2.9.10 限定某个目录禁止解析 PHP

每个网站中都有一个 upload 目录，upload 目录是用来上传图片的，别有用心的人可能会通过一些手段上传 PHP 文件，一旦执行上传的 PHP 文件就可能丢失服务器的 root 权限，非常危险。

为了避免发生上述情况，需要禁止 upload 目录解析 PHP，方法是编辑虚拟主机配置文件，在虚拟主机配置文件中添加如下代码。

```
<Directory /data/www/humingzhe.cn/content/uploadfile>
    php_admin_flag engine off
    <FilesMatch (.*)\.php(.*)>
        Order allow,deny
        Deny from all
    </FilesMatch>
</Directory>
```

2.9.11 限制 user_agent

有时网站会受到 CC 攻击，攻击者通过软件或“肉鸡”（受黑客远程控制的计算机）等手段攻击某个目标站点，如果攻击者手中有一万个“肉鸡”，同时攻击某站点，站点肯定承受不住。攻击者都有一个共同的特征，即 user_agent 是一致的，遇到很多一样的请求时可限制 user_agent 以减轻服务器压力。设置 user_agent 需要编辑虚拟主机配置文件，在虚拟主机配置文件中添加如下代码。

```
<IfModule rewrite.c>
    RewriteEngine on
    RewriteCond %{HTTP_USER_AGENT}  .*curl.* [NC,OR] #指定 user_agent
    RewriteCond %{HTTP_USER_AGENT}  .*baidu.com.* [NC]
    RewriteRule .*  -   [F]
</IfModule>
```

2.10　PHP相关配置

PHP 配置文件路径可以通过浏览器访问 phpinfo 查看，但通过查看 phpinfo 发现并没有加载配置文件路径，需设置配置文件路径，操作命令如下。

```
[root@centos7 ~]# cd /usr/local/src/php-7.2.1/
[root@centos7 php-7.2.1]# cp php.ini-development
/usr/local/php7/etc/php.ini
[root@centos7 php-7.2.1]# /usr/local/apache2.4/bin/apachectl graceful
```

打开 php.ini 配置文件，搜索 disable_functions，禁掉一些相关函数配置，操作命令如下。

```
[root@centos7 php-7.2.1]# vim /usr/local/php7/etc/php.ini
disable_functions =
eval,assert,popen,passthru,escapeshellarg,escapeshellcmd,passthru,exec,s
ystem,chroot,scandir,chgrp,chown,escapenshellcmd,escapshellarg,shell_exe
c,proc_get_status,ini_alter,ini_restore,dl,pfsockopen,openlog,syslog,rea
dlink,symlink,leak,popepassthru,stream_socket_server,popen,proc_open,pro
c_close,phpinfo
```

定义 date.timezone 时区，如果不定义就会报一些语法错误，例如定义亚洲上海时区，操作命令如下。

```
date.timezone = Asia/Shanghai
```

“display_errors = On”表示打开语法错误，打开语法错误会让错误的语法显示在浏览器上，用户会看到，需更改为 Off，即不把语法错误信息显示在浏览器上，操作命令如下。

```
display_errors = Off
```

设置错误日志，在 php.ini 配置文件中搜索 log_error，指定 error_log 目录，刷新浏览器后会在指定目录下生成该错误日志文件，操作命令如下。

```
log_errors = On
error_log = /data/logs/php/php_errors.log
error_reporting = E_NOTICE
```

设置 open_basedir 后可将用户访问文件的活动范围限制在指定的区域，在 php.ini 中设置是针对所有站点进行的，不能指定单一的网站。针对某个站点，可以在虚拟主机配置文件中进行设置，操作命令如下。

```
php_admin_value open_basedir "/data/www/humingzhe.cn:/tmp/"
```

2.11 安装PHP扩展模块Redis

日常工作中会用到 Redis，但是编译安装 PHP 时并没有添加 Redis 这个模块，缺少 Redis 模块也不可能重新编译 PHP，可以把 Redis 编译成以.so 结尾的文件，用该文件编译安装 Redis 扩展模块，操作命令如下。

```
[root@centos7 ~]# cd /usr/local/src/
[root@centos7 src]# wget
https://codeload.github.com/phpredis/phpredis/zip/develop
[root@centos7 src]# mv develop phpredis-develop.zip
[root@centos7 src]# unzip phpredis-develop.zip
[root@centos7 phpredis-develop]# /usr/local/php7/bin/phpize
[root@centos7 phpredis-develop]# ./configure
--with-php-config=/usr/local/php7/bin/php-config
[root@centos7 phpredis-develop]# echo $?
0
[root@centos7 phpredis-develop]# make && make install
[root@centos7 phpredis-develop]# echo $?
0
```

编译并安装完成后，默认不会加载 redis.so 模块，需要在 php.ini 中设置文件，在“;extension=shmop”后增加一行代码，代码如下。

```
[root@centos7 phpredis-develop]# /usr/local/php7/bin/php -m |grep redis
[root@centos7 phpredis-develop]# /usr/local/php7/bin/php -i |grep -i
extension_dir
extension=redis.so
[root@centos7 phpredis-develop]# apachectl -t
Syntax OK
[root@centos7 phpredis-develop]# /usr/local/apache2.4/bin/apachectl
graceful
[root@centos7 phpredis-develop]# /usr/local/php7/bin/php -m |grep redis
```

```
redis
[root@centos7 phpredis-develop]# php -m
```

Redis 和 mongodb 都属于第三方的，PHP 源码包中不自带，在 PHP 源码包中有 ext 目录，ext 目录下有很多模块，系统现在没有 dba 模块，进入 dba 目录，直接执行 phpize 命令即可，如图 2-26 所示。

```
[root@centos7 phpredis-develop]# cd /usr/local/src/php-7.2.1/ext/
[root@centos7 ext]# ls
        ext_skel
        ext_skel_win32.php
[root@centos7 ext]# /usr/local/php7/bin/php -m |grep sqlite
pdo_sqlite
sqlite3
[root@centos7 ext]# /usr/local/php7/bin/php -m |grep dba
[root@centos7 ext]# cd dba/
[root@centos7 dba]# ls
config.m4   dba_cdb.c  dba_db4.c       dba_inifile.c  dba_tcadb.c                php_db3.h  php_flatfile.h  php_ndbm.h
config.w32  dba_db1.c  dba_dbm.c       dba_lmdb.c     install_cdb.sh  php_cdb.h  php_db4.h  php_gdbm.h      php_qdbm.h
CREDITS     dba_db2.c  dba_flatfile.c  dba_ndbm.c                     php_db1.h  php_dba.h  php_inifile.h   php_tcadb.h
dba.c       dba_db3.c  dba_gdbm.c      dba_qdbm.c                     php_db2.h  php_dbm.h  php_lmdb.h      README
[root@centos7 dba]# /usr/local/php7/bin/phpize
Configuring for:
PHP Api Version:         20170718
Zend Module Api No:      20170718
Zend Extension Api No:   320170718
```

图 2-26

执行完毕后设置并编译 dba 模块，运行命令如下。

```
[root@centos7 dba]# ./configure
--with-php-config=/usr/local/php7/bin/php-config
[root@centos7 dba]# echo $?
0
[root@centos7 dba]# make && make install
[root@centos7 dba]# echo $?
0
```

第 3 章

3

LNMP 服务架构

3.1　LNMP服务架构简介

LNMP 架构和 LAMP 架构相似，LAMP 即 Linux+Apache+MySQL+PHP，而 LNMP 只是把 LAMP 中的 Apache 换成 Nginx。Nginx 在架构中所担任的角色和 Apache 是类似的。

在 LAMP 架构中，PHP 是作为 Apache 的一个模块出现的，用户请求 PHP，Apache 会把请求交给 PHP 的模块去处理。

在 LNMP 架构中，PHP 会启动 php-fpm 服务，Nginx 会把用户请求的 PHP 交给 php-fpm 服务去处理，php-fpm 服务和 MySQL 进行交互，数据库验证成功后返回给 php-fpm，php-fpm 再返回给 Nginx，Nginx 返回给用户。

相较 Apache，Nginx 占用的资源和内存更少，在抗并发方面性能是 Apache 的数十倍。Nginx 处理请求是异步阻塞的，而 Apache 是阻塞型的。在高并发情况下，Nginx 能保持低资源、低消耗、高性能、高度模块化的设计，编写模块相对简单，在负载方面也优于 Apache。

3.2　安装MySQL数据库

删除前述 LAMP 章节中安装的 MySQL 数据库，删除 MySQL 前要确保 mysqld 服务处于关闭状态，操作命令如下。

```
[root@centos7 ~]# cd /usr/local/src/
[root@centos7 src]# ps aux |grep mysqld
```

```
    root     122185  0.0  0.0 112660   972 pts/0    S+   10:02   0:00 grep
--color=auto mysqld
    [root@centos7 src]# rm -rf /usr/local/mysql
    [root@centos7 src]# rm -rf /etc/init.d/mysqld
```

下载 MySQL 5.6 版本数据库至/usr/local/src/目录，下载完毕后进行解压，解压至/usr/local/目录，将解压后的 mysql-5.6 目录软链接到 mysql 目录，操作命令如下。

```
    [root@centos7 src]# wget http://mirrors.sohu.com/mysql/MySQL-5.6/
mysql-5.6.36-linux-glibc2.5-x86_64.tar.gz
    [root@centos7 src]# tar xf mysql-5.6.36-linux-glibc2.5-x86_64.tar.gz -C
/usr/local/
    [root@centos7 src]# cd ..
    [root@centos7 local]# ln -sv mysql-5.6.36-linux-glibc2.5-x86_64 mysql
    'mysql' -> 'mysql-5.6.36-linux-glibc2.5-x86_64'
    [root@centos7 local]# ls -l mysql
    lrwxrwxrwx 1 root root 34 Jan 28 11:05 mysql ->
mysql-5.6.36-linux-glibc2.5-x86_64
```

进入 mysql 目录，创建 MySQL 用户和存放数据的/data 目录，因为在前面章节中创建过用户和 data 目录，所以这里不再进行创建，删除 mysql 目录下所有数据后进行 MySQL 设置，操作命令如下。

```
    [root@centos7 local]# cd mysql
    [root@centos7 mysql]# rm -rf /data/mysql/*
    [root@centos7 mysql]# ./scripts/mysql_install_db --user=mysql
--datadir=/data/mysql
    [root@centos7 mysql]# echo $?
    0
```

复制 support-files/my-default.cnf 至/etc/目录下，并重命名为 my.cnf，操作命令如下。

```
    [root@centos7 mysql]# cp support-files/my-default.cnf /etc/my.cnf
```

复制启动脚本 mysql.server 至/etc/init.d/目录下并重命名为 mysql，编辑此配置文件，更改 datadir 和 basedir，操作命令如下。

```
    [root@centos7 mysql]# cp support-files/mysql.server /etc/init.d/mysqld
    [root@centos7 mysql]# vim /etc/init.d/mysqld
    basedir=/usr/local/mysql
    datadir=/data/mysql
```

启动 mysqld 服务，用 ps 命令可查看是否启动成功，设置 MySQL 开机自启动，使用 netstat 命令可查看是否监听 3306 端口，如图 3-1 所示。

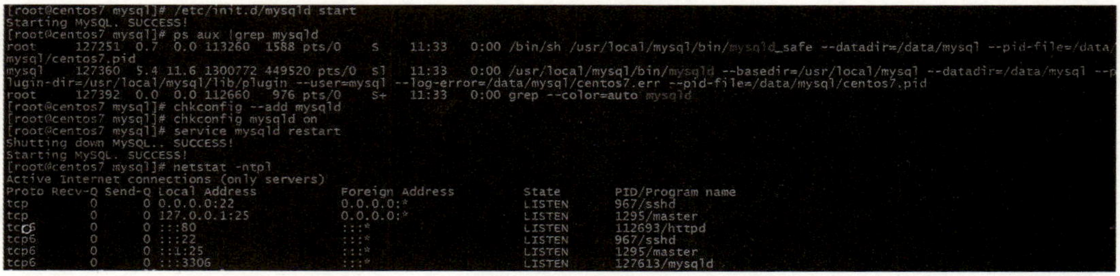

```
[root@centos7 mysql]# /etc/init.d/mysqld start
Starting MySQL. SUCCESS!
[root@centos7 mysql]# ps aux |grep mysqld
root     127251  0.7  0.0 113260  1588 pts/0    S    11:33   0:00 /bin/sh /usr/local/mysql/bin/mysqld_safe --datadir=/data/mysql --pid-file=/data/
mysql/centos7.pid
mysql    127360  5.4 11.6 1300772 449520 pts/0  Sl   11:33   0:00 /usr/local/mysql/bin/mysqld --basedir=/usr/local/mysql --datadir=/data/mysql --p
lugin-dir=/usr/local/mysql/lib/plugin --user=mysql --log-error=/data/mysql/centos7.err --pid-file=/data/mysql/centos7.pid
root     127392  0.0  0.0 112660   976 pts/0    S+   11:33   0:00 grep --color=auto mysqld
[root@centos7 mysql]# chkconfig --add mysqld
[root@centos7 mysql]# chkconfig mysqld on
[root@centos7 mysql]# service mysqld restart
Shutting down MySQL.. SUCCESS!
Starting MySQL. SUCCESS!
[root@centos7 mysql]# netstat -ntpl
Active Internet connections (only servers)
Proto Recv-Q Send-Q Local Address           Foreign Address         State       PID/Program name
tcp        0      0 0.0.0.0:22              0.0.0.0:*               LISTEN      967/sshd
tcp        0      0 127.0.0.1:25            0.0.0.0:*               LISTEN      1295/master
tcp6       0      0 :::80                   :::*                    LISTEN      112693/httpd
tcp6       0      0 :::22                   :::*                    LISTEN      967/sshd
tcp6       0      0 ::1:25                  :::*                    LISTEN      1295/master
tcp6       0      0 :::3306                 :::*                    LISTEN      127613/mysqld
```

图 3-1

3.3 安装PHP

在 LNMP 服务架构中安装 PHP 需要开启 php-fpm 服务，因为在讲解 LAMP 时编译过 PHP 7 版本，所以可以直接进入 PHP 源码包，把之前编译过的文件删除，操作命令如下。

```
    [root@centos7 mysql]# cd /usr/local/src/php-7.2.1/
    [root@centos7 php-7.2.1]# make clean
    find . -name \*.gcno -o -name \*.gcda | xargs rm -f
    find . -name \*.lo -o -name \*.o | xargs rm -f
    find . -name \*.la -o -name \*.a | xargs rm -f
    find . -name \*.so | xargs rm -f
    find . -name .libs -a -type d|xargs rm -rf
    rm -f libphp7.la sapi/cli/php sapi/cgi/php-cgi    libphp7.la modules/*
libs/*
```

在 LAMP 中设置 PHP 时指定的是/usr/local/php7，这里需要把“php7”改为“php-fpm”，设置完毕后进行编译和安装，操作命令如下。

```
    [root@centos7 php-7.2.1]# yum install -y libcurl-devel gcc-c++
    [root@centos7 php-7.2.1]# ./configure --prefix=/usr/local/php-fpm
--with-config-file-path=/usr/local/php-fpm/etc/ --enable-fpm --with-fpm-
user=php-fpm --with-fpm-group=php-fpm --with-mysql=/usr/local/mysql
--with-mysqli=/usr/local/mysql/bin/mysql_config --with-pdo-mysql=/usr/
local/mysql --with-mysql-sock=/tmp/mysql.sock --with-libxml-dir --with-gd
--with-jpeg-dir --with-png-dir --with-freetype-dir --with-iconv-dir
--with-zlib-dir --with-mcrypt --enable-soap --enable-gd-native-ttf
--enable-ftp --enable-mbstring --enable-exif --with-pear --with-curl
--with-openssl
    [root@centos7 php-7.2.1]# echo $?
    0
    [root@centos7 php-7.2.1]# make && make install
```

```
[root@centos7 php-7.2.1]# echo $?
0
```

/usr/local/php-fpm 目录比/usr/local/php7 目录多了一个 sbin 目录，sbin 目录下有一个 php-fpm 文件，该文件是启动 php-fpm 的文件，操作命令如下。

```
[root@centos7 php-7.2.1]# ls /usr/local/php-fpm/
bin  etc  include  lib  php  sbin  var
[root@centos7 php-7.2.1]# ls /usr/local/php7/
bin  etc  include  lib  php  var
```

使用 php-fpm－m 命令可以查看安装了哪些模块，使用-t 选项可以检测语法是否有误，如果觉得路径长，也可以添加至环境变量中，操作命令如下。

```
[root@centos7 php-7.2.1]# /usr/local/php-fpm/sbin/php-fpm -m
[root@centos7 php-7.2.1]# /usr/local/php-fpm/sbin/php-fpm -t
[root@centos7 php-7.2.1]# vim /etc/profile.d/php-fpm.sh
export PATH=/usr/local/php-fpm/sbin:$PATH
[root@centos7 php-7.2.1]# php-fpm -m
[PHP Modules]
cgi-fcgi
Core
```

使用-t 选项检测 php-fpm 语法时提示缺少 php-fpm.conf 文件，先复制 php.ini 文件至/etc/目录（production 适用于生产环境，development 适用于开发环境），然后编辑 php-fpm.conf 配置文件，增加如下代码。

```
[root@centos7 php-7.2.1]# cp php.ini-production
/usr/local/php-fpm/etc/php.ini
[root@centos7 php-7.2.1]# cd /usr/local/php-fpm/etc/
[root@centos7 etc]# ls
pear.conf  php-fpm.conf.default  php-fpm.d  php.ini
[root@centos7 etc]# vim php-fpm.conf
[global]
pid = /usr/local/php-fpm/var/run/php-fpm.pid
error_log = /usr/local/php-fpm/var/log/php-fpm.log
[www]
listen = /tmp/php-fcgi.sock
#listen = 127.0.0.1:9000
listen.mode = 666
user = php-fpm
group = php-fpm
pm = dynamic
pm.max_children = 50
```

```
    pm.start_servers = 20
    pm.min_spare_servers = 5
    pm.max_spare_servers = 35
    pm.max_requests = 500
    rlimit_files = 1024
```

复制 php-fpm 启动脚本文件至/etc/init.d/目录下，操作命令如下。

```
    [root@centos7 etc]# cd /usr/local/src/php-7.2.1/
    [root@centos7 php-7.2.1]# cp sapi/fpm/init.d.php-fpm
/etc/init.d/php-fpm
    [root@centos7 php-7.2.1]# chmod 755 /etc/init.d/php-fpm
    [root@centos7 php-7.2.1]# chkconfig --add php-fpm
    [root@centos7 php-7.2.1]# chkconfig php-fpm on
    [root@centos7 php-7.2.1]# useradd php-fpm
    [root@centos7 php-7.2.1]# service php-fpm start
    Starting php-fpm  done
    [root@centos7 php-7.2.1]# ps aux |grep php-fpm
```

3.4 Nginx简介与安装

3.4.1 Nginx 简介

Nginx 官网是 http://nginx.org/。Nginx 最新版本是 1.13。Nginx 本身自带的模块非常少，但是可以安装第三方模块，如 HTTPS 协议所支持的模块。

Nginx 应用场景有：Web 服务器、反向代理、负载均衡。

Nginx 有一个著名的分支，淘宝网基于 Nginx 开发了 Tengine，Tengine 在使用上和 Nginx 一致，服务名、配置文件名也是一致的。Tengine 和 Nginx 最大的区别在于，Tengine 增加了一些定制化模块，在安全限速方面表现突出，另外，它还支持对 JavaScript、CSS 合并。

3.4.2 Nginx 安装

下载 Nginx 源码包至/usr/local/src/目录，下载完毕后进行解压和参数设置，设置完毕后进行编译和安装，操作命令如下。

```
    [root@centos7 php-7.2.1]# cd /usr/local/src/
    [root@centos7 src]# wget http://nginx.org/download/nginx-1.13.8.tar.gz
```

```
[root@centos7 src]# tar xf nginx-1.13.8.tar.gz
[root@centos7 src]# cd nginx-1.13.8/
[root@centos7 nginx-1.13.8]# ./configure --prefix=/usr/local/nginx
[root@centos7 nginx-1.13.8]# make && make install
```

给 Nginx 添加启动脚本配置文件，Nginx 启动脚本配置文件不是自带的，需要在/etc/init.d/目录下创建配置文件 nginx，由于文件代码过多会占用书中大幅篇章，故把启动脚本代码放到 GitLab 中，读者可访问如下网址进行下载：https://gitlab.com/humingzhe/Linux/blob/master/LAMP/Nginx 启动脚本 For.CentOS7。

```
[root@centos7 nginx-1.13.8]# vim /etc/init.d/nginx
```

更改配置文件 nginx 的权限为 755，开启 Nginx 并设置 Nginx 开机自启动，操作命令如下。

```
[root@centos7 nginx-1.13.8]# chmod 755 /etc/init.d/nginx
[root@centos7 nginx-1.13.8]# chkconfig --add nginx
[root@centos7 nginx-1.13.8]# chkconfig nginx on
```

Nginx 本身自带了 nginx.conf 配置文件，这里不用它自带的，用笔者给大家提供的配置文件代码即可，配置文件链接网址：https://gitlab.com/humingzhe/Linux/blob/master/LAMP/nginx.conf。

```
[root@centos7 nginx-1.13.8]# cd /usr/local/nginx/conf/
[root@centos7 conf]# mv nginx.conf nginx.conf.bak
[root@centos7 conf]# vim nginx.conf
```

保存后使用-t 选项检测文件语法是否有误，无误后即可启动 Nginx 服务。注意，确保 Apache 处于关闭状态后才可开启 Nginx，否则会报错，操作命令如下。

```
[root@centos7 conf]# /usr/local/nginx/sbin/nginx -t
nginx: the configuration file/usr/local/nginx/conf/nginx.conf syntax is ok
nginx: configuration file /usr/local/nginx/conf/nginx.conf test is
successful
[root@centos7 conf]# /etc/init.d/nginx start
Starting nginx (via systemctl):  [  OK  ]
```

启动成功后在浏览器的地址栏中输入 192.168.222.129，即可显示 Nginx 安装成功界面，如图 3-2 所示。

3.4.3　Nginx 测试解析 PHP

存放 PHP 文件的路径是/usr/local/nginx/html/，在该路径下创建 1.php 文件，并在 PHP 文件中输出一段内容，操作代码如下。

```
[root@centos7 conf]# vim /usr/local/nginx/html/1.php
<?php
    echo phpinfo();
```

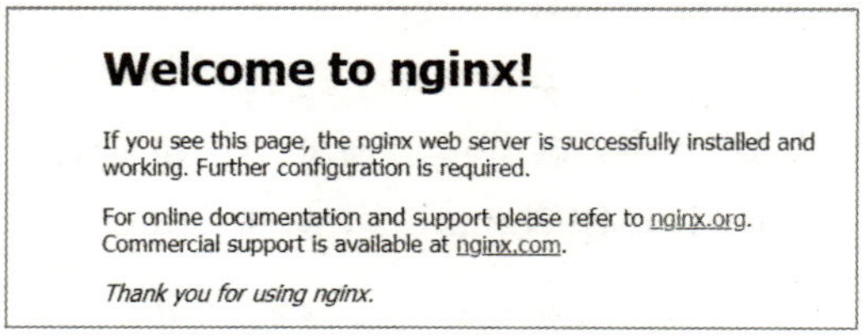

Welcome to nginx!

If you see this page, the nginx web server is successfully installed and working. Further configuration is required.

For online documentation and support please refer to nginx.org.
Commercial support is available at nginx.com.

Thank you for using nginx.

图 3-2

用浏览器访问 1.php 文件，会输出 PHP 的详细配置信息，如图 3-3 所示。

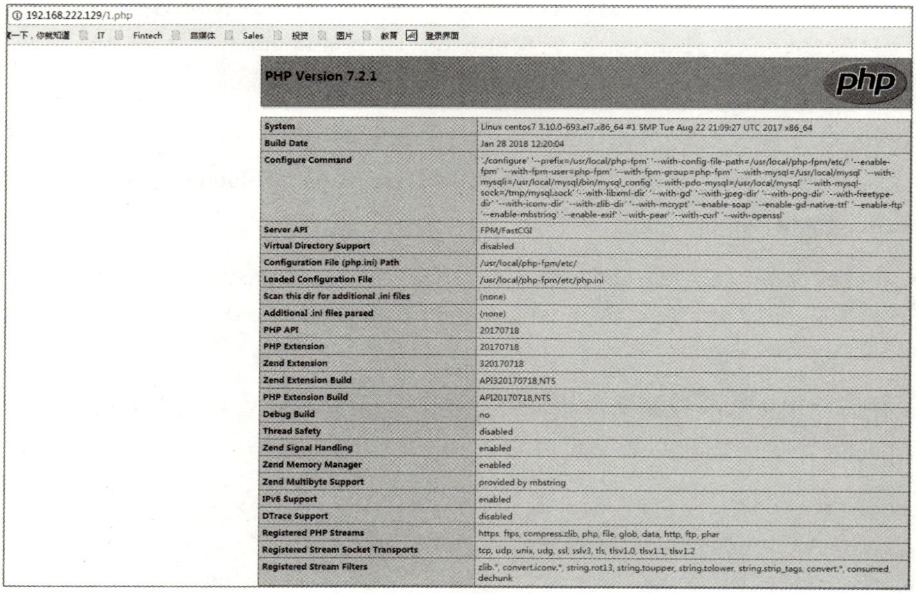

192.168.222.129/1.php

IT　Fintech　自媒体　Sales　投资　图片　教育　登录界面

PHP Version 7.2.1

System	Linux centos7 3.10.0-693.el7.x86_64 #1 SMP Tue Aug 22 21:09:27 UTC 2017 x86_64
Build Date	Jan 28 2018 12:20:04
Configure Command	'./configure' '--prefix=/usr/local/php-fpm' '--with-config-file-path=/usr/local/php-fpm/etc/' '--enable-fpm' '--with-fpm-user=php-fpm' '--with-fpm-group=php-fpm' '--with-mysql=/usr/local/mysql' '--with-mysqli=/usr/local/mysql/bin/mysql_config' '--with-pdo-mysql=/usr/local/mysql' '--with-mysql-sock=/tmp/mysql.sock' '--with-libxml-dir' '--with-gd' '--with-jpeg-dir' '--with-png-dir' '--with-freetype-dir' '--with-iconv-dir' '--with-zlib-dir' '--with-mcrypt' '--enable-soap' '--enable-gd-native-ttf' '--enable-ftp' '--enable-mbstring' '--enable-exif' '--with-pear' '--with-curl' '--with-openssl'
Server API	FPM/FastCGI
Virtual Directory Support	disabled
Configuration File (php.ini) Path	/usr/local/php-fpm/etc/
Loaded Configuration File	/usr/local/php-fpm/etc/php.ini
Scan this dir for additional .ini files	(none)
Additional .ini files parsed	(none)
PHP API	20170718
PHP Extension	20170718
Zend Extension	320170718
Zend Extension Build	API320170718,NTS
PHP Extension Build	API20170718,NTS
Debug Build	no
Thread Safety	disabled
Zend Signal Handling	enabled
Zend Memory Manager	enabled
Zend Multibyte Support	provided by mbstring
IPv6 Support	enabled
DTrace Support	disabled
Registered PHP Streams	https, ftps, compress.zlib, php, file, glob, data, http, ftp, phar
Registered Stream Socket Transports	tcp, udp, unix, udg, ssl, sslv3, tls, tlsv1.0, tlsv1.1, tlsv1.2
Registered Stream Filters	zlib.*, convert.iconv.*, string.rot13, string.toupper, string.tolower, string.strip_tags, convert.*, consumed, dechunk

图 3-3

3.5　Nginx相关配置

3.5.1　Nginx 默认虚拟主机

删除 nginx.conf 配置文件中的 server 段代码，并在 httpd 段中“application/xml;”下面添加一行代码，代码如下。

```
[root@centos7 conf]# vim nginx.conf
application/xml;
include vhost/*.conf;
```

在/usr/local/nginx/conf 目录下创建 vhost 目录，然后在 vhost 目录下创建 mingzhe.com.conf 配置文件，在该配置文件中增加如下代码。

```
[root@centos7 conf]# mkdir vhost
[root@centos7 conf]# cd vhost/
[root@centos7 vhost]# vim mingzhe.com.conf
server
{
   listen 80 default_server;  #default_server表示默认虚拟主机
   server_name humingzhe.com;    #网站域名
   index index.html index.php index.htm;   #指定索引页
   root /data/www/default;     #指定网站路径
}
```

在/data/www/目录下创建 default 目录，然后进入 default 目录创建 index.html 文件，添加如下代码。

```
[root@centos7 vhost]# mkdir -p /data/www/default
[root@centos7 vhost]# cd !$
cd /data/www/default
[root@centos7 default]# vim index.html
<html>
   <head>
      <meta charset="UTF-8">
      <title>胡明哲博客</title>
   </head>
   <body>
      <h3>胡明哲</h3>
   </body>
</html>
```

代码添加完毕后检测是否存在语法错误，无误后重启 Nginx 使之生效，若不重启 Nginx，也可以使用-s 选项进行重载，操作命令如下。

```
[root@centos7 default]# /usr/local/nginx/sbin/nginx -t
nginx: the configuration file /usr/local/nginx/conf/nginx.conf syntax is
ok
nginx: configuration file /usr/local/nginx/conf/nginx.conf test is
successful
[root@centos7 default]# /etc/init.d/nginx restart
```

```
[root@centos7 default]# /usr/local/nginx/sbin/nginx -s reload
```

3.5.2 Nginx 用户认证

1．对整站进行用户认证

创建一个虚拟主机配置文件，名称为 bone.conf.conf，在该配置文件中添加如下代码。

```
[root@centos7 default]# vim /usr/local/nginx/conf/vhost/bone.conf.conf
server
{
    listen 80;
    server_name humingzhe.cn;
    index index.htm index.html index.php;
    root /data/www/humingzhe.cn;

location /
    {
        auth_basic  "Auth";
        auth_basic_user_file /usr/local/nginx/conf/htpasswd;
    }
}
```

Nginx 用户认证也需要用到 Apache 的密码生成文件，如果没有 Apache，使用 yum 命令安装即可，操作命令如下。

```
    [root@centos7 default]# /usr/local/apache2.4/bin/htpasswd -c
/usr/local/nginx/conf/htpasswd bone
    New password:
    Re-type new password:
    Adding password for user bone
    [root@centos7 default]# cat /usr/local/nginx/conf/htpasswd
    bone:$apr1$33gAyJon$nc0zyFuH02fgqZqQoWvb8/
```

用 curl 命令测试 Nginx 用户认证功能，操作命令如下。

```
    [root@centos7 default]# curl -ubone:HMkj8899 -x127.0.0.1:80
humingzhe.cn
    <?php
       echo phpinfo();
    [root@centos7 default]# ls /data/www/humingzhe.cn/
    admin  admin.php  content  index.php
```

```
[root@centos7 default]# cat /data/www/humingzhe.cn/index.php
<?php
    echo phpinfo();
```

2. 对某目录做用户认证

针对某个目录做用户认证，如针对 admin 目录做用户认证，只需要把 bone.conf.conf 配置文件中的“location /”改为“location /admin/”即可，操作命令如下。

```
[root@centos7 default]# vim /usr/local/nginx/conf/vhost/bone.conf.conf
location /admin/
    {
        auth_basic  "Auth";
        auth_basic_user_file /usr/local/nginx/conf/htpasswd;
    }
}
```

在访问 humingzhe.cn 时不需要加用户名和密码就可以访问，但是访问 admin 目录却提示 401 状态码，加上用户名和密码后即可访问，如图 3-4 所示。

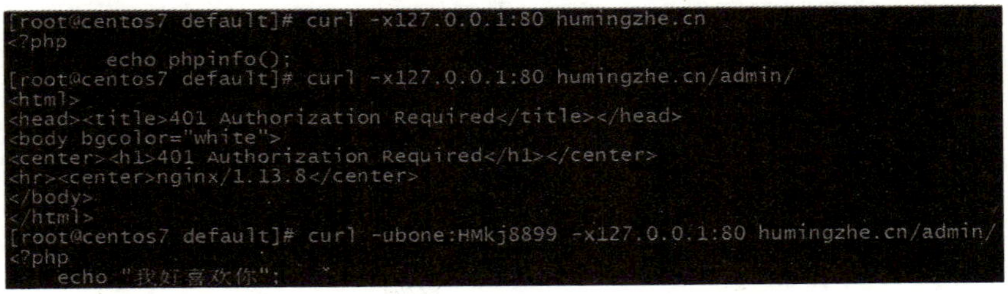

```
[root@centos7 default]# curl -x127.0.0.1:80 humingzhe.cn
<?php
        echo phpinfo();
[root@centos7 default]# curl -x127.0.0.1:80 humingzhe.cn/admin/
<html>
<head><title>401 Authorization Required</title></head>
<body bgcolor="white">
<center><h1>401 Authorization Required</h1></center>
<hr><center>nginx/1.13.8</center>
</body>
</html>
[root@centos7 default]# curl -ubone:HMkj8899 -x127.0.0.1:80 humingzhe.cn/admin/
<?php
    echo "我好喜欢你";
```

图 3-4

3. 对某文件做用户认证

针对某文件做用户认证，只需要编辑虚拟主机配置文件，在配置文件中更改 location 代码行，如针对 admin.php 做用户认证，操作命令如下。

```
[root@centos7 default]# vim /usr/local/nginx/conf/vhost/bone.conf.conf
location ~ admin.php
    {
        auth_basic  "Auth";
        auth_basic_user_file /usr/local/nginx/conf/htpasswd;
    }
}
```

使用 curl 命令访问 admin.php 提示 401 状态码，加上用户名和密码后再去访问显示的是 admin.php 文件中的内容，如图 3-5 所示。

```
[root@centos7 default]# curl -x127.0.0.1:80 humingzhe.cn/admin/
<?php
    echo "我好喜欢你";
[root@centos7 default]# curl -x127.0.0.1:80 humingzhe.cn/admin.php
<html>
<head><title>401 Authorization Required</title></head>
<body bgcolor="white">
<center><h1>401 Authorization Required</h1></center>
<hr><center>nginx/1.13.8</center>
</body>
</html>
[root@centos7 default]# curl -ubone:HMkj8899 -x127.0.0.1:80 humingzhe.cn/admin.php
<?php
    echo "Hello HuMingZhe I Love You~";
```

图 3-5

3.5.3 Nginx 域名跳转

Apache 能实现的功能 Nginx 一样可以实现，域名重定向自然也不例外。编辑虚拟主机配置文件，在配置文件中增加如下代码。

```
[root@centos7 default]# vim /usr/local/nginx/conf/vhost/bone.conf.conf
server
{
    listen 80;
    server_name humingzhe.cn www.humingzhe.com;
    index index.htm index.html index.php;
    root /data/www/humingzhe.cn;
    if ($host != 'humingzhe.cn' ){
        rewrite ^/(.*)$ http://humingzhe.cn/$1 permanent;
    }
}
```

使用 curl 命令访问 www.humingzhe.com 时会自动显示 301 状态码，并告知 location 是 humingzhe.cn，如图 3-6 所示。

```
[root@centos7 default]# curl -x127.0.0.1:80 www.humingzhe.com/admin/index.php -I
HTTP/1.1 301 Moved Permanently
Server: nginx/1.13.8
Date: Sun, 28 Jan 2018 11:30:49 GMT
Content-Type: text/html
Content-Length: 185
Connection: keep-alive
Location: http://humingzhe.cn/admin/index.php

[root@centos7 default]#
```

图 3-6

3.5.4 Nginx 访问日志

Nginx 主配置文件 nginx.conf 中有一段代码是 log_format，log_format 是定义 Nginx 访问日志的格式，代码和格式含义如下。

```
[root@centos7 ~]# vim /usr/local/nginx/conf/nginx.conf
log_format mingzhe '$remote_addr $http_x_forwarded_for [$time_local]' '
$host "$request_uri" $status' ' "$http_referer" "$http_user_agent"';
```

$remote_addr：客户端 IP（公网 IP）。

$http_x_forwarded_for：代理服务器 IP。

$time_local：服务器本地时间。

$host：访问主机名（域名）。

$request_uri：访问的 URL 地址。

$status：状态码。

$httpd_referer：每个页面的引用记录、页面统计。

$http_user_agent：用户代理。

主配置文件定义完日志格式后还需在虚拟主机配置文件中添加代码，代码如下。

```
[root@centos7 vhost]# vim mingzhe.com.conf
server
{
    access_log /data/logs/nginx/humingzhe.com.log mingzhe;
}
```

使用 curl 命令访问 mingzhe.com 站点，在/data/logs/nginx/目录下会生成该日志文件，如图 3-7 所示。

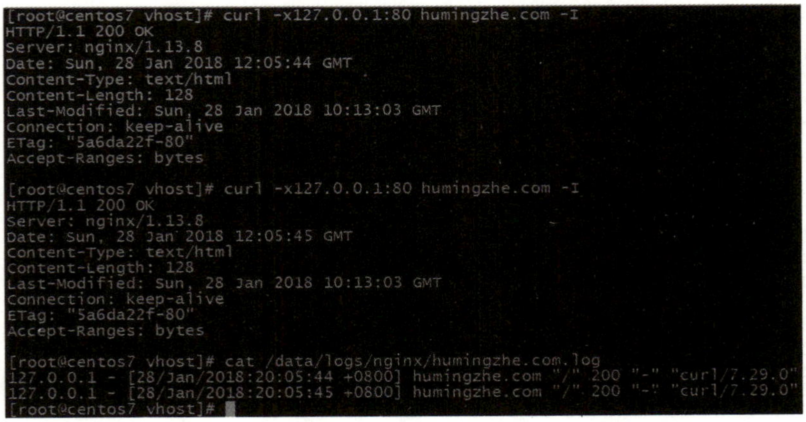

```
[root@centos7 vhost]# curl -x127.0.0.1:80 humingzhe.com -I
HTTP/1.1 200 OK
Server: nginx/1.13.8
Date: Sun, 28 Jan 2018 12:05:44 GMT
Content-Type: text/html
Content-Length: 128
Last-Modified: Sun, 28 Jan 2018 10:13:03 GMT
Connection: keep-alive
ETag: "5a6da22f-80"
Accept-Ranges: bytes

[root@centos7 vhost]# curl -x127.0.0.1:80 humingzhe.com -I
HTTP/1.1 200 OK
Server: nginx/1.13.8
Date: Sun, 28 Jan 2018 12:05:45 GMT
Content-Type: text/html
Content-Length: 128
Last-Modified: Sun, 28 Jan 2018 10:13:03 GMT
Connection: keep-alive
ETag: "5a6da22f-80"
Accept-Ranges: bytes

[root@centos7 vhost]# cat /data/logs/nginx/humingzhe.com.log
127.0.0.1 - [28/Jan/2018:20:05:44 +0800] humingzhe.com "/" 200 "-" "curl/7.29.0"
127.0.0.1 - [28/Jan/2018:20:05:45 +0800] humingzhe.com "/" 200 "-" "curl/7.29.0"
[root@centos7 vhost]#
```

图 3-7

3.5.5　Nginx 日志切割

不像 Apache 自带切割日志的工具，Nginx 若想切割日志文件，需借助系统的日志切割工具，或者去写一个切割脚本。日志切割脚本代码如下。

```
[root@centos7 ~]# vim /usr/local/sbin/nginx_logrotate.sh
#!/bin/bash
##定义 Nginx 日志路径为/data/logs/nginx/
##__author__ is humingzhe
##email admin@humingzhe.com
d=`date -d "-1 day" +%Y%m%d `
logdir="/data/logs/nginx/"
nginx_pid="/usr/local/nginx/logs/nginx.pid"
cd $logdir
for log in `ls *.log`
do
   mv $log $d-$log
done
/bin/kill -HUP `cat $nginx_pid`
```

日志脚本添加完毕后执行该日志脚本，执行过程中可以使用-x 选项查看该脚本文件的执行过程，操作命令如下。

```
[root@centos7 ~]# sh -x /usr/local/sbin/nginx_logrotate.sh
++ date -d '-1 day' +%Y%m%d
+ d=20180127
+ logdir=/data/logs/nginx/
+ nginx_pid=/usr/local/nginx/logs/nginx.pid
+ cd /data/logs/nginx/
++ ls humingzhe.com.log
+ for log in '`ls *.log`'
+ mv humingzhe.com.log 20180127-humingzhe.com.log
++ cat /usr/local/nginx/logs/nginx.pid
+ /bin/kill -HUP 1077
```

日志切割已经设置完毕，还需要加入任务计划，使之每天定点（例如零点）自动进行切割，操作命令如下。

```
[root@centos7 ~]# crontab -e
0 0 * * * /usr/local/sbin/nginx_logrotate.sh
```

切割好的日志会保存在指定目录下，但是长期不删除仍会导致磁盘写满，可以在脚本中添加如下代码，最后放入任务计划中，操作命令如下。

```
find /data/logs/nginx/ -name *.log-* -type f -mtime +30 |xargs rm -rf
```

3.5.6　静态文件不记录日期和时间

Nginx 也可以像 Apache 一样不记录静态文件的日期和时间，编辑虚拟主机配置文件即可，在配置文件中添加如下代码。

```
[root@centos7 ~]# vim /usr/local/nginx/conf/vhost/mingzhe.com.conf
server
{
    location ~ .*\.(gif|jpg|jpeg|png|txt|doc|docx|swf)$
    {
        expires     7d;
        access_log off;
    }
    location ~ .*\.(js|css)$
    {
        expires     12h;
        access_log off;
    }
}
```

3.5.7　Nginx 设置防盗链

Nginx 和 Apache 一样，也可以设置防盗链，编辑 Nginx 虚拟主机配置文件，在配置文件中增加如下代码。

```
[root@centos7 ~]# vim /usr/local/nginx/conf/vhost/mingzhe.com.conf
server
{
    location
~ .*\.(gif|jpg|jpeg|png|txt|doc|docx|swf|flv|rar|zip|gz|bz2|pdf|xls)$
    {
        expires     7d;
        valid_referers none blocked server_name *.mingzhe.com ;
        if ($invalid_referer) {
            return 403;
        }
        access_log off;
    }
}
```

3.5.8 Nginx 进行访问控制

针对/admin/目录做访问控制，只允许指定的 IP 访问/admin/目录，设置代码如下。

```
[root@centos7 ~]# vim /usr/local/nginx/conf/vhost/mingzhe.com.conf
server
{
    location /admin/
    {
        allow 127.0.0.1;
        allow 192.168.222.129;
        deny all;
    }
}
```

Nginx 可以匹配正则表达式进行访问控制，代码如下。

```
[root@centos7 ~]# vim /usr/local/nginx/conf/vhost/mingzhe.com.conf
server
{
    location ~ .*(upload|image)/.*\.php$
    {
        deny all;
    }
}
```

Nginx 可以针对 user_agent 进行限制，代码如下。

```
[root@centos7 ~]# vim /usr/local/nginx/conf/vhost/mingzhe.com.conf
server
{
    if ($http_user_agent ~ 'Spider/3.0|YoudaoBot|Tomtato')
    {
        return 403;
    }
}
```

3.5.9 Nginx 解析 PHP 相关设置

虽然在 mingzhe.com.conf 配置文件中做了很多相关设置，但是还不能进行 PHP 解析操作，需要继续编辑 mingzhe.com.conf 配置文件，增加如下代码。

```
[root@centos7 ~]# vim /usr/local/nginx/conf/vhost/mingzhe.com.conf
server
{
    location ~ \.php$
    {
        include fastcgi_params;
        fastcgi_pass unix:/tmp/php-fcgi.sock;
        fastcgi_index index.php;
        fastcgi_param SCRIPT_FILENAME
/data/www/default$fastcgi_script_name;
    }
}
```

代码添加完毕后先不要重载此配置文件，使用 curl 命令测试，发现返回的是代码，并没有解析 PHP，操作命令如下。

```
[root@centos7 ~]# curl -x127.0.0.1:80 humingzhe.com/test.php
<?php
    echo phpinfo();
```

重载配置文件后，使用 curl 命令测试，返回解析后的结果，如图 3-8 所示。

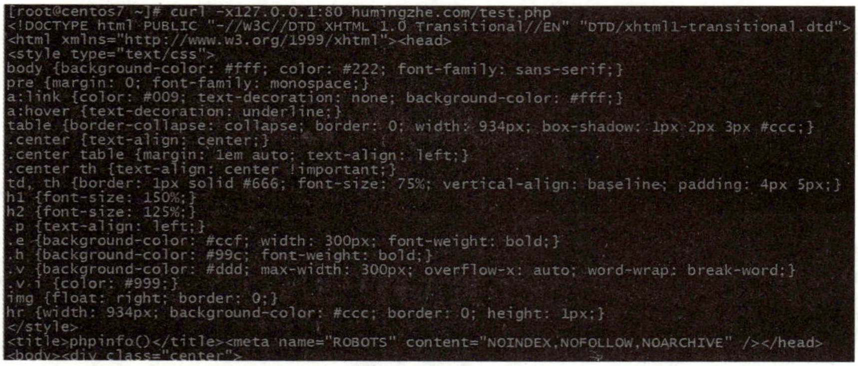

```
[root@centos7 ~]# curl -x127.0.0.1:80 humingzhe.com/test.php
<!DOCTYPE html PUBLIC "-//W3C//DTD XHTML 1.0 Transitional//EN" "DTD/xhtml1-transitional.dtd">
<html xmlns="http://www.w3.org/1999/xhtml"><head>
<style type="text/css">
body {background-color: #fff; color: #222; font-family: sans-serif;}
pre {margin: 0; font-family: monospace;}
a:link {color: #009; text-decoration: none; background-color: #fff;}
a:hover {text-decoration: underline;}
table {border-collapse: collapse; border: 0; width: 934px; box-shadow: 1px 2px 3px #ccc;}
.center {text-align: center;}
.center table {margin: 1em auto; text-align: left;}
.center th {text-align: center !important;}
td, th {border: 1px solid #666; font-size: 75%; vertical-align: baseline; padding: 4px 5px;}
h1 {font-size: 150%;}
h2 {font-size: 125%;}
.p {text-align: left;}
.e {background-color: #ccf; width: 300px; font-weight: bold;}
.h {background-color: #99c; font-weight: bold;}
.v {background-color: #ddd; max-width: 300px; overflow-x: auto; word-wrap: break-word;}
.v i {color: #999;}
img {float: right; border: 0;}
hr {width: 934px; background-color: #ccc; border: 0; height: 1px;}
</style>
<title>phpinfo()</title><meta name="ROBOTS" content="NOINDEX,NOFOLLOW,NOARCHIVE" /></head>
<body><div class="center">
```

图 3-8

3.6 Nginx代理

Nginx 代理是指当用户访问 Web 服务器时，网络不互通，因为 Web 服务器只有私网 IP，若想访问 Web 服务器，只有一个办法，再找一台服务器，这台服务器不仅能够和 Web 服务器进行互通，而且还能和用户进行互通，这台服务器就是代理服务器。

在配置 Nginx 代理前做一个测试，使用 curl 命令访问 www.baidu.com，如图 3-9 所示。

图 3-9

使用 Nginx 代理功能需要在 vhost 目录下创建一个新的配置文件，如 proxy.conf，在该配置文件中增加如下代码。

```
[root@centos7 ~]# cd /usr/local/nginx/conf/vhost/
server
{
    listen 80;
    server_name www.baidu.com;

    location /
    {
        proxy_pass      http://61.135.163.121/;
        proxy_set_header HOST $host;
        proxy_set_header X-Real-IP      $remote_addr;
        proxy_set_header X-Forwarded-For $proxy_add_x_forwarded_for;
    }
}
```

正常情况下不设置代理是不能在本地访问远程站点的，设置完 Nginx 代理后就可以访问 www.baidu.com 了，如图 3-10 所示。

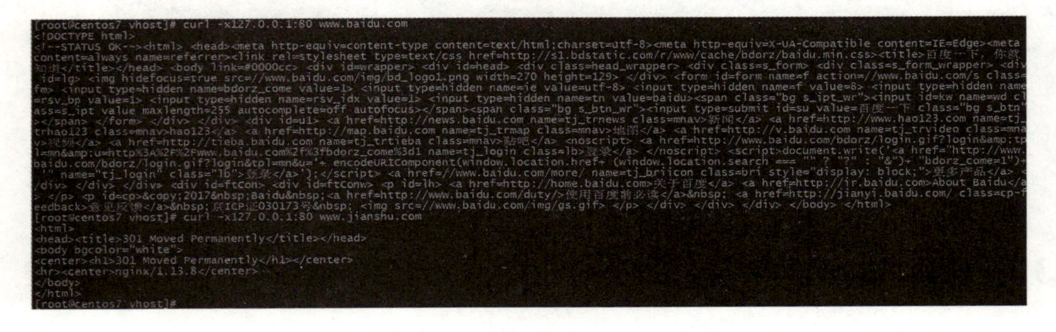

图 3-10

3.7 Nginx负载均衡

代理一台服务器称作代理，代理两台或两台以上的服务器称为负载均衡。代理服务器后面可以是多台 Web 服务器，多台 Web 服务器提供服务时即可实现负载均衡的功能。正常情况下，如果不加代理服务，用户访问 Web 服务器只能一台一台地去请求，或者指定 IP，或者把域名解析到多台服务器上，使得用户一访问 A 服务器、用户二访问 B 服务器，以此类推。虽然这样操作也可以，但不太友好，如果 A 服务器宕机了，用户一因解析到了 A 服务器，就不能够访问该网站。

采用 Nginx 负载均衡后，若 A 服务器宕机，代理服务器就不会把请求发给 A 服务器。Nginx 负载均衡借助 upstream 模块，在 proxy_pass 处不能定义多个 IP，但在 upstream 下面可以定义多个 IP。下面用 qq.com 作为演示对象进行举例，使用 dig 命令可以查看域名解析的 IP 地址，如未安装，使用 yum 命令安装即可，使用 dig qq.com 命令发现 qq.com 解析到了两个 IP 上，可以利用这两个 IP 做负载均衡，操作命令如下。

```
[root@centos7 vhost]# yum install -y bind-utils.x86_64
[root@centos7 vhost]# dig qq.com
;; ANSWER SECTION:
qq.com.                311     IN      A       125.33.240.113
qq.com.                311     IN      A       61.135.157.156
```

创建一个负载均衡配置文件，如命名为 load.conf，在该配置文件中增加如下代码。

```
[root@centos7 vhost]# vim load.conf
upstream qq_com // 自定义名字
{
    ip_hash;    // 让同一个用户始终保持在同一个服务器上
    server  61.135.157.156:80;  // 定义 server 地址
    server  125.33.240.113:80;
}
server
{
    listen 80;  // 监听端口
    server_name www.qq.com; //  网站域名
    location /
    {
        proxy_pass  http://qq_com;
        proxy_set_header Host $host;
        proxy_set_header X-Real-IP  $remote_addr;
        proxy_set_header X-Forwarded-For  $proxy_add_x_forwarded_for;
    }
}
```

正常情况下在本地访问 qq.com 会指向默认页，操作命令如下。

```
[root@centos7 vhost]# curl -x127.0.0.1:80 www.qq.com
<html>
```

```
        <head>
                <meta charset="UTF-8">
                <title>胡明哲博客</title>
        </head>
        <body>
                <h3>胡明哲</h3>
        </body>
</html>
```

重载完配置文件后，使用 curl 命令再访问 qq.com，会显示网站上的内容，如图 3-11 所示。

```
[root@centos7 ~]# curl -x127.0.0.1:80 www.qq.com
<!DOCTYPE html>
<html lang="zh-CN">
<head>
<meta content="text/html; charset=gb2312" http-equiv="Content-Type">
<meta http-equiv="X-UA-Compatible" content="IE=edge">
<meta name="baidu-site-verification" content="cNitg6enc2">
<title></title>
<script type="text/javascript">
if(window.location.toString().indexOf('pref=padindex') != -1){
}else{
        if(/AppleWebKit.*Mobile/i.test(navigator.userAgent) || /\(Android.*Mobile.+\).+Gecko.+Firefox/i.test(navigator.userAgent) || (/MIDP|Symbia
nOS|NOKIA|SAMSUNG|LG|NEC|TCL|Alcatel|BIRD|DBTEL|Dopod|PHILIPS|HAIER|LENOVO|MOT-|Nokia|SonyEricsson|SIE-|Amoi|ZTE/.test(navigator.userAgent))){
      if(window.location.href.indexOf("?mobile")<0){
                try{
                        if(/Android|Windows Phone|webOS|iPhone|iPod|BlackBerry/i.test(navigator.userAgent)){
                                window.location.href="http://xw.qq.com/index.htm";
                        }else if(/iPad/i.test(navigator.userAgent)){
              //window.location.href="http://www.qq.com/pad/"
                        }else{
                                window.location.href="http://xw.qq.com/simple/s/index/"
                        }
                }catch(e){}
        }
        }
}
```

图 3-11

Nginx 不支持代理 HTTPS，server 后面如果写 443 端口是不支持的，Nginx 只能代理 HTTP 和 TCP。若想代理 HTTPS，让用户看到 HTTPS，后端服务器必须是 80 端口，在代理服务器上监听 443 端口，用代理服务器的 443 端口返回给用户。

3.8 Nginx配置SSL

3.8.1 生成 SSL 秘钥对

把公钥和私钥放到/usr/local/nginx/conf/目录下，生成的私钥操作命令如下。

```
[root@centos7 ~]# cd /usr/local/nginx/conf/
[root@centos7 conf]# openssl genrsa -des3 -out tmp.key 2048
Generating RSA private key, 2048 bit long modulus
..............................+++
......................................................................+++
e is 65537 (0x10001)
Enter pass phrase for tmp.key:
Verifying - Enter pass phrase for tmp.key:
```

生成私钥文件时需要输入密码，每次用浏览器访问HTTPS都需要输入密码是不现实的，需要转换秘钥，取消密码，操作命令如下。

```
[root@centos7 conf]# openssl rsa -in tmp.key -out humingzhe.key
Enter pass phrase for tmp.key: #输入tmp.key设置的密码
writing RSA key
[root@centos7 conf]# rm -f tmp.key
```

生成证书请求文件，用该文件和私钥一起生成公钥文件，humingzhe.crt为公钥，操作命令如下。

```
[root@centos7 conf]# openssl req -new -key humingzhe.key -out
humingzhe.csr
[root@centos7 conf]# openssl x509 -req -days 365 -in humingzhe.csr
-signkey humingzhe.key -out humingzhe.crt
Signature ok
subject=/C=11/ST=BeiJing/L=BeiJing/O=Devops/OU=mingzhe/CN=humingzhe/e
mailAddress=admin@humingzhe.com
Getting Private key
[root@centos7 conf]# ls humingzhe.
humingzhe.crt  humingzhe.csr  humingzhe.key
```

3.8.2 Nginx配置SSL

在vhost目录下创建一个新的配置文件，如ssl.conf，在ssl.conf配置文件中添加如下代码。

```
[root@centos7 vhost]# vim ssl.conf
server
{
   listen 443;
   server_name humingzhe.com;
   index index.html index.htm index.php;
   root /data/www/defaul/humingzhe.com;
   ssl on;
   ssl_certificate humingzhe.crt;
   ssl_certificate_key humingzhe.key;
   ssl_protocols TLSv1 TLSV1.1 TLSV1.2;
}
```

检测语法时报错，是因为不支持SSL，需要重新编译Nginx，编译完成后，检测配置文件无误再重启Nginx，操作命令如下。

```
[root@centos7 vhost]# cd /usr/local/src/nginx-1.13.8/
[root@centos7 nginx-1.13.8]# ./configure --prefix=/usr/local/nginx
--with-http_ssl_module
[root@centos7 nginx-1.13.8]# make && make install
[root@centos7 nginx-1.13.8]# /etc/init.d/nginx restart
```

在 default 目录下创建一个测试文件，并把域名写入 hosts 文件中，操作命令如下。

```
[root@centos7 nginx-1.13.8]# vim /data/www/default/test.html
This Is SSL.
[root@centos7 nginx-1.13.8]# vim /etc/hosts
127.0.0.1   humingzhe.com humingzhe.cn
```

配置完毕后打开浏览器，在地址栏中输入域名会提示建立的连接不安全，出现该信息说明 SSL 配置已生效，如图 3-12 所示。

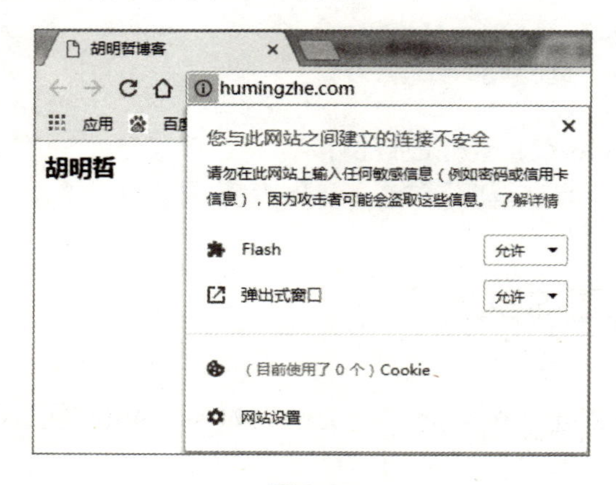

图 3-12

3.9 php-fpm配置

3.9.1 php-fpm 的 pool

php-fpm 的 pool 是 php-fpm 服务的池子，前面章节在 php-fpm.conf 中只定义了一个 www 的 pool。其实 php-fpm 支持定义多个池子，它能够监听多个不同的 sock 或不同的 IP。

如果 Nginx 上运行了多个站点，那么每个站点都可以设置一个 pool，当其中一个站点的 PHP 报 502 状态码或其他错误时，额外的站点不会受到任何影响。如果全部站点都使用同一个 pool，其中一个站点 PHP 出现问题，则所有的站点都会瘫痪，所以需要给每个站点都设置一个 pool，把这些站点隔离开。

定义多个 pool 只需使用 php-fpm.conf 配置文件中的代码，把 www 重命名为其他名称，如更改为 mingzhe，操作命令如下。php-fpm.conf 下载网址：https://gitlab/humingzhe/Linux/blob/master/LNMP/php-fpm.conf。

```
[root@centos7 nginx-1.13.8]# cd /usr/local/php-fpm/etc/
[root@centos7 etc]# vim php-fpm.conf
[mingzhe]
listen = /tmp/mingzhe.sock
listen.mode = 666
user = php-fpm
group = php-fpm
pm = dynamic
pm.max_children = 50
pm.start_servers = 20
pm.min_spare_servers = 5
pm.max_spare_servers = 35
pm.max_requests = 500
rlimit_files = 1024
```

检测语法是否有误，无误后重载 php-fpm，查看设置的 mingzhe 进程池，如图 3-13 所示。

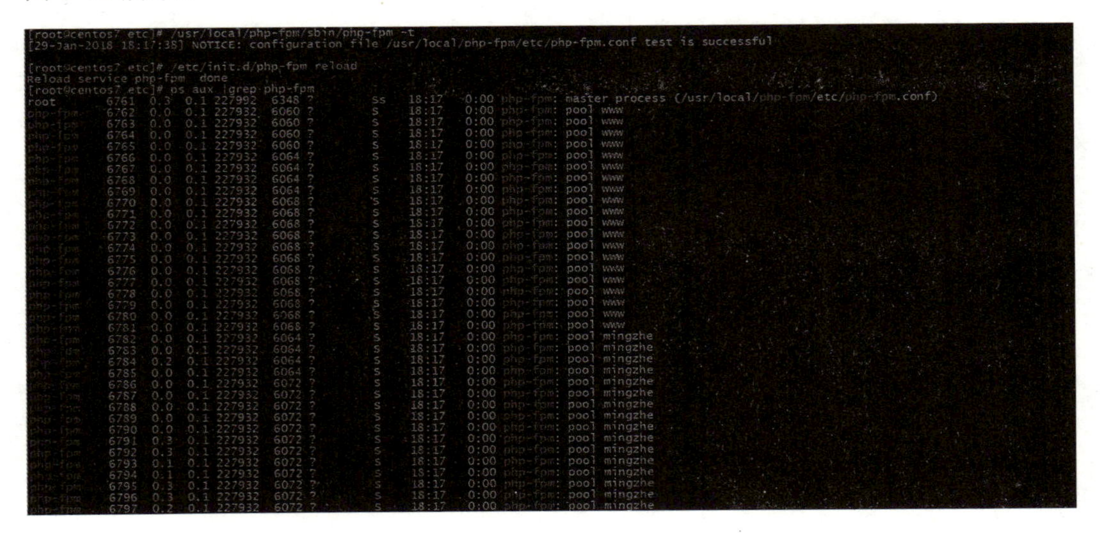

图 3-13

前面设置 Nginx 解析 PHP 时，给 mingzhe.conf 设置的是 php-fcgi.sock，但在 bone.conf.conf 文件中没有进行定义，可以把 mingzhe.sock 定义到 bone.conf.conf 文件中，操作命令如下。

```
[root@centos7 vhost]# ls
bone.conf.conf  load.conf  mingzhe.com.conf  proxy.conf  ssl.conf

server
{
    location ~ \.php$
    {
            include fastcgi_params;
            fastcgi_pass unix:/tmp/mingzhe.sock;
            fastcgi_index index.php;
            fastcgi_param SCRIPT_FILENAME
/data/www//humingzhe_blog$fastcgi_script_name;
    }
}
```

vhost 目录下存在多个.conf 配置文件，这是因为在 nginx.conf 主配置文件中定义了“include vhost/*.conf;”。php-fpm 同样支持这样操作，在 php-fpm.conf 配置文件中增加如下代码，完成后删除下面的所有进程池。

```
[root@centos7 etc]# vim php-fpm.conf
[global]
pid = /usr/local/php-fpm/var/run/php-fpm.pid
error_log = /usr/local/php-fpm/var/log/php-fpm.log
include = etc/php-fpm.d/*.conf
```

创建 php-fpm.d 目录，然后进入 php-fpm.d 目录创建 www.conf 和 mingzhe.conf 文件，把 www 进程池和 mingzhe 进程池分别粘贴到各自的配置文件中，操作命令如下。

```
[root@centos7 etc]# mkdir php-fpm.d
[root@centos7 etc]# cd php-fpm.d/
[root@centos7 php-fpm.d]# vim www.conf
[root@centos7 php-fpm.d]# vim mingzhe.conf
[root@centos7 php-fpm.d]# ls
mingzhe.conf  www.conf  www.conf.default
```

3.9.2 php-fpm 慢执行日志

在日常运维工作中经常会遇到网站访问速度变慢的情况，若想知道速度变慢的原因，可以通过 php-fpm 慢执行日志查看。

针对 www.conf 配置文件来做测试，在 www.conf 配置文件中增加如下代码。添加完毕后检测语法是否有误，无误后重载 php-fpm。

```
[root@centos7 php-fpm.d]# vim www.conf
request_slowlog_timeout = 1
slowlog = /usr/local/php-fpm/var/log/www-slow.log
[root@centos7 php-fpm.d]# ls /usr/local/php-fpm/var/log/www-slow.log
/usr/local/php-fpm/var/log/www-slow.log
```

查看 www-slow.log 文件，没有内容输出，需要模拟慢执行的 PHP。php-fcgi 目前被 mingzhe.com.conf 所用，在 default 目录下创建 sleep.php 文件，在 sleep.php 文件中添加如下代码，让该文件休眠两秒。

```
[root@centos7 php-fpm.d]# vim /data/www/default/sleep.php
<?php
    echo "test slow php";
    sleep(2);
    echo "done";
?>
```

使用 curl 命令执行该休眠文件，执行过程中不会立刻输出信息，而是延迟两秒后再进行输出，操作命令如下。

```
[root@centos7 php-fpm.d]# curl -x127.0.0.1:80 humingzhe.com/sleep.php
test slow phpdone[root@centos7 php-fpm.d]#
```

查看 slow.log 日志文件，该日志文件中也会记录执行过程的信息，而且告知第三行代码执行慢，查看 sleep.php 文件的第三行代码，正是前面添加的“sleep(2);”代码，如图 3-14 所示。

```
[root@centos7 php-fpm.d]# cat /usr/local/php-fpm/var/log/www-slow.log

[29-Jan-2018 20:04:21]  [pool www] pid 1794
script_filename = /data/www/default/sleep.php
[0x00007fd529e1e090] sleep() /data/www/default/sleep.php:3

[29-Jan-2018 20:06:39]  [pool www] pid 1797
script_filename = /data/www/default/sleep.php
[0x00007fd529e1e090] sleep() /data/www/default/sleep.php:3
[root@centos7 php-fpm.d]# cat /data/www/default/sleep.php
<?php
        echo "test slow php";
        sleep(2);
        echo "done";
?>
[root@centos7 php-fpm.d]#
```

图 3-14

3.9.3 open_basedir

在 php-fpm 中可以定义 open_basedir，如果服务器上有多个网站，则在 php.ini 中定义不合适，可以在 Apache 虚拟主机配置文件和 php-fpm 配置文件中进行定义，php-fpm 可以针对不同的进程池设置不同的 open_basedir，如在 www.conf 配置文件中定义，操作命令如下。

```
[root@centos7 php-fpm.d]# vim
/usr/local/php-fpm/etc/php-fpm.d/www.conf
php_admin_value[open_basedir]=/data/www/default/humingzhe.com:/tmp/
```

设置完毕后打开 php.ini 的错误日志并定义日志级别，如果定义的错误日志文件路径不存在，则需自行创建，操作命令如下。

```
[root@centos7 php-fpm.d]# vim /usr/local/php-fpm/etc/php.ini
display_errors = Off
;error_log = syslog
error_log = /usr/local/php-fpm/var/log/php_errors.log
error_reporting = E_ALL
[root@centos7 php-fpm.d]# touch /usr/local/php-fpm/
var/log/php_errors.log
[root@centos7 php-fpm.d]# chmod 777 !$
chmod 777 /usr/local/php-fpm/var/log/php_errors.log
```

3.9.4 php-fpm 管理进程

在 www.conf 配置文件中定义了很多进程，但前面并没有讲过，下面详细讲解。

pm = dynamic：动态进程管理，也可以是静态的（static）。

pm.max_children = 50：最大子进程数，使用 ps aux 命令可查看。

pm.start_servers：启动服务时同时启动的进程数。

pm.min_spare_servers：定义在空闲时段，子进程数的最少数量，如果达到该数值，则 php-fpm 服务会自动派生新的子进程。

pm.max_spare_servers：定义在空闲时段，子进程数的最大数量，如果高于该数值，则会开始清理空闲的子进程。

pm.max_requests：定义一个 php-fpm 的子进程中最多可以处理的请求数，当达到该数值时自动退出。

3.10 部署phpMyAdmin

下载 phpMyAdmin 包到网站根目录下，解压后进入 phpMyAdmin 目录，复制 config.sample.inc.php 配置文件并重命名为 config.inc.php，编辑复制后的 config.inc.php 配置文件，修改“$cfg['blowfish_secret'] = ''”，命令如下。

```
[root@humingzhe ~]# cd /data/www/humingzhe.com
```

```
    [root@humingzhe humingzhe.com]# wget https://files.phpmyadmin.net/
phpMyAdmin/4.7.8/phpMyAdmin-4.7.8-all-languages.tar.gz
    [root@humingzhe humingzhe.com]# tar xf
phpMyAdmin-4.7.8-all-languages.tar.gz
    [root@humingzhe humingzhe.com]# mv phpMyAdmin-4.7.8-all-languages
phpMyAdmin
    [root@humingzhe humingzhe.com]# cd phpMyAdmin
    [root@humingzhe phpMyAdmin]# cp config.sample.inc.php config.inc.php
    [root@humingzhe phpMyAdmin]# vim !$
    $cfg['blowfish_secret'] = 'dhashgdhasgdq3w323hsud312esagdh' #此处字符串
随便输入
```

打开浏览器，输入网站域名/phpMyAdmin 即可访问，如图 3-15 所示。

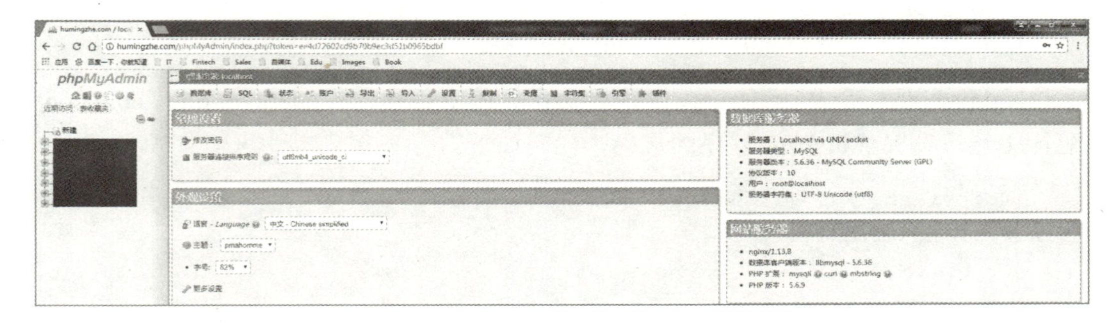

图 3-15

3.11　安装&设置Laravel框架

下载 Laravel 框架源码包到/data/www/目录下，运行命令如下。源码可以在 GitHub 上进行下载，网址是 https://github.com/laravel/laravel。

```
    [root@lanmp ~]# cd /data/www/
    [root@lanmp www]# unzip laravel5.6.zip
    [root@lanmp www]# ls
    laravel56  laravel5.6.zip
    [root@lanmp www]# mv laravel56 laravel
```

新建 laravel.conf 配置文件，在 laravel.conf 中增加如下代码。

```
    [root@lanmp www]# vim /usr/local/nginx/conf/vhost/
laravel.linux.com.conf
    server
    {
        listen 80;
        server_name laravel.linux.com;
```

```
    index index.html index.php index.htm;
    root /data/www/laravel/public/;
    access_log /var/log/nginx/access_laravel.log linux;
}
```

重启 Nginx 服务，运行命令如下。

```
[root@lanmp www]# /etc/init.d/nginx restart
Restarting nginx (via systemctl):                    [  OK  ]
[root@lanmp www]# ps aux|grep nginx
```

在地址栏中输入 laravel.linux.com 并访问会下载一个文件，该文件中所显示的内容和/public/目录下的 index.php 文件所显示的内容是一致的，运行命令如下。

```
[root@lanmp laravel]# cd public/
[root@lanmp public]# cat index.php
```

Nginx 只是一个 Web Server，它并不能解析 PHP 程序，打开 laravel.linux.com.conf 配置文件，设置解析 PHP 程序的代码，代码如下。

```
    [root@lanmp www]# vim /usr/local/nginx/conf/vhost/
laravel.linux.com.conf
    server
    {
        location ~ \.php$
        {
            include fastcgi_params;
            fastcgi_pass unix:/tmp/php-fcgi.sock;
            fastcgi_index index.php;
            try_files $uri =404;
            fastcgi_param SCRIPT_FILENAME
$document_root$fastcgi_script_name;
        }
    }
```

重启 Nginx 服务，操作命令如下。

```
[root@lanmp public]# /etc/init.d/nginx restart
Restarting nginx (via systemctl):                    [  OK  ]
```

再次打开浏览器，在地址栏中输入 laravel.linux.com，显示如图 3-16 所示的界面，提示 logs 目录没有访问权限，所以需要给 logs 目录下的所有文件和目录赋予 777 权限，运行命令如下。

```
[root@lanmp laravel]# cd storage/
[root@lanmp storage]# chmod -R 777 logs/
```

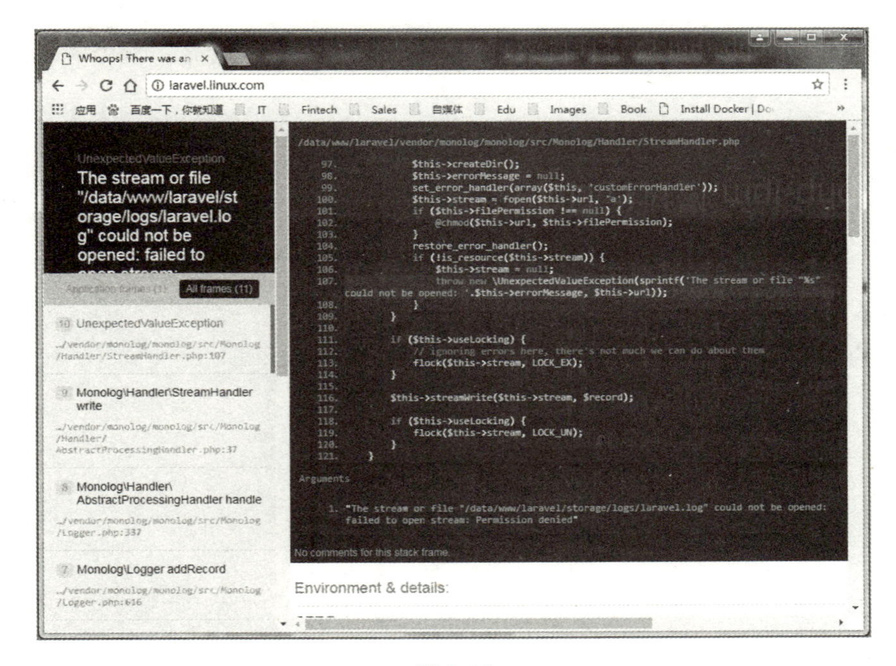

图 3-16

刷新浏览器页面，提示 framework/views/目录没有访问权限，如图 3-17 所示。所以，还需要赋予 framework/views/目录下的所有目录和文件 777 权限，运行命令如下。

```
[root@lanmp storage]# chmod -R 777 framework/views/
```

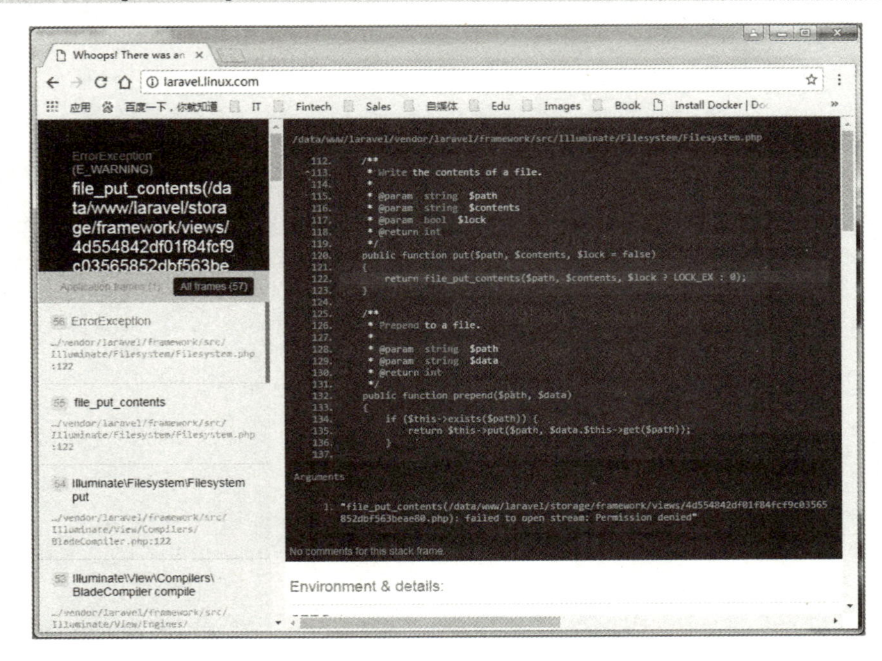

图 3-17

再次刷新浏览器页面，久违的 Laravel 初始界面显示出来了，如图 3-18 所示。

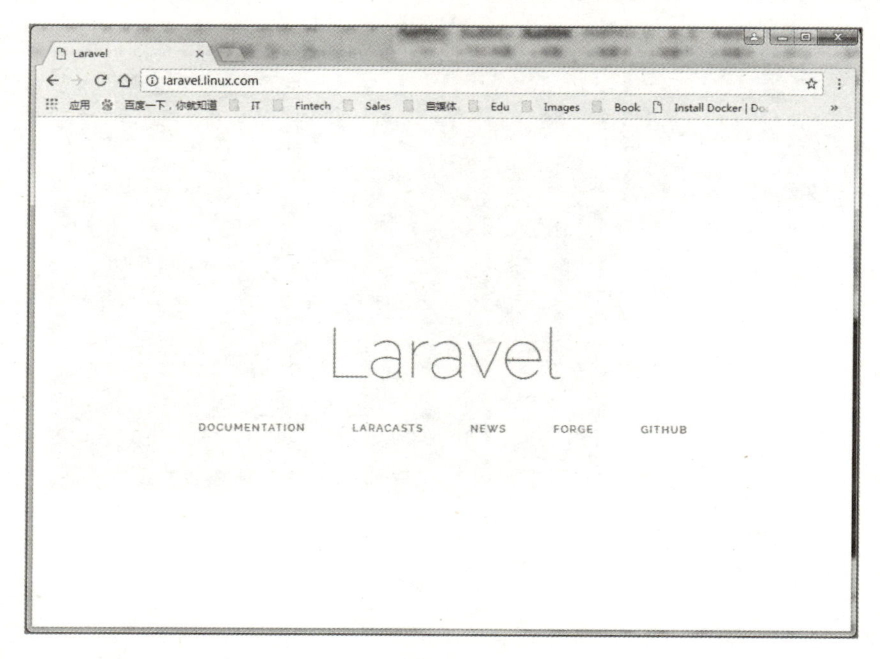

图 3-18

3.12 安装&设置Yii2框架

将 Yii2 框架源码包上传到/data/www/目录下，然后进行解压，解压后将目录重命名为 Yii2，运行命令如下。

```
[root@lanmp ~]# cd /data/www/
[root@lanmp www]# wget https://github.com/yiisoft/yii2/releases/
download/2.0.14/yii-basic-app-2.0.14.tgz
[root@lanmp www]# tar xf yii-basic-app-2.0.14.tgz
[root@lanmp www]# mv basic yii2
```

进入该目录，可以看到 Yii2 框架的结构，运行命令如下。

```
[root@lanmp www]# cd yii2/
[root@lanmp yii2]# ls
assets          commands       composer.lock  controllers
LICENSE.md  models     requirements.php tests    Vagrantfile  views
widgets  yii.bat
codeception.yml  composer.json  config        docker-compose.yml  mail
README.md  runtime          vagrant  vendor      web    yii
```

设置 Yii2 框架的 Nginx 虚拟机配置文件，完成后重载 Nginx 服务，运行命令如下。

```
[root@lanmp yii2]# cd /usr/local/nginx/conf/vhost/
[root@lanmp vhost]# cp laravel.linux.com.conf yii2.linux.com.conf
[root@lanmp vhost]# vim yii2.linux.com.conf
server
{
    listen 80;
    server_name yii2.linux.com;
    index index.html index.php index.htm;
    root /data/www/yii2/web/;
    access_log /var/log/nginx/access_yii2.log;

    location / {
        try_files $uri $uri/ /index.php?$args;
    }
    location ~ \.php$
    {
        include fastcgi_params;
        fastcgi_pass unix:/tmp/php-fcgi.sock;
        fastcgi_index index.php;
        try_files $uri =404;
        fastcgi_param SCRIPT_FILENAME
$document_root$fastcgi_script_name;
    }

}
[root@lanmp vhost]# /usr/local/nginx/sbin/nginx -t
[root@lanmp vhost]# /usr/local/nginx/sbin/nginx -s reload
```

打开浏览器，在地址栏中输入 yii2.linux.com，会显示和前文设置 Laravel 框架时一致的内容，都是没有权限，只需赋予 777 权限即可，运行命令如下。

```
[root@lanmp yii2]# chmod -R 777 runtime/
[root@lanmp yii2]# chmod -R 777 web/assets/
```

再次刷新浏览器页面，会出现如图 3-19 所示的界面，这是 Yii2 框架内部的原因，需要编辑 config/web.php 文件，将“'cookieValidationKey' => '',”改为如下代码即可。

```
[root@lanmp yii2]# vim config/web.php
'cookieValidationKey' => 'lshdjanjsdnka123', # 内容自定义
```

再次刷新浏览器页面就可以看到 Yii2 框架的界面，如图 3-20 所示。

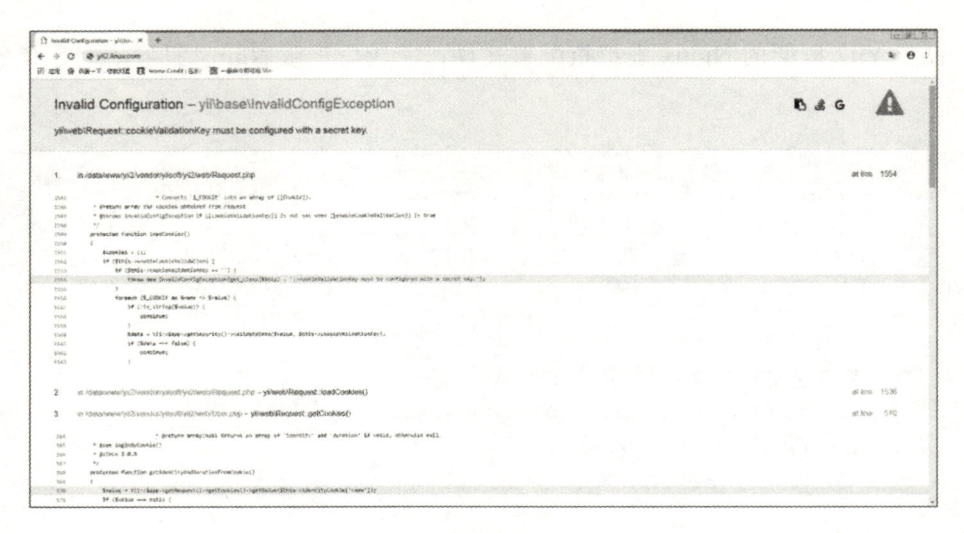

图 3-19

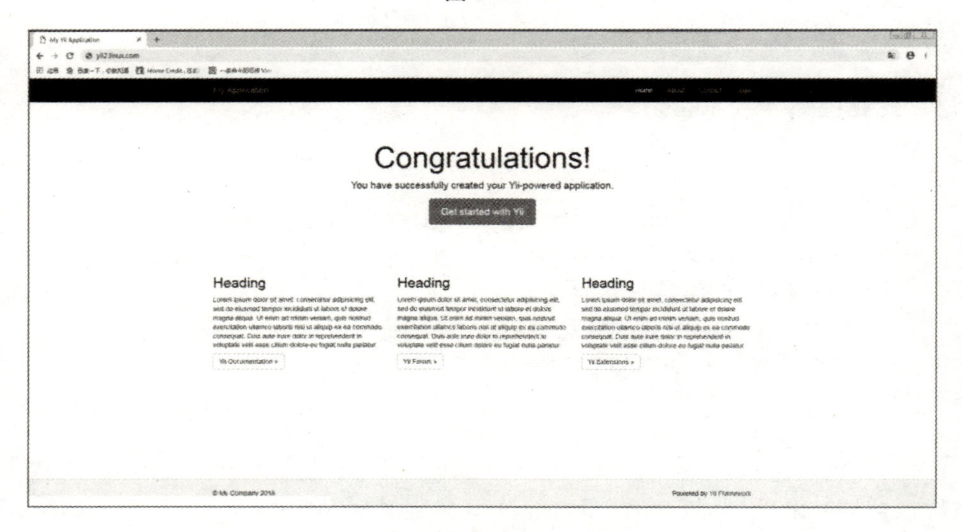

图 3-20

3.13 安装&设置ThinkPHP 5框架

将 ThinkPHP 5 框架的源码包上传到/data/www/目录下，然后进行解压，解压后重命名为 tp5，命令如下。

```
[root@lanmp www]# mkdir tp5
[root@lanmp www]# cd tp5/
[root@lanmp tp5]# unzip thinkphp5.zip
```

设置 ThinkPHP 5 框架的 Nginx 虚拟机配置文件，完成后重载 Nginx 服务，运行命令如下。

```
    [root@lanmp public]# vim
/usr/local/nginx/conf/vhost/tp5.linux.com.conf
    server
    {
        listen 80;
        server_name thinkphp5.linux.com;
        index index.html index.php index.htm;
        root /data/www/tp5/public;
        access_log /var/log/nginx/access_tp5.log;

        location / {
            try_files $uri $uri/ /index.php?$args;
        }
        location ~ \.php$
        {
            include fastcgi_params;
            fastcgi_pass unix:/tmp/php-fcgi.sock;
            fastcgi_index index.php;
            try_files $uri =404;
            fastcgi_param SCRIPT_FILENAME
$document_root$fastcgi_script_name;
        }

    }
    [root@lanmp public]# /usr/local/nginx/sbin/nginx -t
    [root@lanmp public]# /usr/local/nginx/sbin/nginx -s reload
```

打开浏览器，在地址栏中输入 thinkphp5.linux.com 即可访问 ThinkPHP 5 页面，如图 3-21 所示。

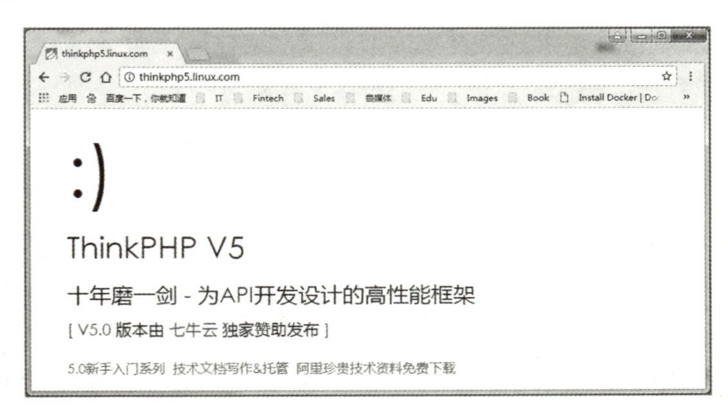

图 3-21

3.14 安装&设置Django框架

3.14.1 安装 Django 框架

安装 Python Web 框架 Django，可以使用 django>=1.11 命令下载 Django 1.11 稳定版。目前最新版本是 Django 2.0，仍然可以使用 django>=1.11 命令下载，会自动下载最新版本的 Django 2.0。若想下载指定的 1.11 版本，可以使用“django>=1.11, <=1.12”命令，运行命令如下。

```
[root@lanmp ~]# source python36env/bin/activate
(python36env) [root@lanmp ~]# pip install "django>=1.11,<=1.12"
Collecting django<=1.12,>=1.11
   100% |████████████████████████████████| 7.0MB 1.2MB/s
Collecting pytz (from django<=1.12,>=1.11)
   100% |████████████████████████████████| 512kB 1.1MB/s
Installing collected packages: pytz, django
Successfully installed django-1.11.12 pytz-2018.4
```

使用 pip list 命令可以查看刚刚安装的 Django Web 框架，运行命令如下。

```
(python36env) [root@lanmp ~]# pip list
Package    Version
---------- -------
Django     1.11.12
pip        10.0.0
PyMySQL    0.8.0
pytz       2018.4
setuptools 33.0.1
wheel      0.31.0
```

创建 Django 数据库，设置字符集为 UTF-8，运行命令如下。

```
    [root@lanmp ~]# mysql -uroot -phumingzhe -e "create database django
CHARACTER SET utf8;"
```

3.14.2 Django runserver

Django 提供了一个命令行的工具，可以用来创建 Django 项目。进入 python36env 环境，在环境下创建项目，创建完成后进入该目录，使用 tree 命令查看该目录的结构，运行代码如下。

```
[root@lanmp ~]# source python36env/bin/activate
(python36env) [root@lanmp ~]# django-admin startproject pyweb
(python36env) [root@lanmp ~]# ls
python36env  pyweb
(python36env) [root@lanmp ~]# tree pyweb/
pyweb/  #最外层的 pyweb 目录是项目的容器
├── manage.py #使用命令行工具，可以让你以各种方式与 Django 项目进行交互
└── pyweb     #内层的 pyweb 目录是项目中的实际 Python 包，通过它可以导入里面的任
何东西
    ├── __init__.py       #空文件，告诉 Python 该目录是一个 Python 包
    ├── settings.py       #Django 项目的配置文件
    ├── urls.py          #Django 项目的 URL 声明
    └── wsgi.py          #WSGI 兼容的 Web 服务器入口
```

Django 项目本身支持的是 SQLite 数据库，但是我们安装的是 MySQL 数据库，修改 settings.py 文件设置 MySQL 数据库，settings.py 文件是一个普通的 Python 模块，每项配置都是一对键值对（key/value），数据库的配置是以 DICT 的形式存放在这个模块中的，key 名为 DATABASES，代码如下。

```
(python36env) [root@lanmp ~]# cd pyweb/
(python36env) [root@lanmp pyweb]# vim pyweb/settings.py
DATABASES = {
    'default': {
        #'ENGINE': 'django.db.backends.sqlite3',
        #'NAME': os.path.join(BASE_DIR, 'db.sqlite3'),
        'ENGINE': 'django.db.backends.mysql',
        'NAME': 'django',
        'USER': 'root',
        'PASSWORD': 'humingzhe',
        'HOST': '127.0.0.1',
        'PORT': '3306',
    }
}
```

设置完成后还是不能用，因为默认调用的是 mysqldb 模块，该模块是 Python 2 版本的，本书使用的是 Python 3.6 版本，需要另行配置。修改 pyweb/__init__.py 文件，添加如下代码。

```
(python36env) [root@lanmp pyweb]# vim pyweb/__init__.py
import pymysql
pymysql.install_as_MySQLdb()
```

启动 Django 项目，如图 3-22 所示，启动成功，但是出现了一个红色的警告信

息，警告信息不妨碍我们启动项目。

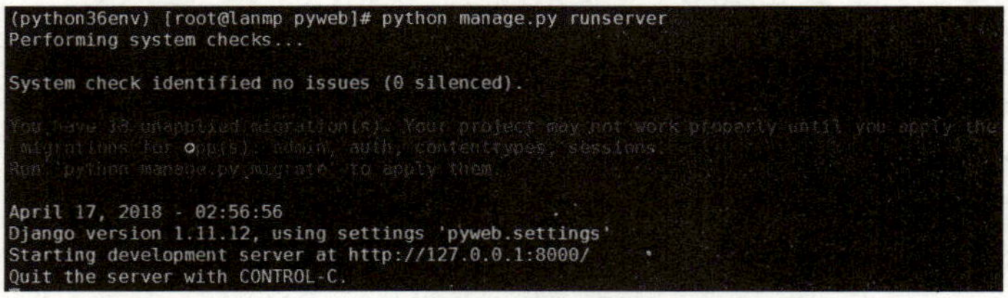

```
(python36env) [root@lanmp pyweb]# python manage.py runserver
Performing system checks...

System check identified no issues (0 silenced).

April 17, 2018 - 02:56:56
Django version 1.11.12, using settings 'pyweb.settings'
Starting development server at http://127.0.0.1:8000/
Quit the server with CONTROL-C.
```

图 3-22

直接运行 python manage.py 命令，后面不加任何选项，会列出如图 3-23 所示的参数。

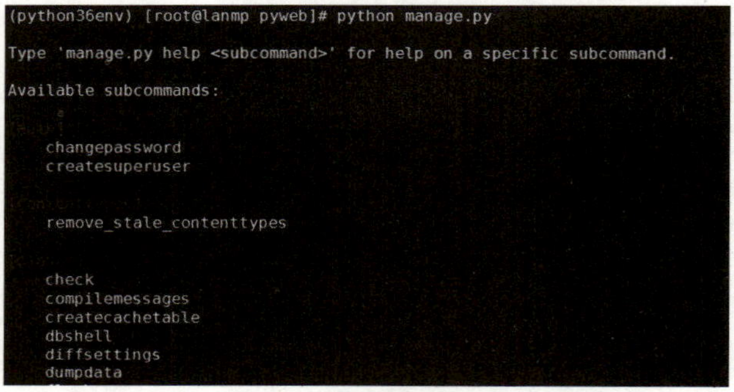

```
(python36env) [root@lanmp pyweb]# python manage.py

Type 'manage.py help <subcommand>' for help on a specific subcommand.

Available subcommands:

    changepassword
    createsuperuser

    remove_stale_contenttypes

    check
    compilemessages
    createcachetable
    dbshell
    diffsettings
    dumpdata
```

图 3-23

runserver 启动项目时，后面不加任何参数，默认监听的 IP 和端口是 127.0.0.1：8000，如果想要更改 IP 和端口，在 Runserver 后面指定即可，如图 3-24 所示。

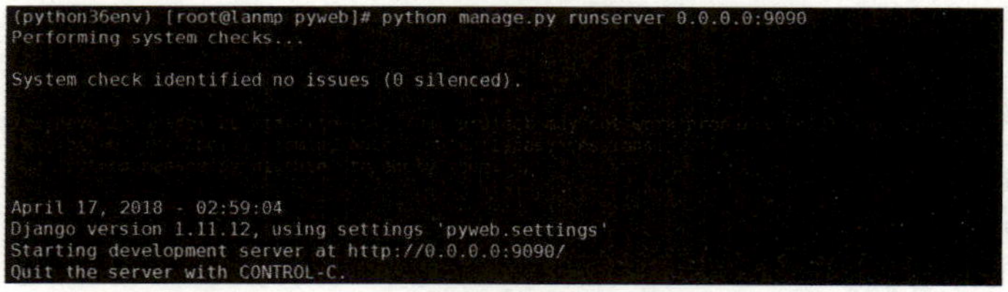

```
(python36env) [root@lanmp pyweb]# python manage.py runserver 0.0.0.0:9090
Performing system checks...

System check identified no issues (0 silenced).

April 17, 2018 - 02:59:04
Django version 1.11.12, using settings 'pyweb.settings'
Starting development server at http://0.0.0.0:9090/
Quit the server with CONTROL-C.
```

图 3-24

Django 项目启动后默认只允许本机访问，若想让所有的 IP 都可以访问，需修改 settings 配置文件，将配置文件中的“ALLOWED_HOSTS = []”更改为如下代码。

```
(python36env) [root@lanmp pyweb]# vim pyweb/settings.py
ALLOWED_HOSTS = ["*"]
```

修改完毕后一定要关闭防火墙，否则不会显示 Django 页面，关闭防火墙后在本地浏览器中输入 IP 地址和端口，即可显示 Django 服务启动页面，如图 3-25 所示。

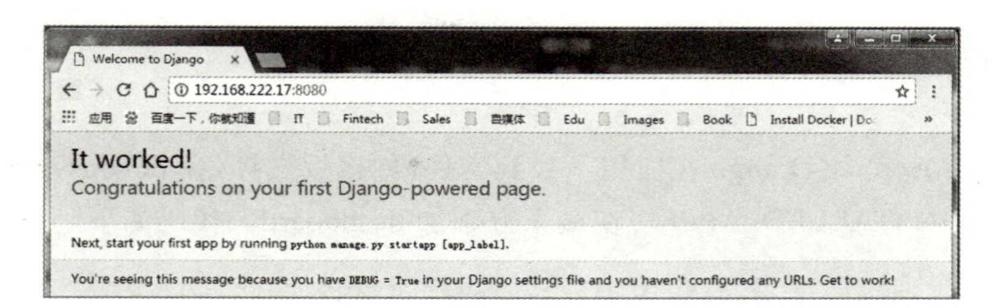

图 3-25

3.14.3　运行 Hello World

新建一个 App 简单展示 Hello World 程序。Django App 和 Django Project 的区别在于，一个 Django 项目（Django Project）可能会有很多子功能，把每个子功能独立成一个 Python 模块，一个模块负责一个功能，这个功能就是一个 Django App。想要成功运行一个 Django App，需要完成如下三个步骤。

（1）创建 Django App，通过 startapp 启动 App，后面加上名称，如 dashboard（指示板），运行命令如下。

```
(python36env) [root@lanmp pyweb]# python manage.py startapp dashboard
(python36env) [root@lanmp pyweb]# ls
dashboard  manage.py  pyweb
(python36env) [root@lanmp pyweb]# tree dashboard/
dashboard/
├── admin.py
├── apps.py
├── __init__.py
├── migrations
│   └── __init__.py
├── models.py
├── tests.py
└── views.py
```

（2）让 Django 通过 URL 找到 dashboard App，设置代码如下。

```
(python36env) [root@lanmp pyweb]# vim pyweb/urls.py
from django.conf.urls import url,include
from django.contrib import admin

urlpatterns = [
    url(r'^admin/', admin.site.urls),
```

```
    url(r'^dashboard/',include("dashboard.urls")),
]
```

（3）激活 App，让 Django 在应用下找到文件，如模板文件，在 devopsweb/settins.py 文件中找到 INSTALLED_APPS，在最下方添加 dashboard，代码如下。

```
(python36env) [root@lanmp pyweb]# vim pyweb/settings.py
INSTALLED_APPS = [
    'django.contrib.admin',
    'django.contrib.auth',
    'django.contrib.contenttypes',
    'django.contrib.sessions',
    'django.contrib.messages',
    'django.contrib.staticfiles',
    'dashboard',
]
```

设置完毕后，编写一个 Hello World 程序，步骤如下。

（1）编写视图文件，运行代码如下。

```
(python36env) [root@lanmp pyweb]# cd dashboard/
(python36env) [root@lanmp dashboard]# ll
total 20
-rw-r--r-- 1 root root 63 Apr 17 11:24 admin.py
-rw-r--r-- 1 root root 93 Apr 17 11:24 apps.py
-rw-r--r-- 1 root root  0 Apr 17 11:24 __init__.py
drwxr-xr-x 2 root root 25 Apr 17 11:24 migrations
-rw-r--r-- 1 root root 57 Apr 17 11:24 models.py
-rw-r--r-- 1 root root 60 Apr 17 11:24 tests.py
-rw-r--r-- 1 root root 63 Apr 17 11:24 views.py
(python36env) [root@lanmp dashboard]# vim views.py
from django.shortcuts import render

# Create your views here.
from django.http import HttpResponse
def index(request):
    return HttpResponse("Hello World,Hello Python")
```

（2）配置 URL，让视图展现在浏览器页面中，每个视图对应一个 URL 地址，如果不对应 URL 地址，则视图没有任何意义，别人访问不到。每个 App 目录下都有一个 urls.py 文件，默认是不存在的，需要手动创建，运行代码如下。

```
(python36env) [root@lanmp dashboard]# vim urls.py
from django.conf.urls import include, url #导入 Django URL 配置相关函数
```

```
from . import views # 导入当前模块下的 views 模块
urlpatterns = [
    url(r'^$', views.index, name='index'),  # URL 声明的列表
]
```

启动 Django 服务器，以便访问刚刚创建的 Hello World 页面，运行代码如下。

```
(python36env) [root@lanmp pyweb]# python manage.py runserver
0.0.0.0:8000
```

在浏览器的地址栏中输入 IP 和端口，显示 Page not found，因为开启了 Debug，所以图 3-26 所示的页面会把相应的路由列出来，一个是 admin，另一个是 dashboard。而这个 admin 是在/pyweb/pyweb/urls.py 文件中的，该 urls.py 文件是 Django 默认调用的 URL 配置文件。

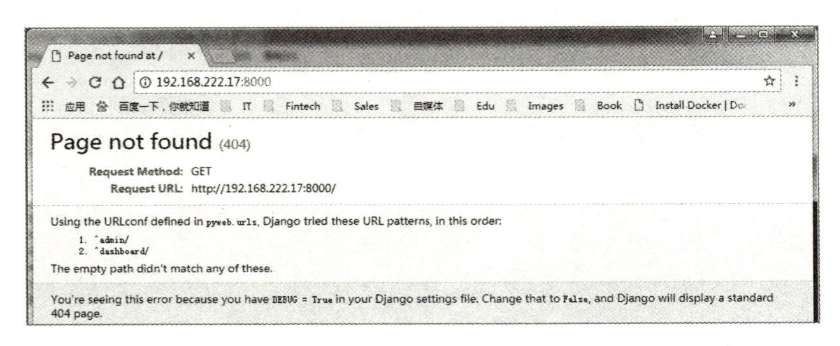

图 3-26

如果不进行任何 URL 设置，则支持的 URL 要么以 admin 开头，要么以 dashboard 开头。现在暂时不用 admin，把它关闭，运行代码如下。

```
(python36env) [root@lanmp pyweb]# vim pyweb/urls.py
urlpatterns = [
    #url(r'^admin/', admin.site.urls),
    url(r'^dashboard/', include("dashboard.urls")),
    ]
```

再次刷新浏览器，现在只显示 dashboard 路由了。进入 dashboard 即可显示刚刚写入视图文件的“Hello World,Hello Python”，如图 3-27 所示。

图 3-27

3.15 安装&设置Flask框架

在 python36env 环境下使用 pip 命令安装 Flask 框架，运行命令如下。

```
(python36env) [root@lanmp ~]# pip install flask
(python36env) [root@lanmp ~]# pip list
Package      Version
------------ -------
click        6.7
Flask        0.12.2
itsdangerous 0.24
Jinja2       2.10
MarkupSafe   1.0
pip          10.0.0
setuptools   39.0.1
Werkzeug     0.14.1
wheel        0.31.0
```

新建一个目录 python，进入该目录，创建一个 index.py 文件，在该文件中写入如下代码。

```
(python36env) [root@lanmp ~]# mkdir python
(python36env) [root@lanmp ~]# cd python
(python36env) [root@lanmp python]# vim index.py
from flask import Flask
app = Flask(__name__)

@app.route('/')
def hello_world():
    return 'Hello World! Hello Flask! Hello HuMingZhe! Hello~'

if __name__ == '__main__':
    app.run()
```

运行 index.py 文件会启动一个服务，提示输入 http://127.0.0.1:5000 可以访问，命令如下。

```
(python36env) [root@lanmp python]# python index.py
 * Running on http://127.0.0.1:5000/ (Press CTRL+C to quit)
```

但是通过虚拟机 IP 地址 192.168.222.18:5000 是不能访问的，所以还需要进行设置。将 index.py 文件中的 app.run()改为 app.run(host='0.0.0.0')即可，代码如下。

```
(python36env) [root@lanmp python]# vim index.py
```

```
    app.run(host='0.0.0.0')
```

再次运行 index.py 文件，会发现下面输出的是 0.0.0.0:5000，打开浏览器，在地址栏中输入虚拟机 IP 地址和端口 192.168.222.18:5000，如图 3-28 所示。

图 3-28

修改 Nginx 虚拟机配置文件，添加如下代码。

```
    (python36env) [root@lanmp python]# vim
/usr/local/nginx/conf/vhost/flask.conf
    server {
        listen 80;
        server_name flask.linux.com;
        location / {
            proxy_pass http://127.0.0.1:5000;
        }
    }
```

检测 flask.conf 虚拟主机配置文件语法是否有误，无误后重载 flask.conf 配置文件，运行命令如下。

```
    (python36env) [root@lanmp python]# /usr/local/nginx/sbin/nginx -t
    nginx: the configuration file /usr/local/nginx/conf/nginx.conf syntax is
ok
    nginx: configuration file /usr/local/nginx/conf/nginx.conf test is
successful
    (python36env) [root@lanmp python]# /usr/local/nginx/sbin/nginx -s
reload
```

再次运行 python index.py 文件，在浏览器的地址栏中输入 flask.linux.com 就可以访问刚才 index.py 输出的信息，如图 3-29 所示。

图 3-29

第4章 MySQL数据库技术实践

4.1 设置和更改root密码

MySQL 数据库和 CentOS 操作系统一样，root 用户均是超级管理员。MySQL 数据库的 root 密码默认为空，无需输入密码即可连接，但是这样不安全，需要给 root 用户设置安全密码。

4.1.1 设置 MySQL 数据库环境变量

在终端窗口输入 mysql –uroot 命令，进入 MySQL 管理界面，但是提示未找到 MySQL 命令，因为安装 MySQL 数据库时的目录在/usr/local/下，环境变量并不存在，所以需要设置环境变量。环境变量设置完毕后，使用 mysql –uroot 命令即可进入 MySQL 管理界面。若想环境变量永久生效，需要编辑/etc/profile 配置文件，在配置文件的末尾添加代码即可，如图 4-1 所示。

4.1.2 设置 MySQL 数据库密码

设置 root 密码的命令是 mysqladmin，在 mysqladmin 命令后指定设置密码的用户和需要设置的密码即可。设置密码后使用 mysql –uroot 命令是不能登录的，需要使用 mysql -uroot –p 命令进行登录，操作命令如下。

```
[root@centos7 ~]# mysqladmin -uroot password 'humingzhe'
```

```
[root@centos7 ~]# mysql -uroot -p
Enter password:
```

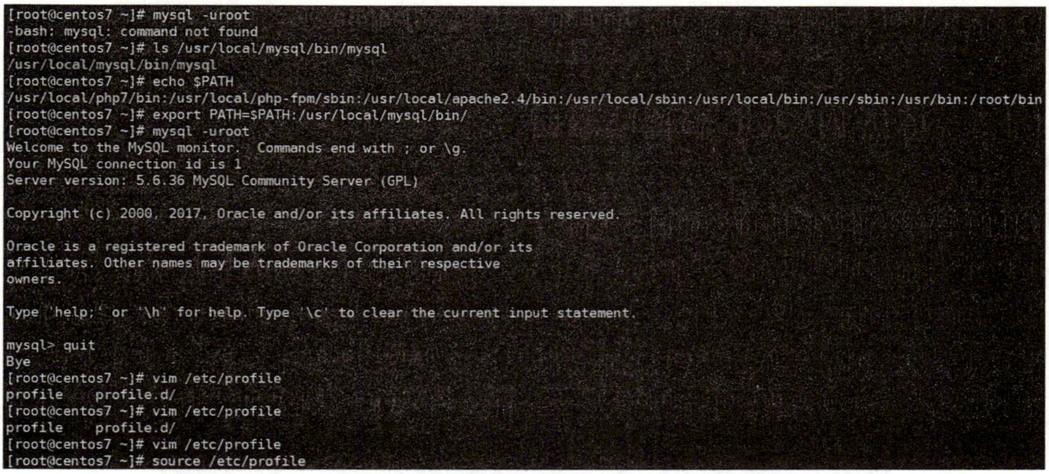

```
[root@centos7 ~]# mysql -uroot
-bash: mysql: command not found
[root@centos7 ~]# ls /usr/local/mysql/bin/mysql
/usr/local/mysql/bin/mysql
[root@centos7 ~]# echo $PATH
/usr/local/php7/bin:/usr/local/php-fpm/sbin:/usr/local/apache2.4/bin:/usr/local/sbin:/usr/local/bin:/usr/sbin:/usr/bin:/root/bin
[root@centos7 ~]# export PATH=$PATH:/usr/local/mysql/bin/
[root@centos7 ~]# mysql -uroot
Welcome to the MySQL monitor.  Commands end with ; or \g.
Your MySQL connection id is 1
Server version: 5.6.36 MySQL Community Server (GPL)

Copyright (c) 2000, 2017, Oracle and/or its affiliates. All rights reserved.

Oracle is a registered trademark of Oracle Corporation and/or its
affiliates. Other names may be trademarks of their respective
owners.

Type 'help;' or '\h' for help. Type '\c' to clear the current input statement.

mysql> quit
Bye
[root@centos7 ~]# vim /etc/profile
profile    profile.d/
[root@centos7 ~]# vim /etc/profile
profile    profile.d/
[root@centos7 ~]# vim /etc/profile
[root@centos7 ~]# source /etc/profile
```

图 4-1

4.1.3　修改 MySQL 数据库密码

修改 MySQL 数据库密码的前提是我们知道 MySQL 数据库的密码，更改密码的操作命令如下。

```
[root@centos7 ~]# mysqladmin -uroot -p'humingzhe' password 'humingzhe123'
Warning: Using a password on the command line interface can be insecure.
[root@centos7 ~]# mysql -uroot -p'humingzhe123'
Warning: Using a password on the command line interface can be insecure.
Welcome to the MySQL monitor.  Commands end with ; or \g.
```

4.1.4　重置 MySQL 数据库密码

不知道 MySQL 数据库密码时使用 mysqladmin 命令是不能更改密码的，需要编辑 MySQL 数据库配置文件 my.cnf，在[mysqld]下方添加 skip-grant（忽略授权），然后保存并重启 MySQL 数据库，代码如下。

```
[root@centos7 ~]# vim /etc/my.cnf
[mysqld]
skip-grant
[root@centos7 ~]# /etc/init.d/mysqld restart
```

```
Shutting down MySQL.. SUCCESS!
Starting MySQL. SUCCESS!
```

重启 MySQL 数据库后，使用 mysql －uroot 命令进入 MySQL 数据库管理界面，在管理界面进行密码的设置，如图 4-2 所示。

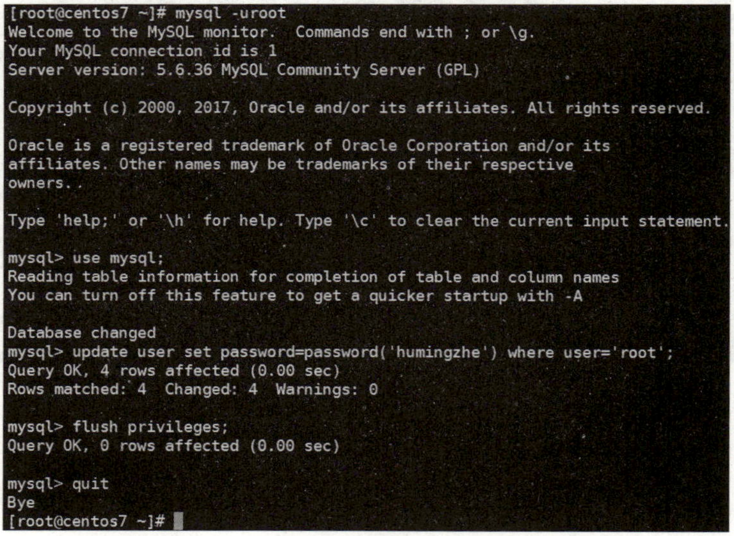

```
[root@centos7 ~]# mysql -uroot
Welcome to the MySQL monitor.  Commands end with ; or \g.
Your MySQL connection id is 1
Server version: 5.6.36 MySQL Community Server (GPL)

Copyright (c) 2000, 2017, Oracle and/or its affiliates. All rights reserved.

Oracle is a registered trademark of Oracle Corporation and/or its
affiliates. Other names may be trademarks of their respective
owners.

Type 'help;' or '\h' for help. Type '\c' to clear the current input statement.

mysql> use mysql;
Reading table information for completion of table and column names
You can turn off this feature to get a quicker startup with -A

Database changed
mysql> update user set password=password('humingzhe') where user='root';
Query OK, 4 rows affected (0.00 sec)
Rows matched: 4  Changed: 4  Warnings: 0

mysql> flush privileges;
Query OK, 0 rows affected (0.00 sec)

mysql> quit
Bye
[root@centos7 ~]#
```

图 4-2

密码设置完毕后需删除 my.cnf 配置文件中的 skip-grant 代码行，然后重启 MySQL 服务，使用 mysql -uroot －p 命令即可登录，如图 4-3 所示。

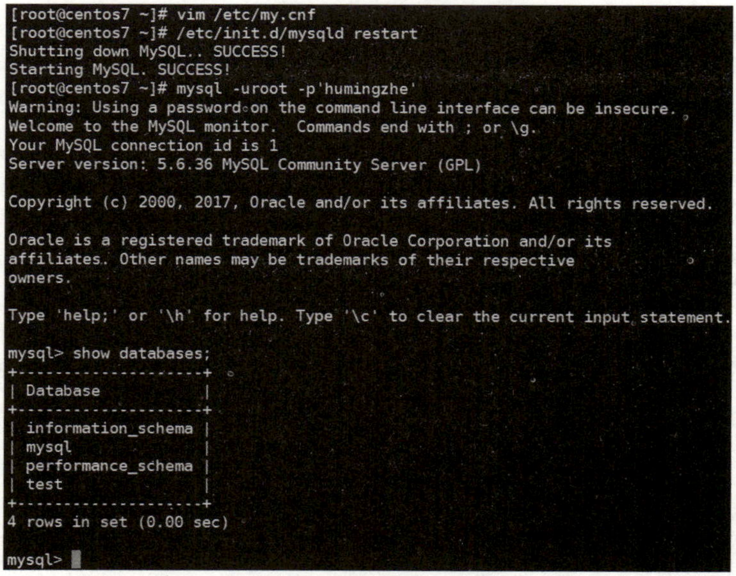

```
[root@centos7 ~]# vim /etc/my.cnf
[root@centos7 ~]# /etc/init.d/mysqld restart
Shutting down MySQL.. SUCCESS!
Starting MySQL. SUCCESS!
[root@centos7 ~]# mysql -uroot -p'humingzhe'
Warning: Using a password on the command line interface can be insecure.
Welcome to the MySQL monitor.  Commands end with ; or \g.
Your MySQL connection id is 1
Server version: 5.6.36 MySQL Community Server (GPL)

Copyright (c) 2000, 2017, Oracle and/or its affiliates. All rights reserved.

Oracle is a registered trademark of Oracle Corporation and/or its
affiliates. Other names may be trademarks of their respective
owners.

Type 'help;' or '\h' for help. Type '\c' to clear the current input statement.

mysql> show databases;
+--------------------+
| Database           |
+--------------------+
| information_schema |
| mysql              |
| performance_schema |
| test               |
+--------------------+
4 rows in set (0.00 sec)

mysql>
```

图 4-3

4.2　连接MySQL的几种方式

连接本机 MySQL 数据库，操作命令如下。

```
[root@centos7 ~]# mysql -uroot -p'humingzhe'
```

连接远程服务器的 MySQL 数据库，比如从 A 服务器连接 B 服务器的 MySQL 数据库，需要输入 B 服务器的 IP 和端口，如图 4-4 所示。

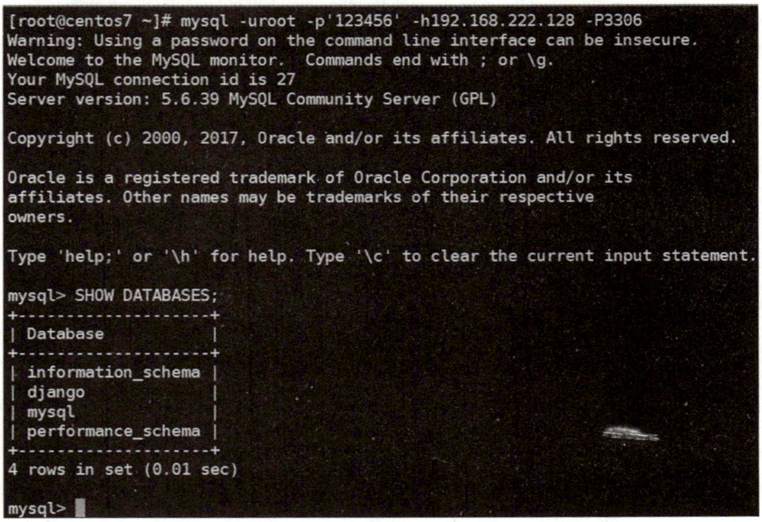

```
[root@centos7 ~]# mysql -uroot -p'123456' -h192.168.222.128 -P3306
Warning: Using a password on the command line interface can be insecure.
Welcome to the MySQL monitor.  Commands end with ; or \g.
Your MySQL connection id is 27
Server version: 5.6.39 MySQL Community Server (GPL)

Copyright (c) 2000, 2017, Oracle and/or its affiliates. All rights reserved.

Oracle is a registered trademark of Oracle Corporation and/or its
affiliates. Other names may be trademarks of their respective
owners.

Type 'help;' or '\h' for help. Type '\c' to clear the current input statement.

mysql> SHOW DATABASES;
+--------------------+
| Database           |
+--------------------+
| information_schema |
| django             |
| mysql              |
| performance_schema |
+--------------------+
4 rows in set (0.01 sec)

mysql>
```

图 4-4

在 UNIX/Linux 操作系统中有一种通信方式是使用 socket，MySQL 数据库不仅监听 3306 端口，还监听 mysql.sock，所以可以使用 socket 连接 MySQL，操作命令如下。

```
[root@centos7 ~]# mysql -uroot -phumingzhe -S/tmp/mysql.sock
Warning: Using a password on the command line interface can be insecure.
Welcome to the MySQL monitor.  Commands end with ; or \g.
```

有时在连接成功 MySQL 数据库后会执行一些命令，这种操作经常用在 shell 脚本中，如查看 MySQL 数据库中有多少个库，如图 4-5 所示。

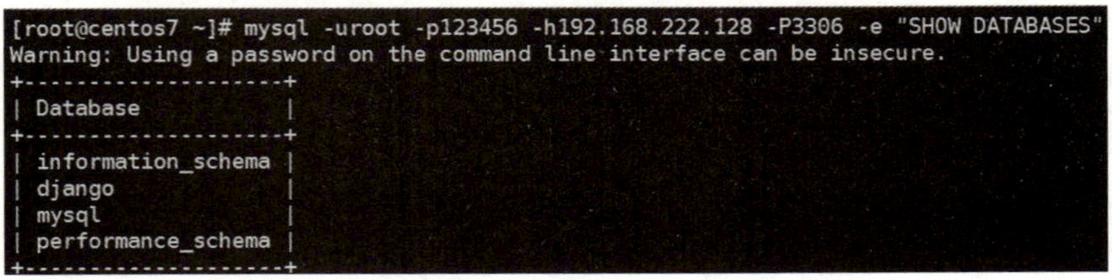

```
[root@centos7 ~]# mysql -uroot -p123456 -h192.168.222.128 -P3306 -e "SHOW DATABASES"
Warning: Using a password on the command line interface can be insecure.
+--------------------+
| Database           |
+--------------------+
| information_schema |
| django             |
| mysql              |
| performance_schema |
+--------------------+
```

图 4-5

4.3 MySQL常用命令

查看 MySQL 数据库中都有哪些数据库，操作命令如下。

```
mysql> SHOW DATABASES;
```

USE 命令表示可以使用某个数据库，如使用 Django 数据库，操作命令如下。

```
mysql> USE django;
Reading table information for completion of table and column names
You can turn off this feature to get a quicker startup with -A

Database changed
```

使用 Django 数据库后可查看 Django 数据库中都有哪些表，操作命令如下。

```
mysql> SHOW TABLES;
+-------------------+
| Tables_in_django  |
+-------------------+
| django_migrations |
+-------------------+
1 row in set (0.00 sec)
```

数据库是由数据表组成的，数据表是由字段组成的。可以查看 django_migrations 表中存在的字段，File 表示字段名字，Type 表示字段类型，如图 4-6 所示。

可以查看数据表是如何创建的，如查看 Django 数据库中的 django_migrations 表，使用\G 选项表示以树型显示，如图 4-7 所示。如果不使用\G 选项，则显示的内容非常乱。

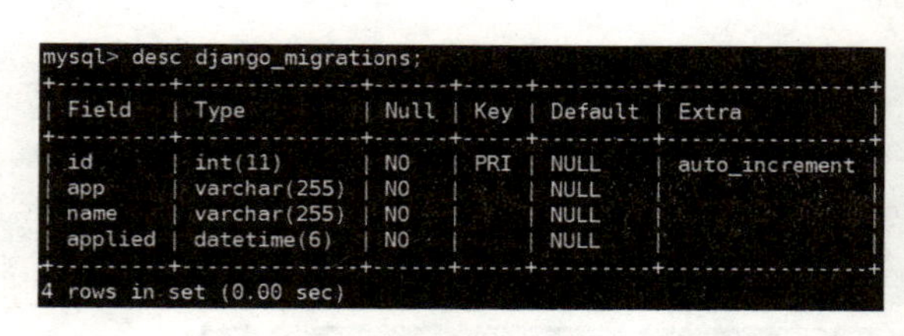

```
mysql> desc django_migrations;
+---------+--------------+------+-----+---------+----------------+
| Field   | Type         | Null | Key | Default | Extra          |
+---------+--------------+------+-----+---------+----------------+
| id      | int(11)      | NO   | PRI | NULL    | auto_increment |
| app     | varchar(255) | NO   |     | NULL    |                |
| name    | varchar(255) | NO   |     | NULL    |                |
| applied | datetime(6)  | NO   |     | NULL    |                |
+---------+--------------+------+-----+---------+----------------+
4 rows in set (0.00 sec)
```

图 4-6

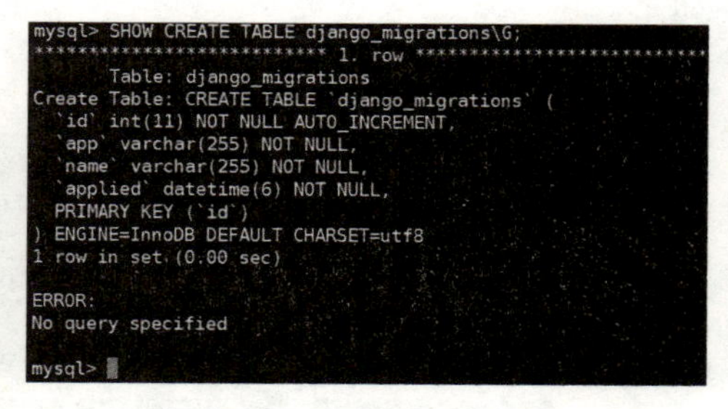

```
mysql> SHOW CREATE TABLE django_migrations\G;
*************************** 1. row ***************************
       Table: django_migrations
Create Table: CREATE TABLE `django_migrations` (
  `id` int(11) NOT NULL AUTO_INCREMENT,
  `app` varchar(255) NOT NULL,
  `name` varchar(255) NOT NULL,
  `applied` datetime(6) NOT NULL,
  PRIMARY KEY (`id`)
) ENGINE=InnoDB DEFAULT CHARSET=utf8
1 row in set (0.00 sec)

ERROR:
No query specified

mysql>
```

图 4-7

查看当前登录的是哪位用户，可以使用 SELECT USER 命令，操作命令如下。

```
mysql> SELECT USER();
+----------------------+
| USER()               |
```

```
+----------------------+
| root@192.168.222.129 |
+----------------------+
1 row in set (0.00 sec)
mysql>
```

可以查看当前使用的是哪个数据库，操作命令如下。

```
mysql> SELECT DATABASE();
+------------+
| DATABASE() |
+------------+
| NULL       |
+------------+
1 row in set (0.00 sec)
mysql>
```

创建 MySQL 数据库，如创建一个 blog 数据库，操作命令如下。

```
mysql> CREATE DATABASE blog;
Query OK, 1 row affected (0.00 sec)
```

切换至 blog 数据库，创建名为 links 的表，第一个字段是 id，第二个字段是 name，定义字段的长度和字符集编码，操作命令如下。

```
mysql> use blog;
Database changed
mysql> CREATE TABLE links(`id` int(4), `name` char(50)) ENGINE=InnoDB
DEFAULT CHARSET=utf8;
Query OK, 0 rows affected (0.02 sec)
```

可以查看当前数据库的版本，操作命令如下。

```
mysql> SELECT VERSION();
+-----------+
| VERSION() |
+-----------+
| 5.6.39    |
+-----------+
1 row in set (0.00 sec)
```

可以查看数据库的状态，操作命令如下。

```
mysql> SHOW STATUS;
```

my.cnf 配置文件中定义的参数可以使用 SHOW VARIABLES 命令进行查看，也可查看指定的参数。例如，只记住了参数前面的 max_connetc，后面的字符记不住了，这时可以用通配符（%）表示，操作命令如下。

```
mysql> SHOW VARIABLES;
mysql> SHOW VARIABLES like 'max_connect%';
+--------------------+-------+
| Variable_name      | Value |
+--------------------+-------+
| max_connect_errors | 100   |
| max_connections    | 151   |
+--------------------+-------+
2 rows in set (0.00 sec)
```

在 my.cnf 配置文件中能修改参数，在命令行界面中一样可以修改参数，如修改“max_connetc_errors=1000”，若想重启 MySQL 数据库后仍保存以上设置，则需编辑配置文件 my.cnf，操作命令如下。

```
mysql> SET GLOBAL max_connect_errors=1000;
Query OK, 0 rows affected (0.00 sec)
```

使用 SHOW PROCESSLIST 命令可以查看队列，但是用该命令查看的最后一列 Info 不会全部显示出来，若想让其全部显示，需要增加“FULL”，如图 4-8 所示。

```
mysql> SHOW PROCESSLIST;
+----+------+----------------------+--------+---------+------+-------+------------------+
| Id | User | Host                 | db     | Command | Time | State | Info             |
+----+------+----------------------+--------+---------+------+-------+------------------+
| 40 | root | 192.168.222.129:59562 | django | Query   |    0 | init  | SHOW PROCESSLIST |
+----+------+----------------------+--------+---------+------+-------+------------------+
1 row in set (0.00 sec)

mysql> SHOW FULL PROCESSLIST;
+----+------+----------------------+--------+---------+------+-------+-----------------------+
| Id | User | Host                 | db     | Command | Time | State | Info                  |
+----+------+----------------------+--------+---------+------+-------+-----------------------+
| 40 | root | 192.168.222.129:59562 | django | Query   |    0 | init  | SHOW FULL PROCESSLIST |
+----+------+----------------------+--------+---------+------+-------+-----------------------+
1 row in set (0.00 sec)

mysql>
```

图 4-8

使用 drop 命令可以删除数据库和数据表，如删除 links 表和 blog 数据库，操作命令如下。

```
mysql> DROP TABLE links;
Query OK, 0 rows affected (0.01 sec)
mysql> DROP DATABASE blog;
Query OK, 0 rows affected (0.01 sec)
```

4.4　MySQL创建用户及授权

MySQL 数据库默认只有 root 用户，root 用户是 MySQL 数据库的超级管理员。某个项目上线时需要连接数据库，并输入用户名和密码，不能都用 root 用户名去连接，会出现误操作。

所以，需要给单独的用户做授权，只需要让它对某个数据库或数据表有执行权限即可。如创建 bone 用户并给它授权只允许 127.0.0.1 的 IP 访问，它只能通过指定 IP 地址登录，如果指定为 localhost 本机，则不指定 IP 地址也可以访问，操作命令如下。

```
mysql> GRANT ALL ON *.* TO 'bone'@'127.0.0.1' IDENTIFIED BY 'bone123';
Query OK, 0 rows affected (0.01 sec)
[root@centos7 ~]# mysql -ubone -pbone123 -h127.0.0.1
mysql> GRANT ALL ON *.* TO bone@'localhost' IDENTIFIED BY 'bone123';
Query OK, 0 rows affected (0.00 sec)
[root@centos7 ~]# mysql -ubone -pbone123
```

指定权限授权用户，如指定 bone 用户只有查询、增加、修改 blog 数据库所有表的权限，并指定通过 192.168.222.128IP 才可以访问，操作命令如下。

```
mysql> GRANT SELECT,UPDATE,INSERT ON blog.* TO bone@'192.128.222.128'
IDENTIFIED BY '123456';
Query OK, 0 rows affected (0.00 sec)
```

可以指定 bone 用户通过所有 IP 执行所有权限，通配符（%）表示所有 IP，操作命令如下。

```
mysql> GRANT ALL ON blog.* TO bone@'%' IDENTIFIED BY '123456';
Query OK, 0 rows affected (0.01 sec)
```

使用 SHOW GRANTS 命令可以查看 MySQL 数据库中的所有授权，也可以查看指定用户的授权，如查看 bone 用户的授权，如图 4-9 所示。

```
mysql> SHOW GRANTS;
+------------------------------------------------------------------------------------------------------------------+
| Grants for root@localhost                                                                                        |
+------------------------------------------------------------------------------------------------------------------+
| GRANT ALL PRIVILEGES ON *.* TO 'root'@'localhost' IDENTIFIED BY PASSWORD '*376A7CB0986C5F1716D014737E28C50F0EE43B4D' WITH GRANT OPTION |
| GRANT PROXY ON ''@'' TO 'root'@'localhost' WITH GRANT OPTION                                                     |
+------------------------------------------------------------------------------------------------------------------+
2 rows in set (0.00 sec)

mysql> SHOW GRANTS FOR bone@'%';
+-----------------------------------------------------------------------------------------------+
| Grants for bone@%                                                                             |
+-----------------------------------------------------------------------------------------------+
| GRANT USAGE ON *.* TO 'bone'@'%' IDENTIFIED BY PASSWORD '*6BB4837EB74329105EE4568DDA7DC67ED2CA2AD9' |
| GRANT ALL PRIVILEGES ON `blog`.* TO 'bone'@'%'                                                |
+-----------------------------------------------------------------------------------------------+
2 rows in set (0.00 sec)

mysql>
```

图 4-9

4.5　MySQL数据备份与恢复

网站迁移或更换服务器时都需要进行数据备份，mysqldump 命令就是用来备份数据的，既可以备份数据库，也可以备份表，如备份 blog 数据库并保存在/tmp/blogbak/目录下的 blog.sql 文件中，操作命令如下。

```
[root@centos7 ~]# mysqldump -uroot -phumingzhe blog > /tmp/blog/blog.sql
Warning: Using a password on the command line interface can be insecure.
[root@centos7 ~]# cat !$
cat /tmp/blog/blog.sql
```

通过备份的 blog.sql 文件也可以恢复 blog 数据库或恢复到其他库，如恢复至 mingzhe 数据库，操作命令如下。

```
[root@centos7 ~]# mysql -uroot -phumingzhe mingzhe < /tmp/blog/blog.sql
[root@centos7 ~]# mysql -uroot -phumingzhe mingzhe
mysql> SELECT DATABASE();
mysql> SHOW TABLES;
```

备份 MySQL 数据库中的 user 表至/tmp/mysqlbak/目录下的 user.sql 文件中，操作命令如下。

```
[root@centos7 ~]# mysqldump -uroot -phumingzhe mysql user >
/tmp/mysqlbak/user.sql
```

恢复 user 表至 mingzhe 数据库中，操作命令如下。

```
[root@centos7 ~]# mysql -uroot -phumingzhe mingzhe <
/tmp/mysqlbak/user.sql
mysql> desc user;
```

MySQL 数据库中有多个库，如果一个库一个库地去备份是比较麻烦的，全部备份可使用-A 选项，如备份 MySQL 数据库中所有的库至/tmp/mysqlbak/mysql.sql 文件中，操作命令如下。

```
[root@centos7 ~]# mysqldump -uroot -p -A > /tmp/mysqlbak/mysql.sql
Enter password:
```

备份数据时可能有这样的需求：不需要表中的数据，只需要表的结构。使用-d 选项可以备份表的结构，操作命令如下。

```
[root@centos7 ~]# mysqldump -uroot -phumingzhe -d mingzhe >
/tmp/mysqlbak/mingzhe.sql
```

4.6　MySQL主从复制监控

MySQL 主从也被称为复制、AB 复制。简单来说，就是 A 和 B 两台服务器做主从后，在 A 服务器上写入数据，B 服务器上也会跟着写入数据，两者之间的数据是实时同步的。

MySQL 主从是基于 binlog 的，主必须开启 binlog 才能进行主从。主从过程大致有如下三个步骤。

（1）主将更改操作记录到 binlog 中。

（2）从将主的 binlog 事件（SQL 语句）同步到本机上并记录在 relaylog 中。

（3）从根据 relaylog 中的 SQL 语句按顺序执行。

主上有一个 log dump 线程，用来和从的 I/O 线程传递 binlog。从上有两个线程，其中 I/O 线程用来同步主的 binlog 并生成 relaylog，另外一个线程用来执行 relaylog 中的 SQL 语句。

4.7　MySQL主从准备工作

MySQL 主从准备工作需要两台服务器，每台服务器上都必须有 MySQL 服务。安装 MySQL 既可使用 yum 命令安装，也可使用二进制免编译安装或源码包编译安装。安装完 MySQL 服务后，设置 my.cnf 并启动脚本。本书在讲解时使用了两台虚拟主机，主机 centos7 为主，主机 devops 为从。

4.8　设置MySQL主

修改 my.cnf 配置文件，增加语句 server-id=129 和 log_bin=centos7，然后重启 MySQL 服务，运行命令如下。

```
[root@centos7 ~]# vim /etc/my.cnf
[mysqld]
server-id=129
log_bin=centos7
[root@centos7 ~]# /etc/init.d/mysqld restart
```

MySQL 服务重启后在/data/mysql 目录下会生成 centos7.index 和 centos7.000001 文件，其中，centos7.index 是索引文件，centos7.000001 是日志文件，运行命令如下。

```
[root@centos7 ~]# ls /data/mysql/
auto.cnf  blog  centos7.000001  centos7.err  centos7.index
```

```
centos7.pid  ibdata1  ib_logfile0  ib_logfile1  mingzhe  mysql
performance_schema  python  test
```

备份 mingzhe 数据库以便后续演示操作，运行命令如下。

```
    [root@centos7 ~]# mysqldump -uroot -phumingzhe mingzhe >
/tmp/mysqlbak/mingzhe.sql
    Warning: Using a password on the command line interface can be insecure.
```

创建新数据库，名称为 humingzhe，然后恢复 mingzhe.sql 数据至 humingzhe 数据库中，运行命令如下。

```
    [root@centos7 ~]# mysqldump -uroot -phumingzhe mingzhe >
/tmp/mysqlbak/mingzhe.sql
    Warning: Using a password on the command line interface can be insecure.
    [root@centos7 ~]# mysql -uroot -phumingzhe humingzhe <
/tmp/mysqlbak/mingzhe.sql
```

创建 MySQL 用户 humingzhe，该用户作为主从数据同步用户，然后进行锁表操作，目的是不让数据继续写，运行命令如下。

```
    mysql> GRANT REPLICATION SLAVE ON *.* TO humingzhe@'192.168.222.128'
IDENTIFIED BY 'master';
    Query OK, 0 rows affected (0.00 sec)
    mysql> FLUSH TABLES WITH READ LOCK;
    Query OK, 0 rows affected (0.00 sec)
```

使用 SHOW MASTER STATUS 命令可以记录主服务器的状态值，在从服务器中设置 master_log_file 和 master_log_pos 时需要用到，如图 4-10 所示。

```
mysql> SHOW MASTER STATUS;
+-----------------+----------+--------------+------------------+-------------------+
| File            | Position | Binlog_Do_DB | Binlog_Ignore_DB | Executed_Gtid_Set |
+-----------------+----------+--------------+------------------+-------------------+
| centos7.000001  |     6501 |              |                  |                   |
+-----------------+----------+--------------+------------------+-------------------+
1 row in set (0.00 sec)
```

图 4-10

主服务器上有 blog、mingzhe、humingzhe、python 库，但不需要同步 MySQL，因为 MySQL 库中有很多用户和权限，除 MySQL 库外，其他库在从服务器上也需要同步这些库，同步库意味着要把这些数据备份至从服务器上。前面备份了 blog.sql 和 mingzhe.sql，其他的数据库也要备份，运行命令如下。

```
    [root@centos7 ~]# mysqldump -uroot -phumingzhe humingzhe >
/tmp/mysqlbak/humingzhe.sql
    [root@centos7 ~]# mysqldump -uroot -phumingzhe python >
/tmp/mysqlbak/python.sql
```

```
[root@centos7 ~]# ls /tmp/mysqlbak/
blog.sql  humingzhe.sql  mingzhe.sql  mysql.sql  python.sql  user.sql
```

4.9　设置MySQL从

编辑 my.cnf 配置文件，增加代码 server-id=128，要求和主不一致，log_bin 无需设置，主服务器需要二进制日志文件，而从服务器不需要，修改后重启 MySQL 服务，运行命令如下。

```
[root@devops ~]# vim /etc/my.cnf
[mysqld]
server-id=128
[root@devops ~]# systemctl restart mysqld.service
```

复制主服务器上备份的库至从服务器上，运行命令如下。

```
[root@devops ~]# scp 192.168.222.129:/tmp/mysqlbak/*.sql /tmp/mybak/
root@192.168.222.129's password:
blog.sql                    100% 1254    66.7KB/s   00:00
humingzhe.sql               100% 6994     2.5MB/s   00:00
mingzhe.sql                 100% 6992     3.2MB/s   00:00
mysql.sql                   100%  648KB   9.5MB/s   00:00
python.sql                  100% 1260   350.5KB/s   00:00
user.sql                    100% 6986   400.9KB/s   00:00
```

登录 MySQL，创建 blog、humingzhe、mingzhe 和 python 库，运行命令如下。

```
[root@devops ~]# mysql -uroot -p
mysql> CREATE DATABASE mingzhe;
mysql> CREATE DATABASE blog;
mysql> CREATE DATABASE python;
mysql> CREATE DATABASE humingzhe;
```

恢复数据至各自的数据库中，运行命令如下。

```
[root@devops ~]# mysql -uroot -p blog < /tmp/mybak/blog.sql
[root@devops ~]# mysql -uroot -p humingzhe < /tmp/mybak/humingzhe.sql
[root@devops ~]# mysql -uroot -p mingzhe < /tmp/mybak/mingzhe.sql
[root@devops ~]# mysql -uroot -p python < /tmp/mybak/python.sql
```

登录 MySQL，关闭同步，运行命令如下。

```
[root@devops ~]# mysql -uroot -p
mysql> STOP SLAVE;
Query OK, 0 rows affected, 1 warning (0.00 sec)
```

下面执行 MySQL 主从同步最关键的一步操作，完毕后开启同步功能，运行命令如下。

```
    mysql> change master to master_host='192.168.222.129',
master_user='humingzhe', master_password='master',
master_log_file='centos7.000001', master_log_pos=6501;
    Query OK, 0 rows affected, 2 warnings (0.03 sec)
    mysql> start slave;
    Query OK, 0 rows affected (0.01 sec)
```

查看主从同步是否设置成功可以使用 show slave status 命令，Slave_IO_Running 和 Slave_SQL_Running 都为 Yes 表示同步成功，运行命令如下。

```
mysql> show slave status\G
    Slave_IO_Running: Yes
    Slave_SQL_Running: Yes
```

注意

主从同步必须关闭防火墙，否则不会显示 Yes。

服务器主从设置成功后恢复主服务器上的写操作，运行命令如下。

```
mysql> unlock tables;
Query OK, 0 rows affected (0.00 sec)
```

4.10 测试MySQL主从同步

同步数据功能支持同步指定的库和忽略指定的库，这些需要在主服务器的 my.cnf 配置文件中进行定义。例如，只想同步 blog 库（多个库以逗号分隔），或者除 python 库外，其他库全部同步，运行命令如下。

```
[mysqld]
binlog-do-db=blog   # 同步指定库，如同步 blog 库
binlog-ignore-db=python     # 忽略指定库，如忽略 python 库
```

在从服务器中也可以进行定义，运行命令如下。

```
[mysqld]
server-id=128
replicate_do_db=
replicate_ignore_db=
replicate_do_table=
replicate_ignore_table=
```

```
replicate_wild_do_table=
replicate_wild_ignore_table=
```

配置 MySQL 主从同步后还需要进行测试，在主服务器上写入数据看能不能同步到从服务器中，或者在主服务器上进行一些操作，如删除 user 表中的数据，运行命令如下。

```
mysql>
mysql> use mingzhe;
mysql> show tables;
mysql> select count(*) user;
mysql> truncate table user;
Query OK, 0 rows affected (0.00 sec)
```

删除 user 表数据后再从服务器上进行查看，如图 4-11 所示，数据已不存在。

```
mysql> use mingzhe;
Reading table information for completion of table and column names
You can turn off this feature to get a quicker startup with -A

Database changed
mysql> select count(*) from user;
+----------+
| count(*) |
+----------+
|        0 |
+----------+
1 row in set (0.00 sec)

mysql>
```

图 4-11

python 库中有多个表，例如删除 user 表，再到从服务器中查看，提示不存在，运行命令如下。

```
mysql> drop table user;
Query OK, 0 rows affected (0.00 sec)
mysql> select * from user;
ERROR 1046 (42S02): Table 'python.user' doesn't exist
```

在主服务器上删除整个 python 库，从服务器中也会删除，运行命令如下。

```
mysql> drop database python;
Query OK, 27 rows affected (0.08 sec)
mysql> select * from python;
ERROR 1046 (42S02): Table 'python.python' doesn't exist
```

温馨提醒

如果想更好地学习 MySQL DBA 相关知识，可以加入读者 QQ 群 99208965 或访问笔者博客 http://humingzhe.com 进行学习与交流。

第 5 章

Tomcat 服务

5.1 Tomcat介绍

Tomcat 是 Apache 软件基金会（Apache Software Foundation）Jakarta 项目中的核心，由 Apache、Sun 和其他一些公司及个人共同开发。使用 Java 语言编写的程序可以用 Tomcat+JDK 来运行。

Tomcat 是一个中间件，真正的起作用的是解析 Java 脚本的 JDK。JDK（Java Development Kit）是整个 Java 的核心，它包含了 Java 的运行环境和一堆与 Java 相关的工具及 Java 基础库。

主流的 JDK 是 Sun 公司发布的 JDK，除此之外，IBM 公司也发布了 JDK，在 CentOS 上可以使用 yum 命令安装 openjdk。

5.2 安装JDK

Tomcat 启动依赖 JDK 环境变量，JDK 目前有 1.6、1.7、1.8、1.9 和 1.10 共 5 个版本，版本简称为 6、7、8、9、10。JDK 官方下载网址是 http://www.oracle.com/technetwork/java/javase/downloads/。

JDK 不能使用 wget 工具进行下载，需要手动打开浏览器下载，下载后上传 JDK 源码包至/usr/local/src/目录下并解压至/usr/local/jdk1.9 目录，运行命令如下。

```
[root@centos7 src]# du -sh jdk-9.0.4_linux-x64_bin.tar.gz
339M    jdk-9.0.4_linux-x64_bin.tar.gz
[root@centos7 src]# tar xf jdk-9.0.4_linux-x64_bin.tar.gz -C /usr/local/
```

```
[root@centos7 src]# du -sh /usr/local/jdk-9.0.4/
556M    /usr/local/jdk-9.0.4/
[root@centos7 src]# mv /usr/local/jdk-9.0.4/ /usr/local/jdk1.9/
```

编辑/etc/profile 配置文件，在下方添加 JDK 环境变量，运行命令如下。

```
[root@centos7 src]# vim /etc/profile
export JAVA_HOME=/usr/local/jdk1.9
export JRE_HOME=${JAVA_HOME}
export CLASSPATH=.:${JAVA_HOME}/lib:${JRE_HOME}/lib
export PATH=${JAVA_HOME}/bin:$PATH
[root@centos7 src]# source /etc/profile
```

使用 java -version 命令可以检测 JDK 是否安装成功，安装成功会显示 JDK 的版本信息，运行命令如下。

```
[root@centos7 src]# java -version
java version "9.0.4"
Java(TM) SE Runtime Environment (build 9.0.4+11)
Java HotSpot(TM) 64-Bit Server VM (build 9.0.4+11, mixed mode)
```

5.3 安装Tomcat

下载 Tomcat 源码包至/usr/local/src/目录下，然后解压至 usr/local/目录并重命名为 tomcat，运行命令如下。

```
    [root@centos7 src]# wget
http://apache.fayea.com/tomcat/tomcat-9/v9.0.4/bin/apache-tomcat-9.0.4.t
ar.gz
    [root@centos7 src]# du -sh apache-tomcat-9.0.4.tar.gz
    9.1M    apache-tomcat-9.0.4.tar.gz
    [root@centos7 src]# tar -xf apache-tomcat-9.0.4.tar.gz -C /usr/local/
    [root@centos7 src]# du -sh /usr/local/apache-tomcat-9.0.4/
    15M /usr/local/apache-tomcat-9.0.4/
    [root@centos7 src]# mv /usr/local/apache-tomcat-9.0.4/
/usr/local/tomcat
    [root@centos7 src]# ls /usr/local/tomcat/
    bin  conf  lib  LICENSE  logs  NOTICE  RELEASE-NOTES  RUNNING.txt  temp
webapps  work
```

执行 Tomcat 脚本启动 tomcat，运行命令如下。

```
[root@centos7 src]# /usr/local/tomcat/bin/startup.sh
Using CATALINA_BASE:   /usr/local/tomcat
```

```
    Using CATALINA_HOME:   /usr/local/tomcat
    Using CATALINA_TMPDIR: /usr/local/tomcat/temp
    Using JRE_HOME:        /usr/local/jdk1.9
    Using CLASSPATH:       /usr/local/tomcat/bin/bootstrap.jar:/usr/local/
tomcat/bin/tomcat-juli.jar
    Tomcat started.
```

查看 8080 端口是否处于监听状态，运行命令如下。Java 共监听了三个端口，8080 端口是提供 Web 服务的端口，8005 端口是管理端口，8009 端口是第三方服务调用端口，如 httpd 和 Tomcat 结合时会用到。

```
    [root@centos7 src]# netstat -ltnp |grep java
    tcp6       0      0 :::8080              :::*      LISTEN      3347/java
    tcp6       0      0 117.0.0.1:8005       :::*      LISTEN      3347/java
    tcp6       0      0 :::8009              :::*      LISTEN      3347/java
```

关闭防火墙后，在浏览器的地址栏中输入 IP 和端口号即可访问 Tomcat 页面，如图 5-1 所示。

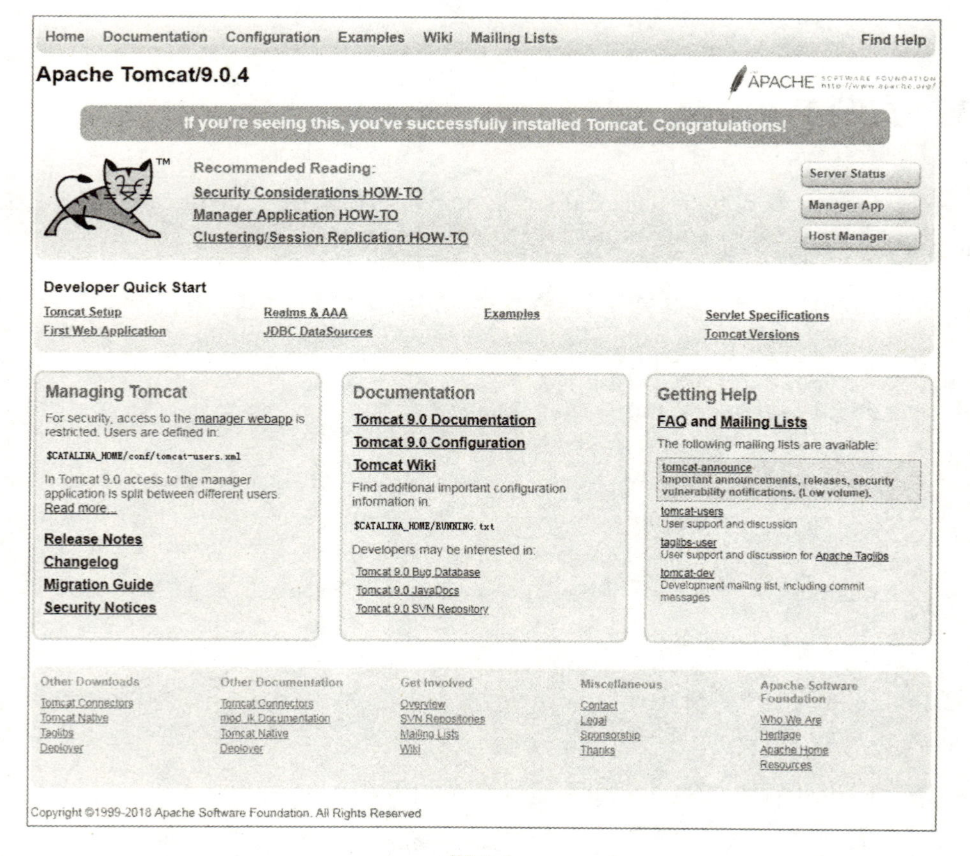

图 5-1

5.4　设置Tomcat监听80端口

Tomcat 默认监听 8080 端口，每次访问还需要在 IP 后面加端口号，比较麻烦。80 端口在浏览器访问时是可以省略的。Tomcat 也支持自定义端口，我们把 8080 端口改为 80 端口需编辑 Tomcat 配置文件 server.xml，搜索 8080，将其改为 80 即可，修改后重启 Tomcat，运行命令如下。

```
[root@centos7 src]# vim /usr/local/tomcat/conf/server.xml
<Connector port="80" protocol="HTTP/1.1"
    connectionTimeout="20000"
    redirectPort="8443" />
[root@centos7 src]# /usr/local/tomcat/bin/shutdown.sh
[root@centos7 src]# /usr/local/tomcat/bin/startup.sh
```

重启后使用 netstat 命令可查看 80 端口是否被 Tomcat 监听，运行命令如下。

```
[root@centos7 src]# netstat -lntp |grep java
tcp6       0      0 :::80                 :::*      LISTEN      3600/java
tcp6       0      0 117.0.0.1:8005        :::*      LISTEN      3600/java
tcp6       0      0 :::8009               :::*      LISTEN      3600/java
```

5.5　Tomcat虚拟主机

5.5.1　设置 Tomcat 虚拟主机

Apache 和 Nginx 都属于 Web Server，都有虚拟主机配置文件，Tomcat 同样有虚拟主机配置文件。Tomcat 配置文件格式是 XML 格式，和 Apache、Nginx 不一样。在 server.xml 配置文件中搜索“/<Host”会定位至如图 5-2 所示的界面。

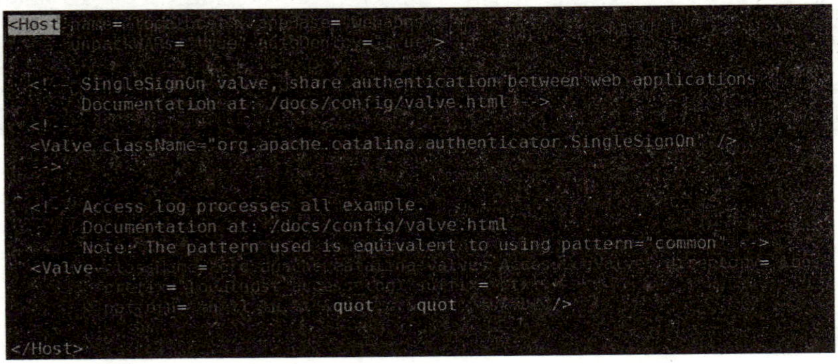

图 5-2

<Host>和</Host>之间是虚拟主机配置部分：name 定义域名；appBase 定义应用的目录。Java 的应用通常是一个 JAR 的压缩包，只要将 JAR 的压缩包放到 appBase 目录下就会自动解压成一个程序目录，刚刚访问的 Tomcat 默认页就在 appBase 目录下，不过是在子目录 ROOT 中。

增加虚拟主机，编辑 server.xml 配置文件，在</Host>下面增加自定义虚拟主机代码，如图 5-3 所示。

docBase 用来定义网站文件的存放路径，如果不定义，默认是在 AppBase/ROOT 目录下，定义 docBase 后就以该目录为主，其中 appBase 和 docBase 可以一致。在这一步操作中如遇到 404 状态码，是因为 docBase 没有被正确定义。

```
<Host name="www.humingzhe.cn" appBase=""
    unpackWARs= "true"  autoDeploy="true"
    xmlValidation="false" xmlNamespaceAware="false">
    <Context path="" docBase="/data/www/wwwhumingzhe.cn/" debug="0" reloadable="true" crossContext="true"/>
</Host>
```

图 5-3

5.5.2 部署一个 Java 项目

本节通过部署一个 Java 应用项目来讲解 appBase 和 docBase 目录的作用。用 PHP 编写的开源 Blog 程序有 WordPress、emblog 和 zblog。Java 也有 Zrlog 博客程序，下载 Zrlog 博客程序至/usr/local/src/目录下，将 Zrlog 程序中的.war 文件复制到 webapps 目录下，用 ls 命令查看，第一次只显示.war 文件，稍后再用 ls 命令查看，多了一个目录，正是解压.war 文件后的目录，如图 5-4 所示。

```
[root@centos7 src]# wget http://dl.zrlog.com/release/zrlog-1.8.0-d1f36bc-release.war
--2018-02-02 03:48:37--  http://dl.zrlog.com/release/zrlog-1.8.0-d1f36bc-release.war
Resolving dl.zrlog.com (dl.zrlog.com)... 61.54.1.130
Connecting to dl.zrlog.com (dl.zrlog.com)|61.54.1.130|:80... connected.
HTTP request sent, awaiting response... 200 OK
Length: 7306860 (7.0M) [application/zip]
Saving to: 'zrlog-1.8.0-d1f36bc-release.war'

100%[=====================================================>] 7,306,860   1.19MB/s   in 6.1s

2018-02-02 03:48:48 (1.14 MB/s) - 'zrlog-1.8.0-d1f36bc-release.war' saved [7306860/7306860]

[root@centos7 src]# ls
apache-tomcat-9.0.4.tar.gz  apr-util-1.6.1.tar.bz2   logs                                         nginx-1.13.8         php-7.2.1             zrlog-1.8.0-d1f36bc-release.war
apr-1.6.3                   httpd-2.4.29             mariadb-10.2.6-linux-glibc_214-x86_64.tar.gz  nginx-1.13.8.tar.gz  php-7.2.1.tar.gz
apr-1.6.3.tar.gz            httpd-2.4.29.tar.gz      mysql                                        php-5.6.9            phpredis-develop
apr-util-1.6.1              jdk-9.0.4_linux-x64_bin.tar.gz  mysql-5.6.36-linux-glibc2.5-x86_64.tar.gz  php-5.6.9.tar.gz     phpredis-develop.zip
[root@centos7 src]#
[root@centos7 src]# ls /usr/local/tomcat/webapps/
docs  examples  host-manager  manager  ROOT
[root@centos7 src]# cp zrlog-1.8.0-d1f36bc-release.war /usr/local/tomcat/webapps/
[root@centos7 src]# !ls
ls /usr/local/tomcat/webapps/
docs  examples  host-manager  manager  ROOT  zrlog-1.8.0-d1f36bc-release.war
[root@centos7 src]# !ls
ls /usr/local/tomcat/webapps/
docs  examples  host-manager  manager  ROOT  zrlog-1.8.0-d1f36bc-release  zrlog-1.8.0-d1f36bc-release.war
[root@centos7 src]#
```

图 5-4

Zrlog 博客程序的目录名字过长，在浏览器的地址栏中输入全称才可以访问安装页面，为了节省时间，可以把目录重命名为 zrlog，运行命令如下。

```
[root@centos7 src]# cd /usr/local/tomcat/webapps/
[root@centos7 webapps]# mv zrlog-1.8.0-d1f36bc-release zrlog
```

```
[root@centos7 webapps]# ls
docs  examples  host-manager  manager  ROOT  zrlog
zrlog-1.8.0-d1f36bc-release.war
```

在浏览器的地址栏中输入 IP+程序名即可跳转到安装向导界面，输入数据库用户名及密码（数据库用户名和密码需要在 MySQL 管理界面创建），如图 5-5 所示。

图 5-5

访问 zrlog 博客时，浏览器的地址栏中会显示 IP/zrlog，若要去掉 zrlog 目录，需要把程序移动至前面在 server.xml 文件中定义 Host 时 docBase 指定的目录下，运行命令如下。

```
[root@centos7 webapps]# mkdir -p /data/www/www.humingzhe.cn
[root@centos7 webapps]# mv /usr/local/tomcat/webapps/zrlog/*
/data/www/www.humingzhe.cn/
[root@centos7 webapps]# /usr/local/tomcat/bin/shutdown.sh
[root@centos7 webapps]# /usr/local/tomcat/bin/startup.sh
```

修改后输入 server.xml 文件中的 humingzhe.cn 域名即可访问该博客站点，如图 5-6 所示。

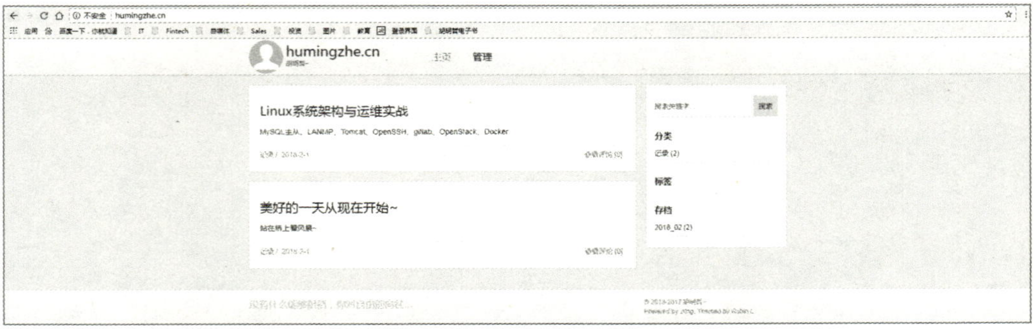

图 5-6

5.6 Tomcat日志

Tomcat 日志在/usr/local/tomcat/logs 目录下，其中以 catalina 开头的日志是 Tomcat 的综合日志文件，它会记录 Tomcat 服务的一些信息，也会记录错误日志。

catalina.2017-02-01-log 和 catalina.out 内容相同，前者会每天生成一个新的日志文件。host-manager 和 manager 为管理相关的日志文件，host-manager 为虚拟主机的管理日志文件。

localhost 和 localhost_access 是与虚拟主机相关的日志文件，带 access 字样的日志为访问日志，不带 access 字样的日志为默认虚拟主机错误日志。

访问日志默认不会自动生成，需要在 server.xml 配置文件中进行定义，如配置 humingzhe.cn 站点的日志，代码如下。

```
    [root@centos7 logs]# vim /usr/local/tomcat/conf/server.xml
    <Host name="humingzhe.cn" appBase=""
        unpackWARs= "true" autoDeploy="true"
        xmlValidation="false" xmlNamespaceAware="false">
        <Context path="" docBase="/data/www/www.humingzhe.cn" debug="0"
reloadable="true" crossContext="true"/>
        <Valve className="org.apache.catalina.valves.AccessLogValve"
directory="logs"
        prefix="humingzhe.cn_access" suffix=".log"
        pattern="%h %l %u %t &quot;%r&quot; %s %b" />
    </Host>
```

注意

prefix 定义访问日志的前缀，suffix 定义日志的后缀，pattern 定义日志格式。新增加的虚拟主机默认不会生成类似默认虚拟主机的“localhost.日期.log”日志，错误日志会统一记录到 catalina.out 中。关于 Tomcat 日志，最需要关注的是 catalina.out 文件，当出现问题时，应该第一时间去查看它。

第 6 章

Linux 集群架构

6.1 Linux集群概述

Linux 集群从功能上可以分为两大类：高可用集群和负载均衡集群。

高可用集群通常为两台服务器，一台工作，另一台冗余。当提供服务的服务器宕机时，冗余服务器将接替宕机的服务器继续提供服务。

实现高可用集群的开源软件有 Heartbeat 和 Keepalived。

负载均衡集群，需要有一台服务器作为分发器，它负责把用户的请求分发给后端的服务器处理。在负载均衡集群中，除分发器外，就是给用户提供服务的服务器了，这些服务器数量至少是两台。

实现负载均衡的开源软件有很多，如 LVS、Keepalived、Haproxy、Nginx，商业版有 F5 和 Netscaler，商业版的使用费用较高，少则几万元，多则数十万元；商业版的优势在于并发量高，拥有很好的稳定性。

6.2 Keepalived简介

Keepalived 是通过 VRRP（Virtual Router Redundancy Protocl）协议来实现高可用的。VRRP 协议会将多台功能相同的路由器组成一个小组，这个小组里有 1 个 master 角色和 N（N>=1）个 backup 角色。

master 会通过组播的形式向各个 backup 发送 VRRP 协议的数据包，当 backup 收不到 master 发来的 VRRP 数据包时，就会认为 master 宕机了。此时就需要根据各个 backup 的优先级来决定谁成为新的 master。

Keepalived 有三个模块，分别是 core、check 和 vrrp。其中，core 模块是 Keepalived 的核心，负责主进程的启动、维护，以及全局配置文件的加载和解析；check 模块负责健康检查；vrrp 模块用来实现 VRRP 协议。

6.3 Keepalived设置高可用集群

6.3.1 集群准备工作

准备两台虚拟主机，IP 为 192.168.222.129 的主机 centos7 作为 master，IP 为 192.168.222.128 的主机 devops 作为 bakup。两台虚拟主机同时安装 Keepalived 和 Nginx，因为 centos7 在前面编译并安装过，所以只需在 devops 上进行 yum 安装，安装服务后关闭防火墙和 selinux，运行命令如下。

```
[root@centos7 ~]# yum install -y keepalived
[root@devops ~]# yum install -y keepalived
[root@devops ~]# yum install -y nginx
[root@centos7 ~]# systemctl stop firewalld.service
[root@centos7 ~]# systemctl disable firewalld.service
[root@devops ~]# systemctl stop firewalld.service
[root@devops ~]# systemctl disable firewalld.service
[root@centos7 ~]# getenforce
Disabled
[root@devops ~]# getenforce
Disabled
```

6.3.2 设置 Keepalived 主服务器

编辑 master 服务器上的 Keepalived 配置文件，清空所有原始配置，将 VIP 设置为 100，代码如下。

```
[root@centos7 ~]# vim /etc/keepalived/keepalived.conf
global_defs {   # 全局定义参数
    notification_email { # 出现问题时给邮箱发邮件
    admin@humingzhe.com
    }
    notification_email_from root@humingzhe.com
    smtp_server 127.0.0.1
    smtp_connect_timeout 30
```

```
    router_id LVS_DEVEL
}
vrrp_script chk_nginx {
    script "/usr/local/sbin/check_ng.sh" #检测服务是否正常，需写 shell 脚本
    interval 3 # 检测间断 3 秒
}
vrrp_instance VI_1 {
    state MASTER    # 定义 master 相关信息
    interface ens33 # 通过 ens33 网卡发送广播
    virtual_router_id 51   # 主从 ID 保持一致
    priority 100    # 权重，主角色和从角色权重不一致
    advert_int 1
    authentication {    # 认证相关信息
        auth_type PASS
        auth_pass humingzhe>com
    }
    virtual_ipaddress { # 定义 vip
        192.168.222.100
    }
    track_script { # 加载脚本
        chk_nginx
    }
}
```

创建检测 Nginx 服务是否正常的脚本，保存后设置为 755 权限，代码如下。

```
[root@centos7 ~]# vim /usr/local/sbin/check_ng.sh
#!/bin/bash
##__author__ is humingzhe
##email admin@humingzhe.com
#时间变量，用于记录日志
d=`date --date today +%Y%m%d_%H:%M:%S`
#计算 Nginx 进程数量
n=`ps -C nginx --no-heading|wc -l`
#如果进程数为 0，则启动 Nginx，并且再次检测 Nginx 进程数
#如果进程数还为 0，则说明 Nginx 无法启动，此时需要关闭 Keepalived
if [ $n -eq "0" ]; then
    /etc/init.d/nginx start
    n2=`ps -C nginx --no-heading|wc -l`
    if [ $n2 -eq "0"  ]; then
        echo "$d nginx down,keepalived will stop" >> /var/log/check_ng.log
        systemctl stop keepalived
    fi
```

```
fi
[root@centos7 ~]# chmod 755 /usr/local/sbin/check_ng.sh
```

启动 Keepalived 服务，使用 ps aux 命令查看 Keepalived 进程，检查 Nginx 是否启动，启动后关闭 Nginx，再次查询它时会自动启动，说明脚本已经生效，如图 6-1 所示。

```
[root@centos7 ~]# systemctl start keepalived
[root@centos7 ~]# ps aux |grep keep
root      3864  0.0  0.0 120740  1404 ?        Ss   06:28   0:00 /usr/sbin/keepalived -D
root      3865  0.0  0.0 127476  3272 ?        S    06:28   0:00 /usr/sbin/keepalived -D
root      3866  0.0  0.0 131780  3128 ?        S    06:28   0:00 /usr/sbin/keepalived -D
root      4180  0.0  0.0 112660   976 pts/1    S+   06:30   0:00 grep --color=auto keep
[root@centos7 ~]# ps aux |grep ng nx
root      4083  0.0  0.0  46000  1304 ?        Ss   06:30   0:00 nginx: master process /usr/local/nginx/sbin/nginx -c /usr/local/nginx/conf/nginx.conf
nobody    4084  0.0  0.1  48488  4192 ?        S    06:30   0:00 nginx: worker process
nobody    4085  0.0  0.1  48488  3936 ?        S    06:30   0:00 nginx: worker process
root      4210  0.0  0.0 112660   976 pts/1    S+   06:30   0:00 grep --color=auto nginx
[root@centos7 ~]# /etc/init.d/nginx stop
Stopping nginx (via systemctl):                            [  OK  ]
[root@centos7 ~]# ps aux |grep nginx
root      4252  0.0  0.0 112660   976 pts/1    S+   06:31   0:00 grep --color=auto nginx
[root@centos7 ~]# ps aux |grep nginx
root      4275  0.0  0.0  46000  1304 ?        Ss   06:31   0:00 nginx: master process /usr/local/nginx/sbin/nginx -c /usr/local/nginx/conf/nginx.conf
nobody    4276  0.0  0.1  48488  3936 ?        S    06:31   0:00 nginx: worker process
nobody    4277  0.0  0.1  48488  3936 ?        S    06:31   0:00 nginx: worker process
root      4282  0.0  0.0 112660   976 pts/1    S+   06:31   0:00 grep --color=auto nginx
[root@centos7 ~]#
```

图 6-1

Keepalived 服务日志在/var/log/目录下的 message 文件中，运行命令如下。

```
[root@centos7 ~]# less /var/log/messages
```

设置的 VIP 使用 ifconfig 命令查看不到，要使用 ip addr 命令查看，如图 6-2 所示。

```
[root@centos7 ~]# ip addr
1: lo: <LOOPBACK,UP,LOWER_UP> mtu 65536 qdisc noqueue state UNKNOWN qlen 1
    link/loopback 00:00:00:00:00:00 brd 00:00:00:00:00:00
    inet 127.0.0.1/8 scope host lo
       valid_lft forever preferred_lft forever
    inet6 ::1/128 scope host
       valid_lft forever preferred_lft forever
2: ens33: <BROADCAST,MULTICAST,UP,LOWER_UP> mtu 1500 qdisc pfifo_fast state UP qlen 1000
    link/ether 00:0c:29:d6:dd:d9 brd ff:ff:ff:ff:ff:ff
    inet 192.168.222.129/24 brd 192.168.222.255 scope global ens33
       valid_lft forever preferred_lft forever
    inet 192.168.222.100/32 scope global ens33
       valid_lft forever preferred_lft forever
    inet6 fe80::6be1:bf5e:d449:d3a5/64 scope link
       valid_lft forever preferred_lft forever
3: ens37: <BROADCAST,MULTICAST,UP,LOWER_UP> mtu 1500 qdisc pfifo_fast state UP qlen 1000
    link/ether 00:0c:29:d6:dd:e3 brd ff:ff:ff:ff:ff:ff
    inet 192.168.222.129/24 brd 192.168.222.255 scope global ens37
       valid_lft forever preferred_lft forever
    inet6 fe80::24b6:3812:4da9:80f1/64 scope link
       valid_lft forever preferred_lft forever
[root@centos7 ~]#
```

图 6-2

6.3.3 设置 Keepalived 从服务器

编辑 backup 服务器中的 Keepalived 配置文件，清空所有原始配置，将 VIP 设置为 90，代码如下。

```
[root@devops ~]# > /etc/keepalived/keepalived.conf
global_defs {
```

```
    notification_email {
    admin@humingzhe.com
    }
    notification_email_from root@humingzhe.com
    smtp_server 127.0.0.1
    smtp_connect_timeout 30
    router_id LVS_DEVEL
}
vrrp_script chk_nginx {
    script "/usr/local/sbin/check_ng.sh"
    interval 3
}
vrrp_instance VI_1 {
    state BACKUP
    interface ens33
    virtual_router_id 51
    priority 90
    advert_int 1
    authentication {
        auth_type PASS
        auth_pass humingzhe>com
    }
    virtual_ipaddress {
        192.168.222.100
    }
    track_script {
    chk_nginx
    }
}
```

创建检测 Nginx 服务是否正常的脚本，保存后设置为 755 权限，代码如下。

```
#!/bin/bash
##__author__ is humingzhe
##email admin@humingzhe.com
#时间变量，用于记录日志
d=`date --date today +%Y%m%d_%H:%M:%S`
#计算 Nginx 进程数量
n=`ps -C nginx --no-heading|wc -l`
#如果进程数为 0，则启动 Nginx，并且再次检测 Nginx 进程数
#如果进程数还为 0，则说明 Nginx 无法启动，此时需要关闭 Keepalived
if [ $n -eq "0" ]; then
    systemctl start nginx
```

```
    n2=`ps -C nginx --no-heading|wc -l`
    if [ $n2 -eq "0"  ]; then
        echo "$d nginx down,keepalived will stop" >> /var/log/check_ng.log
        systemctl stop keepalived
    fi
fi
[root@devops ~]# chmod 755 /usr/local/sbin/check_ng.sh
```

启动 Keepalived 服务，使用 ps aux 命令查看 Keepalived 进程，运行命令如下。

```
[root@devops ~]# systemctl restart keepalived.service
[root@devops ~]# ps aux |grep keep
```

6.3.4 区分主从 Nginx 服务

主服务器和从服务器都安装了 Nginx，主服务器是在前面编译并安装的，从服务器是刚刚使用 yum 命令安装的。主服务器的默认虚拟主机路径是/usr/local/nginx/conf/vhost/目录下的 mingzhe.com.conf 配置文件中定义的 default_server，网站目录在/data/www/default/目录下，如图 6-3 所示。

```
[root@centos7 ~]# cat /usr/local/nginx/conf/vhost/mingzhe.com.conf
server
{
        listen 80 default_server;
        server_name humingzhe.com;
        index index.html index.php index.htm;
        root /data/www/default;
        access_log /data/logs/nginx/humingzhe.com.log mingzhe;
```

图 6-3

编辑/data/www/default/目录下的 index.html 文件，添加“Keepalived Master”代码，如图 6-4 所示。在浏览器的地址栏中输入 IP 地址 192.168.222.129，就会显示刚刚定义的内容，如图 6-5 所示。

图 6-4

图 6-5

从服务器使用 yum 命令安装了 Nginx，它的网站根目录在/usr/share/nginx/html/目录下，清空原有的 index.html 文件内容，添加“Keepalived Backup”代码，如图

6-6 所示。在浏览器的地址栏中输入 IP 就会显示 index.html 文件中添加的内容，如图 6-7 所示。

```
[root@devops html]# vim /usr/share/nginx/html/index.html

<html>
    <head>
        <meta charset="UTF-8">
        <title>Keepalived Backup</title>
    </head>
    <body>
        <h1>Keepalived Backup</h1>
        <h3>Host:192.168.222.128</h3>
    </body>
</html>
```

图 6-6

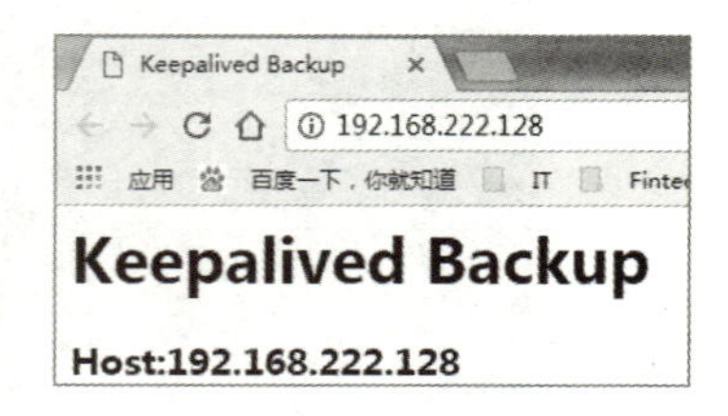

图 6-7

在浏览器的地址栏中输入 VIP 地址 192.168.222.100 就会访问主服务器的内容，因为 VIP 目前在 master 上，如图 6-8 所示。

图 6-8

6.3.5　测试 Keepalived 高可用

在主服务器上添加一条防火墙规则，把主服务器上 VRRP 协议出去的包封掉，如图 6-9 所示。但是把 VRRP 包封掉也不能达到切换资源的目的，用浏览器访问 VIP 地址依然停留在 master 上，所以还要把 iptables 规则清空。

```
[root@centos7 ~]# iptables -I OUTPUT -p vrrp -j DROP
[root@centos7 ~]# iptables -nvL
Chain INPUT (policy ACCEPT 49 packets, 2688 bytes)
 pkts bytes target     prot opt in     out     source               destination

Chain FORWARD (policy ACCEPT 0 packets, 0 bytes)
 pkts bytes target     prot opt in     out     source               destination

Chain OUTPUT (policy ACCEPT 16 packets, 1432 bytes)
 pkts bytes target     prot opt in     out     source               destination
   14   560 DROP       112  --  *      *       0.0.0.0/0            0.0.0.0/0
[root@centos7 ~]#
```

图 6-9

模拟生产环境中主机宕机的场景最简答的做法是把 master 上的 Keepalived 服务停掉，停掉 Keepalived 也就意味着该服务器宕机了，自然 VIP 地址也会被释放，如图 6-10 所示。

```
[root@centos7 ~]# systemctl stop keepalived.service
[root@centos7 ~]# ip addr
1: lo: <LOOPBACK,UP,LOWER_UP> mtu 65536 qdisc noqueue state UNKNOWN qlen 1
    link/loopback 00:00:00:00:00:00 brd 00:00:00:00:00:00
    inet 127.0.0.1/8 scope host lo
       valid_lft forever preferred_lft forever
    inet6 ::1/128 scope host
       valid_lft forever preferred_lft forever
2: ens33: <BROADCAST,MULTICAST,UP,LOWER_UP> mtu 1500 qdisc pfifo_fast state UP qlen 1000
    link/ether 00:0c:29:d6:dd:d9 brd ff:ff:ff:ff:ff:ff
    inet 192.168.222.129/24 brd 192.168.222.255 scope global ens33
       valid_lft forever preferred_lft forever
    inet6 fe80::6be1:bf5e:d449:d3a5/64 scope link
       valid_lft forever preferred_lft forever
3: ens37: <BROADCAST,MULTICAST,UP,LOWER_UP> mtu 1500 qdisc pfifo_fast state UP qlen 1000
    link/ether 00:0c:29:d6:dd:e3 brd ff:ff:ff:ff:ff:ff
    inet 192.168.222.129/24 brd 192.168.222.255 scope global ens37
       valid_lft forever preferred_lft forever
    inet6 fe80::24b6:3812:4da9:80f1/64 scope link
       valid_lft forever preferred_lft forever
[root@centos7 ~]#
```

图 6-10

在从服务器上运行 ip addr 命令会发现 VIP 地址绑定到了从服务器上，如图 6-11 所示。

```
[root@devops html]# ip addr
1: lo: <LOOPBACK,UP,LOWER_UP> mtu 65536 qdisc noqueue state UNKNOWN qlen 1
    link/loopback 00:00:00:00:00:00 brd 00:00:00:00:00:00
    inet 127.0.0.1/8 scope host lo
       valid_lft forever preferred_lft forever
    inet6 ::1/128 scope host
       valid_lft forever preferred_lft forever
2: ens33: <BROADCAST,MULTICAST,UP,LOWER_UP> mtu 1500 qdisc pfifo_fast state UP qlen 1000
    link/ether 00:0c:29:e4:4c:2b brd ff:ff:ff:ff:ff:ff
    inet 192.168.222.128/24 brd 192.168.222.255 scope global ens33
       valid_lft forever preferred_lft forever
    inet 192.168.222.100/32 scope global ens33
       valid_lft forever preferred_lft forever
    inet6 fe80::b6ed:d94b:215d:19ad/64 scope link
       valid_lft forever preferred_lft forever
[root@devops html]#
```

图 6-11

在浏览器的地址栏中输入 VIP 地址 192.168.222.100，会访问从服务器上的内容，如图 6-12 所示。

图 6-12

6.4 负载均衡集群介绍

负载均衡集群由 LVS、Keepalived、Haporxy、Nginx 等开源软件实现。LVS 基于 4 层（OSI 网络 7 层模型），Nginx 基于 7 层，Haporxy 既可以用作 4 层，也可以用作 7 层。Keepalived 的负载均衡功能就是 LVS。

LVS（4 层）负载均衡可以分发 TCP 协议，Web 服务是 80 端口，除分发 80 端口外，它还可以分发其他端口通信。比如，MySQL 负载均衡也可以用 LVS 去实现，而 Nginx 则不支持这个功能，仅支持 HTTP、HTTPS、Mail；而 Haporxy 也支持像 MySQL 这种 TCP 负载均衡。

相比较而言，LVS 这种 4 层的结构更稳定，能承载的并发量更高，而 Nginx 这种 7 层的结构更加灵活，能实现更多的个性化需求。

6.5　LVS介绍

LVS 是我国的章文嵩开发的，可以说是世界知名的软件，因为开源，所以很多企业在用，流行度不亚于 Apache 的 htppd。LVS 是基于 TCP/IP 做的路由转发，稳定性和效率非常高。

LVS 最新版是基于 Linux 内核 2.6 的，其实已经很多年没有进行更新了。LVS 有 3 种常见的模式，分别是 NAT 模式、DR 模式和 IP Tunnel 模式。

LVS 架构中有一个核心的角色叫作分发器（Load Balance），用来分发用户的请求，还有诸多处理用户请求的服务器（Real Server，简称 RS）。

LVS NAT 模式是借助于 iptables 的 nat 表来实现的，用户的请求到分发器后，通过预设的 iptables 规则，把请求的数据包转发到后端的 RS 上。RS 设定网关为分发器的内网 IP，用户请求的数据包全部经过分发器，所以分发器成为瓶颈。在 NAT 模式中，分发器有公网 IP 即可，所以比较节省公网 IP 资源。

LVS IP Tunnel 模式需要有一个公网 IP 配置在分发器和所有 RS 上，我们把它称之为 VIP。客户端请求的目标 IP 为 VIP，分发器接收到请求数据包会对数据包做加工处理，把目标 IP 改为 RS 的 IP，这样数据包就到了 RS 上。RS 接收数据包后，会还原原始数据包，这样目标 IP 就为 VIP，因为所有 RS 上配置了这个 VIP，所以它的任务就是它自己。

LVS DR 模式也需要有一个公共 IP 配置在分发器和所有 RS 上，也就是 VIP。和 IP Tunnel 不同的是，它会把数据包的 MAC 地址修改为 RS 的 MAC 地址。RS 接收数据包后，会还原原始数据包，这样目标 IP 就为 VIP，因为所有 RS 上配置了这个 VIP，所以它会认为是它自己。

6.6　LVS的调度算法

LVS 支持的算法包括 8 种：轮询（Round-Robin，简称 rr）、加权轮询（Weight Round-Robin，简称 wrr）、最小连接（Least-Connection，简称 lc）、加权最小连接（Weight Least-Connection，简称 wlc）、基于局部性的最小连接（Locality-Based Least Connections，简称 lblc）、带复制的基于局部性最小连接（Locality-Based Least Connections with Replication，简称 lblcr）、目标地址散列调度（Destination Hashing，简称 dh）和源地址散列调度（Source Hashing，简称 sh）。

6.7 NAT模式的LVS搭建

6.7.1 NAT 模式搭建准备工作

准备 3 台服务器，其中一台作为分发器（也叫作调度器，简称 Dir），另外两台是 Real Server，用作处理用户请求的服务器。

分发器需要两个网卡，IP 为 192.168.222.129 是内网，IP 为 192.168.17.3 是外网（VMWare 仅主机模式），主机名为 centos7。

Real Server 1 IP 为 192.168.222.128，主机名为 devops，网关为 222.129。

Real Server 2 IP 为 192.168.222.130，主机名为 python，网关为 222.129。

为 3 台服务器分别设置好 IP 地址和主机名，并关闭防火墙，如图 6-13 所示。

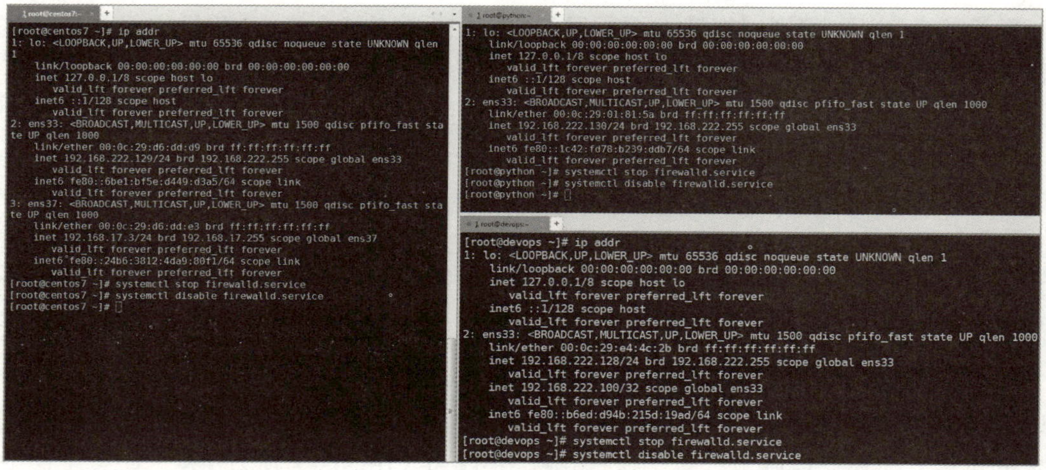

图 6-13

给 3 台机器同时安装 iptables 工具，安装后启动 iptables 服务并添加开机自启动功能，运行命令如下。

```
[root@centos7 ~]# yum install -y iptables-services
[root@python ~]# yum install -y iptables-services
[root@devops ~]# yum install -y iptables-services
[root@centos7 ~]# systemctl start iptables.service
[root@python ~]# systemctl start iptables.service
[root@devops ~]# systemctl start iptables.service
[root@centos7 ~]# systemctl enable iptables.service
[root@python ~]# systemctl enable iptables.service
[root@devops ~]# systemctl enable iptables.service
```

清空 iptables 规则，并保存空规则，运行命令如下。

```
[root@python ~]# iptables -F
[root@python ~]# service iptables save
iptables: Saving firewall rules to /etc/sysconfig/iptables:[  OK  ]
[root@centos7 ~]# iptables -F
[root@centos7 ~]# service iptables save
iptables: Saving firewall rules to /etc/sysconfig/iptables:[  OK  ]
[root@devops ~]# iptables -F
[root@devops ~]# service iptables save
iptables: Saving firewall rules to /etc/sysconfig/iptables:[  OK  ]
```

关闭 Selinux，运行命令如下。

```
[root@centos7 ~]# getenforce
Disabled
[root@python ~]# getenforce
Disabled
[root@devops ~]# getenforce
Disabled
```

设置两台内网服务器的网关为分发器的内网 IP，如图 6-14 所示。

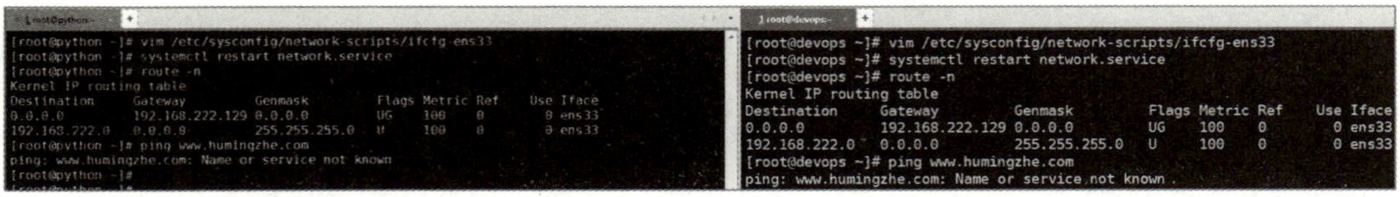

图 6-14

6.7.2　设置分发器

在 Dir 服务器上安装 ipvsadm，ipvsadm 是实现 LVS 功能的重要工具，运行命令如下。

```
[root@centos7 ~]# yum install -y ipvsadm
```

LVS 全部以脚本的形式执行，方便维护。在 Dir 服务器上编写一个脚本，代码如图 6-15 所示。

执行该脚本，运行命令如下。

```
[root@centos7 ~]# sh /usr/local/sbin/lvs_nat.sh
```

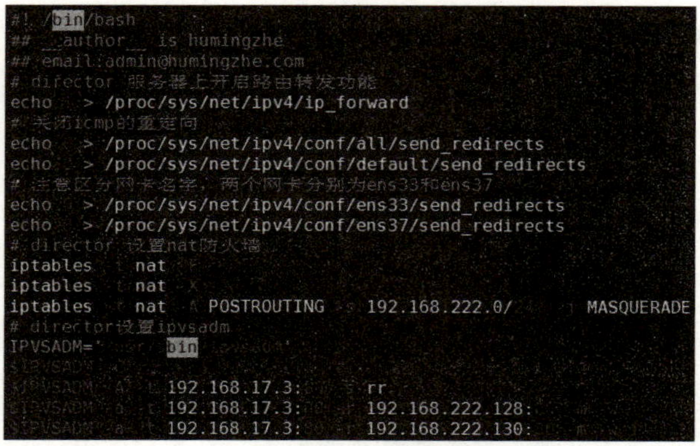

```
#! /bin/bash
## __author__ is humingzhe
## email:admin@humingzhe.com
# director 服务器上开启路由转发功能
echo  > /proc/sys/net/ipv4/ip_forward
# 关闭icmp的重定向
echo  > /proc/sys/net/ipv4/conf/all/send_redirects
echo  > /proc/sys/net/ipv4/conf/default/send_redirects
# 注意区分网卡名字，两个网卡分别为ens33和ens37
echo  > /proc/sys/net/ipv4/conf/ens33/send_redirects
echo  > /proc/sys/net/ipv4/conf/ens37/send_redirects
# director 设置nat防火墙
iptables  nat
iptables  nat
iptables  nat  POSTROUTING  192.168.222.0/  MASQUERADE
# director设置ipvsadm
IPVSADM='  bin
$IPVSADM
$IPVSADM  192.168.17.3:  rr
$IPVSADM  192.168.17.3:  192.168.222.128:
$IPVSADM  192.168.17.3:  192.168.222.130:
```

图 6-15

6.7.3 Real Server 安装 Nginx 服务

安装 Nginx 前要配置 epel 源，运行命令如下。

```
[root@devops ~]# yum install -y epel-release.noarch
[root@devops ~]# yum install -y nginx
[root@devops ~]# systemctl start nginx.service
[root@devops ~]# ps aux |grep nginx
[root@python ~]# yum install -y epel-release.noarch
[root@python ~]# yum install -y nginx
[root@python ~]# systemctl start nginx.service
[root@python ~]# ps aux |grep nginx
```

在两台 Real Server 上分别设置不同的主页内容，方便区分，如 Real Server 1 主页内容为 Real Server 1，Real Server2 主页内容为 Real Server 2，运行命令如下。

```
[root@python ~]# vim /usr/share/nginx/html/index.html
Real Server 1
[root@devops ~]# vim /usr/share/nginx/html/index.html
<html>
    <head>
        <meta charset="UTF-8">
        <title>Real Server 2</title>
    </head>
    <body>
        <h1>Real Server 2</h1>
        <h3>Host:192.168.222.128</h3>
    </body>
</html>
```

使用 curl 命令访问外网 IP，会显示刚刚编辑的 index.html 文件中的内容，而且每次访问显示的结果都不一致，如图 6-16 所示。

```
[root@centos7 ~]# curl 192.168.17.3
Real Server 1
[root@centos7 ~]# curl 192.168.17.3
<html>
    <head>
        <meta charset="UTF-8">
        <title>Real Server 2</title>
    </head>
    <body>
        <h1>Real Server 2</h1>
        <h3>Host:192.168.222.128</h3>
    </body>
</html>

[root@centos7 ~]# curl 192.168.17.3
Real Server 1
```

图 6-16

6.8　DR模式的LVS搭建

网站规模不是很大且服务器在 10 台以内时可以选择用 NAT 模式，因为 NAT 模式节省公网 IP 资源，对于小企业来说，公网 IP 也是需要花钱的，所以用的 IP 少非常占优势。反之，如果几十台、几百台服务器，每一台都设置公网 IP 就会非常浪费 IP 资源，尤其是现在公网 IP 越来越少。实际上还有一种方法可搭建一个内部的 LVS，全部都用内网，VIP 也用内网，用一个公网 IP 做映射即可（公网 IP 的 80 端口映射到内网 VIP 的 80 端口）。

6.8.1　DR 模式的准备工作

依然准备 3 台服务器，3 台服务器中只需要一个网卡、一个 IP，把 192.168.222 网段设置为公网 IP，VIP 为 222.200。更改前面配置的两台 Real Server 网关为原始的 192.168.222.2，运行命令如下。

```
[root@python ~]# vim /etc/sysconfig/network-scripts/ifcfg-ens33
GATEWAY=192.168.222.2
[root@devops ~]# vim /etc/sysconfig/network-scripts/ifcfg-ens33
GATEWAY=192.168.222.2
[root@devops ~]# systemctl restart network.service
[root@python ~]# systemctl restart network.service
```

6.8.2　设置 LVS 的 DR 模式

在 Dir 服务器上创建一个脚本，设置一些规则，代码如下。

```
[root@centos7 ~]# vim /usr/local/sbin/lvs_dr.sh
#! /bin/bash
## __author__ is humingzhe
## Email:admin@humingzhe.com
echo 1 > /proc/sys/net/ipv4/ip_forward
ipv=/usr/sbin/ipvsadm
vip=192.168.222.200
rs1=192.168.222.128
rs2=192.168.222.130
#设置虚拟网卡，网卡名为 ens33:2，IP 为 192.168.222.200
ifdown ens33
ifup ens33
ifconfig ens33:2 $vip broadcast $vip netmask 255.255.255.255 up
#设置网关
route add -host $vip dev ens33:2
$ipv -C
$ipv -A -t $vip:80 -s rr
$ipv -a -t $vip:80 -r $rs1:80 -g -w 1
$ipv -a -t $vip:80 -r $rs2:80 -g -w 1
```

执行 lvs_dr 脚本，使脚本生效，运行命令如下。

```
[root@centos7 ~]# sh /usr/local/sbin/lvs_dr.sh
Device 'ens33' successfully disconnected.
Connection successfully activated (D-Bus active path:
/org/freedesktop/NetworkManager/ActiveConnection/7)
```

两台 Real Server 服务器也需要写脚本，代码如图 6-17 所示。

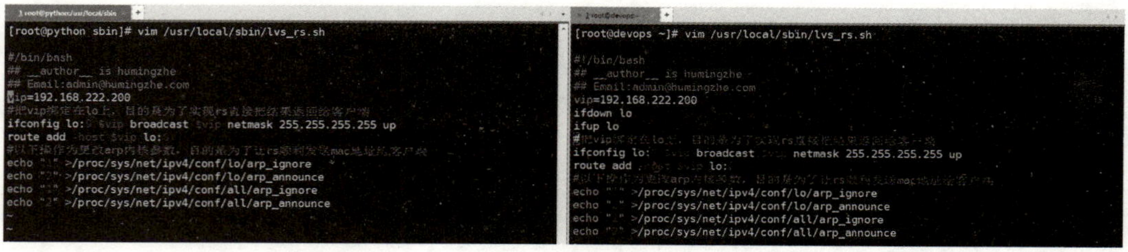

图 6-17

执行 Real Server 服务器中的 lvs_rs 脚本，运行命令如下。

```
[root@python ~]# sh /usr/local/sbin/lvs_rs.sh
[root@devops ~]# sh /usr/local/sbin/lvs_rs.sh
```

使用 route -n 命令可以查看路由表，其中有一个是 VIP 的路由，查看 IP 发现 Dir 服务器中 ens33 网卡上多了 222.200，Real Server 中的两个 lo 上多了 222.200，如图 6-18 所示。

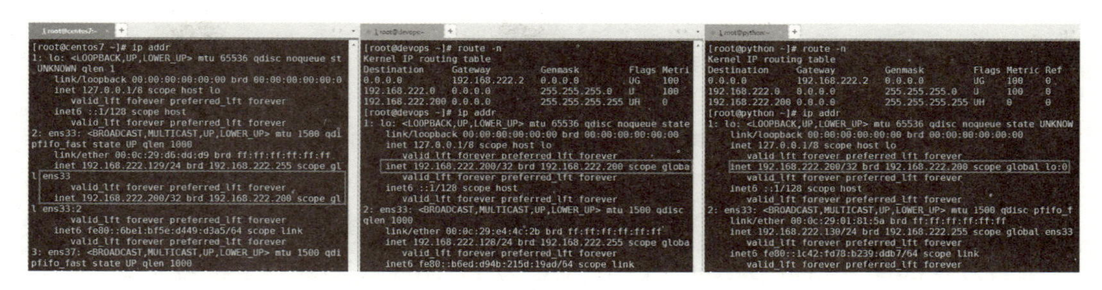

图 6-18

6.8.3　测试 LVS 的 DR 模式

在浏览器的地址栏中输入 VIP 地址 192.168.222.200，会显示和输入内容不同的内容，如图 6-19 所示。

图 6-19

6.9　Keepalived+LVS

把 Keepalived 加入到 LVS 中的原因有以下两个。

- LVS 拥有一个很关键的角色 Dir（分发器），如果分发器宕掉，所有的服务和访问都会被中断。因为入口全部在 Dir 上，所以需要把分发器做高可用，Keepalived 可以实现高可用，放在这个位置最合适不过，并且它具有负载均衡的功能。
- 在使用 LVS 时，如果没有其他额外的操作，把其中一个 Real Server 关机，再去访问是会出问题的，LVS 并不聪明，即使后端的一个服务器宕机，它也会把请求转发过去。Keepalived 的出现就是为了解决这个问题，即使有一台 Real Server 宕机了，它也能够正常提供服务，当请求分发过来时它能够自动检测到后端的 Real Server 宕机，就不会再把请求继续转发到有问题的 Real Server 上。

完整架构需要两台服务器（角色为 Dir），分别安装 Keepalived 软件，目的是实现高可用，但 Keepalived 本身也有负载均衡的功能，所以只安装一台 Keepalived 也是没问题的。

Keepalived 内置了 ipvsadm 功能，所以不需要再安装 ipvsadm 包，也不用编写和执行 lvs_dir 脚本。

准备 3 台机器，分发器（Dir）安装 Keepalived 服务，IP 地址为 192.168.222.129，主机名为 centos7。Real Server 1 的 IP 地址为 192.168.222.128，主机名为 devops。Real Server 2 的 IP 地址为 192.168.222.130，主机名为 python。VIP 地址为 192.168.222.200。

编辑 Keepalived 配置文件，删除原有内容，添加新内容，代码如下。

```
[root@centos7 ~]# vim /etc/keepalived/keepalived.conf
vrrp_instance VI_1 {
    #备用服务器上为 backup
    state MASTER
    #绑定 VIP 的网卡为 ens33
    interface ens33
    virtual_router_id 51
    #备用服务器上为 90
    priority 100
    advert_int 1
    authentication {
        auth_type PASS
        auth_pass humingzhe
    }
    virtual_ipaddress {
        192.168.222.200
    }
}
virtual_server 192.168.222.200 80 {
    #(每隔 10 秒查询 Real Server 状态)
    delay_loop 10
    #(LVS 算法)
    lb_algo wlc
    #(DR 模式)
    lb_kind DR
    #(同一个 IP 的连接 60 秒内被分配到同一台 Real Server)
    persistence_timeout 0
    #(用 TCP 协议检查 Real Server 状态)
    protocol TCP
    real_server 192.168.222.128 80 {
        #(权重)
```

```
        weight 100
        TCP_CHECK {
        #(10 秒无响应超时)
        connect_timeout 10
        nb_get_retry 3
        delay_before_retry 3
        connect_port 80
        }
    }
real_server 192.168.222.130 80 {
        weight 100
        TCP_CHECK {
        connect_timeout 10
        nb_get_retry 3
        delay_before_retry 3
        connect_port 80
        }
    }
}
```

先关闭 Keepalived 服务，这一步操作是为了关闭之前的虚拟 IP，使用 **ip addr** 命令查看没有虚拟 IP 后再启动 Keepalived 服务，并查看 Keepalived 规则，如图 6-20 所示。

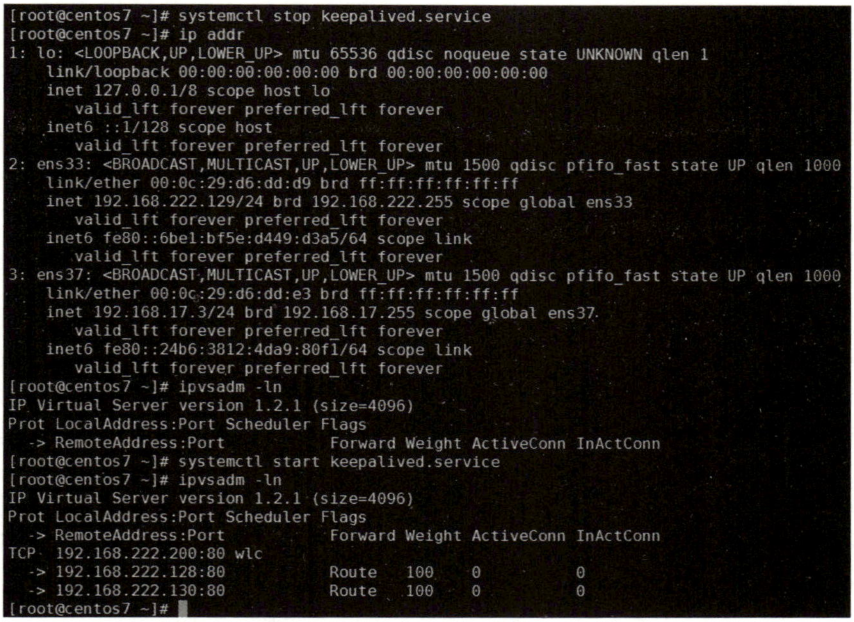

```
[root@centos7 ~]# systemctl stop keepalived.service
[root@centos7 ~]# ip addr
1: lo: <LOOPBACK,UP,LOWER_UP> mtu 65536 qdisc noqueue state UNKNOWN qlen 1
    link/loopback 00:00:00:00:00:00 brd 00:00:00:00:00:00
    inet 127.0.0.1/8 scope host lo
       valid_lft forever preferred_lft forever
    inet6 ::1/128 scope host
       valid_lft forever preferred_lft forever
2: ens33: <BROADCAST,MULTICAST,UP,LOWER_UP> mtu 1500 qdisc pfifo_fast state UP qlen 1000
    link/ether 00:0c:29:d6:dd:d9 brd ff:ff:ff:ff:ff:ff
    inet 192.168.222.129/24 brd 192.168.222.255 scope global ens33
       valid_lft forever preferred_lft forever
    inet6 fe80::6be1:bf5e:d449:d3a5/64 scope link
       valid_lft forever preferred_lft forever
3: ens37: <BROADCAST,MULTICAST,UP,LOWER_UP> mtu 1500 qdisc pfifo_fast state UP qlen 1000
    link/ether 00:0c:29:d6:dd:e3 brd ff:ff:ff:ff:ff:ff
    inet 192.168.17.3/24 brd 192.168.17.255 scope global ens37
       valid_lft forever preferred_lft forever
    inet6 fe80::24b6:3812:4da9:80f1/64 scope link
       valid_lft forever preferred_lft forever
[root@centos7 ~]# ipvsadm -ln
IP Virtual Server version 1.2.1 (size=4096)
Prot LocalAddress:Port Scheduler Flags
  -> RemoteAddress:Port           Forward Weight ActiveConn InActConn
[root@centos7 ~]# systemctl start keepalived.service
[root@centos7 ~]# ipvsadm -ln
IP Virtual Server version 1.2.1 (size=4096)
Prot LocalAddress:Port Scheduler Flags
  -> RemoteAddress:Port           Forward Weight ActiveConn InActConn
TCP  192.168.222.200:80 wlc
  -> 192.168.222.128:80           Route   100    0          0
  -> 192.168.222.130:80           Route   100    0          0
[root@centos7 ~]#
```

图 6-20

注意

两台 Real Server 依然要执行/usr/local/sbin/lvs_rs.sh 脚本。Keepalived 有一个较好的功能，当有一台 Real Server 宕机时，就不会再把请求转发过去。如关闭 192.168.222.130IP 地址的 Nginx 服务，它就会自动把 222.130IP 地址“踢”出去，当开启 Nginx 服务后又会自动加载。

温馨提醒

如果想更好地学习 Linux 集群的相关内容，可以加入读者 QQ 群 99208965 或访问笔者博客 http://humingzhe.com 进行学习与交流。

第7章 Zabbix运维监控

7.1 Linux监控平台简介

监控是一项非常重要的运维工作，尤其对于一些比较重要的业务，如果没有监控，就只能等着用户反馈。

常见的开源监控软件有 Cacti、Nagios、Zabbix、Smokeping 和 Open-falcon 等。Cacti 和 Smokeping 倾向于基础监控，成图非常漂亮。Cacti、Nagios 和 Zabbix 服务端监控中心需要 PHP 环境支持，其中 Zabbix 和 Cacti 需要安装 MySQL 作为存储数据库。Nagios 不用存储历史数据，注重服务或监控项的状态。Zabbix 会获取服务或监控项目的数据，把数据记录到数据库中，可以成图查看。

Open-falcon 是小米公司开发的一款监控软件，开源后受到诸多大公司和运维工程师的追捧，滴滴出行、360 软件、新浪微博和京东等知名互联网公司都在使用此款监控软件。

7.2 Zabbix监控介绍

Zabbix 是 C/S 架构，抓取数据是通过客户端抓取的，在客户端必须有服务启动，该服务负责采集数据，数据会主动上报给服务端，也可让服务端连接客户端去抓取数据。客户端分为两种模式，即主动模式和被动模式。

Zabbix 是基于 C++开发的，所以其性能还不错。监控中心需要 PHP 环境开启 Web 界面，单台 Server 节点在理论上是可以支持上万台客户端的。

Zabbix 目前最新的版本是 3.4，更新速度较快。Zabbix 官方文档网址是

https://www.zabbix.com/manuals。在 Zabbix 架构中包含如下 5 个组件。

- zabbix-server 监控中心：接收客户端上报的信息，负责设置、统计、操作数据。
- 数据存储：Zabbix 数据需要存储到数据库中，如 MySQL、PgSQL 和 MariaDB。
- Web 界面：可以在 Web 界面中操作是 Zabbix 简单易用的主要原因。
- Zabbix-Proxy：可选组件，可以代替 zabbix-server 的功能，减轻服务器的压力。
- Zabbix-agent：客户端软件，负责采集各个监控服务或项目的数据并上报。

Zabbix 监控流程图如图 7-1 所示。想监控一台服务器，首先要在监控中心添加主机，让两者进行通信，然后添加监控项目，设置好监控项目就可以和服务端进行通信了，然后就可以采集数据并存储到数据库中。Zabbix Server 采集过程可以是主动的，也可以是被动的，需要设置报警规则，采集的数据值达到报警线就会报警，否则只进行存储。除设置报警规则外，还需要设置报警机制，检测到数据不正常也需要报警。Zabbix 是通过 E-mail 报警，还是通过短信报警，以及发给谁都需要在服务端的 Server 中心进行设置，而这些设置过程都在 Zabbix 的 Web 界面中进行。

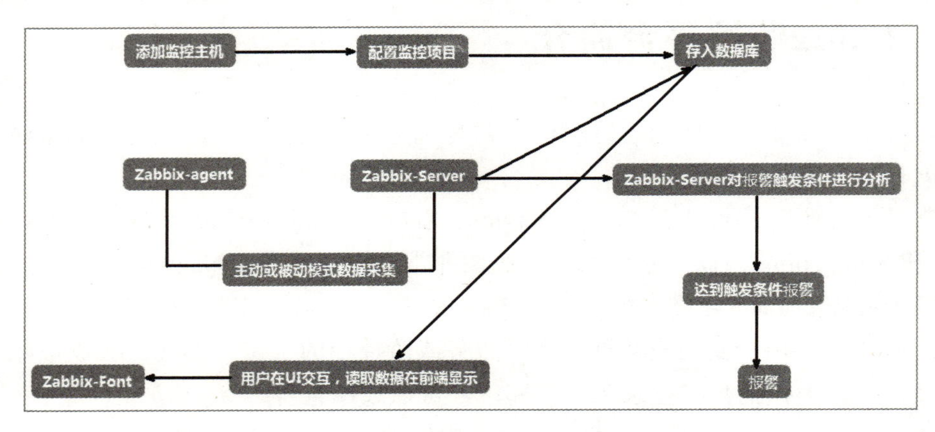

图 7-1

7.3 安装Zabbix监控

7.3.1 安装准备工作

准备两台服务器，主机名 zabbix-1 作为 Server（服务端，监控中心），主机名 zabbix-2 作为 Agent（客户端）。

Zabbix 是可以使用 yum 命令进行安装的，但是需要安装 epel 扩展源，在 epel 扩展源中安装的是 2.2 版本的 Zabbix，推荐使用官方提供的对应版本的 yum 源。官方下载网址是 www.zabbix.com/download，该页面中会提供对应的版本下载方式。本书选择 Centos 7 X86_64 位版本进行演示，选择完毕后页面下方会提供对应的 yum 源下载地址，如图 7-2 所示。

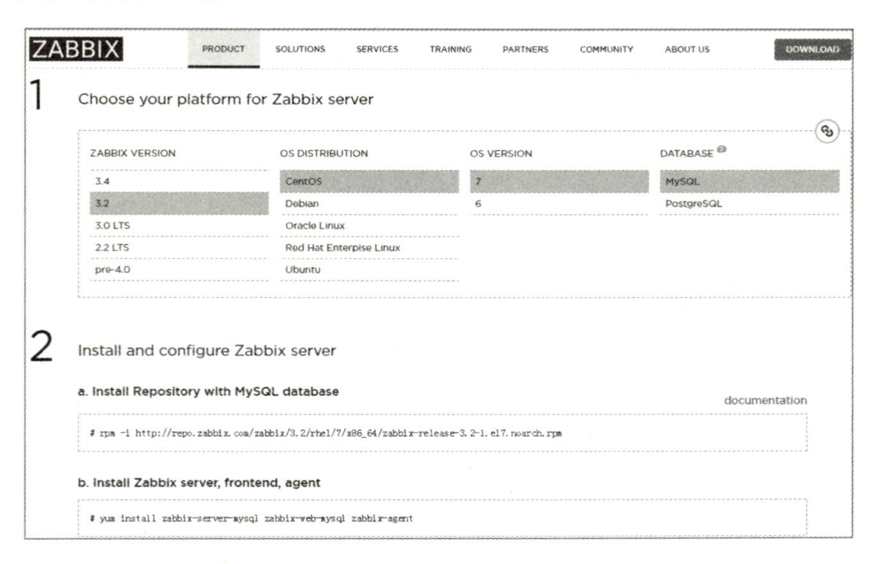

图 7-2

安装自动补全命令工具包，命令如下。

```
[root@zabbix-1 ~]# yum install -y bash-completion-extras.noarch
bash-completion.noarch
```

7.3.2　ntpdate 设置时间同步

要确保时间是同步的，否则发出报警时无法知道具体的报警时间。同步时间的命令是 ntpdate，还可以设置时间同步任务计划，命令如下。

```
[root@centos7 ~]# yum install -y ntpdate
[root@centos7 ~]# echo '* * * * * /usr/sbin/ntpdate ntp1.aliyun.com
&>/dev/null/' >> /var/spool/cron/root
[root@centos7 ~]# crontab -l
* * * * * /usr/sbin/ntpdate ntp1.aliyun.com &>/dev/null/
```

7.3.3　安装 Zabbix 服务器端

安装 Zabbix 的 yum 源，安装后 yum.repos.d 目录下会生成一个 zabbix.repo 文件，命令如下。

```
    [root@zabbix-1 ~]# wget
http://repo.zabbix.com/zabbix/3.2/rhel/7/x86_64/zabbix-release-3.2-1.el7
.noarch.rpm
    [root@zabbix-1 ~]# rpm -ive zabbix-release-3.2-1.el7.noarch.rpm
    [root@zabbix-1 ~]# ls /etc/yum.repos.d/zabbix.repo
    /etc/yum.repos.d/zabbix.repo
```

更改 Zabbix yum 源为阿里云以提高下载速度，命令如下。

```
    [root@zabbix-1 ~]# vim /etc/yum.repos.d/zabbix.repo
    baseurl=http://mirrors.aliyun.com/zabbix/zabbix/3.0/rhel/7/$basearch/
    baseurl=http://mirrors.aliyun.com/zabbix/non-supported/rhel/7/$basear
ch/
```

有了 zabbix.repo 源后就可以使用 yum 命令安装 Zabbix 了，需要安装 zabbix-agent、zabbix-server-mysql、zabbix-get 等一系列的包，默认会把 Apache 和 PHP 服务一起安装，命令如下。

```
    [root@zabbix-1 ~]# yum install -y zabbix-agent zabbix-get
zabbix-server-mysql zabbix-web zabbix-web-mysql
```

安装 MariaDB 数据库并启动，命令如下。

```
    [root@zabbix-1 ~]# yum install -y mariadb-server
    [root@zabbix-1 ~]# systemctl enable mariadb.service
    [root@zabbix-1 ~]# systemctl start mariadb.service
    [root@zabbix-1 ~]# netstat -ltnp
```

执行 MySQL 安全设置向导并设置密码，命令如下。

```
    [root@zabbix-1 ~]# mysql_secure_installation
```

编辑 MariaDB 主配置文件 my.cnf，设置字符集编码为 UTF-8，命令如下。

```
    [root@zabbix-1 ~]# vim /etc/my.cnf
    [mysqld]
    character_set_server = utf8
    [root@zabbix-1 ~]# systemctl restart mariadb.service
```

创建 Zabbix 数据库，设置字符集编码为 UTF-8，并给 Zabbix 数据库授权，命令如下。

```
    [root@zabbix-1 ~]# mysql -uroot -p123456
    MariaDB [(none)]> create database zabbix character set utf8 collate
utf8_bin;
    MariaDB [(none)]> grant all on zabbix.* to zabbix@'127.0.0.1' identified
by '123456';
```

```
MariaDB [(none)]> flush privileges;
MariaDB [(none)]> quit
```

导入初始化数据到 Zabbix 数据库中，导入后进入数据库查看是否导入成功，命令如下。

```
[root@zabbix-1 ~]# zcat
/usr/share/doc/zabbix-server-mysql-3.2.11/create.sql.gz |mysql -uroot
-p123456 zabbix
[root@zabbix-1 ~]# mysql -uroot -p123456
MariaDB [(none)]> use zabbix;
MariaDB [zabbix]> show tables;
```

编辑 httpd.conf 配置文件，将 ServerName 修改为 127.0.0.1:80，命令如下。

```
[root@zabbix-1 ~]# vim /etc/httpd/conf/httpd.conf
ServerName 127.0.0.1:80
```

配置 Zabbix Server，让 Zabbix 服务运行起来，编辑配置文件 zabbix_server.conf，修改 DBHost 和 DBPassword，修改完毕后启动 Zabbix 服务，Zabbix 默认监听 10051 端口，命令如下。

```
[root@zabbix-1 ~]# vim /etc/zabbix/zabbix_server.conf
DBHost=127.0.0.1  #前面数据库定义的哪个 IP，在此就输入哪个 IP。
DBPassword=123456 # 前面设置的什么数据库密码，在此就输入什么密码。
```

启动 zabbix-server 和 httpd 服务，并设置开机自启动，命令如下。

```
[root@zabbix-1 ~]# systemctl start httpd.service
[root@zabbix-1 ~]# systemctl enable httpd.service
[root@zabbix-1 ~]# systemctl enable zabbix-server.service
[root@zabbix-1 ~]# systemctl start zabbix-server.service
[root@zabbix-1 ~]# netstat -lntp
```

7.3.4　Web 界面安装 Zabbix

在浏览器的地址栏中输入 IP 地址“/zabbix”，进入 Zabbix 安装向导页面，如图 7-3 所示。

单击 Next step 按钮，提示 Time zone for PHP is not set (configuration parameter "date.timezone")，意思是需要修改时区，在 php.ini 配置文件中修改时区为“亚洲/上海”，修改后重启 httpd 服务，命令如下。

```
[root@zabbix-1 ~]# vim /etc/php.ini
```

```
date.timezone = Asia/Shanghai
[root@zabbix-1 ~]# systemctl restart httpd.service
```

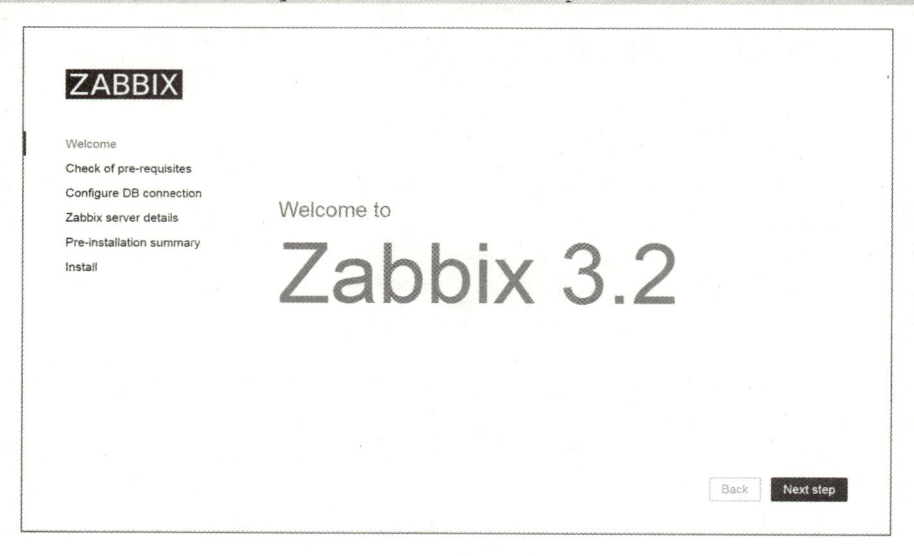

图 7-3

修改后按 F5 键刷新页面，timezone 就会显示 OK 标识，单击 Next step 按钮打开数据库设置界面，将 Database host 修改为 127.0.0.1，Database port 默认为 0 表示 3306 端口，其他的选项正常设置即可，如图 7-4 所示。

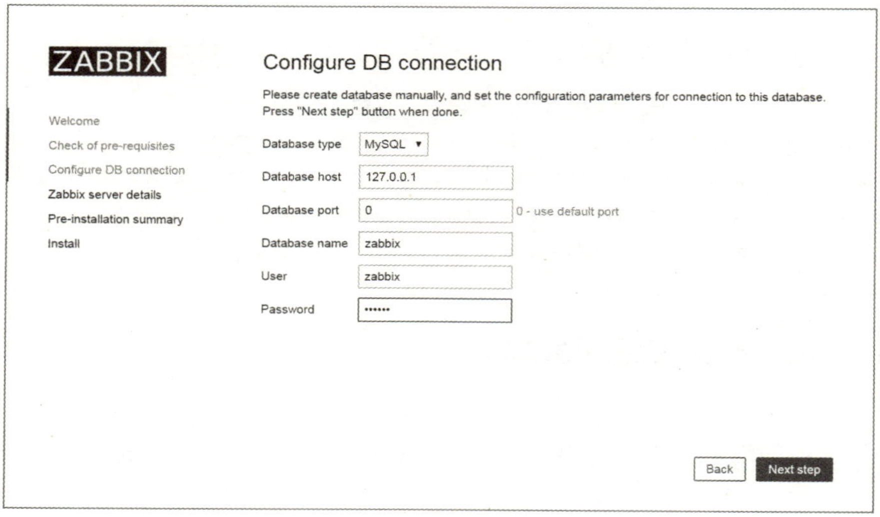

图 7-4

设置完成后单击 Next step 按钮会显示登录页面，输入用户名和密码（官方默认用户名是 Admin，密码是 zabbix），登录成功后即可看到如图 7-5 所示的页面。

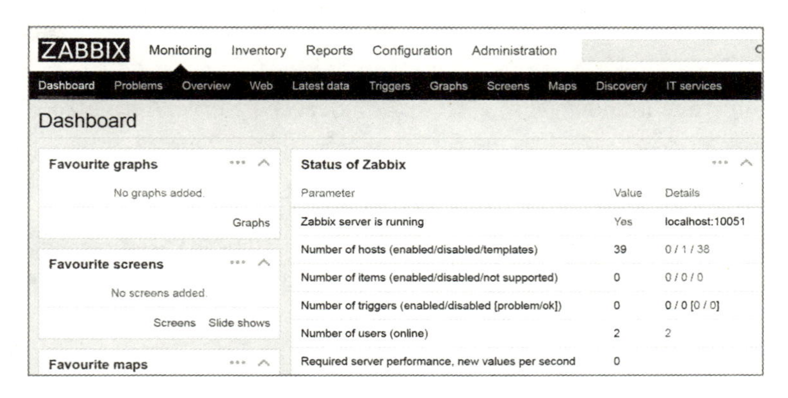

图 7-5

7.3.5　修改 Admin 管理员密码

Zabbix 的初始密码非常不安全，需要更改 Admin 管理员密码。如果在线上监控中心没有更改管理员密码，就很容易被别人拿到权限，进行一些恶意操作。依次选择 Administration→Users→Admin→Password，单击 Change password 按钮即可修改密码。界面下方的 Language 选项用于修改语言，默认是英文，可以更改成中文（Chinese (zh_CN)），然后单击 Update（更新）按钮。

7.3.6　重置 Admin 管理员密码

许久不登录 Zabbix，管理员账号的密码可能会忘记，这时可以进入 MySQL 数据库命令行界面，选择 zabbix 库（密码存储在 users 表中），重置 Admin 管理员密码，命令如下。

```
    [root@zabbix-1 ~]# mysql -uroot -p123456
    MariaDB [(none)]> use zabbix;
    MariaDB [zabbix]> update users set passwd=md5('newpasswd') where 
alias='Admin';
```

7.3.7　安装 Zabbix 客户端

安装 Zabbix 的 yum 源和 zabbix-agent 客户端服务，命令如下。

```
    [root@zabbix-2 ~]# wget http://repo.zabbix.com/zabbix/3.2/rhel/7/
x86_64/zabbix-release-3.2-1.el7.noarch.rpm
    [root@zabbix-2 ~]# rpm -ivh zabbix-release-3.2-1.el7.noarch.rpm
```

```
[root@zabbix-2 ~]# yum install -y zabbix-agent.x86_64
```

编辑 zabbix_agentd.conf 配置文件，修改 Server、ServerActive 和 Hostname。其中，Server 是指定白名单 IP，如果不指定白名单 IP，则所有人都可以访问，所以需要在客户端指定服务端的 IP。ServerActive 涉及主动模式和被动模式，客户端有可能用主动模式，也有可能用被动模式，如果只写上面的 Server 表示只能用被动模式，等着服务端过来采集数据。如果客户端想主动上报服务端，需要设置 ServerActive，命令如下。

```
[root@zabbix-2 ~]# vim  /etc/zabbix/zabbix_agentd.conf
Server=192.168.222.132
ServerActiv=192.168.222.132
Hostname=zabbix-2
```

启动 zabbix-agent 客户端服务，zabbix-agent 默认监听 10050 端口，而 zabbix-server 默认监听 10051 端口，如图 7-6 所示。

```
[root@zabbix-2 zabbix]# systemctl restart zabbix-agent.service
[root@zabbix-2 zabbix]# netstat -lntp
Active Internet connections (only servers)
Proto Recv-Q Send-Q Local Address           Foreign Address         State       PID/Program name
tcp        0      0 0.0.0.0:22              0.0.0.0:*               LISTEN      752/sshd
tcp        0      0 127.0.0.1:25            0.0.0.0:*               LISTEN      834/master
tcp        0      0 0.0.0.0:10050           0.0.0.0:*               LISTEN      13374/zabbix_agentd
tcp6       0      0 :::22                   :::*                    LISTEN      752/sshd
tcp6       0      0 ::1:25                  :::*                    LISTEN      834/master
tcp6       0      0 :::10050                :::*                    LISTEN      13374/zabbix_agentd
[root@zabbix-2 zabbix]# systemctl enable zabbix-agent.service
Created symlink from /etc/systemd/system/multi-user.target.wants/zabbix-agent.service to /usr/lib/systemd/system/zabbix-agent.service.
[root@zabbix-2 zabbix]# ps aux |grep zabbix
zabbix    13374  0.0  0.1  80664  1260 ?        S    20:17   0:00 /usr/sbin/zabbix_agentd -c /etc/zabbix/zabbix_agentd.conf
zabbix    13375  0.0  0.1  80664  1340 ?        S    20:17   0:00 /usr/sbin/zabbix_agentd: collector [idle 1 sec]
zabbix    13376  0.0  0.1  80664  1840 ?        S    20:17   0:00 /usr/sbin/zabbix_agentd: listener #1 [waiting for connection]
zabbix    13377  0.0  0.1  80664  1840 ?        S    20:17   0:00 /usr/sbin/zabbix_agentd: listener #2 [waiting for connection]
zabbix    13378  0.0  0.1  80664  1840 ?        S    20:17   0:00 /usr/sbin/zabbix_agentd: listener #3 [waiting for connection]
zabbix    13379  0.0  0.2  80796  2200 ?        S    20:17   0:00 /usr/sbin/zabbix_agentd: active checks #1 [idle 1 sec]
root      13440  0.0  0.0 112660   968 pts/1    R+   20:22   0:00 grep --color=auto zabbix
[root@zabbix-2 zabbix]#
```

图 7-6

7.4 添加监控主机

Configuration 菜单项下面有很多二级菜单，其中，主机群组是用来给机器建主机组的，创建主机时可选择组来创建。模板是监控项目的集合，可预设监控 CPU、内存、磁盘等，把这些监控的项目集合在一起就可以组成一个模板，后面添加第一台主机用模板 1，添加第二台主机用模板 2。

7.4.1 Web 界面添加 Host 主机

在客户端已经安装了 zabbix-agent 服务并启动了服务，现在创建一个新的监控主机。依次选择 Configuration→Hosts，单击 Create host 按钮，即可进入添加主机的界面，如图 7-7 所示。

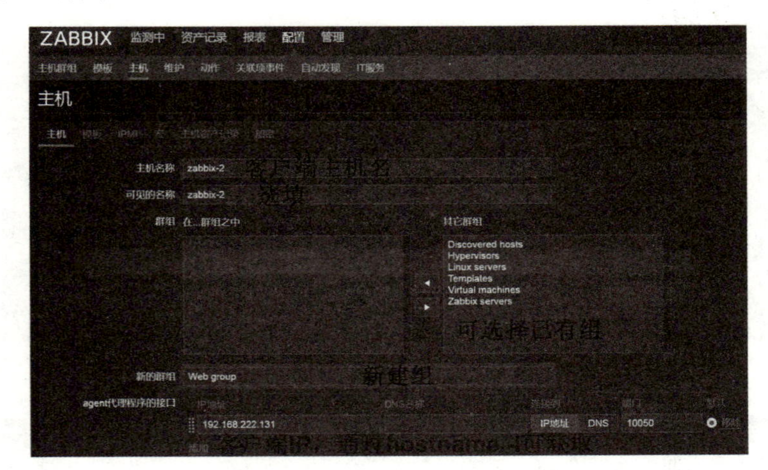

图 7-7

单击 Templates（模板）子菜单，在 Link new template 文本框中搜索 Linux，选择 Template OS Linux 模板，如图 7-8 所示。单击 Add 按钮后，如果 Template 处没有加载模板，则说明没有链接成功，需要再链接一次。

图 7-8

添加主机成功后，状态为已启用状态，Enabled 后面的 ZBX 按钮为绿色表示处于监控状态。一台主机中有应用级（Applications）、监控项（Items）、触发器（Triggers）、图形（Graphs）、自动发现（Discovery）和 Web 监测（Web Interface）六项，如图 7-9 所示。

定位至新主机 Items 处，可以查看默认模板有哪些 Items，如图 7-10 所示。

Items 中的 Key 可以通过命令行的方式进行取值，它的类型是 Zabbix Agent，使用 zabbix_get 命令只能得到 Zabbix Agent 类型的监控项。使用 agent.ping 命令可以判断这台机器是否可以正常通信，带有[]的 Key 表示只定义了一个 Key，[]中是传的参数，逗号表示前面第一个参数为空，多个参数以逗号分隔，这些 Key 都属于内置 Key，使用 zabbix_agentd –p 命令可查看所有的内置 Key，命令如下。

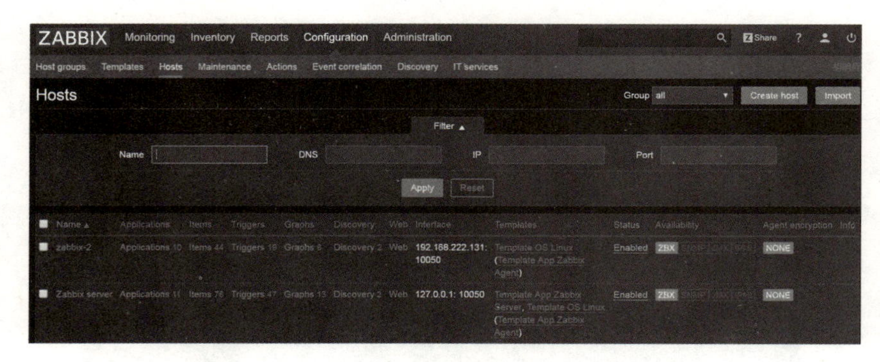

图 7-9

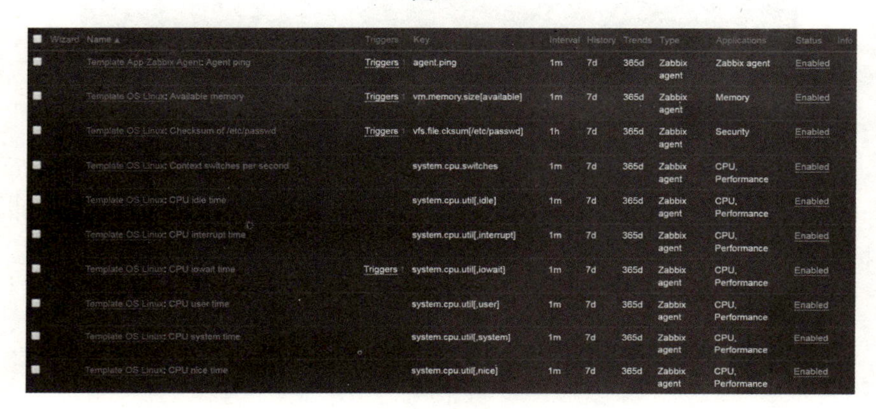

图 7-10

```
    [root@zabbix-1 ~]# zabbix_get -s 192.168.222.131 -k agent.ping
    1
    [root@zabbix-1 ~]# zabbix_get -s 192.168.222.131 -k
system.cpu.util[,system]
    0.300903
    [root@zabbix-1 ~]# zabbix_get -s 192.168.222.131 -k
system.cpu.util[,iowait]
    0.000000
    [root@zabbix-1 ~]# zabbix_agentd -p
```

7.4.2 解决 Zabbix 页面乱码

查看自定义的监听 80 端口图形时下方显示乱码，安装插件解决乱码问题或修改 PHP 源码文件都是非常麻烦的事情。其实，Android、iOS 、Linux 和 Windows 等操作系统的字体都是.ttf 格式的文件，Windows 字体路径是 C:\Windows\Fonts，可以复制当前 Windows 系统中的字体到 Linux 系统中，如复制方正书宋简体。

在/etc/httpd/conf.d/zabbix.conf 配置文件中可以找到 Zabbix 的 Web 安装路径在/usr/share/zabbix/目录下，该目录下有一个 fonts 目录，进入该目录，使用 ls 命令列出字体文件，将本地 Windows 系统中 C:\Windows\Fonts 路径下的字体上传至当前目录，并重命名为原字体的名字，命令如下。

```
[root@zabbix-1 ~]# cat /etc/httpd/conf.d/zabbix.conf
[root@zabbix-1 ~]# cd /usr/share/zabbix/
[root@zabbix-1 zabbix]# cd fonts/
[root@zabbix-1 fonts]# ll -l graphfont.ttf
lrwxrwxrwx 1 root root 33 Feb  5 17:20 graphfont.ttf ->
/etc/alternatives/zabbix-web-font
[root@zabbix-1 fonts]# rz -E
[root@zabbix-1 fonts]# ls
graphfont.ttf  msyh.ttf
[root@zabbix-1 fonts]# \mv msyh.ttf graphfont.ttf
[root@zabbix-1 fonts]# ll graphfont.ttf
-rw-r--r-- 1 root root 21767952 Jun 11  2009 graphfont.ttf
```

替换字体文件后刷新浏览器页面，乱码问题已经解决了，如图 7-11 所示。

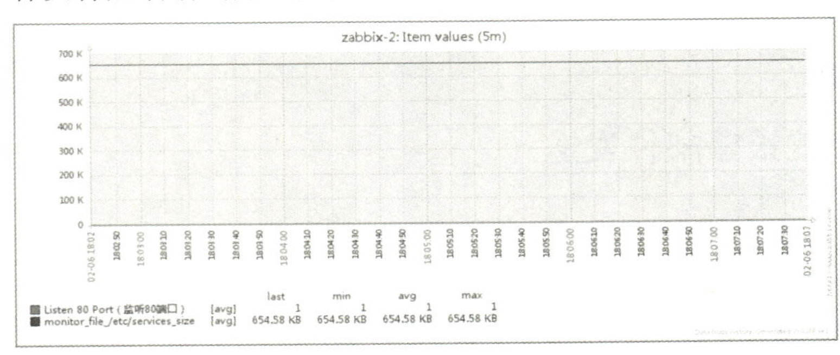

图 7-11

7.5　使用SMTP发送邮件报警及定制报警

7.5.1　添加 Triggers 触发器

在 Zabbix Web 界面中依次单击 Configuration→Hosts→Triggers，即可添加触发器。Name 可自定义，如 HTTP Port Down；可设置报警级别（设置报警级别的好处是，可以根据报警级别发送给不同的人，比如领导只接收高级（High）报警，其他级别的员工接收高级以下的报警），此处选择 Average（一般严重）；单击 Expression

文本框右侧的 Add 按钮可添加表达式，在打开的对话框（如图 7-12 所示）的 Function 下拉列表框中选择最新的 T 值，如果 T 值不是 1，就报警，表达式输入完毕后单击对话框下方的 Insert 按钮完成添加，返回 Zabbix Web 界面，如图 7-13 所示。

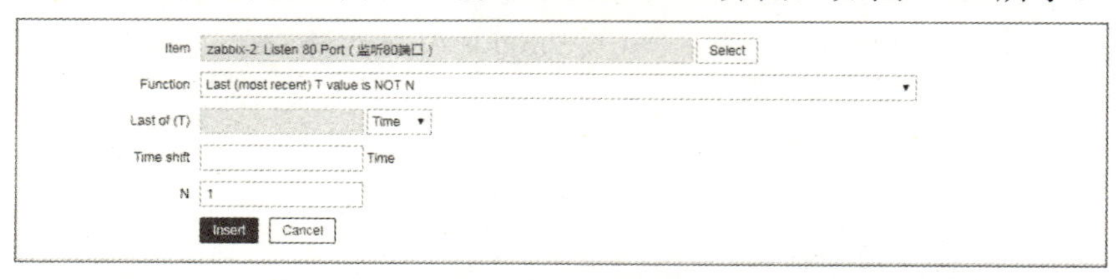

图 7-12

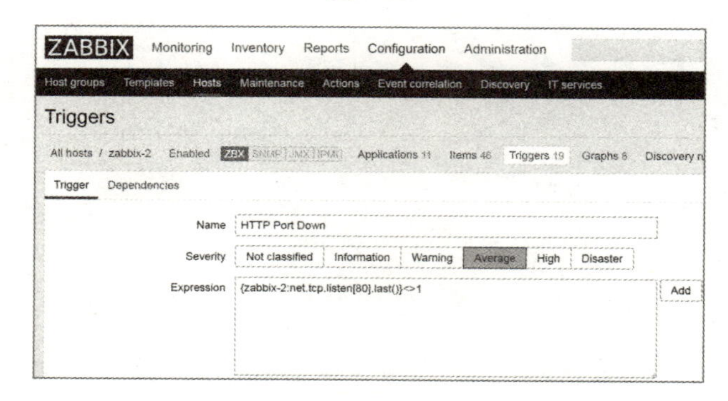

图 7-13

7.5.2 设置报警邮件发送

触发器添加成功后，还需要定义发送动作，如发邮件或执行远程命令。依次选择 Configuration→Actions，该菜单下面有一个 Action 选项，但这个 Action 默认是关闭状态的，需要单击 Disabled 按钮启动，如图 7-14 所示。启动后单击 Name 下方的 Report problems to Zabbix administrators 链接进入设置页面，选择 Recovery operations 选项卡，单击 Operations 选项后面的 Edit 按钮，并选择发送 E-mail，完成后单击 Update 按钮，如图 7-15 所示。

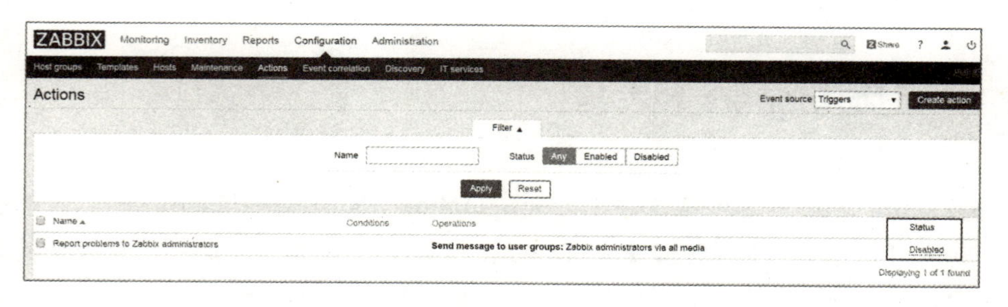

图 7-14

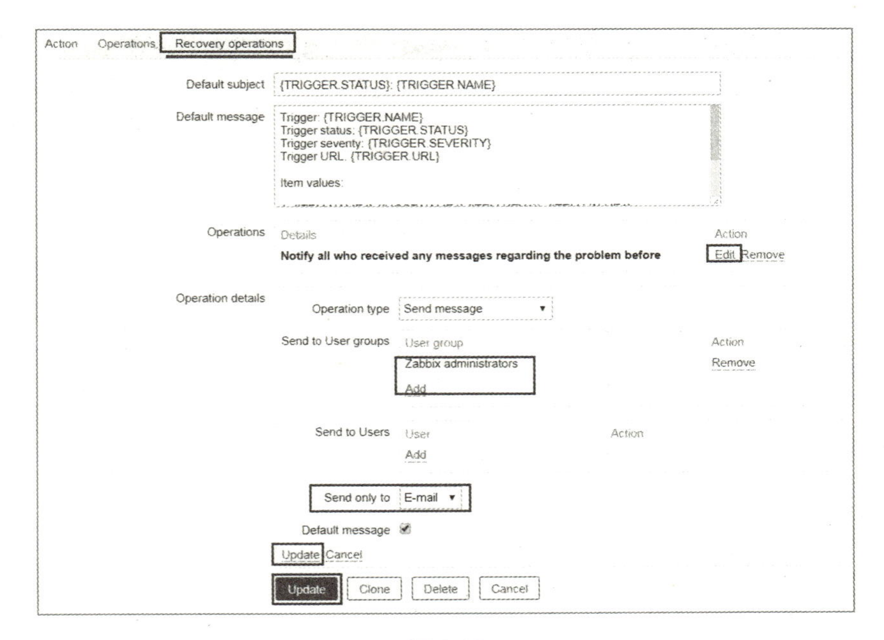

图 7-15

上述操作只定义了发送 E-mail 功能，没有定义使用哪个 E-mail 去发送。在界面中依次选择 Administration→Media types，在打开的界面中可定义发件人，单击 E-mail 选项可以进行设置，如图 7-16 所示。

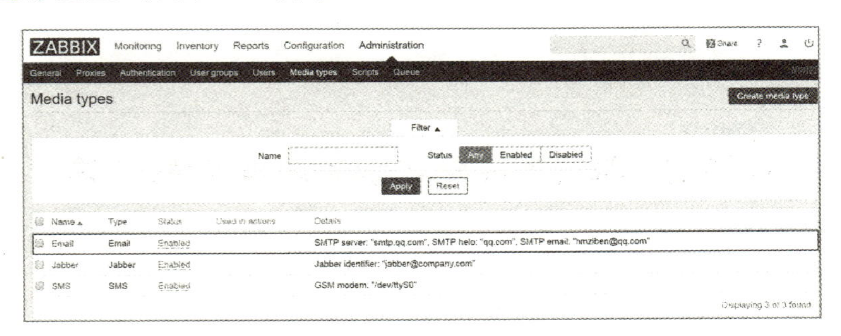

图 7-16

在 Password 文本框中填写 SMTP 授权码，设置内容如图 7-17 所示。授权码在 email.qq.com 账户设置界面中生成，如图 7-18 所示。

7.5.3　设置报警邮件接收

在 Zabbix Web 界面依次单击 Administration→Users，选择 Admin 用户进入设置页面。选择 Media 选项卡，单击 Add 选项进行设置，如图 7-19 所示。

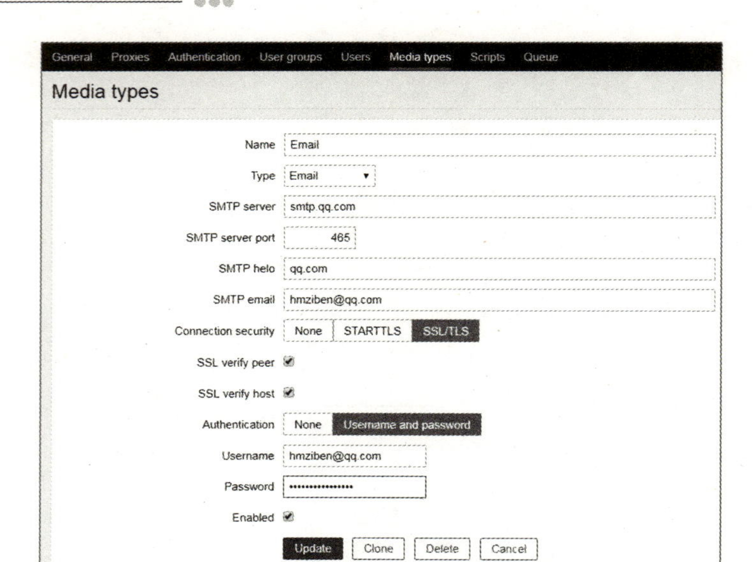

图 7-17

POP3/IMAP/SMTP/Exchange/CardDAV/CalDAV服务

开启服务： POP3/SMTP服务（如何使用 Foxmail 等软件收发邮件？） 已开启 | 关闭

IMAP/SMTP服务（什么是 IMAP，它又是如何设置？） 已关闭 | 开启

Exchange服务（什么是Exchange，它又是如何设置？） 已关闭 | 开启

CardDAV/CalDAV服务（什么是CardDAV/CalDAV，它又是如何设置？） 已关闭 | 开启

（POP3/IMAP/SMTP/CardDAV/CalDAV服务均支持SSL连接。如何设置？）

温馨提示：登录第三方客户端时，密码框请输入"授权码"进行验证。生成授权码

图 7-18

ZABBIX Monitoring Inventory Reports Configuration Administration

General Proxies Authentication User groups Users Media types Scripts Queue

Users

User Media Permissions

Media Type Send to When active Use if severity Status Action

Add 第一步

Update Delete Cancel

第三步

Media - Google Chrome

不安全 | 192.168.222.132/zabbix/popup_media.php?dstfrm=userForm

Media

Type Email

Send to admin@humingzhe.com

When active 1-7,00:00-24:00

Use if severity Not classified

Information

Warning

Average

High

Disaster

Enabled

第二步

Add Cancel

图 7-19

7.5.4　添加报警动作

关闭 Apache 服务，添加报警动作，命令如下。

```
[root@zabbix-2 ~]# systemctl stop httpd.service
```

在 Zabbix Web 界面依次单击 Monitoring→Dashboard，在 Last 20 issues 下面会显示 HTTP Port Done，如图 7-20 所示。邮件随之也接收到了，如图 7-21 所示。

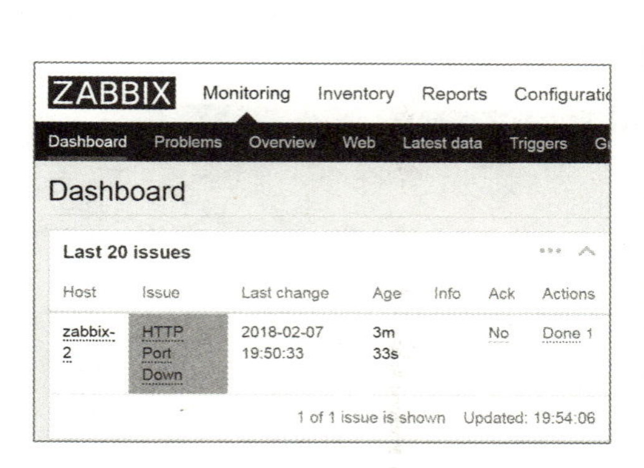

图 7-20

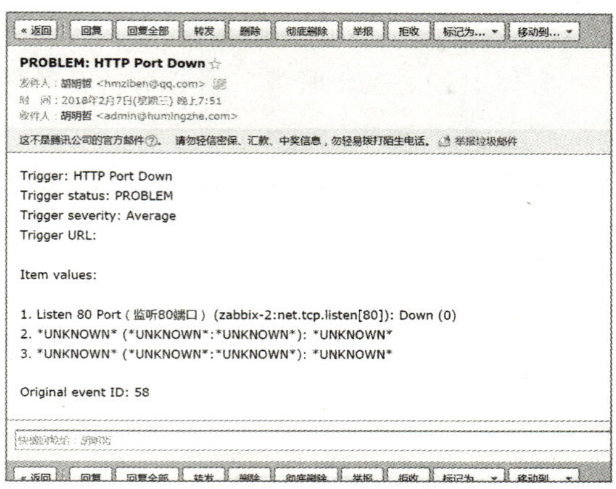

图 7-21

接收到的告警邮件中字数过多，如果发送短信告警的话，字数会超限，可在 Zabbix Web 界面中依次单击 Configuration→Actions→Report problems to Zabbix administrators，在 Operations 选项卡中进行自定义，如图 7-22 所示。

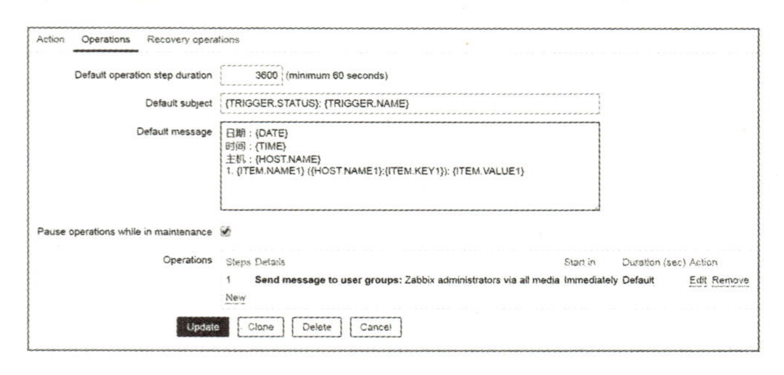

图 7-22

7.5.5　设置邮件报警铃声

设置报警铃声，可以单击管理员 Logo，再选择 Messaging 选项卡，启动 Frontend messaging 后单击 Update 按钮，如图 7-23 所示。

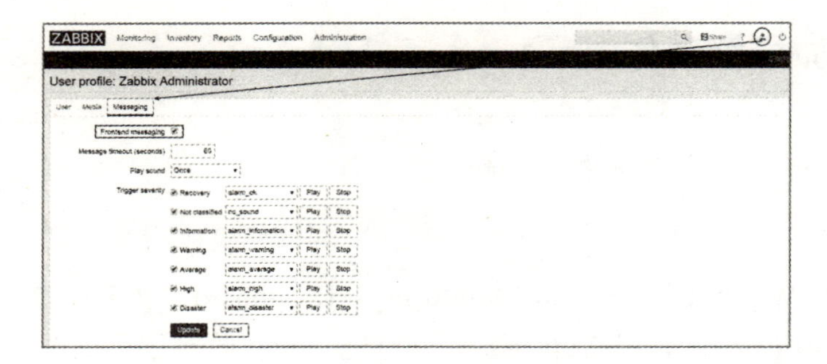

图 7-23

宕掉 HTTP Port 服务，耐心等待几秒钟即可听到报警铃声，如图 7-24 所示。

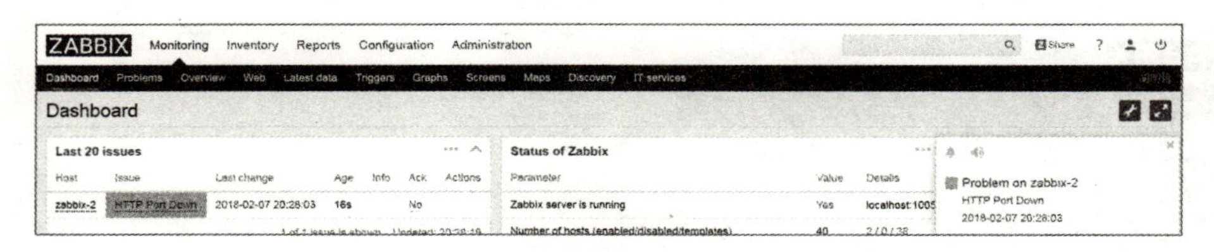

图 7-24

7.5.6 设置微信报警

企业版微信注册网址是 https://work.weixin.qq.com/。注册企业微信界面如图 7-25 所示。

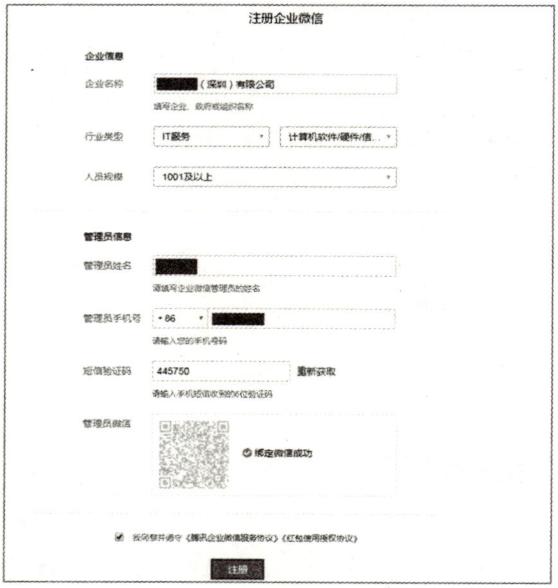

注册企业微信

企业信息

企业名称 （深圳）有限公司

填写企业、政府或组织名称

行业类型 IT服务 计算机软件/硬件/信...

人员规模 1001及以上

管理员信息

管理员姓名

请填写企业微信管理员的姓名

管理员手机号 +86

请输入您的手机号码

短信验证码 445750 重新获取

请输入手机短信收到的6位验证码

管理员微信 绑定微信成功

我同意并遵守《腾讯企业微信服务协议》《红包使用授权协议》

注册

图 7-25

注册完成后进入企业微信后台管理系统，单击“我的企业”选项卡，在“企业信息”设置界面的下方会显示 CorpID，如图 7-26 所示。我们需要记录 CorpID，以便后续操作。

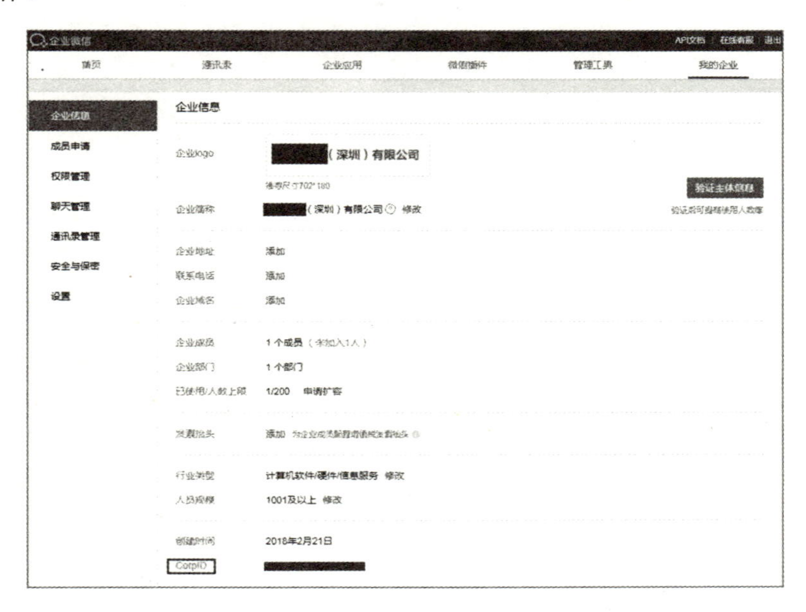

图 7-26

单击“通讯录”选项卡，添加组或成员。添加成员时需要记录账号（ID），添加后单击“微信邀请”按钮进行邀请，邀请时需扫描二维码，添加成员结果如图 7-27 所示。

图 7-27

单击“企业应用”选项卡，再单击“创建应用”按钮，创建的应用如图 7-28 所示。创建完成后需记录 AgentId 和 Secret 的值，如图 7-29 所示。

图 7-28

图 7-29

企业微信注册成功并添加成员后，需要编辑 zabbix_server.conf 配置文件，指定 AlertScriptsPath 的路径，然后在 Web 端就可以获取脚本了，命令如下。

```
[root@zabbix-1 ~]# vim /etc/zabbix/zabbix_server.conf
AlertScriptsPath=/usr/lib/zabbix/alertscripts
```

安装 simplejson 并设置 Python 脚本，命令如下。

```
[root@zabbix-1 ~]# wget https://pypi.python.org/packages/f0/07/
26b519e6ebb03c2a74989f7571e6ae6b82e9d7d81b8de6fcdbfc643c7b58/simplejson-
3.8.2.tar.gz
[root@zabbix-1 ~]# tar xf simplejson-3.8.2.tar.gz
[root@zabbix-1 ~]# cd simplejson-3.8.2/
[root@zabbix-1 simplejson-3.8.2]# python setup.py build
[root@zabbix-1 simplejson-3.8.2]# python setup.py install
```

下载微信告警脚本，下载后复制到/usr/lib/zabbix/alertscripts 目录下，并赋予 x 执行权限，更改属主和属组为 zabbix，命令如下。下载后要修改 Python 脚本中的代码，代码示例文件在本书读者 QQ 群文件中，文件名是 wechat.py。

```
[root@zabbix-1 ~]# git clone https://github.com/X-Mars/
Zabbix-Alert-WeChat.git
[root@zabbix-1 ~]# cp Zabbix-Alert-WeChat/wechat.py
/usr/lib/zabbix/alertscripts/
[root@zabbix-1 ~]# cd /usr/lib/zabbix/alertscripts/
[root@zabbix-1 alertscripts]# chmod +x wechat.py && chown zabbix:zabbix
wechat.py
```

执行 wechat 脚本进行测试，命令如下。如果设置成功，在微信中会接收到告警测试信息，如图 7-30 所示。

```
[root@zabbix-1 alertscripts]# ./wechat.py humingzhe 微信告警测试 微信告警
测试
https://qyapi.weixin.qq.com/cgi-bin/gettoken?corpid=ww28885c7c08116e3
b&corpsecret=N4sVA2bvNZU98jd85TkgqIlavrXjLmJu2Ug_KeCm9z8
{u'invaliduser': u'\u5fae\u4fe1\u544a\u8b66\u6d4b\u8bd5', u'errcode':
0, u'errmsg': u'ok'}
```

图 7-30

登录 Zabbix Web 客户端，依次单击 Administration→Media types，进行微信脚本设置，如图 7-31 所示。

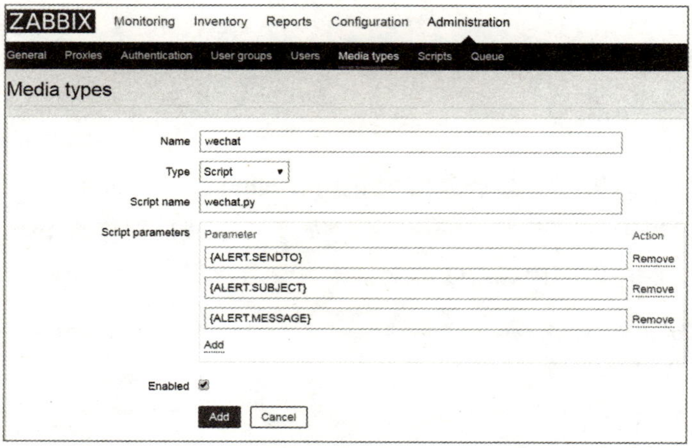

ZABBIX Monitoring Inventory Reports Configuration Administration
General Proxies Authentication User groups Users Media types Scripts Queue
Media types
Name wechat
Type Script
Script name wechat.py
Script parameters Parameter Action
{ALERT.SENDTO} Remove
{ALERT.SUBJECT} Remove
{ALERT.MESSAGE} Remove
Add
Enabled
Add Cancel

图 7-31

依次单击 Administration→Users，分别设置 User 和 Media 选项卡中的内容，如图 7-32 所示。

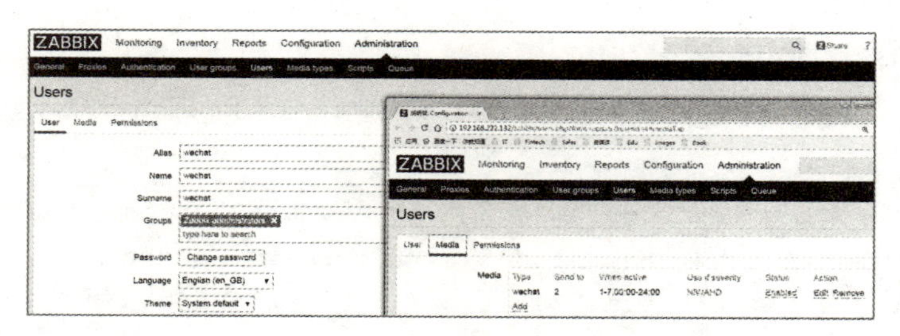

图 7-32

依次单击 Configuration→Actions，进行告警设置，Action 选项卡中的设置内容如图 7-33 所示，Operations 选项卡中的设置内容如图 7-34 所示，Recovery operations 选项卡中的设置内容如图 7-35 所示。

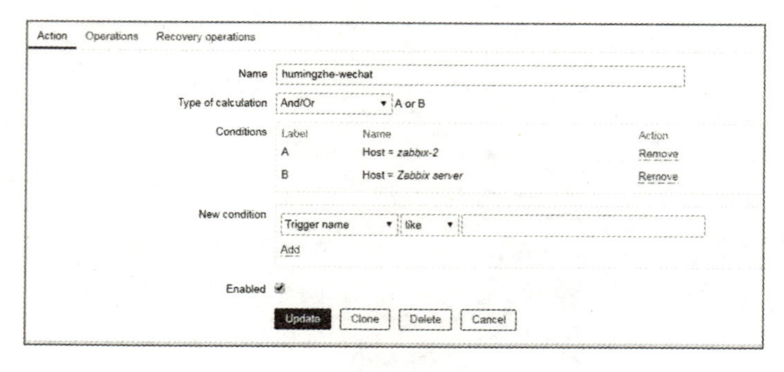

图 7-33

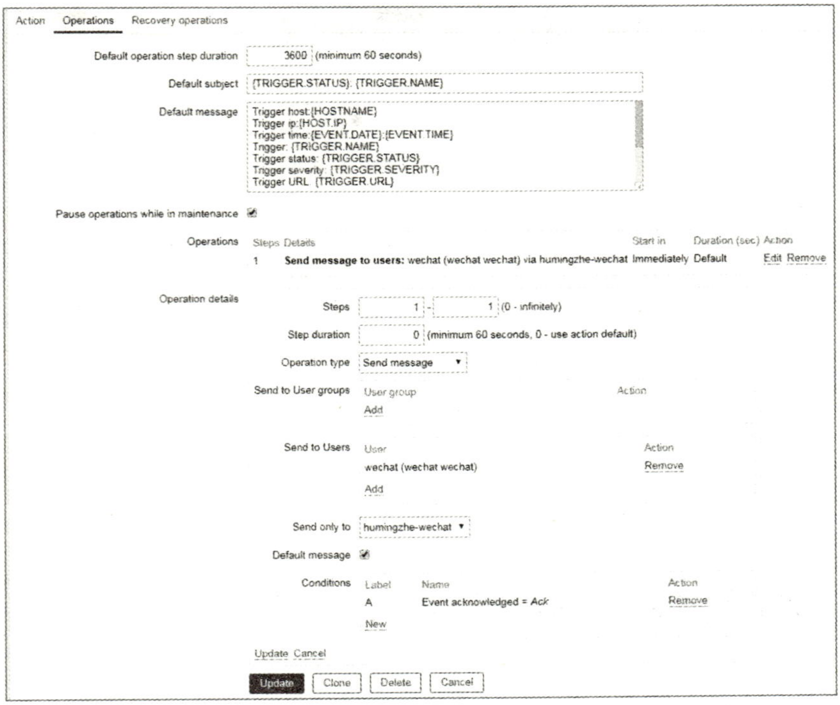

Action　Operations　Recovery operations

Default operation step duration　3600 (minimum 60 seconds)

Default subject　{TRIGGER.STATUS}: {TRIGGER.NAME}

Default message　Trigger host:{HOSTNAME}
Trigger ip:{HOST.IP}
Trigger time:{EVENT.DATE}:{EVENT.TIME}
Trigger: {TRIGGER.NAME}
Trigger status: {TRIGGER.STATUS}
Trigger severity: {TRIGGER.SEVERITY}
Trigger URL: {TRIGGER.URL}

Pause operations while in maintenance ☑

Operations　Steps Details　Start in　Duration (sec)　Action
1　**Send message to users:** wechat (wechat wechat) via humingzhe-wechat　Immediately　Default　Edit Remove

Operation details
Steps　1 - 1 (0 - infinitely)
Step duration　0 (minimum 60 seconds, 0 - use action default)
Operation type　Send message
Send to User groups　User group　Action
Add
Send to Users　User　Action
wechat (wechat wechat)　Remove
Add
Send only to　humingzhe-wechat
Default message　☑
Conditions　Label　Name　Action
A　Event acknowledged = *Ack*　Remove
New
Update Cancel

Update　Clone　Delete　Cancel

图 7-34

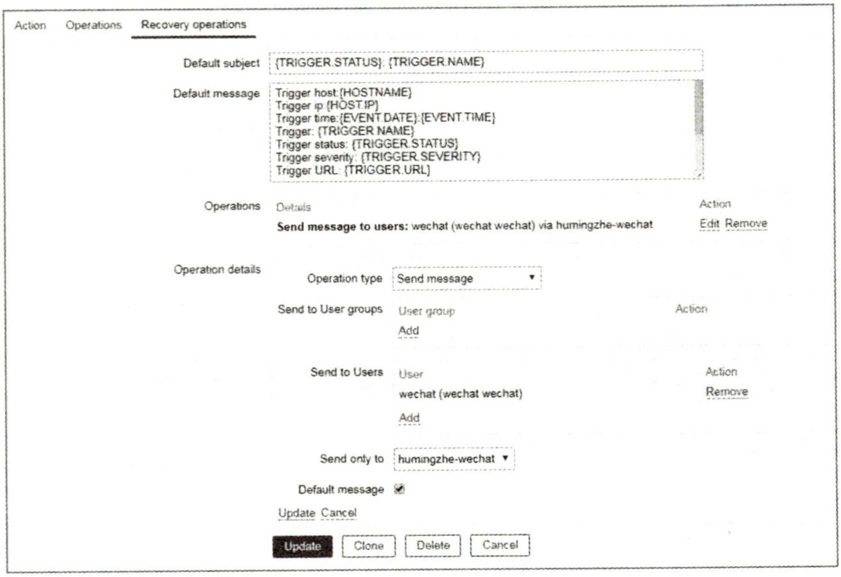

Action　Operations　Recovery operations

Default subject　{TRIGGER.STATUS}: {TRIGGER.NAME}

Default message　Trigger host:{HOSTNAME}
Trigger ip:{HOST.IP}
Trigger time:{EVENT.DATE}:{EVENT.TIME}
Trigger: {TRIGGER.NAME}
Trigger status: {TRIGGER.STATUS}
Trigger severity: {TRIGGER.SEVERITY}
Trigger URL: {TRIGGER.URL}

Operations　Details　Action
Send message to users: wechat (wechat wechat) via humingzhe-wechat　Edit Remove

Operation details
Operation type　Send message
Send to User groups　User group　Action
Add
Send to Users　User　Action
wechat (wechat wechat)　Remove
Add
Send only to　humingzhe-wechat
Default message　☑
Update Cancel

Update　Clone　Delete　Cancel

图 7-35

全部设置完毕后就可以进行测试了，例如，在宕掉 httpd 服务和启动 httpd 服务时，微信都会发送告警消息，如图 7-36 所示。

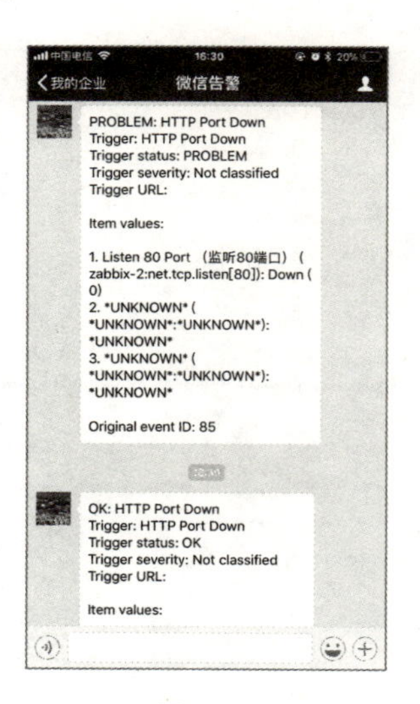

图 7-36

7.6 Web监控和MySQL监控

7.6.1 Web 监控

在 Zabbix Web 界面依次单击 Configuration→Hosts，选择 Web 选项，可以监控 Web 的响应速度，返回状态码和下载速度。如监控 https://www.toutiao.com/i6519259428263297539/页面，选择 Zabbix server 选项卡，因为只有 Zabbix server 这一台计算机目前可以连接外网。

选择 Web 选项后，在页面右上角单击 Create web scenario 按钮进行创建，打开的 Scenario 选项设置如图 7-37 所示。

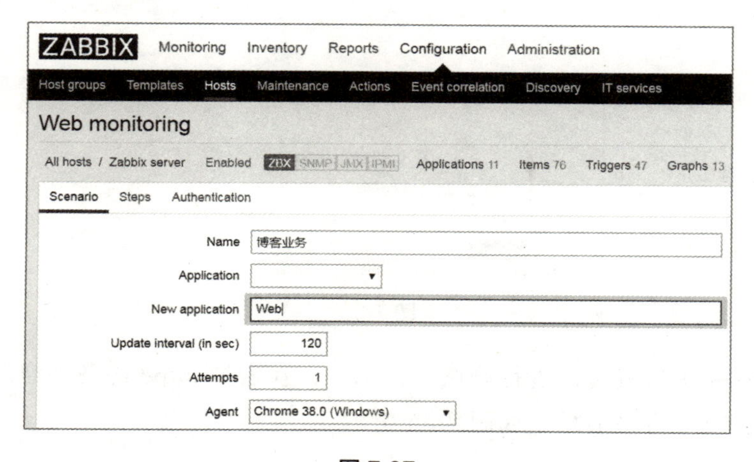

图 7-37

单击 Steps（步骤）选项卡，可以实现登录，登录后再去查看页面上的内容，如有关键字，过滤关键字。如果正常，再去测试退出。登录和退出的验证有一个要求，登录时不能有验证码，因为它无法自动识别验证码。单击 Add 按钮可添加设置内容，如图 7-38 所示。

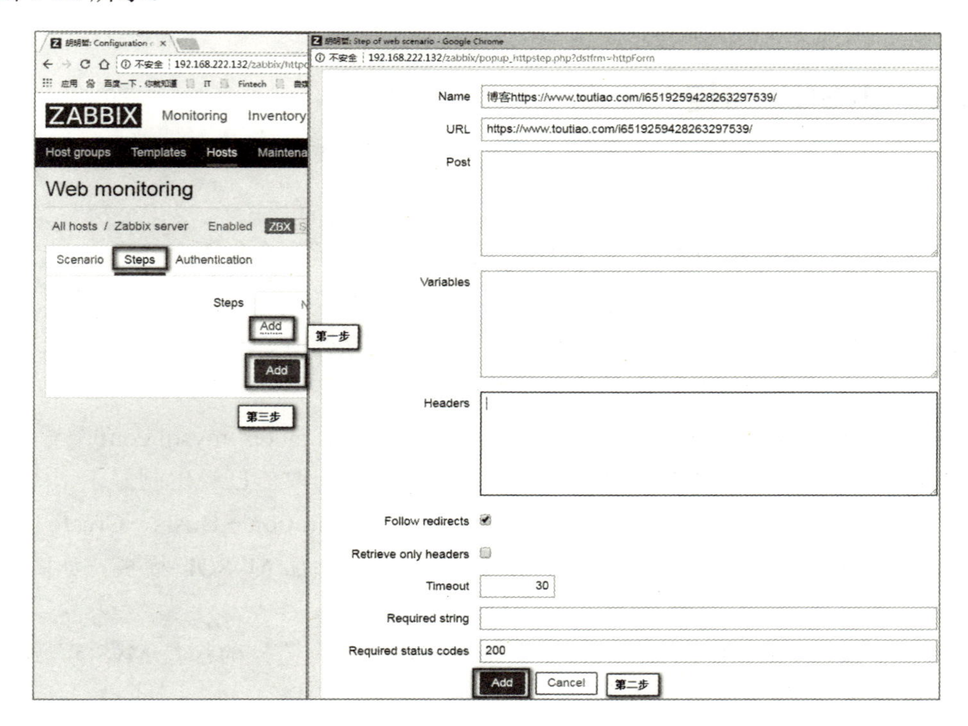

图 7-38

添加成功后在 Monitoring→Web 下方可查看刚刚添加的项目，单击即可查看图形，如图 7-39 所示。

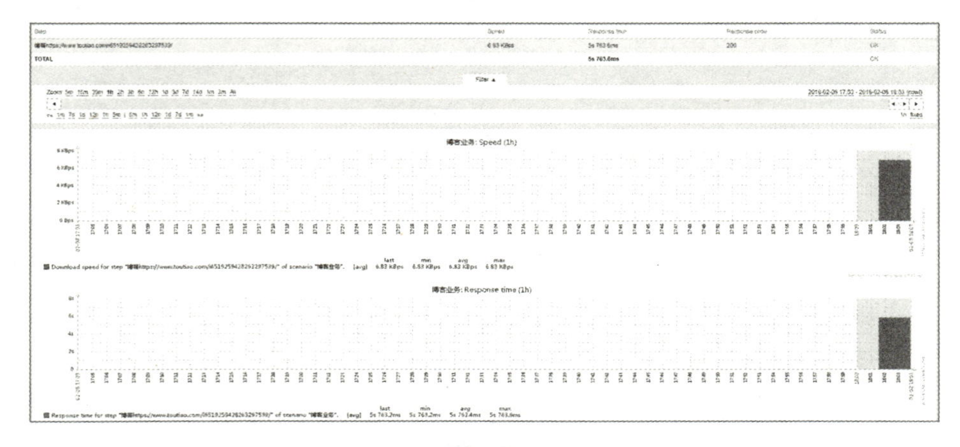

图 7-39

使用 curl 命令可以查看天气，过滤 rain（下雨）关键字，0 表示天晴，不是 0 就添加到触发器中，使其报警提醒自己要带伞，命令如下。

```
[root@zabbix-2 ~]# curl http://www.weather.com.cn/data/sk/
101010100.html 2>/dev/null |python -m json.tool
```

7.6.2 MySQL 监控

1. 无密码监控

在 zabbix-2 服务器中安装 MariaDB 数据库并启动，命令如下。

```
[root@zabbix-2 ~]# yum install -y mariadb-server
[root@zabbix-2 zabbix_agentd.d]# systemctl start mariadb.service
[root@zabbix-2 zabbix_agentd.d]# systemctl enable mariadb.service
```

在/etc/zabbix/zabbix_agentd.d/目录下有一个 userparameter_mysql.conf 配置文件，该配置文件就是用来监控 MySQL 的，该配置文件中的取值已经定制好了，无须进行额外操作，打开 Zabbix Web 页面，依次单击 Configuration→Hosts→Create host，打开如图 7-40 所示的页面。单击 Templates 选项卡，添加 MySQL 模板，如图 7-41 所示。

添加后的 MySQL Host 默认有 14 个监控项，大部分监控的都是 MySQL 数据库的状态信息，如图 7-42 所示。

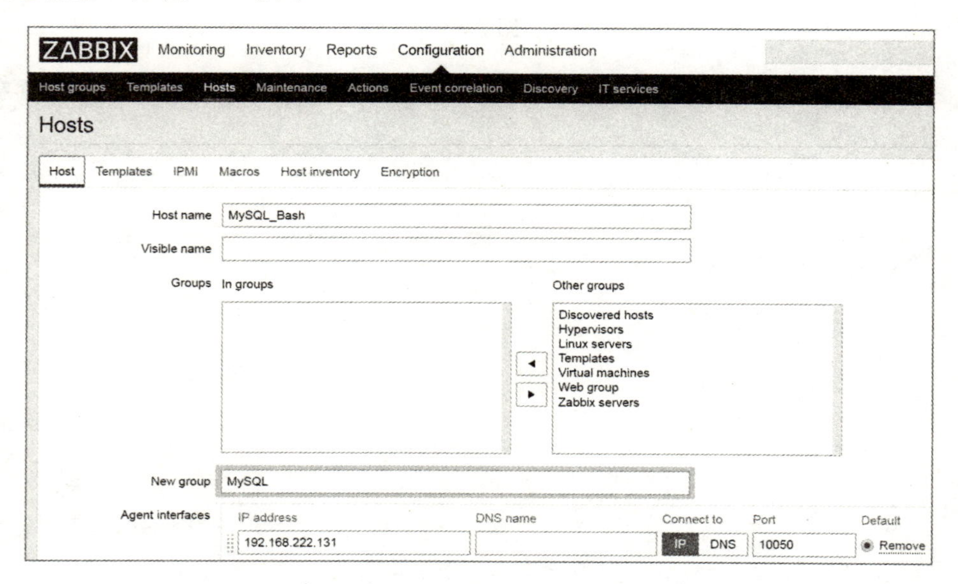

图 7-40

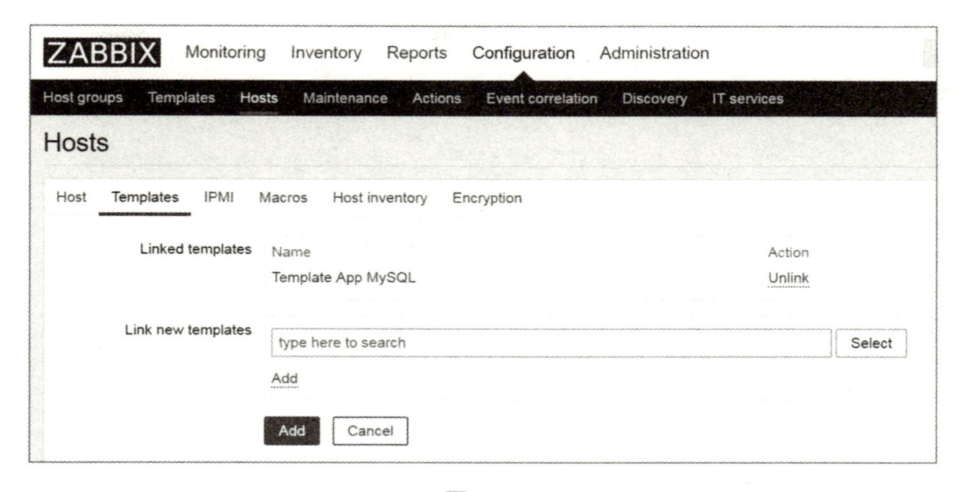

图 7-41

	Wizard	Name ▲	Triggers	Key	Interval	History	Trends	Type	Applications	Status	Info
		Template App MySQL: MySQL begin operations per second		mysql.status[Com_begin]	1m	7d	365d	Zabbix agent	MySQL	Enabled	
		Template App MySQL: MySQL bytes received per second		mysql.status[Bytes_received]	1m	7d	365d	Zabbix agent	MySQL	Enabled	
		Template App MySQL: MySQL bytes sent per second		mysql.status[Bytes_sent]	1m	7d	365d	Zabbix agent	MySQL	Enabled	
		Template App MySQL: MySQL commit operations per second		mysql.status[Com_commit]	1m	7d	365d	Zabbix agent	MySQL	Enabled	
		Template App MySQL: MySQL delete operations per second		mysql.status[Com_delete]	1m	7d	365d	Zabbix agent	MySQL	Enabled	
		Template App MySQL: MySQL insert operations per second		mysql.status[Com_insert]	1m	7d	365d	Zabbix agent	MySQL	Enabled	
		Template App MySQL: MySQL queries per second		mysql.status[Questions]	1m	7d	365d	Zabbix agent	MySQL	Enabled	
		Template App MySQL: MySQL rollback operations per second		mysql.status[Com_rollback]	1m	7d	365d	Zabbix agent	MySQL	Enabled	
		Template App MySQL: MySQL select operations per second		mysql.status[Com_select]	1m	7d	365d	Zabbix agent	MySQL	Enabled	
		Template App MySQL: MySQL slow queries		mysql.status[Slow_queries]	1m	7d	365d	Zabbix agent	MySQL	Enabled	
		Template App MySQL: MySQL status	Triggers 1	mysql.ping	1m	7d	365d	Zabbix agent	MySQL	Enabled	
		Template App MySQL: MySQL update operations per second		mysql.status[Com_update]	1m	7d	365d	Zabbix agent	MySQL	Enabled	
		Template App MySQL: MySQL uptime		mysql.status[Uptime]	1m	7d	365d	Zabbix agent	MySQL	Enabled	
	...	Template App MySQL: MySQL version		mysql.version	1h	7d		Zabbix agent	MySQL	Enabled	

图 7-42

验证 MySQL 取值方法，如查看 MySQL 的版本，以及 14 个监控项 Item 中的值，命令如下。

```
[root@zabbix-2 ~]# mysql -V
mysql  Ver 15.1 Distrib 5.5.56-MariaDB, for Linux (x86_64) using readline
5.1
[root@zabbix-2 ~]# mysqladmin ping | grep -c alive
1
[root@zabbix-2 ~]# mysqladmin ping
mysqld is alive
[root@zabbix-2 ~]# echo "show global status where Variable_name=
'Com_begin';" | HOME=/var/lib/zabbix mysql -N | awk '{print $$2}'
Com_begin    0
```

依次单击 Monitoring→Latest data，可查看 MySQL 监控到的数据，如图 7-43 所示。

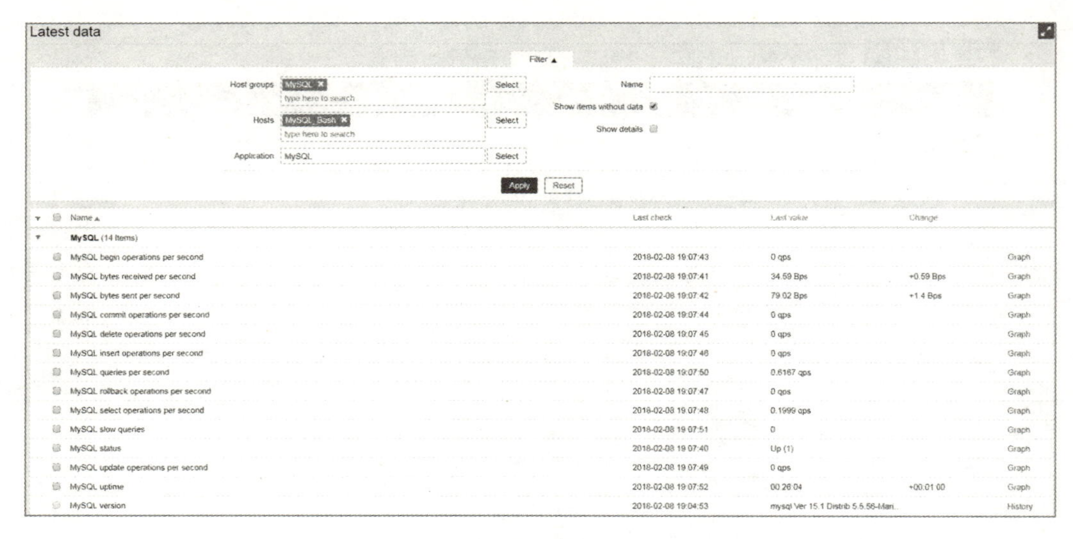

图 7-43

2. 有密码监控

没有密码的监控是非常不安全的，可以设置密码后再进行监控，需要编辑 zabbix-2 服务器中的 userparameter_mysql.conf 配置文件，在如下代码中添加用户名和密码，注意 zabbix-2 服务器中没有 zabbix 用户，需要先创建用户并授权，命令如下。

```
[root@zabbix-2 zabbix_agentd.d]# vim userparameter_mysql.conf
UserParameter=mysql.status[*],echo "show global status where Variable_name='$1';" | HOME=/var/lib/zabbix mysql -uzabbix -p123456 -N | awk '{print $$2}'
UserParameter=mysql.ping,HOME=/var/lib/zabbix mysqladmin -uzabbix -p123456 ping | grep -c alive
[root@zabbix-2 zabbix_agentd.d]# mysql
MariaDB [(none)]> grant all on *.* to zabbix@'127.0.0.1' identified by '123456';
MariaDB [(none)]> flush privileges;
MariaDB [(none)]> quit
[root@zabbix-2 zabbix_agentd.d]# systemctl restart zabbix-agent.service
```

设置密码后，Configuration→Hosts→Items 页面下方的 Key 值会出现红色标识，这是设置密码的缘故，依然可以取到值，命令如下。

```
[root@zabbix-1 ~]# zabbix_get -s 192.168.222.131 -k mysql.status[Uptime]
2929
```

3. 更加灵活的带密码监控

当数据库非常多，而且每个数据库的密码都不一致导致无法自动化时，更加灵

活的带密码监控方法就派上用场了。Zabbix 可以自定义变量，编辑 zabbix-2 服务器中的 userparameter_mysql.conf 配置文件，把配置文件中所有的用户名和密码都换为 $1 和$2，命令如下。

```
    [root@zabbix-2 zabbix_agentd.d]# vim userparameter_mysql.conf
    UserParameter=mysql.status[*],echo "show global status where
Variable_name='$3';" | HOME=/var/lib/zabbix mysql -u$1 -p$2 -N | awk '{print
$$2}'
    UserParameter=mysql.ping[*],HOME=/var/lib/zabbix mysqladmin -u$1 -p$2
ping | grep -c alive
    [root@zabbix-2 zabbix_agentd.d]# systemctl restart zabbix-agent.servic
```

再去获取 Key 值时就需要加上用户名和密码了，命令如下。

```
    [root@zabbix-1 ~]# zabbix_get -s 192.168.222.131 -k
mysql.status[zabbix,123456,Uptime]
    4085
```

Zabbix Web 界面的模板中还没有定义用传参的方式去取值，可以依次单击 Configuration→Templates→Template App MySQL→Macros，在页面中定义变量，如图 7-44 所示。

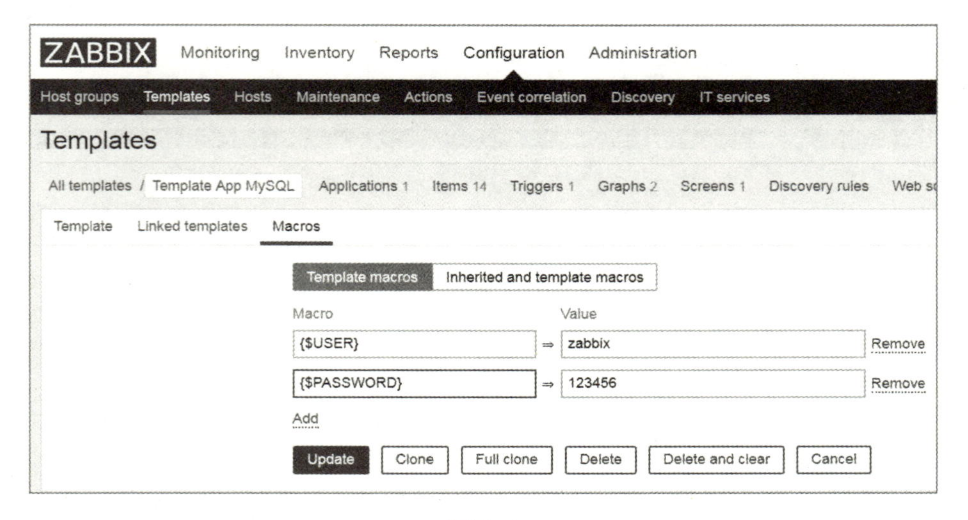

图 7-44

依次单击 Configuration→Hosts→Items，然后选择 MySQL 主机，将 Key 值为 mysql.status 的变量全部添加如图 7-45 所示的参数。添加后的显示结果如图 7-46 所示，在 userparameter_mysql.conf 配置文件中定义的$1 对应$USER、$2 对应$PASSWORD。

Items

All templates / Template App MySQL　Applications 1　Items 14　Triggers 1　Graphs 2　Screens 1　Discovery rules

Name: MySQL begin operations per second

Type: Zabbix agent

Key: mysql.status[{USER},{PASSWORD},Com_begin]　Select

图 7-45

Wizard	Name ▲	Triggers	Key	Interval	History	Trends	Type	Applications	Status
	MySQL begin operations per second		mysql.status[{USER},{PASSWORD},Com_begin]	1m	7d	365d	Zabbix agent	MySQL	Enabled
	MySQL bytes received per second		mysql.status[{USER},{PASSWORD},Bytes_received]	1m	7d	365d	Zabbix agent	MySQL	Enabled
	MySQL bytes sent per second		mysql.status[{USER},{PASSWORD},Bytes_sent]	1m	7d	365d	Zabbix agent	MySQL	Enabled
	MySQL commit operations per second		mysql.status[{USER},{PASSWORD},Com_commit]	1m	7d	365d	Zabbix agent	MySQL	Enabled
	MySQL delete operations per second		mysql.status[{USER},{PASSWORD},Com_delete]	1m	7d	365d	Zabbix agent	MySQL	Enabled
	MySQL insert operations per second		mysql.status[{USER},{PASSWORD},Com_insert]	1m	7d	365d	Zabbix agent	MySQL	Enabled
	MySQL queries per second		mysql.status[{USER},{PASSWORD},Questions]	1m	7d	365d	Zabbix agent	MySQL	Enabled
	MySQL rollback operations per second		mysql.status[{USER},{PASSWORD},Com_rollback]	1m	7d	365d	Zabbix agent	MySQL	Enabled
	MySQL select operations per second		mysql.status[{USER},{PASSWORD},Com_select]	1m	7d	365d	Zabbix agent	MySQL	Enabled
	MySQL slow queries		mysql.status[{USER},{PASSWORD},Slow_queries]	1m	7d	365d	Zabbix agent	MySQL	Enabled
	MySQL status	Triggers 1	mysql.ping[{USER},{PASSWORD}]	1m	7d	365d	Zabbix agent	MySQL	Enabled
	MySQL update operations per second		mysql.status[{USER},{PASSWORD},Com_update]	1m	7d	365d	Zabbix agent	MySQL	Enabled
	MySQL uptime		mysql.status[{USER},{PASSWORD},Uptime]	1m	7d	365d	Zabbix agent	MySQL	Enabled
...	MySQL version		mysql.version	1h	7d		Zabbix agent	MySQL	Enabled

Displaying 14 of 14 found

图 7-46

依次单击 Configuration→Hosts→MySQL_Bash→Macros，在 Inherited and host macros 中即可看到设置的变量参数，如图 7-47 所示。

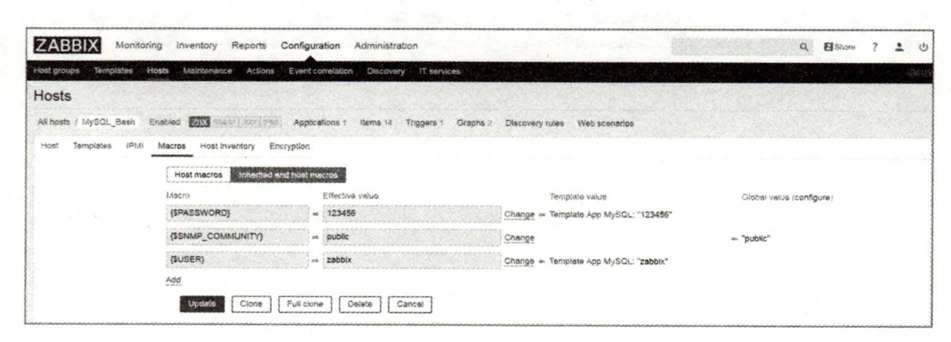

图 7-47

重新授权一个 MariaDB 用户，用户名为 monitor，密码为 humingzhe，命令如下。

```
[root@zabbix-2 zabbix_agentd.d]# mysql
MariaDB [(none)]> grant all on *.* to monitor@'localhost' identified by 'humingzhe';
MariaDB [(none)]> flush privileges;
```

依次单击 Configuration→Hosts→MySQL_Bash→Macros，在 Host macros 中添加新的用户名和密码，如图 7-48 所示。

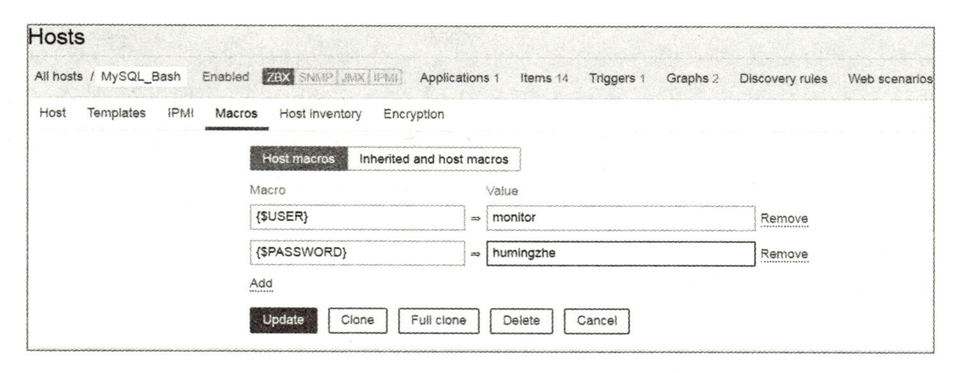

图 7-48

7.7　TCP状态监控和Nginx监控

7.7.1　TCP 状态监控

当有人对服务器进行 DDOS 攻击时，可以在 SYN-RECEIVED 值变大时采取报警措施。监控会涉及取值，使用 man netstat 命令可以查看 TCP 的 12 种状态，分别是 ESTABLISHED、SYN_SENT、SYN_RECV、FIN_WAIT1、FIN_WAIT2、TIME_WAIT、CLOSE、CLOSE_WAIT、LAST_ACK、LISTEN、CLOSING、UNKNOWN。

使用 netstat 和 grep 命令进行取值，如取 UNKNOWN 的值，命令如下。后边的 UNKNOWN 可以写成$1 来传参。配置文件只需要一行，即可将 12 种监控状态写成模板。

```
[root@zabbix-2 ~]# netstat -ant |grep -c UNKNOWN
0
```

在/etc/zabbix/zabbix_agentd.d/目录下创建一个监控 TCP 状态的配置文件，命令如下。

```
[root@zabbix-2 ~]# cd /etc/zabbix/zabbix_agentd.d/
[root@zabbix-2 zabbix_agentd.d]# vim TCP_Status.conf
UserParameter=TCP_Status[*],netstat -ant |grep -c $1
[root@zabbix-2 zabbix_agentd.d]# systemctl restart zabbix-agent.service
```

在 zabbix-1（Zabbix Server）服务器中进行测试取值操作，命令如下。

```
[root@zabbix-1 ~]# zabbix_get -s 192.168.222.131 -k
TCP_Status[TIME_WAIT]
54
```

需要注意自定义的取值方法存在一个问题，在命令行中执行时，是以 root（超级管理员）身份执行的，但是写进配置文件后，让 Zabbix 取值时，它是以 Zabbix 身份执行的，很多时候会面临 Zabbix 没有这个命令权限的问题，如使用 netstat -p 命令就无法查看。针对这个问题，可以使用 sudo 和 suid 两个方法。

7.7.2 Nginx 服务监控

Nginx 监控利用 Nginx 自带的 stub_status，将 stub_status 开启即可，关闭 zabbix-2（Agent）客户端的 Apache 服务，下载安装 Nginx 服务，命令如下。

```
[root@zabbix-2 ~]# systemctl stop httpd.service
[root@zabbix-2 ~]# yum install -y nginx.x86_64 nginx-all-modules.noarch
```

编辑 nginx.conf 配置文件，开启 Nginx 监控功能，关闭 IPv6（listen [::]:80 default_server;）功能。创建一个新的 location，并添加如下内容。

```
[root@zabbix-2 ~]# vim /etc/nginx/nginx.conf
server {
    location /nginx_status {
        stub_status on;
        access_log off;
        allow 192.168.222.0/24;
        allow 127.0.0.1;
        deny all;
    }
}
[root@zabbix-2 ~]# systemctl start nginx.service
```

在/etc/zabbix/zabbix_agentd.d/目录下创建自定义的.conf 配置文件，命令如下。

```
[root@zabbix-2 zabbix_agentd.d]# vim nginx-params.conf
UserParameter=nginx[*],/usr/local/zabbix-agent/scripts/nginx-check.sh
"$1"
```

创建 nginx-check.sh 脚本文件，由于监控 Nginx 脚本配置文件代码过多，所以笔者上传到 QQ 群文件中，可以在读者 QQ 群中下载对应文件，导入成功后赋予脚本执行权限，命令如下。

```
[root@zabbix-2 zabbix_agentd.d]# mkdir -p
/usr/local/zabbix-agent/scripts
[root@zabbix-2 zabbix_agentd.d]# cd /usr/local/zabbix-agent/scripts
[root@zabbix-2 scripts]# vim nginx-check.sh
```

```
[root@zabbix-2 scripts]# chmod +x nginx-check.sh
[root@zabbix-2 zabbix_agentd.d]# systemctl restart zabbix-agent.service
```

依次单击 Configuration→Templates→Import，导入自定义的模板内容，由于自定义的模板是 XML 文件，代码有数百行，所以也上传到读者 QQ 群 99394384 中，文件名是 nginx-template.xml。导入自定义模板后创建一个新主机，如图 7-49 所示。

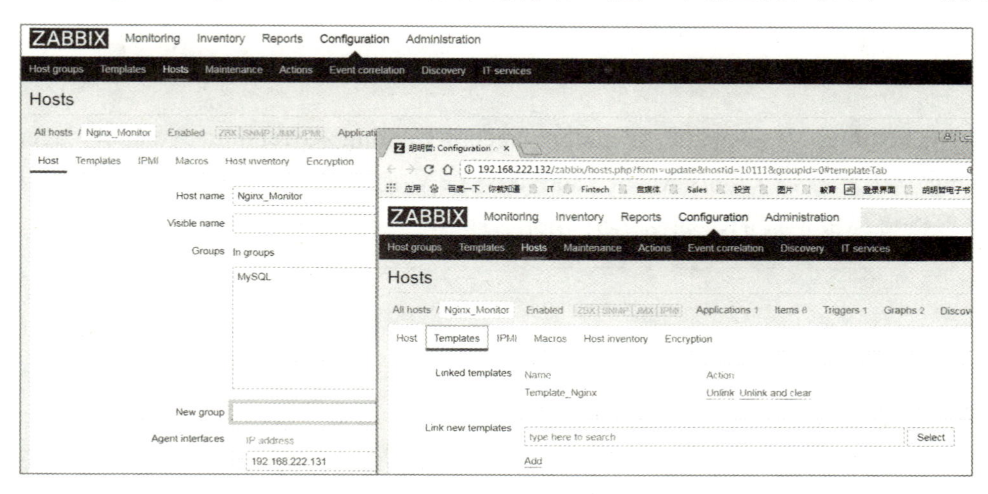

图 7-49

7.8　Zabbix主动模式

7.8.1　添加主动模式模板

Zabbix 默认没有主动模式模板，需要手动制作主动模式模板。模板也可以复制，为了提高工作效率，此处选择复制。依次单击 Configuration→Templates，复制 Template OS Linux 模板。将 Template name（目标名）改为 Template OS Linux Active，其余参数不变，在界面最下方单击 Add 按钮进行添加操作。

选择刚添加的 Template OS Linux Active 模板，定位至 Items 处，它的 Type 类型均是 Zabbix Agent 模式，将所有的类型全选后在界面底部单击 Mass update 按钮进行批量更新。进入批量更新界面后，勾选 Type 选项，选择 Zabbix Agent（active）模式，在界面底部单击 Update 按钮完成更新操作。更新完后会变成 Zabbix Agent（active）模式，如图 7-50 所示。

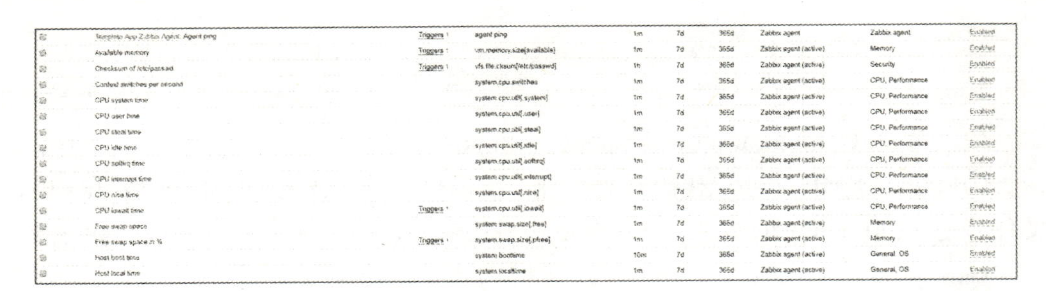

图 7-50

Agent ping、Host name of zabbix_agentd running 和 Version of zabbix_agent(d) running 三项并不是 Zabbix Agent（active）模式，而是 Zabbix Agent 模式，这是模板还可以链接模板造成的，如图 7-51 所示。

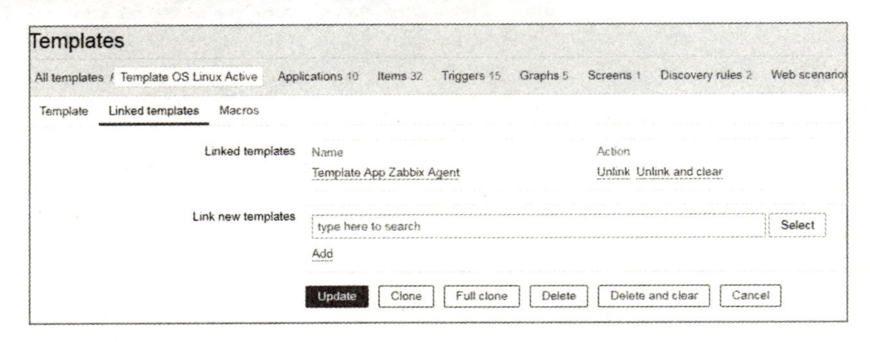

图 7-51

要完全实现主动模式，还需要复制一个 Template App Zabbix Agent 模板。单击 Template App Zabbix Agent 模板，在界面底部选择 Full clone，将 Template name 改为 Template App Zabbix Agent Active 即可，其余参数不变。同样也需要到 Items 中进行批量更新，选择 Zabbix Agent（active）。

依次单击 Configuration→Templates→Template OS Linux Active→Linked templates，单击 Unlink and clear 按钮，然后链接新模板 Template App Zabbix Agent Active，如图 7-52 所示。

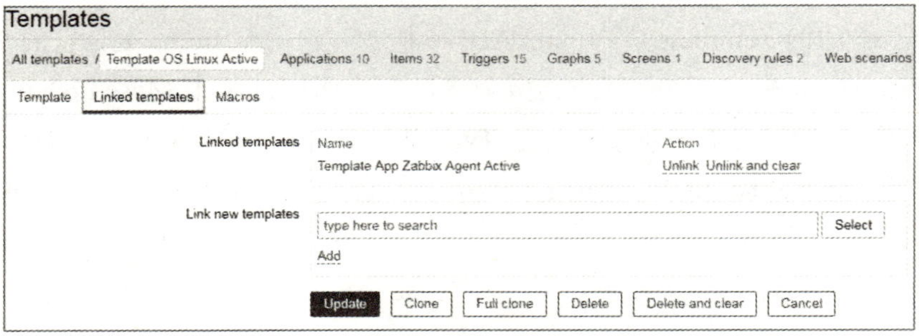

Templates
All templates / Template OS Linux Active　Applications 10　Items 32　Triggers 15　Graphs 5　Screens 1　Discovery rules 2　Web scenarios
Template　Linked templates　Macros
Linked templates　Name　Action
Template App Zabbix Agent Active　Unlink　Unlink and clear
Link new templates　type here to search　Select
Add
Update　Clone　Full clone　Delete　Delete and clear　Cancel

图 7-52

上述步骤操作完成后，Type 类型都会更改为 Zabbix Agent（active）模式。

7.8.2　添加主动模式主机

编辑 zabbix_agentd.conf 配置文件，将被动模式改为主动模式，命令如下。

```
[root@zabbix-2 ~]# cd /etc/zabbix/zabbix_agentd.d/
#Server=192.168.222.132
StartAgents=0
[root@zabbix-2 zabbix_agentd.d]# grep "^[a-Z]"
/etc/zabbix/zabbix_agentd.conf
PidFile=/var/run/zabbix/zabbix_agentd.pid
LogFile=/var/log/zabbix/zabbix_agentd.log
LogFileSize=0
StartAgents=0
ServerActive=192.168.222.132
HostnameItem=system.hostname
Timeout=15
Include=/etc/zabbix/zabbix_agentd.d/*.conf
[root@zabbix-2 zabbix_agentd.d]# systemctl restart zabbix-agent.service
```

回到 Zabbix Web 界面，依次单击 Configuration→Hosts→Create host 添加新的主机，输入 Host name（zabbix-2，由于使用的是 system.hostname，所以一定要使用 hostname 命令获取主机名，不要乱填写），接着选择一个主机组（Groups In groups），最后选择 Template OS Linux Active 即可完成添加主机操作。

默认使用纯主动模式时，ZBX 图标很难变成绿色，笔者曾经多次尝试，即便变成绿色也需要一小时左右的时间，不如被动模式时变绿色图标快。

温馨提醒

如果想更好地学习监控这一章节，可以加入读者 QQ 群 99208965 或访问笔者博客 http://humingzhe.com 进行学习与交流。

第 8 章

8

NoSQL 非关系型数据库

8.1 NoSQL非关系型数据库简介

NoSQL 即 Not only SQL，被称为非关系型数据库，是用来存储数据的，类似于关系型数据库。典型的关系型数据库有 MySQL、Oracle 和 SQL Server。关系型数据库有一个共同的特点是可以使用 SQL 语句，而非关系型数据库没有 SQL 语句。

关系型数据库需要把数据存储到库、表、行、字段，最小的单元是字段，如创建一张表，表中包含用户名、ID 和密码，查询时需要根据条件一行行地匹配查询。当数据量非常多时，匹配会非常耗费时间和资源，因为数据存储在服务器的磁盘中，数据量非常多时需要遍历整个存储空间。

NoSQL 的存储相对来讲是非常简单的，没有错综复杂的关系，不存在行和字段等数据结构，而是以典型的 Key/Value（键值）对的方式存在。如存储用户，对应的名称是 username；存储一个值，名称是 age，对应的年龄是 20。在查询时，用户可直接匹配出来，无需一行一行地匹配，而且 NoSQL 不存在多个表之间的关联。

NoSQL 数据可以存储在内存中，查询时速度非常快。NoSQL 在性能上虽然优于关系型数据库，但是它并不能完全替代关系型数据库，因为一些非常复杂的需求不能被满足，原因是 NoSQL 的设计非常简单。NoSQL 只能满足一般的应用，当网站访问量非常大时，可以加一层缓存，缓存就可以使用 NoSQL。缓存必然是在内存中的，比如在查询一件商品时，商品查询命令先被放入内存的 NoSQL 中，商品都有唯一的 ID，商品的 ID 对应商品的内容，这样就可以理解为商品 ID 是 Key，商品内容是 Value，可以把一条条的商品信息缓存在 NoSQL 中，当用户访问时，不用查询 MySQL 数据库中的内容，而是直接查询缓存中的内容，这样就减轻了数据的查询压力，这是最典型的用法。

NoSQL 因为没有复杂的数据结构，所以扩展也非常容易，而且还支持分布式。

常用的 NoSQL 数据库有 Memcached、Redis 和 MongoDB。Memcached 和 Redis 属于 Key/Value 的形式，比如前面讲到的商品的缓存就可以用 Memcached 和 Redis 来实现，还可以把它们作为用户会话（Session，用户登录信息），在集群中 Session 共享是非常关键的，所以用 Memcached 和 Redis 存储 Session 非常合适。

MongoDB 属于文档数据库，它将数据以文档的形式存储。每个文档都是一系列数据项的集合。每个数据项都有一个名称对应的值，值既可以是简单的数据类型，如字符串、数字和日期等，也可以是复杂的类型，如有序列表和关联对象。数据存储的最小单位是文档，同一个表中存储的文档属性可以是不同的，数据可以使用 XML、JSON、JSONB 等多种形式存储。

HBase 是 Hadoop Database 的简称，它是一个分布式的、面向列存储的开源数据库。

图存储数据库有 Neo4J、Infinite Graph 和 OrientDB。

8.2 Memcached

8.2.1 Memcached 简介

Memcached 是以 Key/Value 形式存在的，和 LNMP 服务一起使用，被称为 LNMMP，增加的那个 M 就是 Memcached，它扮演的是缓存的角色。Memcached 是国外社区网站 LiveJournal 团队开发的，目的是为了通过缓存数据库查询结果，减少数据库访问的次数，从而提高动态 Web 站点的性能。

Memcached 官方网址是 http://www.memcached.org，最新版本是 v1.5.5。Memcached 数据在内存中不支持落地，也不支持持久化，这就意味着服务器一旦重启或重启 Memcached 服务，原先存储的数据就会丢失，所以重要的数据一定要注意保存。为了保证数据的安全，可以不定期地把数据落地，存入硬盘中。比如，每次重启服务前把数据存到磁盘中，当 Memcached 服务启动后再把磁盘中的数据导入 Memcached 中。

Memcached 是多线程的服务，当服务器 CPU 较多时，使用 Memcached 可以明显感觉到速度很快。Memcached 是基于 C/S 架构的，协议很简单。它基于 Libevent 事件处理，支持自主内存存储处理（Slab 分配，Slab Alloaction）。Memcached 的数据过期方式有两种，即 Lazy Expiration 和 LRU。

Slab Allocation 原理是将分配的内存分割成各种尺寸的块（Chunk），并把尺寸相同的块分成组（Chunk 集合），每个 Chunk 集合被称为 Slab。Memcached 的内存分配以 Page 为单位，Page 默认值为 1MB，可以在启动时通过-I 参数指定。Slab 是由多个 Page 组成的，Page 按照指定大小切割成多个 Chunk。

Chunk 的大小是由 Growth factor 来决定的。Memcached 在启动时通过-f 选项可以指定 Growth Factor 因子，该值控制 Chunk 大小的差异，默认值为 1.25。通过 memcached-tool 命令可以查看指定 Memcached 实例的不同 Slab 状态，可以看到各 Item 所占大小（Chunk 的大小）的差距为 1.25，命令如下。

```
[root@nosql~]#memcached-tool 127.0.0.1:11211 display
```

Lazy Expiration 是一种比较懒惰的方式，在创建 Key/Value 时需要给 Key 指定过期时间，如一个小时后过期，过期后就不会在内存中保存。Memcached 不会监视记录是否过期，在 get 请求时查看记录的时间戳，检查记录是否过期。Lazy Expiration 的优势在于不会在过期监视上耗费 CPU 资源。

Memcached 会优先使用已超时的记录的空间，但即便如此，也会发生追加新记录时空间不足的情况，此时就要使用 Least Recently Used（LRU）机制来分配空间。顾名思义，这是删除“最近最少使用”的记录的机制。因此，当内存空间不足时（无法从 Slab class 获取到新的空间时），就从最近未被使用的记录中搜索，并将其空间分配给新的记录。从缓存的使用角度来看，该模型十分理想。

假设有一台数据库 MySQL 服务器，还有一台 Web 服务器（Nginx+PHP+MySQL），正常情况下是 PHP 和 MySQL“打交道”，Nginx 调用 PHP 进行数据交互，比如查询帖子内容，一般是用户发起请求给 Nginx，Nginx 把 PHP 请求交给 PHP 脚本，PHP 脚本把请求交给 php-fpm 服务，php-fpm 服务解析完脚本后发现是查询某个帖子，它就去 MySQL 数据库中查询数据，查询完数据最终由 Nginx 反馈给用户，这是正常的数据流向。但是在 MySQL 服务器中查数据比较慢，而且并发量很大，这时可以增加一个缓存层，把查询到的数据缓存在内存中，用户再次查询时就可以到内存中获取数据结果并反馈给用户，不用再去 MySQL 数据库中查数据，从而减少了查询 MySQL 数据库的次数。

Memcached 其实就是作为这样一个缓存服务存在的，如图 8-1 所示，首先在 MySQL 服务器中查询数据，查询到数据后通过 Web 服务器（PHP 客户端）把数据存储到 Memcached 中，再次查询时就无需到 MySQL 数据库服务器中查询，直接从 Memcached 服务器中反馈给用户即可。

8.2.2 安装 Memcached

安装 Memcached 无需下载官方源码包，使用 yum 命令即可安装。如果想安装官方最新版，可以在官网下载源码包进行编译安装。本节使用 yum 命令进行安装，然后启动并设置开机自启动，命令如下。

```
[root@nosql ~]# yum install -y memcached libmemcached libevent
[root@nosql ~]# systemctl enable memcached.service
[root@nosql ~]# systemctl start memcached.service
[root@nosql ~]# ps aux |grep memcached
```

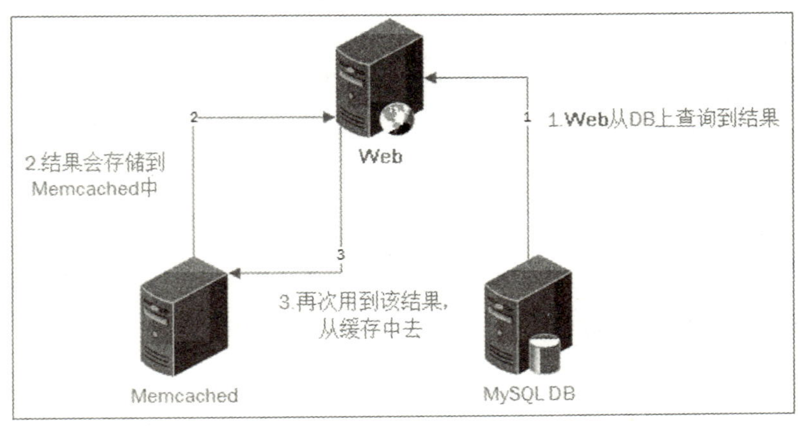

图 8-1

使用 ps aux 命令查看 Memcached 服务时，使用-p 选项可以指定它的监听端口是 11211；使用-m 选项可以指定给 Memcached 分配多大的内存，默认是 64MB；使用-c 选项可以指定最大的并发数，默认是 1024。这些参数都可以在/etc/sysconfig/memcached 文件中进行更改，命令如下。

```
[root@nosql ~]# netstat -lntp
tcp    0     0 0.0.0.0:11211    0.0.0.0:*      LISTEN     1483/memcached
[root@nosql ~]# vim /etc/sysconfig/memcached
PORT="11211"
USER="memcached"
MAXCONN="1024"
CACHESIZE="64"
OPTIONS=""
```

8.2.3　查看 Memcached 状态

使用 memcached-tool 命令可以查看 Memcached 的状态信息，查看方法是 memcached-tool+IP+Port，命令如下。

```
[root@nosql ~]# memcached-tool 127.0.0.1:11211 stats
```

查看 Memcached 状态时主要关注 get_hits 和 cmd_get，get_hits 表示 get 请求命

中总次数，cmd_get 表示总的 get 次数。用 get_hits 和 cmd_get 可计算出命中率有多高，计算命中率的目的是确认 Memcached 中缓存了数据和请求，若只有缓存而没有请求，那么做缓存也就没有任何意义。

使用 nc 命令也可以查看 Memcached 的状态信息，nc 命令默认不存在，需要使用 yum 命令安装，安装后即可查看 Memcached 的状态，命令如下。

```
[root@nosql ~]# yum install -y nc
[root@nosql ~]# echo stats |nc 127.0.0.1 11211
```

如果安装了 libmemcached，可以使用 memstat –servers 命令查看 Memcached 服务的状态，命令如下。

```
[root@nosql ~]# memstat --servers=127.0.0.1:11211
```

8.2.4 Memcache 命令行

MySQL 数据库支持创建表、插入数据和查看数据，Memcached 同样也支持。使用 telnet 命令进入 Memcached 中，命令如下。

```
[root@nosql ~]# yum install -y telnet
[root@nosql ~]# telnet 127.0.0.1 11211
```

注意

进入 telnet 后不要乱按键盘，因为 telnet 不像 Shell 终端那样好用，敲错命令是比较麻烦的。

set 命令可用来存储数据，例如下面的命令，key2 表示 key 的名字，数字 30 表示过期时间，数字 2 表示存储的数值是 2 字节，如果指定 3 字节，只输入两位时不会产生任何反应，再输入一位时就会报错，命令如下。

```
set key2 0 30 2
12
STORED
set key1 0 30 3
12
1
CLIENT_ERROR bad data chunk
ERROR
set key1 0 20 3
abc
STORED
```

虽然终端界面显示 STORED 表示存储成功，但是为了保证万无一失，还要用“get+key 名”来查看，命令如下。

```
get key3
VALUE key3 0 3
abc
END
```

注意

使用 get 命令获取 key 存储的数据时没有返回相应的数据，这意味着服务器中不存在这些项，有可能过期了，也有可能被删除了。

Memcached 语法规则：<command name> <key> <flags> <exptime> <bytes>\r\n <data block>\r\n。

<command name> 可以是 set、add 和 replace。set 表示按照相应的<key>存储该数据，没有数据的时候增加，有数据的时候覆盖。add 表示按照相应的<key>添加该数据，如果该<key>已经存在，则操作失败。replace 表示按照相应的<key>替换数据，如果该<key>不存在，则操作失败。<key>客户端需要保存数据的 key。

<flags> 是一个 16 位的无符号的整数（以十进制的方式表示）。它将和需要存储的数据一起存储，并在客户端获取数据时返回。客户端可以将此标志用做特殊用途，此标志对服务器来说是不透明的。“set key2 0 30 2”中的 0 即 flags。

<exptime> 为过期的时间。若为 0，表示存储的数据永远不过期（但可被服务器算法 LRU 等替换）。如果非 0（UNIX 时间或距离此时的秒数），当过期后，服务器可以保证用户得不到该数据（以服务器时间为标准）。

<bytes>表示需要存储的字节数，当用户希望存储空数据时，<bytes>可以为 0。

<data block>表示需要存储的内容，输入完成后，最后客户端需要加上\r\n（直接按 Enter 键）作为结束标志。

注意

\r\n 在 Windows 操作系统中是 Enter 键。

replace 命令可以用来替换数据，如存储一条数据，数值是 1234，可以通过 replace 命令进行替换，如替换成 abcde，命令如下。

```
set key4 1 100 4
1234
STORED
replace key4 1 0 5
abcde
```

```
STORED
get key4
VALUE key4 1 5
abcde
END
```

使用 delete 命令可以删除一个 key，如删除 key4，命令如下。

```
delete key4
DELETED
get key4
END
```

注意

按 Ctrl+]组合键，再输入 quit 即可退出 Telnet。

8.2.5 Memcached 数据的导入和导出

重启 Memcached 服务时需要备份数据，将内存中的数据导出至服务器的磁盘中，启动 Memcached 服务后再将备份到磁盘中的数据导入内存中。

memcached-tool 工具支持导出数据，命令如下。

```
[root@nosql ~]# memcached-tool 127.0.0.1:11211 dump > data.txt
Dumping memcache contents
  Number of buckets: 1
  Number of items  : 6
Dumping bucket 1 - 6 total items
```

注意

导出的数据是带有时间戳的，这个时间戳是该条数据过期的时间点，如果当前时间已经超过该时间戳的时间，就不能导回。

nc 命令支持把导出的 Memcached 数据导回，Memcached 的数据有过期时间限制，可以修改 data.txt 文件，将导出数据文件的时间戳增加一小时，再用 nc 命令导回，命令如下。

```
[root@nosql ~]# date -d "+1 hour " +%s
1518791192
[root@nosql ~]# vim data.txt
add passwd 1 1518791192 8
```

```
shdasaaa
add key6 1 1518791192 12
123sadadsada
[root@nosql ~]# nc 127.0.0.1 11211 < data.txt
```

注意

提示 NOT_STORED 是因为 Memcached 数据库中已经存储了该条数据，若想导入，需要重启 Memcached 服务，重启后即可成功导入。

8.2.6　PHP 连接 Memcached

在 LNMP 架构中若想让 PHP 连接 Memcached 服务，需要安装 PHP 的 Memcached 扩展模块 memcache（软件包可在读者 QQ 群文件中下载），命令如下。

```
[root@nosql ~]# cd /usr/local/src/
[root@nosql src]# ls
memcache-3.0.8.tgz
[root@nosql src]# tar xf memcache-3.0.8.tgz
[root@nosql src]# cd memcache-3.0.8/
[root@nosql memcache-3.0.8]# /usr/local/php-fpm/bin/phpize
[root@nosql memcache-3.0.8]# ./configure
--with-php-config=/usr/local/php-fpm/bin/php-config
[root@nosql memcache-3.0.8]# make && make install
```

编辑 php.ini 配置文件，在代码“;extension=shmop”下面添加 extension=memcache.so 模块，命令如下。

```
[root@nosql memcache-3.0.8]# vim /usr/local/php-fpm/etc/php.ini
extension=memcache.so
```

使用/usr/local/php-fpm/bin/php -m 命令可以查看是否成功安装了 memcache 扩展模块，命令如下。

```
[root@nosql memcache-3.0.8]# /usr/local/php-fpm/bin/php -m
memcache
```

下载 PHP 脚本测试 PHP 是否支持 memcache 扩展（脚本在读者 QQ 群文件中，文件名是 memcache.php），命令如下。

```
[root@nosql ~]# /usr/local/php-fpm/bin/php memcache.php
Get key1 value: This is first value<br>Get key1 value: This is replace
value<br>Get key2 value: Array
```

```
(
    [0] => aaa
    [1] => bbb
    [2] => ccc
    [3] => ddd
)
<br>Get key1 value: <br>Get key2 value: <br>[root@nosql ~]#
```

8.2.7 Memcached 中存储 Session

在多台 Web 服务器的场景下，用户第一次可能在 A 服务器上登录，第二次可能在 B 服务器上登录，这样就无法保证用户登录状态始终在一台服务器上。针对以上情况，可以不把 Session 存储到服务器的磁盘上，而是存储到 Memcached 中。Memcached 作为公共的服务器，访问时使用其中一个内网 IP，而不是 127.0.0.1。

在 php-fpm.conf 配置文件对应的 pool 中添加如下代码。

```
[root@nosql ~]# vim /usr/local/php-fpm/etc/php-fpm.d/mingzhe.conf
[mingzhe]
php_value[session.save_handler] = memcache
php_value[session.save_path] = " tcp://192.168.222.129:11211"
[root@nosql vhost]# /etc/init.d/php-fpm restart
```

在根目录中创建 memcached_session.php 文件，添加如下代码。

```
[root@nosql default]# vim memcache_session.php
<?php
session_start();
if (!isset($_SESSION['TEST'])) {
$_SESSION['TEST'] = time();
}
$_SESSION['TEST3'] = time();
print $_SESSION['TEST'];
print "<br><br>";
print $_SESSION['TEST3'];
print "<br><br>";
print session_id();
?>
```

添加后使用 curl 命令生成 Memcached 的 Session 文件，再通过 Telnet 获取 key 值，命令如下。

```
[root@nosql default]# curl localhost/memcache_session.php
```

```
1518973112<br><br>1518973112<br><br>aqfeiesefvanti4oijkd483ij5
get aqfeiesefvanti4oijkd483ij5
VALUE aqfeiesefvanti4oijkd483ij5 0 36
TEST|i:1518973112;TEST3|i:1518973112;
END
```

8.3 Redis

8.3.1 Redis 简介

Redis 和 Memcached 类似，但是功能比 Memcached 丰富很多。Redis 官方网址是 https://redis.io/，目前最新版本是 v4.0.8。Redis 支持多种 value 类型，除 string 类型外，还支持 hash（字典）、list（链表）、set（集合）和 sort set（有序集合）。

Redis 使用了两种文件格式：全量数据（RDB）和增量请求（AOF）。全量数据格式是把内存中的数据写入磁盘，便于下次读取文件时加载。增量请求文件则是把内存中的数据序列化为操作请求，用于读取文件并重新执行，从而得到数据，增量请求类似于 mysql binlog。Redis 的存储分为内存存储、磁盘存储和 log 文件三部分。

8.3.2 Redis 安装

下载最新稳定版 Redis，然后编译安装，命令如下。

```
[root@nosql ~]# cd /usr/local/src/
[root@nosql src]# wget
http://download.redis.io/releases/redis-4.0.8.tar.gz
[root@nosql src]# tar xf redis-4.0.8.tar.gz
[root@nosql src]# cd redis-4.0.8/
[root@nosql redis-4.0.8]# make && make install
```

将 Redis 的配置文件 redis.conf 复制到/etc/目录下，命令如下。

```
[root@nosql redis-4.0.8]# cp redis.conf /etc/
```

复制后打开/etc/redis.conf 配置文件，修改如下代码。

```
[root@nosql redis-4.0.8]# vim /etc/redis.conf
daemonize no 改为 daemonize yes
logfile "" 改为 logfile "/var/log/redis.log"
dir ./ 改为 dir /data/redis
```

```
appendonly no 改为 appendonly yes
[root@nosql redis-4.0.8]# mkdir /data/redis
```

启动 Redis 服务，其方式有些特别，通过 redis-server 指定 redis.conf 文件启动，默认监听端口是 6379，命令如下。

```
[root@nosql redis-4.0.8]# redis-server /etc/redis.conf
[root@nosql redis-4.0.8]# ps aux |grep redis
root       8911  0.2  0.1 145260  7520 ?        Ssl  20:11   0:00
redis-server 127.0.0.1:6379
```

查看 Redis 日志时发现有两个关于内核参数的警告：WARNING overcommit_memory is set to 0 和 WARNING you have Transparent Huge Pages。虽然不修改不会有影响，但还是建议修改，命令如下。

```
[root@nosql redis-4.0.8]# less /var/log/redis.log
[root@nosql redis-4.0.8]# sysctl vm.overcommit_memory=1
vm.overcommit_memory = 1
[root@nosql redis-4.0.8]# echo never >
/sys/kernel/mm/transparent_hugepage/enabled
```

添加至 rc.local 配置文件中可以让服务器启动时直接执行这两条命令，命令如下。

```
[root@nosql redis-4.0.8]# vim /etc/rc.local
sysctl vm.overcommit_memory=1
echo never > /sys/kernel/mm/transparent_hugepage/enabled
```

修改完系统内核参数后，重启 Redis 服务即可。

8.3.3 Redis 持久化

Redis 提供了两种持久化的方式，分别是 RDB（Redis DataBase）和 AOF（Append Only File）。RDB 是在不同的时间点，将 Redis 存储的数据生成快照并存储到磁盘等介质上。AOF 则换了一个角度来实现持久化，将 Redis 执行过的所有写指令记录下来，在下次 Redis 重启时，只要把这些指令从前到后重复执行一遍，就可以实现数据恢复。

RDB 和 AOF 两种方式可以同时使用，在这种情况下如果重启 Redis 服务，则会优先采用 AOF 方式进行数据恢复，这是因为 AOF 方式的数据恢复完整度更高。

如果没有数据持久化的需求，就可以关闭 RBD 和 AOF。这样 Redis 将变成一个纯内存数据库，和 Memcached 一样。

在/etc/redis.conf 配置文件中有如下几个关于持久化的参数。

- save 900 1、save 300 10 和 save 60 10000：这三个参数是关于 RDB 持久化的。save 900 1 表示每 15 分钟且至少有 1 个 key 改变，就触发一次持久化；save 300 10 表示每 5 分钟且至少有 10 个 key 改变，就触发一次持久化；save 60 10000 表示每 60 秒且至少有 10000 个 key 改变，就触发一次持久化。若想关闭 RDB 持久化，可以注释掉这三行参数，打开上面的"save"即可。
- appendonly yes：如果是 yes，则表示开启 AOF 持久化，appendfilename "appendonly.aof"表示指定 AOF 文件名称。
- appendfsync everysec：表示指定 fsync()调用模式，有 no（不调用 fsync）、always（每次写都会调用 fsync）和 everysec（每秒调用一次 fsync）三种。第一种最快；第二种数据最安全，但性能差一些；第三种在性能和持久化方面做了很好的折中，推荐使用，默认为第三种方案。

8.3.4　Redis 数据类型

Redis 共有五种数据类型，第一种数据类型是 string，它是最简单的类型，与 Memcached 类型一样，一个 key 对应一个 value。支持的操作和 Memcached 支持的操作类似，但 Redis 的 string 功能更加丰富，可以存储二进制的对象。

Redis 连接的命令是 redis-cli，redis 可以设置密码，没有密码也可以进入 Redis 的命令行。在 Redis 命令行界面存储 string 类型的数据（当输入 set 命令时后面会跟随参数，非常人性化），命令如下。

```
[root@nosql ~]# redis-cli
127.0.0.1:6379> set name 223
OK
127.0.0.1:6379> get name
"223"
```

string 类型支持同时设置多个键值对，可以使用 MSET 命令进行设置，命令如下。

```
127.0.0.1:6379> MSET name mingzhe age 20 passwd 123
OK
127.0.0.1:6379> MGET name age passwd
1) "mingzhe"
2) "20"
3) "123"
```

注意

Redis 比 Memcached 要人性化很多，支持用 Tab 键补全，支持↑（上翻）键和↓（下翻）键。

第二种数据类型是 list，它是一个链表结构，主要功能是增加/插入（push）、移除/取出（pop）、获取一个范围的所有值等。操作中 key 表示链表的名字。

使用 list 结构可以轻松实现最新消息排行等功能（比如新浪微博的 TimeLine）。list 的另外一个应用是消息队列，可以利用 list 的 push 操作，将任务存在 list 中，工作线程用 pop 操作将任务取出并执行。push 和 pop 是一对反义词，意思是压进去、挤出来。它们拥有先进先出、后进后出的特性。示例代码如下。

```
127.0.0.1:6379> LPUSH list "humingzhe"
(integer) 1
127.0.0.1:6379> LPUSH list "mingzhe"
(integer) 2
127.0.0.1:6379> LPUSH list "bone"
(integer) 3
127.0.0.1:6379> LRANGE list 0 -1
1) "bone"
2) "mingzhe"
3) "humingzhe"
```

使用 LPOP 命令可以取出数据，如取出 bone，取出后再查看会发现它不存在了，命令如下。

```
127.0.0.1:6379> LPOP list
"bone"
127.0.0.1:6379> LRANGE list 0 -1
1) "mingzhe"
2) "humingzhe"
```

第三种数据类型是 set，表示集合，和数学中的集合的概念是相似的。集合的操作有添加、删除元素，还有对多个集合求交集、并集、差集等操作。操作中 key 表示集合的名字。比如，在微博应用中，可以将一个用户所有的关注对象存放在一个集合中，将其所有的“粉丝”存放在一个集合中。因为 Redis 非常人性化地为集合提供了求交集、并集、差集等操作，所以就可以非常方便地实现如共同关注、共同喜好、二度好友等功能。对于上面的所有集合操作，还可以使用不同的命令将结果返回给客户端或将结果存到一个新的集合中。

用 SADD 命令创建两个集合，如 set1 和 set2，命令如下。

```
127.0.0.1:6379> SADD set1  a
(integer) 1
127.0.0.1:6379> SADD set1  b
(integer) 1
```

```
127.0.0.1:6379> SADD set1  c
(integer) 1
127.0.0.1:6379> SMEMBERS set1
127.0.0.1:6379> SADD set2 a
(integer) 1
127.0.0.1:6379> SADD set2 2
(integer) 1
127.0.0.1:6379> SADD set2 3
(integer) 1
127.0.0.1:6379> SADD set2 c
(integer) 1
127.0.0.1:6379> SMEMBERS set2
```

前面提到 set 数据类型可以求并集、交集和差集，它是通过 SUNION 命令求解的，如求并集，把 set1 和 set2 的元素取出来，将重复的数据去重，命令如下。

```
127.0.0.1:6379> SUNION set1 set2
1) "a"
2) "b"
3) "3"
4) "c"
5) "2"
```

使用 SINTER 命令可以求交集，命令如下。

```
127.0.0.1:6379> SINTER set1 set2
1) "c"
2) "a"
```

使用 SDIFF 命令可以求差集，所谓差集就是两者相减，命令如下。

```
127.0.0.1:6379> SDIFF set1 set2
1) "b"
127.0.0.1:6379> SDIFF set2 set1
1) "2"
2) "3"
```

使用 SREM 命令可以删除某个元素，如删除 set1 元素，命令如下。

```
127.0.0.1:6379> SREM set1 c
(integer) 1
127.0.0.1:6379> SMEMBERS set1
1) "b"
2) "a"
```

第四种数据类型是 sort set（有序集合），它比 set 多了一个权重参数 score，使得集合中的元素能够按 score 进行有序排序。比如，一个存储全班学生成绩的 Sorted Sets，其集合 value 可以是学生的学号，score 可以是学生的考试得分，这样在将数据插入集合的时候，就已经进行了天然的排序操作。

添加几个有序集合，使用 ZRANGE 命令可以查看有序集合的值，在查看值时会发现它是从小到大进行排序的，命令如下。

```
127.0.0.1:6379> ZADD set6 12 aaa
(integer) 1
127.0.0.1:6379> ZADD set6 11 bbb
(integer) 1
127.0.0.1:6379> ZADD set6 11 "bbb ccc"
(integer) 1
127.0.0.1:6379> ZADD set6 24 "bone"
(integer) 1
127.0.0.1:6379> ZADD set6 5 "hank"
(integer) 1
127.0.0.1:6379> ZRANGE set6 0 -1
1) "hank"
2) "bbb"
3) "bbb ccc"
4) "aaa"
5) "bone"
```

使用 ZREVRANGE 命令可以进行倒序排序，和上面排序的结果相反，倒序排序是从大到小进行排序的，命令如下。

```
127.0.0.1:6379> ZREVRANGE set6 0 -1
1) "bone"
2) "aaa"
3) "bbb ccc"
4) "bbb"
5) "hank"
```

第五种数据类型是 hash，在 Memcached 中我们经常将一些结构化的信息打包成 hashmap，在客户端序列化后存储为一个字符串的值（一般是 JSON 格式），比如用户的昵称、年龄、性别、积分等。

使用 HSET 命令可以添加 hash 形式的数据类型，命令如下。

```
127.0.0.1:6379> HSET hash name humingzhe
(integer) 1
127.0.0.1:6379> HSET hash age 18
(integer) 1
127.0.0.1:6379> HSET hash job Financial_Qunt
```

```
(integer) 1
```

使用 HGET 命令可以获取 hash 的元素，刚才添加的 hash 中还存在子元素，所以 hash 后面还需要加上子元素，命令如下。

```
127.0.0.1:6379> HGET hash name
"humingzhe"
127.0.0.1:6379> HGET hash age
"18"
127.0.0.1:6379> HGET hash job
"Financial_Qunt"
```

使用 HGETALL 命令可以显示 hash 的所有元素，奇数行是 key，偶数行是 value，命令如下。

```
127.0.0.1:6379> HGETALL hash
1) "name"
2) "humingzhe"
3) "age"
4) "18"
5) "job"
6) "Financial_Qunt"
```

8.3.5　Redis 常用操作

1. string 类型常用操作

使用 set key1 命令可以创建 key1，如果 key1 已经存在，再次使用 key1 去创建值会覆盖之前的值，命令如下。

```
127.0.0.1:6379> set key1 mingzhe
OK
127.0.0.1:6379> GET key1
"mingzhe"
127.0.0.1:6379> SET key1 humingzhe
OK
127.0.0.1:6379> GET key1
"humingzhe"
```

使用 SETNX 命令可以检测 key 是否存在，如检测 key1 aaa，会返回 0，表示 key 已经存在了，不能覆盖。如果创建一个不存在的 key，会返回 1，返回 1 的同时会把定义的 key 和 value 写进去，命令如下。

```
127.0.0.1:6379> SETNX key1 aaa
```

```
(integer) 0
127.0.0.1:6379> SETNX key2 aaa
(integer) 1
127.0.0.1:6379> GET key2
"aaa"
```

使用 SETEX 命令可以给某个 key 设置过期时间；如设置 key1 过期时间为 10 秒，等待 10 秒后再获取值，会返回不存在，命令如下。

```
127.0.0.1:6379> SETEX key1 10 1 #创建后设置过期时间
OK
127.0.0.1:6379> GET key1
(nil)
127.0.0.1:6379> set key1 humingzhe ex 10 # 创建时设置过期时间
OK
127.0.0.1:6379> GET key1
(nil)
```

2. list 类型常用操作

LPUSH 命令是从左边加入一个元素，LPOP 命令是从左边取出第一个元素，RPUSH 命令是从右边加入一个元素，RPOP 命令是从右边取出第一个元素，命令如下。

```
127.0.0.1:6379> LPUSH list3 aaa
(integer) 1
127.0.0.1:6379> LPUSH list3 bbb
(integer) 2
127.0.0.1:6379> LRANGE list3 0 -1
1) "bbb"
2) "aaa"
127.0.0.1:6379> LPOP list3
"bbb"
127.0.0.1:6379> RPOP list3
"aaa"
```

LINSERT 命令用来插入元素，如 LINSERT list2 before bbb 3，表示在 bbb 的前面插入一个元素 3，命令如下。

```
127.0.0.1:6379> LINSERT list2 before bbb 3
(integer) 4
127.0.0.1:6379> LRANGE list2 0 -1
1) "ccc"
2) "3"
```

```
3) "bbb"
4) "aaa"
```

LSET 命令用来修改某个元素，如将 list2 中的元素 3 修改为元素 ddd，命令如下。

```
127.0.0.1:6379> LSET list2 1 ddd
OK
127.0.0.1:6379> LRANGE list2 0 -1
1) "ccc"
2) "ddd"
3) "bbb"
4) "aaa"
```

使用 LINDEX 命令可以查看元素，如查看第 4 个元素，命令如下。

```
127.0.0.1:6379> LINDEX list2 3
"aaa"
```

使用 LLEN 命令可以查看链表中有多少个元素，命令如下。

```
127.0.0.1:6379> LLEN list2
(integer) 4
```

3. set 类型常用操作

使用 sadd set1 aaa 命令可以向集合 set1 中加入 aaa 元素，使用 SMEMBERS set1 命令可以查看 set1 集合中所有的元素，命令如下。

```
127.0.0.1:6379> sadd set1 aaa
(integer) 1
127.0.0.1:6379> SMEMBERS set1
1) "aaa"
2) "b"
3) "a"
```

使用 srem set1 aaa 命令可以删除 set1 集合中的 aaa 元素，使用 spop set1 命令可以随机抽取 set1 集合中的一个元素并删除，命令如下。

```
127.0.0.1:6379> srem set1 aaa
(integer) 1
127.0.0.1:6379> spop set1
"b"
127.0.0.1:6379> SMEMBERS set1
1) "a"
```

使用 sdiff 命令可以求差集，如 sdiff set1 set2，set1 和 set2 谁在前面就以谁为标准求差集；使用 sdiffstore 可以求差集并存储，如 sdiffstore set1 set2 set3，表示把求到的差集存储到 set1 中，命令如下。

```
127.0.0.1:6379> sdiff set1 set2
1) "aaa"
2) "bbb"
3) "ccc"
127.0.0.1:6379> sdiffstore set1 set2 set3
(integer) 6
127.0.0.1:6379> SMEMBERS set1
```

SINTER 命令表示求交集，SINTERSTORE 命令表示求交集并存储，命令如下。

```
127.0.0.1:6379> SINTER set1 set2
127.0.0.1:6379> SINTERSTORE set1 set2 set3
```

SUNION 命令表示求并集，如 sunion set1 set2；SUNIONSTORE 命令表示求并集并存储，命令如下。

```
127.0.0.1:6379> SUNION set1 set2
127.0.0.1:6379> SUNIONSTORE set3 set1 set2
```

SISMEMBER 命令用来判断一个元素是否属于一个集合，命令如下。

```
127.0.0.1:6379> SISMEMBER set3 aaa
(integer) 0
127.0.0.1:6379> SISMEMBER set3 4412
(integer) 1
```

SRANDMEMBER 命令表示取出一个随机的元素，但不进行删除操作，也可以指定数字，如指定数字 4 表示取出 4 个元素，命令如下。

```
127.0.0.1:6379> SRANDMEMBER set3
"eee"
127.0.0.1:6379> SRANDMEMBER set3 4
1) "213"
2) "a"
3) "3"
4) "eee"
```

4. Zset 类型常用操作

ZADD 命令用于创建有序集合，如 ZADD zset1 11 23；ZRANGE 命令用于显示有序集合元素，如 ZRANGE zset1 0 -1，表示显示所有有序集合元素，命令如下。

```
127.0.0.1:6379> ZADD zset1 11 23
(integer) 1
127.0.0.1:6379> ZADD zset1 11 aaa
```

```
(integer) 1
127.0.0.1:6379> ZRANGE zset1 0 -1
1) "23"
2) "aaa"
```

使用 ZREM 命令可以删除指定元素；使用 ZRANK 命令可以返回元素的索引值，索引值从 0 开始，按 score 值正向排序；ZREVRANK 命令的作用和 ZRANK 命令的作用不同，它是按 score 值反向排序的，命令如下。

```
127.0.0.1:6379> ZREM zset1 aaa
(integer) 1
127.0.0.1:6379> ZRANGE zset1 0 -1
127.0.0.1:6379> ZRANGE zset1 0 -1
1) "23"
2) "aaaa"
3) "bbbb"
4) "dddd"
5) "dddf"
6) "dddsadf"
127.0.0.1:6379> ZRANK zset1 dddd
(integer) 3
127.0.0.1:6379> ZREVRANK zset1 dddd
(integer) 2
```

使用 ZREVRANGE 命令可以反序显示所有元素，并带分值；使用 ZCARD 命令可以返回集合中所有元素的个数，命令如下。

```
127.0.0.1:6379> ZREVRANGE zset1 0 -1
1) "dddsadf"
2) "dddf"
3) "dddd"
4) "bbbb"
5) "aaaa"
6) "23"
127.0.0.1:6379> ZCARD zset1
(integer) 6
```

使用 ZCOUNT 命令可以返回分值范围内的个数，如 ZCOUNT zset1 10 30，命令如下。

```
127.0.0.1:6379> ZCOUNT zset1 10 30
(integer) 3
```

使用 ZRANGEBYSCORE 命令可以返回分值范围内的元素，如 zrangebyscore

zset1 10 30，命令如下。

```
127.0.0.1:6379> ZRANGEBYSCORE zset1 10 30
```

使用 ZREMRANGEBYRANK 命令可以删除指定索引范围内的元素，按 score 值正向排序，命令如下。

```
127.0.0.1:6379> ZREMRANGEBYRANK zset1 1 10
(integer) 5
```

使用 ZREMRANGEBYSCORE 命令可以删除指定分值范围内的元素，命令如下。

```
127.0.0.1:6379> ZREMRANGEBYSCORE zset1 10 30
(integer) 1
```

5. Hash 类型常用操作

使用 HSET 命令可以创建 hash，使用 HMSET 命令可以批量创建键值对，命令如下。

```
127.0.0.1:6379> HSET user name humingzhe
(integer) 1
127.0.0.1:6379> HSET user age 16
(integer) 1
127.0.0.1:6379> HSET user job Financial_Quant
(integer) 1
127.0.0.1:6379> HGETALL  user
127.0.0.1:6379> HMSET user1 name humingzhe age 18 job it
OK
127.0.0.1:6379> HMGET user1 name age job
1) "humingzhe"
2) "18"
3) "it"
```

使用 HDEL 命令可以删除指定的区域，如删除 name，命令如下。

```
127.0.0.1:6379> HDEL user1 name
(integer) 1
127.0.0.1:6379> HGETALL user1
1) "age"
2) "18"
3) "job"
4) "it"
```

使用 HKEYS 命令可以获取哈希表中的字段，使用 HVALS 命令可以获取哈希表中所有的值，使用 HLEN 命令可以获取哈希表中字段的数量，命令如下。

```
127.0.0.1:6379> HKEYS user1
1) "age"
2) "job"
127.0.0.1:6379> HVALS user1
1) "18"
2) "it"
127.0.0.1:6379> HLEN user1
(integer) 2
```

8.3.6 Redis 操作键值

使用 KEYS *命令可以删除 Redis 中存储的所有 key；KEYS 命令也支持模糊匹配，比如匹配以 list 开头的 key，命令如下。

```
127.0.0.1:6379> KEYS *
127.0.0.1:6379> KEYS list*
1) "list1"
2) "list"
3) "list2"
```

使用 EXISTS 命令可以判断一个 key 是否存在，如果存在，则返回 1，否则返回 0，命令如下。

```
127.0.0.1:6379> EXISTS name
(integer) 1
127.0.0.1:6379> EXISTS namea
(integer) 0
```

使用 DEL 命令可以删除指定的 key，命令如下。

```
127.0.0.1:6379> DEL name
(integer) 1
```

使用 EXPIRE 命令可以给 key 设置过期时间，如给 set1 设置 10 秒的过期时间，命令如下。

```
127.0.0.1:6379> EXPIRE set1 10
(integer) 1
127.0.0.1:6379> GET set1
(nil)
```

使用 ttl 命令可以查看 key 还有多长时间过期，默认单位是秒，命令如下。

```
127.0.0.1:6379> EXPIRE age 20
(integer) 1
```

```
127.0.0.1:6379> ttl age
(integer) 16
127.0.0.1:6379> ttl age
(integer) -2
127.0.0.1:6379> GET age
(nil)
```

注意

当 key 不存在时，返回-2；当 key 存在但没有设置剩余生存时间时，返回-1，否则返回 key 的剩余生存时间。

使用 SELECT 命令可以选择具体的数据库，select 0 表示选择当前数据库，默认进入的是“0 数据库”，命令如下。

```
127.0.0.1:6379> SELECT 0
OK
127.0.0.1:6379> SELECT 1
OK
127.0.0.1:6379[1]> KEYS *
(empty list or set)
```

使用 MOVE 命令可以将 key 移动到指定的数据库中，如将 user 移动到“1 数据库”中，命令如下。

```
127.0.0.1:6379> MOVE user 1
(integer) 1
127.0.0.1:6379> SELECT 1
OK
127.0.0.1:6379[1]> KEYS *
1) "user"
```

使用 PERSIST 命令可以取消 key 的过期时间，命令如下。

```
127.0.0.1:6379> EXPIRE list 3000
(integer) 1
127.0.0.1:6379> ttl list
(integer) 2996
127.0.0.1:6379> PERSIST list
(integer) 1
127.0.0.1:6379> ttl list
(integer) -1
```

使用 RANDOMKEY 命令可以随机返回一个 key，命令如下。

```
127.0.0.1:6379> RANDOMKEY
```

```
"list1"
127.0.0.1:6379> RANDOMKEY
"key2"
```

使用 RENAME 命令可以重命名一个 key，如把 list1 重命名为 list11，命令如下。

```
127.0.0.1:6379> RENAME list1 list11
OK
```

使用 TYPE 命令可以返回 key 的数据类型，如查看 list11、set2 的数据类型，命令如下。

```
127.0.0.1:6379> TYPE list11
list
127.0.0.1:6379> TYPE set2
set
127.0.0.1:6379> TYPE user1
hash
```

使用 DBSIZE 命令可以返回当前数据库中 key 的数目，使用 INFO 命令可以返回 Redis 数据库的状态信息，命令如下。

```
127.0.0.1:6379> DBSIZE
(integer) 10
127.0.0.1:6379> KEYS *
127.0.0.1:6379> INFO
```

使用 FLUSHALL 命令可以清空数据库中的所有 key；使用 FLUSHDB 命令可以清空当前数据库中的所有 key，命令如下。

```
127.0.0.1:6379> SELECT 1
OK
127.0.0.1:6379[1]> KEYS *
1) "user"
127.0.0.1:6379[1]> FLUSHDB
OK
127.0.0.1:6379[1]> KEYS *
(empty list or set)
```

使用 BGSAVE 命令可以保存数据到 RDB 文件中，并在后台运行。SAVE 命令的作用和 BGSAVE 命令的作用一致，SAVE 命令在前台运行，命令如下。

```
127.0.0.1:6379[1]> BGSAVE
Background saving started
127.0.0.1:6379[1]> SAVE
OK
```

使用 CONFIG GET *命令可以获取所有配置的参数；使用 CONFIG GET 命令可以获取配置的参数，使用 CONFIG SET 命令可以更改配置的参数，命令如下。

```
127.0.0.1:6379[1]> CONFIG GET *
127.0.0.1:6379[1]> CONFIG GET port
1) "port"
2) "6379"
127.0.0.1:6379[1]> CONFIG SET timeout 100
OK
127.0.0.1:6379[1]> CONFIG GET timeout
1) "timeout"
2) "100"
```

Redis 的数据恢复首先要定义或确定 dir 目录和 dbfilename，然后把备份的 RDB 文件放到 dir 目录下面，重启 Redis 服务即可恢复数据，命令如下。

```
127.0.0.1:6379> SAVE
OK
127.0.0.1:6379> CONFIG GET dir
1) "dir"
2) "/data/redis"
[root@nosql ~]# ls /data/redis/
appendonly.aof  dump.rdb
[root@nosql ~]# redis-cli -p 6379 shutdown
[root@nosql ~]# redis-server /etc/redis.conf
```

8.3.7 Redis 安全设置

Redis 服务默认监听的是公网 IP，而不是内网 IP。如果服务器没有设置 iptables 规则，6379 端口没有限制，公网 IP 也开放着，那么就可以远程连接 Redis 服务器。前面讲过 Redis 服务默认是没有密码的，所以人人都可以连接，最终会导致系统被“黑”。解决该问题的方法是设置 Redis 密码、监听内网 IP 加防火墙或以普通用户的身份启动。

编辑 redis.conf 配置文件，设置 Redis 服务密码，修改完毕后重启 Redis 服务，命令如下。

```
[root@nosql ~]# vim /etc/redis.conf
requirepass humingzhe+-.com
[root@nosql ~]# killall redis-server
[root@nosql ~]# redis-server /etc/redis.conf
```

使用 redis-cli 命令依然可以进入 Redis 终端窗口，但是执行命令时会报错。若想登录，使用 redis-cli -a 命令加密码即可，命令如下。

```
[root@nosql ~]# redis-cli
127.0.0.1:6379> KEYS *
(error) NOAUTH Authentication required.
[root@nosql ~]# redis-cli -a 'humingzhe+-.com'
127.0.0.1:6379> KEYS *
1) "zset1"
```

将 Redis 服务中的 config 命令重命名，如重命名为 financial，或者禁掉 config 命令，命令如下。

```
[root@nosql ~]# vim /etc/redis.conf
rename-command CONFIG Financial
[root@nosql ~]# killall redis-server
[root@nosql ~]# redis-server /etc/redis.conf
127.0.0.1:6379> CONFIG GET dir
(error) ERR unknown command 'CONFIG'
127.0.0.1:6379> financial get dir
1) "dir"
2) "/data/redis"
[root@nosql ~]# vim /etc/redis.conf
rename-command CONFIG ""
127.0.0.1:6379> CONFIG GET dir
(error) ERR unknown command 'CONFIG'
127.0.0.1:6379> financial get dir
(error) ERR unknown command 'financial'
```

8.3.8　Redis 慢查询日志

MySQL 有慢查询日志功能，同样的，Redis 也有该功能，编辑 redis.conf 配置文件就可以开启慢查询日志功能，命令如下。

```
[root@nosql ~]# vim /etc/redis.conf
slowlog-log-slower-than 10000 #单位是微秒，表示慢于 10000 微秒就记录日志
slowlog-max-len 128 #定义日志长度，最多存 128 条
```

注意

慢查询日志可以设置两个参数，一个是执行时长，单位是微秒；另外一个是慢查询日志的长度。当一个新的命令被写入日志时，最早的一条命令会从日志队列中移除。

使用 SLOWLOG get 命令可以列出所有的慢查询日志。SLOWLOG get 命令后面加数字，表示列出几条命令，如 SLOWLOG get 3 表示列出 3 条命令、SLOWLOG get 1 表示列出最新的一条命令。使用 SLOWLOG len 命令可以查看慢查询日志条数，命令如下。

```
127.0.0.1:6379> SLOWLOG get
127.0.0.1:6379> SLOWLOG get  3
127.0.0.1:6379> SLOWLOG len
(integer) 7
```

8.3.9 PHP 安装 Redis 扩展模块

下面讲解如何在 PHP 中使用 Redis 服务。进入到/usr/local/src/目录，安装 Redis 扩展模块（模块在读者 QQ 群文件中，文件名是 phpredis.zip），命令如下。

```
[root@nosql ~]# cd /usr/local/src/
[root@nosql src]# unzip phpredis.zip
[root@nosql src]# cd phpredis-develop/
[root@nosql phpredis-develop]# /usr/local/php-fpm/bin/phpize
[root@nosql phpredis-develop]# ./configure
--with-php-config=/usr/local/php-fpm/bin/php-config
[root@nosql phpredis-develop]# make && make install
```

编辑 php.ini 配置文件，在 extension=memcache.so 下面添加一行代码 extension=redis.so，使 php-fpm 识别 Redis 模块，命令如下。

```
[root@nosql phpredis-develop]# vim /usr/local/php-fpm/etc/php.ini
extension=memcache.so
extension=redis.so
[root@nosql phpredis-develop]# php-fpm -m
redis
[root@nosql phpredis-develop]# /etc/init.d/php-fpm restart
```

8.3.10 Redis 存储 session

Redis 存储 session 的操作和 Memcached 存储 session 的操作是一样的，在对应的 pool 池中添加存储 session 的代码即可，命令如下。

```
[root@nosql ~]# vim /usr/local/php-fpm/etc/php-fpm.d/mingzhe.conf
[mingzhe]
```

```
php_value[session.save_handler] = redis
php_value[session.save_path] = " tcp://127.0.0.1:6379 "
[root@nosql ~]# /etc/init.d/php-fpm restart
```

将读者 QQ 群文件中的 redis_session.php 文件上传到对应目录，命令如下。

```
[root@nosql ~]# curl localhost/redis_session.php
1518972701<br><br>1518972701<br><br>3h5kps5mgukg0vstdvbahnsni7
[root@nosql ~]# curl localhost/redis_session.php
1518972703<br><br>1518972703<br><br>98pjv7bqbug7rksmkrafvnsgd1
[root@nosql ~]# redis-cli
127.0.0.1:6379> KEYS *
1) "PHPREDIS_SESSION:98pjv7bqbug7rksmkrafvnsgd1"
2) "PHPREDIS_SESSION:3h5kps5mgukg0vstdvbahnsni7"
127.0.0.1:6379> GET "PHPREDIS_SESSION:98pjv7bqbug7rksmkrafvnsgd1"
"TEST|i:1518972703;TEST3|i:1518972703;"
```

8.3.11　Redis 主从配置

Redis 主从配置和 MySQL 主从配置类似，而且配置起来比 MySQL 主从配置简单得多。为了方便测试，在一台服务器上启动两个端口来模拟两台 Redis 服务器。

复制 redis.conf 配置文件，然后编辑 redis2.conf 配置文件，修改端口号、日志文件路径、pid、主 IP 和端口号，如果在主上设置了 Redis 密码，那么在从上就需要开启密码功能，命令如下。

```
[root@nosql ~]# cp /etc/redis.conf /etc/redis2.conf
[root@nosql ~]# vim /etc/redis2.conf
port 9520
pidfile /var/run/redis_9520.pid
logfile "/var/log/redis2.log"
dir /data/redis2
masterauth humingzhe+-.com
slaveof 127.0.0.1 6379
```

在/data/目录下创建 redis2 目录，创建后启动 redis2 服务，命令如下。

```
[root@nosql ~]# mkdir /data/redis2
[root@nosql ~]# redis-server /etc/redis2.conf
[root@nosql ~]# netstat -lntp
tcp     0    0 127.0.0.1:6379     0.0.0.0:*     LISTEN    124693/redis-server
tcp     0    0 0.0.0.0:11211      0.0.0.0:*     LISTEN    2231/memcached
tcp     0    0 127.0.0.1:9520     0.0.0.0:*     LISTEN    130638/redis-server
```

注意

Redis 主从无需事先同步数据，会自动同步主上的数据。

登录主 Redis 服务，创建一些数据，再登录从 Redis 服务会自动同步数据，命令如下。

```
[root@nosql ~]# redis-cli -h 127.0.0.1 -p 6379
127.0.0.1:6379> KEYS *
1) "job"
2) "age"
3) "name"
127.0.0.1:6379> quit
[root@nosql ~]# redis-cli -h 127.0.0.1 -p 9520
127.0.0.1:9520> KEYS *
1) "name"
2) "age"
3) "job"
```

8.3.12 Redis 集群简介

Redis 集群官方称为 Redis Cluster，是从 Redis 3.0 版本才开始支持的架构。当数据量非常大时，单台机器就不能够满足存储和查询了，需要用多台服务器构成集群来解决存储空间、查询速度、负载过高等瓶颈问题。

Redis Cluster 是一个分布式集群，支持横向扩展，服务器加入集群中很容易，就好比 LVS 做负载均衡去运行 Nginx、PHP 服务，每增加一个节点，仅需要在节点上编写代码并部署好环境。Redis 集群解决方案除官方的 Redis Cluster 外，还有 Codis。Codis 是第三方公司提供的一种 Redis 集群解决方案，原理类似于 LVS 做负载均衡，用的是路由转发的机制。

官方的 Redis Cluster 支持把所有的节点进行互联，能够实现数据的共享。所有的节点可以是一主一从，也可以是一主多从。其中，从不提供服务，仅作为备用。它不支持同时处理多个键，如 mget/mget，因为 Redis 需要把键均分在各个节点上，在并发量很高的情况下同时创建键值会降低性能并导致不可预测的事情发生。

Redis Cluster 支持在线增加和删除节点。客户端可以连接任何一个主节点进行读写操作。

8.3.13　Redis 集群搭建与配置

准备两台服务器，分别开启 3 个 Redis 服务（Redis Port），并关闭 iptables 和 selinux。NoSQL 服务器上的 3 个端口分别是 9191、9193 和 9195，全部为主。Node 服务器上的 3 个端口分别是 9192、9194 和 9196，全部为从。两台服务器上都需要编译安装 Redis 服务，然后复制并编辑 3 个不同的 redis.conf 配置文件，分别设置不同的端口号、dir、logfile、pid 等参数，还需要增加 Cluster 相关参数，最后分别启动 6 个 Redis 服务。

搭建 Redis Cluster 需要 Ruby 环境支持，yum 源中虽然有 Ruby，但是版本是 2.0，不能给最新版的 Redis 提供支持。Ruby 环境在一台机器上运行即可。

安装开发工具 Development Tools 和所支持的依赖环境，命令如下。

```
    [root@nosql ~]# yum groupinstall -y "Development Tools" "Server Platform
Development"
    [root@nosql ~]# yum -y install gdbm-devel libdb4-devel libffi-devel
libyaml libyaml-devel ncurses-devel openssl-devel readline-devel tcl-deve
```

在主上和从上分别创建 redis_9191、redis_9192、redis_9193、redis_9194、redis_9195、redis_9193 这 6 个配置文件，命令如下。

```
    [root@nosql ~]# touch /etc/redis_{9191,9193,9195}.conf
    [root@node ~]# touch /etc/redis_{9192,9194,9196}.conf
```

在 6 个 redis_.conf 配置文件中设置集群端口和 pid，配置文件代码如下（6 个配置文件的具体代码可参考读者 QQ 群文件中的相应配置文件）。

```
    [root@node ~]# vim /etc/redis_9196.conf
    port 9196
    bind 192.168.222.132
    daemonize yes
    pidfile /var/run/redis_9196.pid
    dir /data/redis_data/9196
    cluster-enabled yes
    cluster-config-file nodes_9196.conf
    cluster-node-timeout 10100
    appendonly yes
```

分别创建 6 个 redis-server 服务的启动目录，命令如下。启动 6 个 Redis 服务，如图 8-2 所示。

```
    [root@nosql ~]# mkdir -p /data/redis_data/{9191,9193,9195}
    [root@nosql ~]# ls /data/redis_data/919
```

```
    9191/ 9193/ 9195/
    [root@node ~]# mkdir -p /data/redis_data/{9192,9194,9196}
    [root@node ~]# ls /data/redis_data/919
    9192/ 9194/ 9196/
```

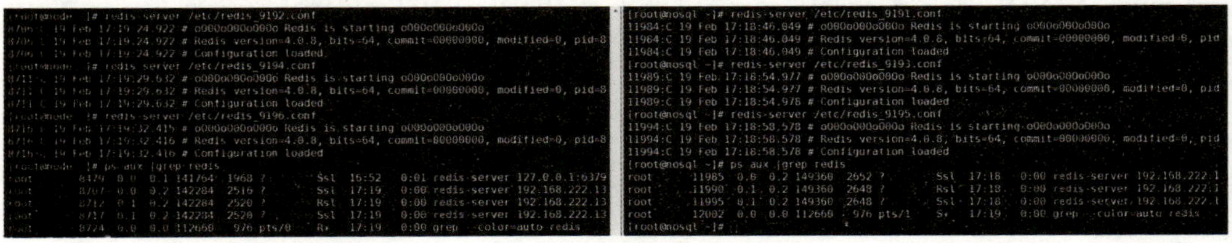

图 8-2

创建 Ruby.rpm 包所需要的目录，创建后下载所需的软件包，命令如下。

```
    [root@nosql ~]# mkdir -p
rpmbuild/{BUILD,BUILDROOT,RPMS,SOURCES,SPECS,SRPMS}
    [root@nosql ~]# wget
http://cache.ruby-lang.org/pub/ruby/2.2/ruby-2.2.3.tar.gz -P
rpmbuild/SOURCES
    [root@nosql ~]# wget
https://raw.githubusercontent.com/tjinjin/automate-ruby-rpm/master/ruby2
2x.spec -P rpmbuild/SPECS
    [root@nosql ~]# yum -y install rpm-build
```

将 Ruby 源码包制作成 RPM 包，命令如下。

```
    [root@nosql ~]# rpmbuild -bb rpmbuild/SPECS/ruby22x.spec
    [root@nosql ~]# ls
rpmbuild/RPMS/x86_64/ruby-2.2.3-1.el7.centos.x86_64.rpm
    rpmbuild/RPMS/x86_64/ruby-2.2.3-1.el7.centos.x86_64.rpm
    [root@nosql ~]# du -sh !$
    du -sh rpmbuild/RPMS/x86_64/ruby-2.2.3-1.el7.centos.x86_64.rpm
    10M rpmbuild/RPMS/x86_64/ruby-2.2.3-1.el7.centos.x86_64.rpm
    [root@nosql ~]# yum -y localinstall
rpmbuild/RPMS/x86_64/ruby-2.2.3-1.el7.centos.x86_64.rpm
    [root@nosql ~]# ruby --version
    ruby 2.2.3p173 (2015-08-18 revision 51636) [x86_64-linux]
    [root@nosql ~]# gem install redis
```

将 redis-trib.rb 文件复制到/usr/bin/目录下，命令如下。

```
    [root@nosql ~]# cp /usr/local/src/redis-4.0.8/src/redis-trib.rb
/usr/bin/
    [root@nosql ~]# redis-trib.rb
```

将 redis-trib.rb 文件复制到/usr/bin/目录下后可直接运行，但需要 Ruby 环境的支持。

使用 redis-trib.rb 命令创建 Redis Cluster，命令如下。

```
[root@nosql ~]# redis-trib.rb create --replicas 1 192.168.222.130:9191
192.168.222.130:9193 192.168.222.130:9195 192.168.222.132:9192
192.168.222.132:9194 192.168.222.132:9196
```

在提示“Can I set the above configuration? (type 'yes' to accept):”时输入 yes，当下方出现两个 OK 时表示集群设置成功，如图 8-3 所示。

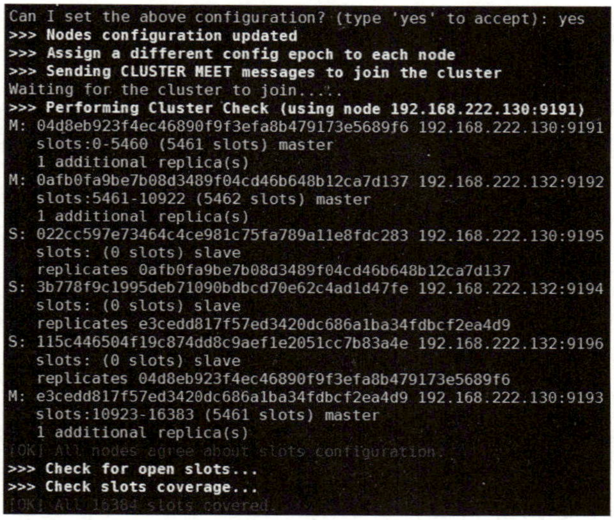

```
Can I set the above configuration? (type 'yes' to accept): yes
>>> Nodes configuration updated
>>> Assign a different config epoch to each node
>>> Sending CLUSTER MEET messages to join the cluster
Waiting for the cluster to join....
>>> Performing Cluster Check (using node 192.168.222.130:9191)
M: 04d8eb923f4ec46890f9f3efa8b479173e5689f6 192.168.222.130:9191
   slots:0-5460 (5461 slots) master
   1 additional replica(s)
M: 0afb0fa9be7b08d3489f04cd46b648b12ca7d137 192.168.222.132:9192
   slots:5461-10922 (5462 slots) master
   1 additional replica(s)
S: 022cc597e73464c4ce981c75fa789a11e8fdc283 192.168.222.130:9195
   slots: (0 slots) slave
   replicates 0afb0fa9be7b08d3489f04cd46b648b12ca7d137
S: 3b778f9c1995deb71090bdbcd70e62c4ad1d47fe 192.168.222.132:9194
   slots: (0 slots) slave
   replicates e3cedd817f57ed3420dc686a1ba34fdbcf2ea4d9
S: 115c446504f19c874dd8c9aef1e2051cc7b83a4e 192.168.222.132:9196
   slots: (0 slots) slave
   replicates 04d8eb923f4ec46890f9f3efa8b479173e5689f6
M: e3cedd817f57ed3420dc686a1ba34fdbcf2ea4d9 192.168.222.130:9193
   slots:10923-16383 (5461 slots) master
   1 additional replica(s)
[OK] All nodes agree about slots configuration.
>>> Check for open slots...
>>> Check slots coverage...
[OK] All 16384 slots covered.
```

图 8-3

8.3.14　Redis 集群操作

因为是分布式集群，所以可以连接任何一个端口，使用-c 选项可以以集群的方式登录，登录后可以创建 key，命令如下。

```
[root@nosql ~]# redis-cli -c -h 192.168.222.130 -p 9191
192.168.222.130:9191> SET name humingzhe
-> Redirected to slot [5798] located at 192.168.222.132:9192
OK
```

注意

每次存储的节点都不一致，会提示重定向到了其他的节点上，当不提示重定向时，表示存储到本地。

使用 redis-trib.rb check 命令可以检测集群的状态，命令如下。

```
[root@nosql ~]# redis-trib.rb check 192.168.222.130:9195
[root@nosql ~]# redis-trib.rb check 192.168.222.130:9193
```

登录集群后，使用 CLUSTER NODES 命令可以列出节点，使用 CLUSTER INFO 命令可以列出集群信息，使用 CLUSTER MEET ip port 命令可以添加节点，使用 CLUSTER FORGET node_id 命令可以移除某个节点，使用 CLUSTER REPLICATE node_id 命令可以将当前节点设置为指定节点的从，使用 CLUSTER SAVECONFIG 命令可以保存配置文件，示例命令如下。

```
[root@nosql ~]# redis-cli -c -h 192.168.222.130 -p 9191
192.168.222.130:9191> CLUSTER NODES
192.168.222.130:9191> CLUSTER INFO
192.168.222.130:9191> CLUSTER MEET 192.168.222.132 32283
OK
192.168.222.130:9191> CLUSTER FORGET
e8fd8e5b80f8adfb4c2078c21cd17ac385158964
OK
192.168.222.130:9195> CLUSTER REPLICATE
0afb0fa9be7b08d3489f04cd46b648b12ca7d137
OK
192.168.222.130:9195> CLUSTER SAVECONFIG
OK
```

使用 CLUSTER SAVECONFIG 命令保存的配置文件路径是/data/redis_data/9195/nodes_9194.conf，命令如下。

```
[root@nosql ~]# ls /data/redis_data/9195/nodes_9194.conf
/data/redis_data/9195/nodes_9194.conf
```

8.4 MongoDB

8.4.1 MongoDB 简介

MongoDB 的官方网址是 https://www.mongodb.com/，MongoDB 由 CPP 编写，最新版本是 3.6。MongoDB 属于文档型数据库，将数据存储在类似 JSON 的灵活文档中，这意味着字段可能因文档而异，数据结构可随时间而改变。文档模型映射到应用程序代码中的对象，使数据易于使用。即席查询、索引和实时聚合提供了访问和分析数据的强大方法。MongoDB 是一个分布式数据库，它的核心是高可用性、横向扩展和地理分布，并且易于使用。MongoDB 是免费且开源的，在 GNU Affero General Public License 上发布。

JSON 是 JavaScript 对象表示法，类似于 XML，但是 JSON 比 XML 更小、更快、更易解析。JSON 示例代码如下。

```
{
    "employees": [
        { "firstName":"Bill" , "lastName":"Gates" },
        { "firstName":"George" , "lastName":"Bush" },
        { "firstName":"Thomas" , "lastName":"Carter" }
    ]
}
```

8.4.2　安装 MongoDB

epel 源中自带的 MongoDB 版本是 2.6，目前最新版本是 3.6，所以安装最新的 3.6 版本进行演示。

官方安装文档的网址是 https://docs.mongodb.com/manual/tutorial/install-mongodb-on-red-hat/。

创建 MongoDB 官方所提供的 mongodb-org-3.6.repo 文件，在该文件中添加如下代码。

```
[root@nosql ~]# vim /etc/yum.repos.d/mongodb-org-3.6.repo
[mongodb-org-3.6]
name=MongoDB Repository
baseurl=https://repo.mongodb.org/yum/redhat/$releasever/mongodb-org/3
.6/x86_64/
gpgcheck=1
enabled=1
gpgkey=https://www.mongodb.org/static/pgp/server-3.6.asc
```

设置好 MongoDB 3.6 版本的 yum 源后，使用 yum list|grep mongodb 命令即可列出最新的 MongoDB 3.6 版本的 RPM 包，用 yum 命令安装 MongoDB 数据库，安装成功后启动 MongoDB 服务，命令如下。

```
[root@nosql ~]# yum list|grep mongodb
collectd-write_mongodb.x86_64      5.8.0-1.el7       epel
mongodb.x86_64                     2.6.12-6.el7      epel
mongodb-org.x86_64                 3.6.2-1.el7       mongodb-org-3.6
mongodb-org-mongos.x86_64          3.6.2-1.el7       mongodb-org-3.6
```

```
mongodb-org-server.x86_64          3.6.2-1.el7          mongodb-org-3.6
mongodb-org-shell.x86_64           3.6.2-1.el7          mongodb-org-3.6
mongodb-org-tools.x86_64           3.6.2-1.el7          mongodb-org-3.6
[root@nosql ~]# yum install -y mongodb-org
[root@nosql ~]# systemctl start mongod.service
```

8.4.3 连接 MongoDB

MongoDB 默认监听 27017 端口。使用 mongo 命令可以进入 MongoDB 终端管理界面，如果默认监听的不是 27017 端口，则需在 mongo 命令后面加--port 指定端口号；连接远程 MongoDB，需加--host 选项指定 IP 地址，如果设置了密码验证，在连接时需要输入用户名和密码，命令如下。

```
[root@nosql ~]# mongo
[root@nosql ~]# mongo --port 27017
[root@nosql ~]# mongo --port 27017 --host 127.0.0.1
[root@nosql ~]# mongo --port 27017 -u "admin" -p "humingzhe"
--authenticationDatabase "admin"
```

8.4.4 MongoDB 用户管理

进入 MongoDB 终端管理界面，切换到 admin 库，创建用户名及密码，命令如下。

```
[root@nosql ~]# mongo
> use admin
switched to db admin
> db.createUser( { user: "admin", customData: {description: "superuser"},
pwd: "humingzhe", roles: [ { role:"root", db: "admin" } ] } )
Successfully added user: {
    "user" : "admin",
    "customData" : {
        "description" : "superuser"
    },
    "roles" : [
        {
            "role" : "root",
            "db" : "admin"
        }
```

```
    ]
}
```

注意

use admin 命令表示切换到 admin 库。user 表示用户；customDate 表示字段说明，此处可省略；pwd 表示用户密码；roles 表示指定用户的角色；db 表示指定库名。

使用 db.system.users.find()命令可以列出所有的用户，在使用该命令前需切换到 admin 库，否则不会输出任何信息，命令如下。

```
> use admin
switched to db admin
> db.system.users.find()
{ "_id" : "test.admin", "user" : "admin", "db" : "test", "credentials" : { "SCRAM-SHA-1" : { "iterationCount" : 10000, "salt" : "OnIMY+O9sG2Ez/ZeflQ+HQ==", "storedKey" : "oTot6uzUnCbtBudoNktvFRBrb8Y=", "serverKey" : "YR/AkPUKaE59F5urIRJ5/lut5QU=" } }, "customData" : { "description" : "superuser" }, "roles" : [ { "role" : "root", "db" : "admin" } ] }
```

使用 show users 命令可以查看当前库中的所有用户，命令如下。

```
> show users
{
    "_id" : "admin.admin",
    "user" : "admin",
    "db" : "admin",
    "customData" : {
        "description" : "superuser"
    },
    "roles" : [
        {
            "role" : "root",
            "db" : "admin"
        }
    ]
}
```

使用 db.dropUser()命令可以删除用户，如删除 admin 用户，命令如下。

```
> db.dropUser("admin")
true
```

授权用户后若想生效，需编辑 mongodb.service 启动脚本，在“OPTIONS=”后面添加“--auth”，编辑后重启 MongoDB 服务即可使用用户名和密码登录，命令如下。

```
[root@nosql ~]# vim /usr/lib/systemd/system/mongod.service
Environment="OPTIONS=--auth -f /etc/mongod.conf"
[root@nosql ~]# systemctl daemon-reload
[root@nosql ~]# systemctl restart mongod.service
[root@nosql ~]# mongo --host 192.168.222.130 --port 27017 -u "bone" -p
"humingzhe" --authenticationDatabase "admin"
```

切换到 dbone 数据库，授权一个用户 test，使 test 用户对 dbone 数据库有读写权限、对 dbtwo 库只有读权限，命令如下。

```
[root@nosql ~]# mongo --host 192.168.222.130 --port 27017 -u "bone" -p
"humingzhe" --authenticationDatabase "admin"
> use dbone
switched to db dbone
> db.createUser( { user: "test", pwd: "humingzhe", roles: [ { role:
"readWrite", db: "dbone" }, {role: "read", db: "dbtwo" } ] } )
> db.auth('test','humingzhe')
1
```

注意

use dbone 命令表示用户在 dbone 库中进行创建，那么就一定要在 dbone 库中验证身份，即用户的信息跟随数据库。比如，上述 test 用户虽然有 dbtwo 库的读权限，但是一定要先在 dbone 库中进行身份验证，直接访问会提示验证失败。

MongoDB 用户角色管理相关命令如下。

- Read：允许用户读取指定数据库。
- readWrite：允许用户读写指定数据库。
- dbAdmin：允许用户在指定数据库中执行管理函数，如索引的创建和删除、查看统计、访问 system.profile。
- userAdmin：允许用户对 system.users 集合进行写入操作，可以在指定数据库中创建、删除和管理用户。
- clusterAdmin：只在 admin 数据库中可用，赋予用户所有分片和复制集相关函数的管理权限。
- readAnyDatabase：只在 admin 数据库中可用，赋予用户所有数据库的读权限。
- readWriteAnyDatabase：只在 admin 数据库中可用，赋予用户所有数据库的读写权限。

- userAdminAnyDatabase：只在 admin 数据库中可用，赋予用户所有数据库的 userAdmin 权限。
- dbAdminAnyDatabase：只在 admin 数据库中可用，赋予用户所有数据库的 dbAdmin 权限。
- root：只在 admin 数据库中可用。超级账号，超级权限。

MongoDB 对库进行管理的相关命令如下。

- db.version()：查看版本。
- use userdb：如果库存在，就切换，否则就创建。
- show dbs：查看库，此时 userdb 并没有出现，这是因为该库是空的，还没有任何集合，创建一个集合就能看到了。
- db.createCollection('clo1')：创建集合 clo1，在当前库下面创建。
- db.dropDatabase()：删除当前库，要想删除某个库，必须切换到那个库下面。
- db.stats()：查看当前库的信息。
- db.serverStatus()：查看 MongoDB 服务器的状态。

8.4.5　MongoDB 创建集合和数据管理

db.createCollection 命令用来创建集合，语法格式是 db.createCollection(name, options)，命令如下。

```
> use dbone
switched to db dbone
> db.createCollection("mycol", { capped : true, size : 6142800, max : 10000 } )
{ "ok" : 1 }
```

注意

name 是集合的名字，如 mycol。Options 是可选项，用来设置集合的参数，参数有 4 个：capped true/false，如果为 true，则启用封顶集合，封顶集合是固定大小的集合，当它达到最大时，会自动覆盖最早的条目，如果指定 true，那么也要指定尺寸参数；autoindexID true/false，如果为 true，则自动创建索引_id 字段，默认值是 false；size，指定最大字节封顶集合，如果封顶集合是 true，那么还需要指定这个字段，单位是 B；max，指定封顶集合允许在文件中的最大数量。

使用 show tables 命令和 show collections 命令可以查看存在哪些集合，命令如下。

```
> show tables
humingzhe
mycol
> show collections
humingzhe
mycol
```

db.Account.insert()命令用来在集合中插入数据，如果集合不存在，则直接插入数据，MongoDB 会自动创建该集合，命令如下。

```
> db.Account.insert({AccountID:1,UserName:"bone",password:"123456"})
WriteResult({ "nInserted" : 1 })
```

db.Account.update()命令用来更新数据，例如在 AccountID 为 1 的数据中加入 age 为 20，命令如下。

```
> db.Account.update( {AccountID:1},{"$set":{"Age":20}} )
WriteResult({ "nMatched" : 1, "nUpserted" : 0, "nModified" : 1 })
```

db.Account.find()命令用来查看文档中的结果，命令如下。

```
> db.Account.find()
{ "_id" : ObjectId("5a8ae5c3d0e9426982bc3162"), "AccountID" : 1,
"UserName" : "bone", "password" : "123456", "Age" : 20 }
```

db.Account.find({})命令表示根据条件去查询，{}中填写查询的条件，命令如下。

```
> db.Account.find({boneID:1})
{ "_id" : ObjectId("5a8ae5fbd0e9426982bc3163"), "boneID" : 1, "UserName" :
"bone", "password" : "123456" }
{ "_id" : ObjectId("5a8ae608d0e9426982bc3164"), "boneID" : 1, "UserName" :
"boneeli", "password" : "123456" }
```

db.Account.remove({})命令表示根据条件去删除，{}中填写删除的条件，命令如下。

```
> db.Account.remove({boneID:1})
WriteResult({ "nRemoved" : 2 })
```

db.Account.drop()命令用来删除所有的文档（集合），命令如下。

```
> db.Account.drop()
true
> db.Account.find()
> db.mycol.drop()
true
> show tables
```

```
humingzhe
```

db.printCollectionStats()命令用来查看集合的状态，命令如下。

```
> db.printCollectionStats()
```

8.4.6 PHP 的 MongoDB 扩展

PHP 官方有两个 Mongo 扩展，一个是 mongodb.so，另一个是 mongo.so。mongo.so 扩展针对的是 PHP 5.X 版本，而且是老扩展，以后不再使用，而是使用 MongoDB 扩展。现在还有很多线上项目使用的是 PHP 5.X 版本的代码，所以本小节和 8.4.7 小节对两个扩展分别进行讲解。

安装 MongoDB 扩展有两种方法，一种是使用 git 命令直接把 mongo-php-driver 复制到服务器中，另一种是在 Pecl 官方网站下载 MongoDB 的扩展源码包。

使用 git 命令复制到本地服务器需要耗费大量的时间，复制后进入该目录执行 submodule update 命令（执行该命令也需要耗费大量时间），然后使用 phpize 命令生成./configure 命令，命令如下。

```
[root@nosql ~]# cd /usr/local/src/
[root@nosql src]# git clone https://github.com/mongodb/mongo-php-driver
[root@nosql src]# cd mongo-php-driver/
[root@nosql mongo-php-driver]# git submodule update --init
[root@nosql mongo-php-driver]# /usr/local/php-fpm/bin/phpize
[root@nosql mongo-php-driver]# ./configure
--with-php-config=/usr/local/php-fpm/bin/php-config
[root@nosql mongo-php-driver]# make && make install
```

在 Pecl 的官方网站下载最新 1.4.0 版本的 MongoDB 扩展源码包，然后进行编译安装，命令如下。

```
[root@nosql ~]# cd /usr/local/src/
[root@nosql src]# wget https://pecl.php.net/get/mongodb-1.4.0.tgz
[root@nosql src]# tar xf mongodb-1.4.0.tgz
[root@nosql src]# cd mongodb-1.4.0/
[root@nosql mongodb-1.4.0]# /usr/local/php-fpm/bin/phpize
[root@nosql mongodb-1.4.0]# ./configure
--with-php-config=/usr/local/php-fpm/bin/php-config
[root@nosql mongodb-1.4.0]# make && make install
```

编辑 php.ini 配置文件，添加 mongodb.so，重启 php-fpm 使之生效，命令如下。

```
[root@nosql mongodb-1.4.0]# vim /usr/local/php-fpm/etc/php.ini
```

```
extension=mongodb.so
[root@nosql mongodb-1.4.0]# php-fpm -m
[PHP Modules]
mongodb
[root@nosql mongodb-1.4.0]# /etc/init.d/php-fpm restart
```

8.4.7 PHP 的 Mongo 扩展

在 Pecl 的官方网站下载最新 1.6.16 版本的 Mongo 扩展源码包，然后进行编译安装，命令如下。

```
[root@nosql src]# wget https://pecl.php.net/get/mongo-1.6.16.tgz
[root@nosql src]# tar xf mongo-1.6.16.tgz
[root@nosql src]# cd mongo-1.6.16/
[root@nosql mongo-1.6.16]# /usr/local/php-fpm/bin/phpize
[root@nosql mongo-1.6.16]# ./configure
--with-php-config=/usr/local/php-fpm/bin/php-config
[root@nosql mongo-1.6.16]# make && make install
```

编辑 php.ini 配置文件，添加 mongo.so，重启 php-fpm 使之生效，命令如下。

```
[root@nosql mongo-1.6.16]# vim /usr/local/php-fpm/etc/php.ini
extension=mongo.so
[root@nosql mongo-1.6.16]# php-fpm -m
[PHP Modules]
mongo
[root@nosql mongo-1.6.16]# /etc/init.d/php-fpm restart
```

8.4.8 测试 Mongo 扩展

Mongo 安装完毕后，剩下的工作就是测试了。在网站根目录下创建 mongo.php 测试脚本文件，在脚本中添加测试代码，命令如下。

```
[root@nosql ~]# vim /data/www/default/mongo.php
<?php
$m = new MongoClient(); // 连接
$db = $m->test; // 获取名称为 "test" 的数据库
$collection = $db->createCollection("runoob");
echo "集合创建成功";
```

用 curl 命令访问网站根目录下的 mongo.php 脚本文件，如果输出“集合创建成功”，则表示 Mongo 扩展没问题。更准确的方法是去 Mongo 的库中查看 runoob 表是否创建成功，命令如下。

```
[root@nosql ~]# curl localhost/mongo.php
集合创建成功[root@nosql ~]# curl localhost/mongo.php
集合创建成功[root@nosql ~]# curl localhost/mongo.php
[root@nosql ~]# mongo --port 27017 --host 127.0.0.1
> use test
switched to db test
> show tables
runoob
```

注意

必须关闭--auth 用户认证才能够查询到 runoob 表。

8.4.9　MongoDB 副本集简介

MongoDB 的架构类似 MySQL 主从的架构，MongoDB 早期的版本也支持一个主、一个从的模式，和 MySQL 主从架构基本上一致。后期 MongoDB 推出了一个新的架构副本集，目的也是为了实现 MongoDB 的高可用，可以把副本集当作 MySQL 主从一样去用。一主一从或一主二从，目的是为了保证数据的安全，读写仅仅是在主（Primary）上进行，从（Secondary）不需要提供服务，是只读的。

还有一种方式是手动指定读库的目标 Server，假如在代码中实现读写分离，读的时候去其中一个 Secondary（从库）上读，这样也可以实现负载均衡。对于多个从也可以增加 LVS 代理。

8.4.10　MongoDB 副本集搭建

准备 3 台服务器，主机名为 nosql、IP 为 192.168.222.130 的服务器为主（Primary），主机名 node-1、IP 为 192.168.222.131 和主机名为 node-2、IP 为 192.168.222.132 的服务器均为从（Secondary）。

3 台服务器需要全部安装 MongoDB，因为前面的章节已经在 nosql 服务器上安装过 MongoDB，现在只需在其余两台 Secondary 服务器上安装 MongoDB，命令如下。

```
[root@nosql ~]# scp /etc/yum.repos.d/mongodb-org-3.6.repo
```

```
root@192.168.222.131:/etc/yum.repos.d/
    [root@nosql ~]# scp /etc/yum.repos.d/mongodb-org-3.6.repo
root@192.168.222.132:/etc/yum.repos.d/
    [root@node-1 ~]# yum install -y mongodb-org
    [root@node-2 ~]# yum install -y mongodb-org
    [root@node-1 ~]# systemctl start mongod.service
    [root@node-2 ~]# systemctl start mongod.service
```

编辑 3 台服务器的 MongoDB 配置文件，开启 replication 功能，并在下方定义日志大小和副本集名称，命令如下。

```
    [root@nosql ~]# vim /etc/mongod.conf
    replication:
      oplogSizeMB: 20
      replSetName: humingzhe
    bindIp: 127.0.0.1,192.168.222.130
    bindIp: 127.0.0.1,192.168.222.131
    bindIp: 127.0.0.1,192.168.222.132
    [root@nosql ~]# systemctl restart mongod.service
```

连接主，在主上运行 Mongo 命令，执行配置副本集命令，并进行初始化，如果显示 OK 就表示初始化成功，命令如下。

```
    > config={_id:"humingzhe",members:[{_id:0,host:"192.168.222.130:
27017"},{_id:1,host:"192.168.222.131:27017"},{_id:2,host:"192.168.222.13
2:27017"}]}
    > rs.initiate(config)
    {
       "ok" : 1,
    }
```

执行 rs.status()命令可查看副本集的状态，命令如下。

```
    humingzhe:SECONDARY> rs.status()
```

如果两个从服务器的 stateStr 状态是 STARUP，说明没有正常启动，需要重新执行一次 rs.reconfig(config)命令。

8.4.11 MongoDB 副本集测试

在主上创建一个库，然后创建一个集合并进行相应操作，命令如下。

```
    humingzhe:PRIMARY> use mydb
    switched to db mydb
```

```
    humingzhe:PRIMARY>
db.acc.insert({AccountID:1,UserName:"bone",password:"123456"})
    WriteResult({ "nInserted" : 1 })
```

查询创建的 mydb 库，查看库中的表，命令如下。

```
humingzhe:PRIMARY> show dbs
admin   0.000GB
config  0.000GB
dbone   0.000GB
local   0.000GB
mydb    0.000GB
test    0.000GB
humingzhe:PRIMARY> use mydb
switched to db mydb
humingzhe:PRIMARY> show tables
acc
```

登录从服务器，使用 show dbs 命令查看库，若提示“Error: listDatabases failed:{ "ok" : 0, "errmsg" : "not master and slaveOk=false",}”，是因为它不是 Master，并且 slaveOK=false，这样的状态是无法收集 dbs 的，需执行 rs.slaveok()命令，执行后即可查看，命令如下。

```
humingzhe:SECONDARY> rs.slaveOk()
humingzhe:SECONDARY> show dbs
admin   0.000GB
config  0.000GB
dbone   0.000GB
local   0.000GB
mydb    0.000GB
test    0.000GB
humingzhe:SECONDARY> use mydb
switched to db mydb
humingzhe:SECONDARY> show tables
acc
```

使用 rs.config()命令可以看到 3 台服务器的权重（Priority）都为 1，如果要模拟 Master 服务器宕机，则设置的第一台 Secondary 服务器会自动切换为主，命令如下。

```
[root@nosql ~]# iptables -I INPUT -p tcp --dport 27017 -j DROP
```

由于服务器的权重都是 1，是相互平等的，所以也有可能是第二台 Secondary 服务器变为 Primary。可以给其中的一台服务器设置高权重，从而到新的 Primary 服务器上进行操作，命令如下。

```
[root@nosql ~]# iptables -D INPUT -p tcp --dport 27017 -j DROP
humingzhe:PRIMARY> cfg=rs.config()
humingzhe:PRIMARY> cfg.members[0].priority = 3
3
humingzhe:PRIMARY> cfg.members[1].priority = 2
2
humingzhe:PRIMARY> cfg.members[2].priority = 1
1
```

使用 rs.reconfig(cfg)命令使配置生效，然后使用 rs.config()命令查看权重，会发生改变，命令如下。

```
humingzhe:PRIMARY>
2018-02-20T19:36:45.913+0800 I NETWORK  [thread1] trying reconnect to
127.0.0.1:27017 (127.0.0.1) failed
2018-02-20T19:36:45.915+0800 I NETWORK  [thread1] reconnect
127.0.0.1:27017 (127.0.0.1) ok
humingzhe:SECONDARY>
humingzhe:PRIMARY> rs.reconfig(cfg)
humingzhe:PRIMARY> rs.config()
```

8.4.12 MongoDB 分片介绍

在中小型企业的日常工作中使用副本集足矣满足日常所需，因为平时数据量并不大，一旦有很庞大的使用场景，就可以把 MongoDB 多做几个副本集，如像 MySQL 那样，多做几个主从，每个主从中运行的库不一致就可以了。

在 BAT 这类大型企业中，它们的一个库非常庞大，数据量很多，此时就无法进行分库分表。MongoDB 有类似于分库分表的架构，称之为分片。分片类似于 Redis Cluster，属于分布式，它会把数据库的数据分别存储到各个分片中。分片有一个比较小的单元，该单元就是副本集。

分片就是将数据库进行拆分，将大型的集合分割到不同的服务器上，比如 A 服务器存 A 集合，B 服务器存 B 集合。还可以将一个集合中的数据分割为两部分，第一部分存储到 A 服务器中，其余部分存储到 B 服务器中。例如，有 100GB 的数据，可以将数据分割成 10 份存储到 10 台服务器上，这样每台服务器上就有 10GB 的数据。

MongoDB 分片通过 mongos 进程（路由）实现分片后的数据存储与访问，也就是说，mongos 是整个分片架构的核心，对客户端而言是不知道是否有分片的，客户端把读写操作转送给 mongos 即可。

虽然分片会把数据分割到很多台服务器上，但是每个节点都需要有一个备用角色，这样才能保证数据的高可用。当系统需要更多的空间或资源时，分片可以让我们按需扩展，只需把提供 MongoDB 服务的服务器加入到分片集群中即可。MongoDB 分片架构如图 8-4 所示。

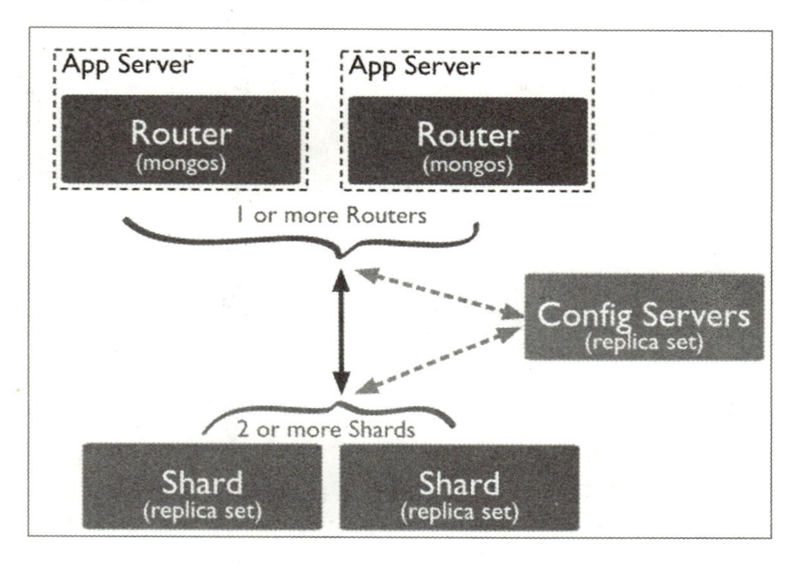

图 8-4

8.4.13　MongoDB 分片重要角色

mongos：数据库集群请求的入口，所有的请求都通过 mongos 进行协调，不需要在应用程序中添加路由选择器，mongos 自己就是一个请求分发中心，它负责把对应的数据请求转发到对应的 Shard 服务器上。在生产环境中通常有多个 mongos 作为请求的入口，其中一个“挂掉”后所有的 mongodb 请求都没有办法操作。

Config Servers：配置服务器，存储所有数据库元信息（路由、分片）的配置。mongos 本身没有物理存储分片服务器和数据路由信息，只是缓存在内存里，配置服务器实际存储这些数据。mongos 第一次启动或关掉重启就会从 Config Servers 加载配置信息，然后配置服务器信息发生变化，会通知所有的 mongos 更新自己的状态，这样 mongos 就能继续准确路由。在生产环境中通常有多个 Config Servers 配置服务器，因为它存储了分片路由的元数据，防止数据丢失。

Shard：存储了一个集合部分数据的 MongoDB 实例，每个分片是单独的 MongoDB 服务或副本集，在生产环境中，所有的分片都应该是副本集。

8.4.14 MongoDB 分片搭建

1. 分片搭建准备工作

准备 3 台服务器，A 服务器搭建 mongos、Config Servers、副本集 1 主节点、副本集 2 仲裁、副本集 3 从节点；B 服务器搭建 mongos、Config Servers、副本集 1 从节点、副本集 2 主节点、副本集 3 仲裁；C 服务器搭建 mongos、Config Servers、副本集 1 仲裁、副本集 2 从节点、副本集 3 主节点。

端口分配：mongos 20000、config 21000、副本集 1 27001、副本集 2 27002、副本集 3 27003。

3 台服务器全部关闭 Selinux 服务和 Firewalld 服务，或者增加对应端口规则。

分别在 3 台服务器上创建各自角色所需要的目录，3 台服务器分别执行如下命令。

```
[root@nosql ~]# mkdir -p /data/mongodb/mongos/log
[root@nosql ~]# mkdir -p /data/mongodb/config/{data,log}
[root@nosql ~]# mkdir -p /data/mongodb/shard1/{data,log}
[root@nosql ~]# mkdir -p /data/mongodb/shard2/{data,log}
[root@nosql ~]# mkdir -p /data/mongodb/shard3/{data,log}
```

2. 配置 Config Servers

注意

MongoDB 3.4 版本后需要对 Config Servers 创建副本集。

分别在 3 台服务器上创建 Config Servers 配置文件所需的 mongod 目录，命令如下。

```
[root@nosql ~]# mkdir /etc/mongod/
[root@node-1 ~]# mkdir /etc/mongod/
[root@node-2 ~]# mkdir /etc/mongod/
```

分别在 3 台服务器的/etc/mongod/目录下创建 config 配置文件，在该配置文件中添加如下代码。

```
[root@nosql ~]# vim /etc/mongod/config.conf
pidfilepath = /var/run/mongodb/configsrv.pid
dbpath = /data/mongodb/config/data
logpath = /data/mongodb/config/log/congigsrv.log
logappend = true
bind_ip = 192.168.222.130
port = 21000
```

```
fork = true
configsvr = true #declare this is a config db of a cluster;
replSet=configs #副本集名称
maxConns=20000 #设置最大连接数
```

注意

3 台服务器上的 bind_ip 不一致，此处 3 台服务器的 IP 分别是 192.168.222.130、192.168.222.131 和 192.168.222.132。

启动 3 台服务器的 Config Servers 服务，使用 ps aux 命令查看 mongod 进程，检查是否监听了 21000 端口，命令如下。

```
[root@nosql ~]# mongod -f /etc/mongod/config.conf
[root@nosql ~]# ps aux|grep mongo
[root@node-1 ~]# mongod -f /etc/mongod/config.conf
[root@node-1 ~]# ps aux|grep mongo
[root@node-2 ~]# mongod -f /etc/mongod/config.conf
[root@node-2 ~]# ps aux|grep mongo
[root@node-2 ~]# netstat -lntp|grep mongod
tcp   0   0 192.168.222.130:21000   0.0.0.0:*    LISTEN    100197/mongod
[root@node-1 ~]# netstat -lntp|grep mongod
tcp   0   0 192.168.222.131:21000   0.0.0.0:*    LISTEN    32724/mongod
[root@node-2 ~]# netstat -lntp|grep mongod
tcp   0   0 192.168.222.132:21000   0.0.0.0:*    LISTEN    28950/mongod
```

登录任意一台服务器的 21000 端口，初始化副本集，命令如下。

```
[root@nosql ~]# mongo --host 192.168.222.130 --port 21000
> config = { _id: "configs", members: [ {_id : 0, host :
"192.168.222.130:21000"},{_id : 1, host : "192.168.222.131:21000"},{_id :
2, host : "192.168.222.132:21000"}] }
> rs.initiate(config)
configs:PRIMARY> rs.status()
```

3. 配置 Shard

分别在 3 台服务器的/etc/mongod/目录下创建 shard1、shard2、shard3 配置文件，在该配置文件中添加如下代码。

```
[root@nosql ~]# vim /etc/mongod/shard1.conf
pidfilepath = /var/run/mongodb/shard1.pid
dbpath = /data/mongodb/shard1/data
logpath = /data/mongodb/shard1/log/shard1.log
logappend = true
```

```
    bind_ip = 192.168.222.130
    port = 27001
    fork = true
    replSet=shard1 #副本集名称
    shardsvr = true #declare this is a shard db of a cluster;
    maxConns=20000 #设置最大连接数
    [root@nosql ~]# cp /etc/mongod/shard1.conf /etc/mongod/shard2.conf
    [root@nosql ~]# cp /etc/mongod/shard1.conf /etc/mongod/shard3.conf
    [root@nosql ~]# sed -i 's/shard1/shard2/g' /etc/mongod/shard2.conf
    [root@nosql ~]# sed -i 's/shard1/shard3/g' /etc/mongod/shard3.conf
    [root@nosql ~]# scp -r /etc/mongod/shard*.conf
root@192.168.222.131:/etc/mongod/
100%  365   123.6KB/s   00:00
    [root@nosql ~]# scp -r /etc/mongod/shard*.conf
root@192.168.222.132:/etc/mongod/
```

注意

3 台服务器的 shard.conf 配置文件中的 pidfilepath、dbpath、logpath、bind_ip、port 和 replSet 也不一致。

分别在 3 台服务器上启动 shard1，命令如下。

```
    [root@nosql ~]# mongod -f /etc/mongod/shard1.conf
    about to fork child process, waiting until server is ready for connections.
    forked process: 100818
    child process started successfully, parent exiting
    [root@node-1 ~]# mongod -f /etc/mongod/shard1.conf
    [root@node-2 ~]# mongod -f /etc/mongod/shard1.conf
```

登录 130 或 131 服务器，以 27001 端口初始化副本集，命令如下。132 服务器之所以不能用，是因为 132 服务器 shard1 的 27001 端口作为了仲裁点。

```
    [root@node-2 ~]# mongo --port 27001 --host 192.168.222.130
    > use admin
    switched to db admin
    > config = { _id: "shard1", members: [ {_id : 0, host :
"192.168.222.130:27001"}, {_id: 1, host : "192.168.222.131:27001"},{_id :
2, host : "192.168.222.132:27001",arbiterOnly:true}] }
    > rs.initiate(config)
    { "ok" : 1 }
```

分别在 3 台服务器上启动 shard2，命令如下。

```
    [root@nosql ~]# mongod -f /etc/mongod/shard2.conf
```

```
[root@node-1 ~]# mongod -f /etc/mongod/shard2.conf
[root@node-2 ~]# mongod -f /etc/mongod/shard2.conf
```

登录 131 或 132 服务器，以 27002 端口初始化副本集，命令如下。130 服务器之所以不能用，是因为 130 服务器 shard2 的 27002 端口作为了仲裁点。

```
[root@nosql ~]# mongo --port 27002 --host 192.168.222.132
> use admin
switched to db admin
> config = { _id: "shard2", members: [ {_id : 0, host : "192.168.222.130:27002" ,arbiterOnly:true}, {_id: 1, host : "192.168.222.131:27002"},{_id : 2, host : "192.168.222.132:27002"}] }
> rs.initiate(config)
{ "ok" : 1 }
```

分别在 3 台服务器上启动 shard3，命令如下。

```
[root@nosql ~]# mongod -f /etc/mongod/shard3.conf
[root@nosql ~]# mongod -f /etc/mongod/shard3.conf
[root@nosql ~]# mongod -f /etc/mongod/shard3.conf
```

登录 130 或 132 服务器，以 27003 端口初始化副本集，命令如下。131 服务器之所以不能用，是因为 131 服务器 shard3 的 27003 端口作为了仲裁点。

```
[root@node-2 ~]# mongo --port 27003 --host 192.168.222.132
> use admin
switched to db admin
> config = { _id: "shard3", members: [ {_id : 0, host : "192.168.222.130:27003"}, {_id: 1, host : "192.168.222.131:27003", arbiterOnly:true},{_id : 2, host : "192.168.222.132:27003"}] }
> rs.initiate(config)
{ "ok" : 1 }
shard3:PRIMARY>
```

4. 配置 mongos

分别在 3 台服务器的/etc/mongod/目录下创建 mongos.conf 配置文件，在该配置文件中添加如下代码。

```
[root@nosql ~]# vim /etc/mongod/mongos.conf
pidfilepath = /var/run/mongodb/mongos.pid
logpath = /data/mongodb/mongos/log/mongos.log
logappend = true
bind_ip = 192.168.222.130
port = 20000
fork = true
```

```
    configdb = configs/192.168.222.130:21000, 192.168.222.131:21000,
192.168.222.132:21000 #监听的配置服务器,只能有 1 个或 3 个, configs 为配置服务器的副本集名字
    maxConns=20000 #设置最大连接数
```

注意

3 台服务器的 mongos.conf 配置文件中的 bind_ip 不一致，在 configdb 处需要改为当前服务器的 IP，切勿照抄。

分别在 3 台服务器上启动 mongos 服务，命令如下。

```
[root@nosql ~]# mongos -f /etc/mongod/mongos.conf
[root@nosql ~]# mongos -f /etc/mongod/mongos.conf
[root@node-2 ~]# mongos -f /etc/mongod/mongos.conf
```

登录任意一台服务器的 20000 端口，把所有的分片和路由进行串联操作，命令如下。

```
    [root@node-2 ~]# mongo --host 192.168.222.130 --port 20000
    mongos> use admin
    switched to db admin
    mongos>
sh.addShard("shard1/192.168.222.130:27001,192.168.222.131:27001,192.168.222.132:27001")
    mongos>
sh.addShard("shard2/192.168.222.130:27002,192.168.222.131:27002,192.168.222.132:27002")
    mongos>
sh.addShard("shard3/192.168.222.130:27003,192.168.222.131:27003,192.168.222.132:27003")
```

使用 sh.status()命令可以查看集群的状态，命令如下。

```
mongos> sh.status()
```

8.4.15 MongoDB 分片测试

登录任意一台服务器的 20000 端口，然后使用 admin 库指定要分片的数据库，命令如下。db.runCommand()命令等同于 sh.enableSharding()命令。

```
[root@node-2 ~]# mongo --host 192.168.222.130 --port 20000
mongos> use admin
```

```
    switched to db admin
    mongos> db.runCommand({ enablesharding : "testdb"})
    mongos> sh.enableSharding("testdb")
```

指定数据库中需要分片的集合和片键，命令如下。db.runCommand()命令等同于 sh.shardCollection()命令。

```
    mongos> db.runCommand( { shardcollection : "testdb.table1",key : {id:
1} } )
    mongos> sh.shardCollection("testdb.table1",{"id":1} )
```

进入 testdb 数据库，在其中插入测试数据，命令如下。

```
    mongos> use testdb
    switched to db testdb
    mongos> for (var i = 1; i <= 10000; i++)
db.table1.save({id:i,"test1":"testval1"})
    WriteResult({ "nInserted" : 1 })
    mongos> show dbs
    mongos> sh.status()
    mongos> db.table1.stats()
```

8.4.16　MongoDB 备份与恢复

1. MongoDB 备份

因为使用了分片，所以备份时需要到分片的端口中进行备份，MongoDB 备份的命令是 mongodump。例如备份指定的 testdb 库，备份后 mongobak 目录下会生成 testdb 目录，testdb 目录下又会生成 table1.bson 和 table1.metadata.json 两个文件，命令如下。

```
    [root@nosql ~]# mkdir /tmp/mongobak
    [root@nosql ~]# mongodump --host 192.168.222.130 --port 20000 -d testdb
-o /tmp/mongobak/
    2018-02-20T23:10:03.157+0800    writing testdb.table1 to
    2018-02-20T23:10:03.283+0800    done dumping testdb.table1 (10000
documents)
    [root@nosql ~]# ls /tmp/mongobak/testdb/table1.
    table1.bson          table1.metadata.json
```

备份所有库只需要把-d 选项去掉，其余参数保持不变，命令如下。

```
    [root@nosql ~]# mongodump --host 192.168.222.130 --port 20000 -o
```

```
/tmp/mongobak/
    [root@nosql ~]# ls /tmp/mongobak/
    admin/  config/ testdb/
```

备份指定集合，使用-d 选项指定库，使用-c 选项指定集合，命令如下。

```
    [root@nosql ~]# mongodump --host 192.168.222.130 --port 20000 -d testdb
-c table1 -o/tmp/mongobackup
    2018-02-20T23:19:25.430+0800    writing testdb.table1 to
    2018-02-20T23:19:25.526+0800    done dumping testdb.table1 (10000
documents)
    [root@nosql ~]# ls /tmp/mongobackup/
    testdb
    [root@nosql ~]# ls /tmp/mongobackup/testdb/table1.
    table1.bson           table1.metadata.json
```

除备份 MongoDB 默认的两种文件外，还可以导出 JSON 格式的文件，命令如下。

```
    [root@nosql ~]# mongoexport --host 192.168.222.130 --port 20000 -d testdb
-c table1 -o /tmp/table/table1.json
    2018-02-20T23:24:05.433+0800    connected to: 192.168.222.130:20000
    2018-02-20T23:24:05.786+0800    exported 10000 records
    [root@nosql ~]# ls /tmp/table/table1.json
    /tmp/table/table1.json
    [root@nosql ~]# less /tmp/table/table1.json
```

2. MongoDB 恢复

mongorestore 是恢复库的命令，即恢复所有的库，命令如下。

```
    [root@nosql ~]# mongorestore --host 192.168.222.131 --port 20000 --drop
/tmp/mongobak/
```

注意

其中/tmp/mongobak 是备份所有库的目录名字；--drop 选项可选，意思是把恢复之前的数据删除，不建议使用。

恢复指定库，如恢复 testdb 库，命令如下。

```
    [root@nosql ~]# mongorestore -d testdb --drop /tmp/mongobak/
```

恢复集合，命令如下。

```
    [root@nosql ~]# mongorestore -d testdb -c table1
```

```
/tmp/mongobackup/testdb/table1.bson
```

注意

-c 选项后面跟要恢复的集合名称，/tmp/mongobackup 是备份 testdb 库时生成的文件的所在路径，即 BSON 文件的所在路径。

导入 JSON 格式的集合，命令如下。

```
    [root@nosql ~]# mongoimport -d testdb -c table1 --file
/tmp/table/table1.json
```

第9章

Jenkins 持续化集成

9.1 Jenkins介绍

随着软件开发复杂度的不断提高，团队开发成员之间如何更好地协同工作以确保软件开发的质量已经成为开发过程中不可回避的问题。尤其是近些年来，敏捷（Agile）在软件工程领域越来越“火”，如何在不断变化的需求中快速适应和保证软件的质量显得尤其重要。

持续化集成（Continuous Integration，简称 CI）正是针对这类问题的一种软件开发实践。它倡导团队开发成员必须经常集成他们的工作，甚至每天都可能进行多次集成。而每次的集成都是通过自动化的构建来验证的，包括自动编译、发布和测试，从而尽快地发现集成错误，让团队成员能够更快地开发内聚的软件。

下面以笔者经历的项目为例进行描述。首先，解释一下集成。我们所有项目的代码都是托管在 SVN 服务器上的。每个项目都要有若干个单元测试，并有一个所谓的集成测试。集成测试就是把所有的单元测试运行一遍，以及进行其他一些能自动完成的测试。只有在本地计算机上通过了集成测试的代码才能上传到 SVN 服务器上，这样可以保证上传的代码没有问题。所以，集成就是指集成测试。

再说持续。不言而喻，持续就是指长期对项目代码进行集成测试。既然是长期，那么肯定是自动执行的，否则人工执行没有保证，而且耗费人力。对此，需要有一台服务器，它会定期地从 SVN 中检出代码并编译，然后运行集成测试，每次集成测试结果都会记录在案。完成这个工作的就是下面要介绍的 Jenkins 软件。当然，它的功能远不止这些。在我们的项目中，执行这个工作的周期是 1 天。也就是说，服务器每天都会准时地对 SVN 上的最新代码自动进行一次集成测试。

持续交付（Continuous Delivery）指的是频繁地将软件的新版本交付给质量团队或用户以供评审。如果评审通过，代码就进入生产阶段。持续交付可以看作持续集成的下一步。它强调的是不管怎么更新，软件是随时随地可以交付的。

持续部署（Continuous Deployment）是持续交付的下一步，指的是代码通过评审以后，自动部署到生产环境。持续部署的目标是，代码在任何时候都是可部署的，可以进入生产阶段。持续部署的前提是能自动化完成测试、构建、部署等步骤。

Jenkins，原名 Hudson，2011 年改为现在的名字，它是一个基于 Web 界面平台开源的、可扩展的持续集成、交付、部署（软件/代码的编译、打包、部署）的工具。Jenkins 能实时监控集成中存在的错误，提供详细的日志文件和提醒功能，还能用图表的形式形象地展示项目构建的趋势和稳定性。

Jenkins 官方网站的网址是 https://jenkins.io/。Jenkins 官网文档的网址是 https://jenkins.io/doc/。

9.2　Jenkins安装

安装 Jenkins 需要安装 JDK，JDK 版本必须高于或等于 1.8，可使用源码包安装 JDK，也可使用 yum 命令安装 openjdk，此处使用 yum 命令安装 openjdk，命令如下。

```
[root@jenkins ~]# yum install -y java-1.8.0-openjdk
```

安装 Jenkins 工具的 yum 源，设置好 yum 源后使用 rpm 命令安装 Jenkins 的 key，因为在 Jenkins 的 yum 源配置文件中 gpgcheck=1，所以需要验证 Jenkins 的 key，命令如下。

```
    [root@jenkins ~]# wget -O /etc/yum.repos.d/jenkins.repo
https://pkg.jenkins.io/redhat/jenkins.repo
    [root@jenkins ~]# ls /etc/yum.repos.d/jenkins.repo
    /etc/yum.repos.d/jenkins.repo
    [root@jenkins ~]# rpm --import
https://pkg.jenkins.io/redhat/jenkins.io.key
```

使用 yum 命令安装 Jenkins，安装后启动 Jenkins 服务，命令如下。

```
[root@jenkins ~]# yum install -y jenkins
[root@jenkins ~]# systemctl start jenkins
[root@jenkins ~]# ps aux |grep jenkins
[root@jenkins ~]# netstat -lntp|grep 8080
tcp6      0      0 :::8080         :::*        LISTEN      48932/java
```

Jenkins 的日志文件是/var/log/jenkins/jenkins.log，通过该日志文件可查看 admin

密码，也可以到/var/lib/jenkins/secrets/initialAdminPassword 文件中查看该密码，命令如下。

```
[root@jenkins ~]# less /var/log/jenkins/jenkins.log|grep -A5 password
Jenkins initial setup is required. An admin user has been created and a
password generated.
Please use the following password to proceed to installation:

0bd44e9c1e984c82bde7da72e9b8c619    #初始化 Jenkins 密码

This may also be found at: /var/lib/jenkins/secrets/initialAdminPassword
```

打开浏览器，在地址栏中输入 IP 和 Port 进行 Jenkins 的安装，如图 9-1 所示。

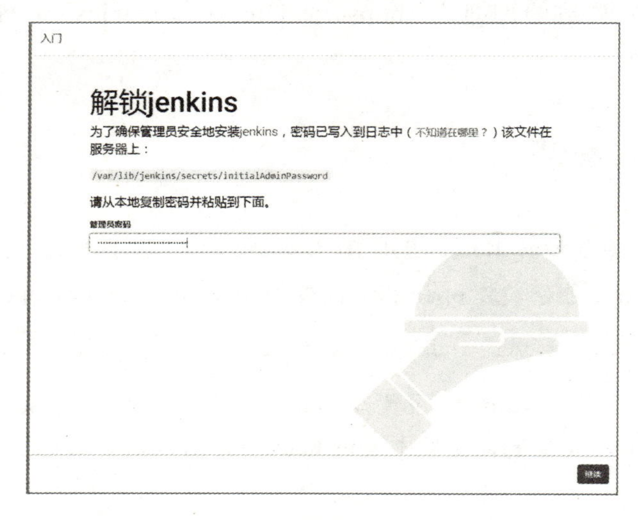

图 9-1

单击“继续”按钮选择要安装的插件，默认安装官方提供的插件，无需自定义。插件安装完成后要求设置管理员账号和密码，与搭建开源 Web 网站一样。安装完成后会看到如图 9-2 所示的 Jenkins 后台管理界面。

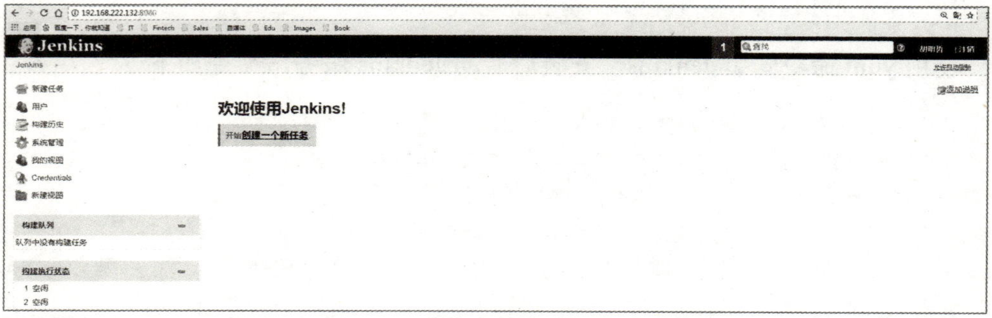

图 9-2

在/etc/sysconfig/目录下有一个 jenkins 文件，该文件是 Jenkins 的配置文件，命令如下。

```
[root@jenkins ~]# ls /etc/sysconfig/jenkins
/etc/sysconfig/jenkins
```

Jenkins 程序主目录在/var/lib/jenkins/目录下，jobs 目录下存放的是在 Jenkins 浏览器界面中创建的任务。例如，在 Jenkins 后台 Web 界面中创建一个“胡明哲”的任务，在 jobs 目录下就会生成一个“胡明哲”的目录，命令如下。

```
[root@jenkins ~]# ls /var/lib/jenkins/jobs/
胡明哲
[root@jenkins ~]# ls /var/lib/jenkins/jobs/胡明哲/
builds  config.xml
```

logs 目录是 Jenkins 日志相关的目录；nodes 是多节点时用到的目录；plugins 是 Jenkins 插件所在的目录，该目录下有很多插件，如新建一个插件，插件会自动保存在该目录下，命令如下。

```
[root@jenkins ~]# ls /var/lib/jenkins/logs/
tasks
[root@jenkins ~]# ls /var/lib/jenkins/nodes/
[root@jenkins ~]# ls /var/lib/jenkins/plugins/
ace-editor     email-ext      github     github-api
```

secrets 是 Jenkins 密码、秘钥存放的目录；users 是与用户相关的目录，命令如下。

```
[root@jenkins ~]# ls /var/lib/jenkins/users/
$80e1$660e$54f2
```

如果需要备份 Jenkins，直接把/var/lib/jenkins/目录下的文件或目录打包到新服务器上即可，Jenkins 无需借助数据库存储数据，它的配置全部存在 XML 格式的文件中。

9.3　Jenkins发布PHP代码

打开 Jenkins 后台 Web 界面，选择“系统管理”→“管理插件”→“已安装”，检查是否有 Git plugin 和 Publish Over SSH 两个插件。如果没有，则需要单击“可选插件”按钮进行安装。安装后重启 Jenkins 服务，命令如下。重启 Jenkins 后需要重新在 Web 界面中进行登录。

```
[root@jenkins ~]# systemctl restart jenkins.service
```

生成一对秘钥用来远程登录服务器，然后将公钥追加到远程服务器的/root/.ssh/authorized_keys 文件中，命令如下。

```
[root@jenkins ~]# ssh-keygen -f /root/.ssh/jenkins
[root@jenkins ~]# ls /root/.ssh/
authorized_keys  jenkins  jenkins.pub
[root@jenkins ~]# cat .ssh/jenkins.pub
[root@node-1 ~]# vim .ssh/authorized_keys
```

安装好两个插件后，选择“系统管理”→“系统设置”，在界面底部选择 Publish Over SSH 插件进行相关设置，如图 9-3 所示。

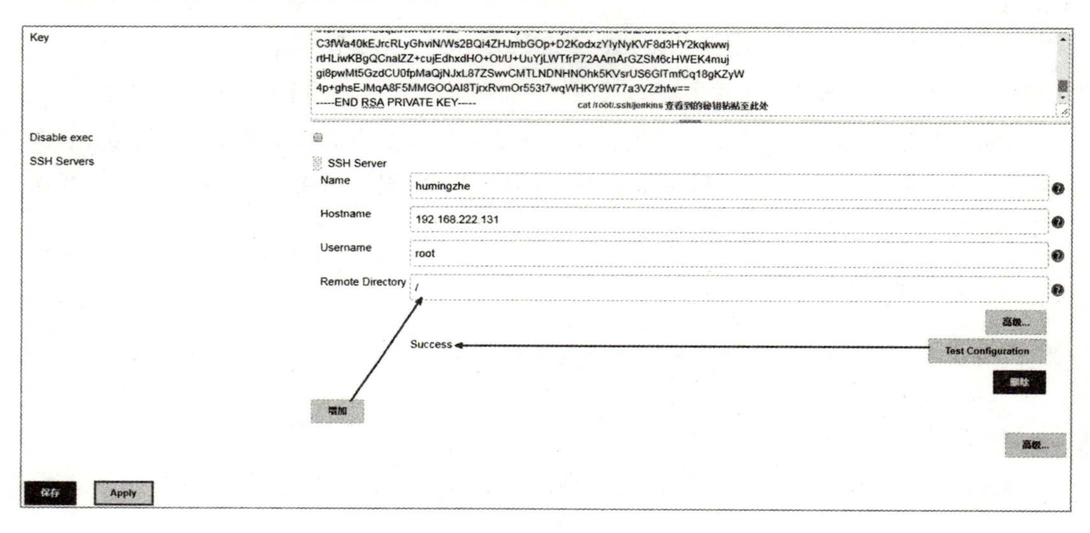

图 9-3

注意

在 Passphrase 文本框中输入密码，Path to key 留空，在 key 文本框中粘贴/root/.ssh/jenkins 文件中的内容。单击左下角的“增加”按钮增加 SSH Server，name 可自定义，在 Hostname 文本框中输入线上 Web 服务器 IP，在 Username 文本框中输入 root，在 Remote Directory 文本框中输入“/”（根）。如果是多台 Web Server，继续单击“添加”按钮重复上述操作即可。操作完毕后单击 Apply 按钮或“保存”按钮。

设置完插件后创建新任务，任务名可自定义，如 Publish PHP code，选择构建一个自由风格的软件项目，最后单击“确定”按钮进入下一页。在“描述”文本框中输入“发布 PHP 代码”（自定义即可），在“源码管理”页面可根据个人或公司情况选择 Git 或 SVN，此处选择 Git 作为演示，如图 9-4 所示。

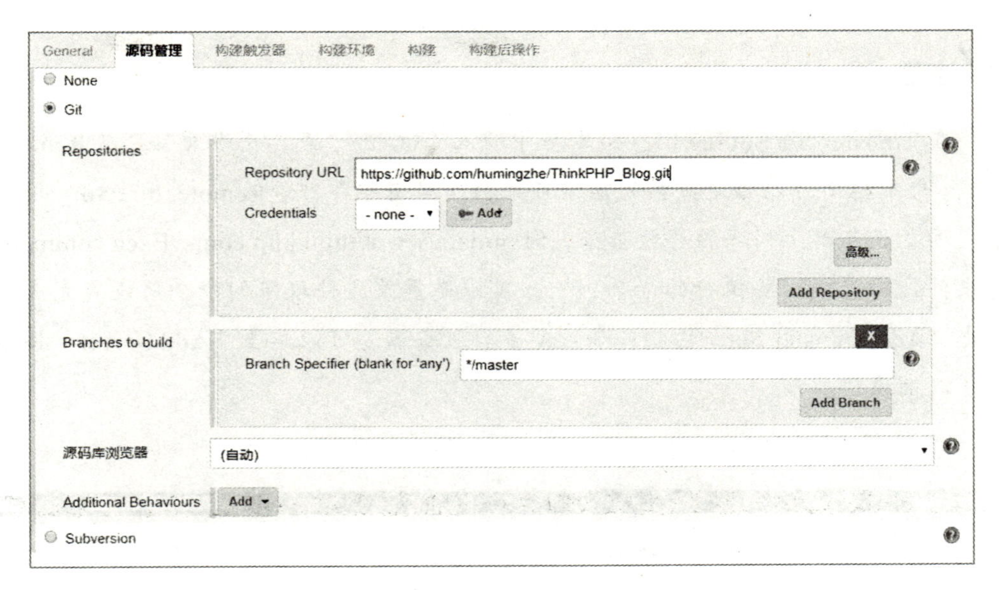

图 9-4

“构建触发器”和“构建环境”处留空，不做任何选择。增加构建步骤 Send files or execute commands over SSH，设置如图 9-5 所示。

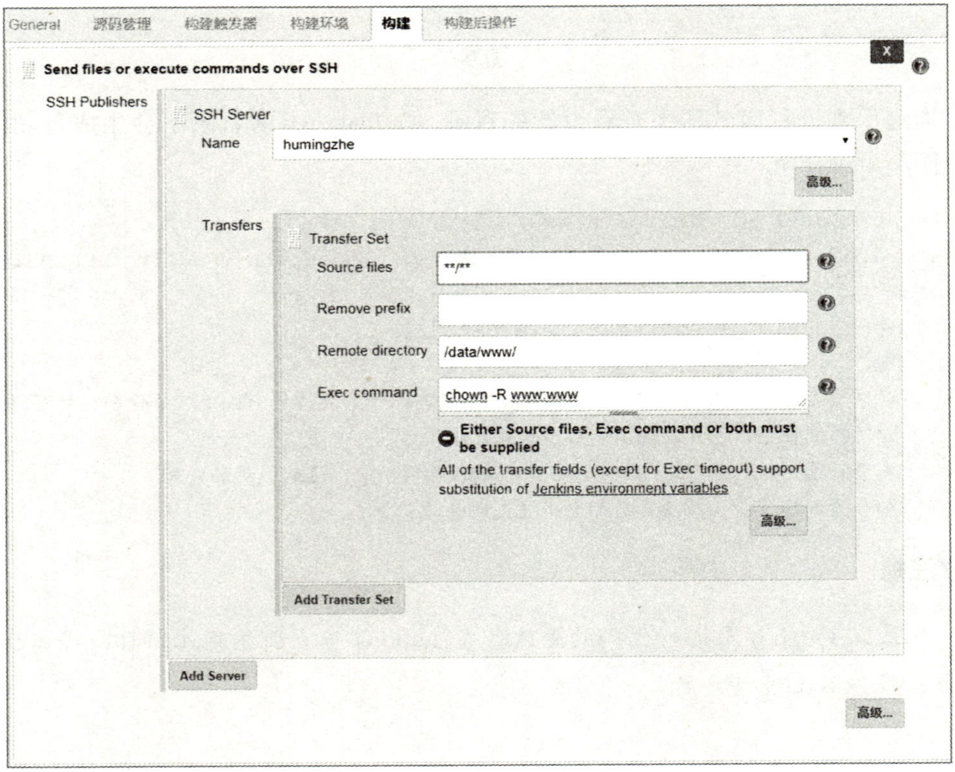

图 9-5

注意

在 Transfer Set Source files 文本框中输入“**/**”，表示全部文件。在 Remove prefix 文本框中可以指定截掉的前缀目录，这里留空即可。Remote directory 用于指定远程服务器上代码的存放路径，例如/data/www/thinkphp.com。Exec command 为文件传输完成后要执行的命令，例如可以是更改文件权限的命令。设置完成后单击“Add Transfer Set”按钮；如果是多台服务器，可以单击“Add Server”按钮重复以上操作。

单击 Publish PHP code 进行项目构建，如图 9-6 所示。

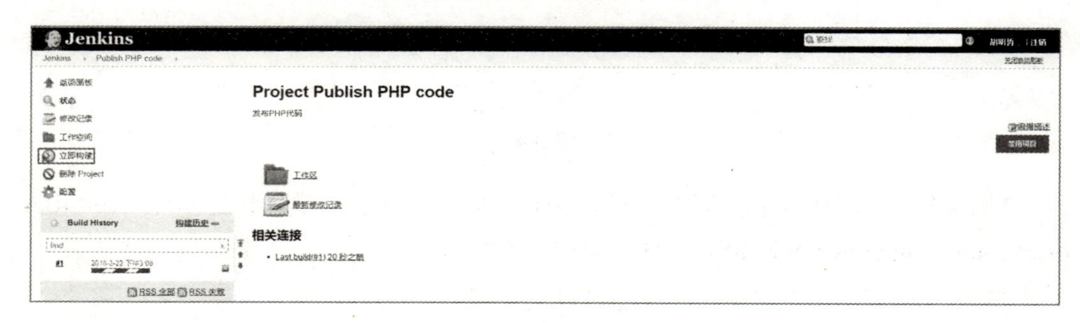

图 9-6

构建后到远程服务器中查看设置的目录，如/data/www/目录下会生成复制的源码文件，命令如下。

```
    [root@node-1 ~]# ls /data/www/
    admin.php  App  composer.json  hank.sql  index.php  Public  README.md
ThinkPHP
    [root@node-1 ~]# cat /data/www/README.md
    ## 简介
        此项目是笔者利用业余时间用 ThinkPHP 编写的一套开源的 Blog 综合程序，既可以自主
改为 cms 管理系统程序，也可以运用到众多行业之中。
        本项目采用 Layer 三方插件美化后台 Web 各项功能以达到美观的效果。
        如今开源给读者，此项目后期会一直更新并维护下去。
```

注意

如果在 GitHub 上进行代码的更改，在 Jenkins 中重新构建此项目，那么代码也会在服务器上进行更新。

9.4　Jenkins邮件设置

前面在讲解 Zabbix 时介绍过邮件告警机制，无论是服务停止、启动，还是出现故障，都会用 E-mail 发送给用户，Jenkins 也不例外，发布代码成功或失败都可以通过 E-mail 告诉用户。

Jenkins 邮件设置在“系统管理”→“系统设置”→Jenkins Location→“系统管理员邮件地址”中，在此处可设置发送 E-mail 的地址。在页面中找到“邮件通知”，输入 SMTP 服务器，单击右侧“高级选项”按钮进行选择，选中“使用 SMTP 认证”，输入用户名和密码，如果使用的是 SSL，则 SSL 协议也需要被选中，SMTP 默认端口号是 25，勾选“通过发送测试邮件测试配置”复选框，输入接收邮件的地址，单击右侧的 Test configuration 按钮可进行邮件发送测试，如图 9-7 所示。

邮件通知

SMTP服务器	smtp.163.com
用户默认邮件后缀	
☑ 使用SMTP认证	
用户名	██████@163.com
密码	••••••••
使用SSL协议	○
SMTP端口	25
Reply-To Address	
字符集	UTF-8
☑ 通过发送测试邮件测试配置	
Test e-mail recipient	██████@qq.com
	Email was successfully sent　　Test configuration

图 9-7

完成上述操作后，在收件箱中会收到 Jenkins 发送的 E-mail 邮件，如图 9-8 所示。

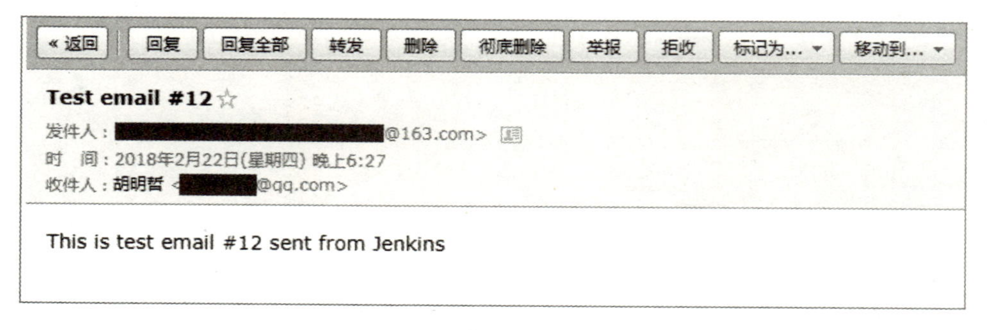

图 9-8

测试成功后，进入到发布 PHP 代码的任务中进行邮件发布设置，单击“构建后操作”下拉列表框，选择 E-mail Notification 选项，如图 9-9 所示。

图 9-9

注意

默认自带的邮件发送功能只在 Job 构建失败时才能发送邮件。

9.5 插件Email-ext

9.4 节 Jenkins 邮件设置中提到默认自带的邮件发送功能只有在 Job 构建失败时才会发送 E-mail，而 Email-ext 插件无论 Job 构建成功与否都可以发送 E-mail。插件名字是 Email Extension Plugin，默认存在；如果不存在，则需要手动安装。使用 Email-ext 插件需要关闭前面章节中设置的邮件发送规则。

单击“系统管理”→“系统设置”→Extended E-mail Notification，在打开的页面中进行 E-mail 插件的设置，如图 9-10 所示。

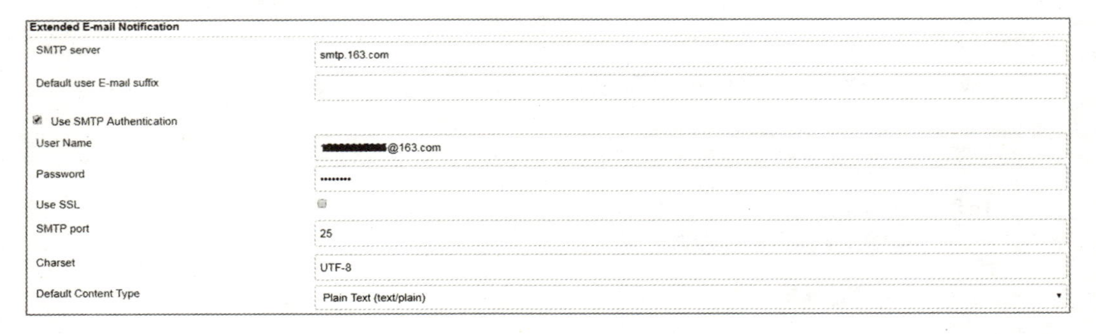

图 9-10

单击 Default Triggers 下拉按钮，选择 Always 选项，意思是无论什么情况下都发送 E-mail。完成后到对应的任务中修改设置，下拉页面找到“构建后操作”选项，删除 E-mail Notification。删除后单击“增加构建后操作”按钮，选择 Editable Email Notification。其中，Project Recipient List 为接收邮件的收件人邮箱，可以在默认内容后面增加额外的收件人邮箱，用逗号分隔，如图 9-11 所示。

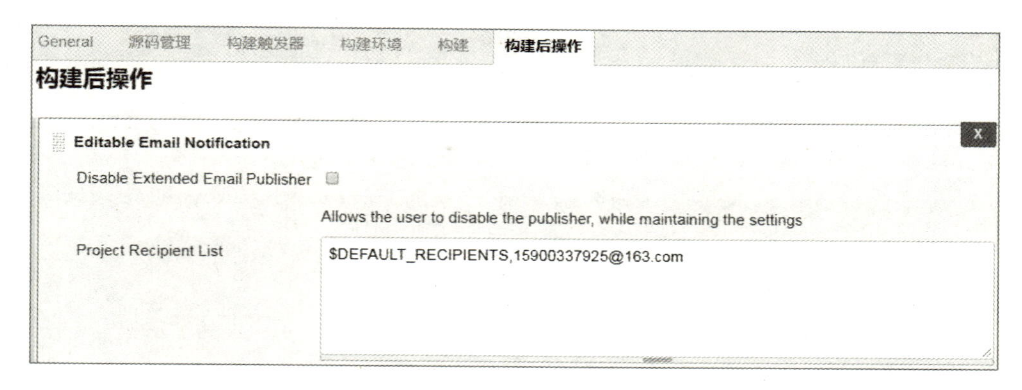

图 9-11

单击右下角的“Advanced settings”下拉按钮，选择“Triggers”，再单击下方的“Add Trigger”按钮，可以添加发邮件的条件，如图 9-12 所示。

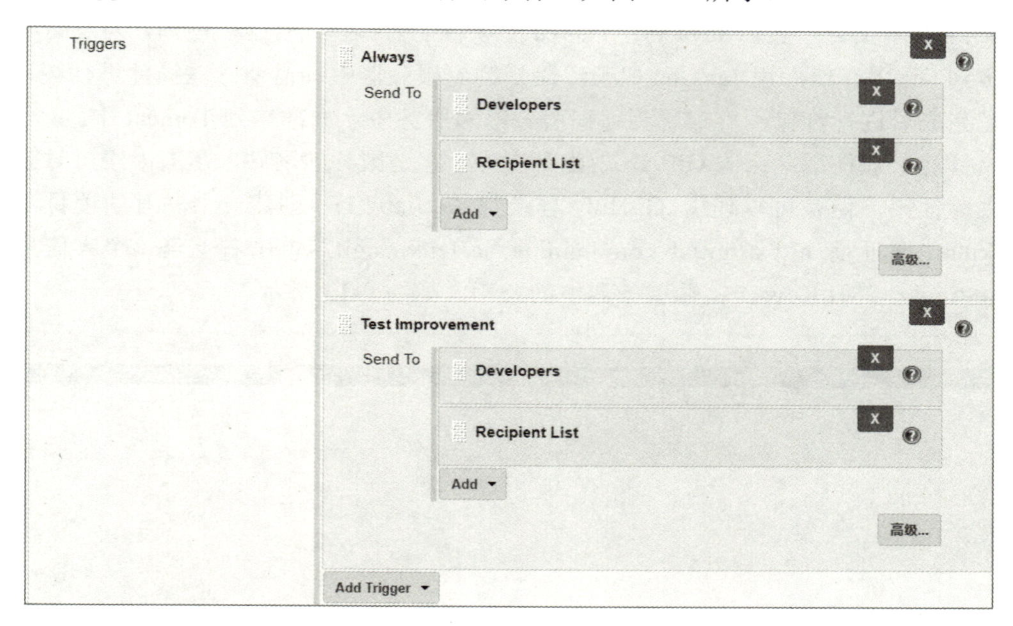

图 9-12

9.6　管理员密码破解

如果忘记了 Jenkins 后台管理员密码，可以编辑 config.xml 配置文件，删除<passwordHash>这行代码，粘贴笔者提供的代码，修改后重启 Jenkins 服务，命令如下。

```
[root@jenkins ~]# cd /var/lib/jenkins/users/\$80e1\$660e\$54f2/
[root@jenkins $80e1$660e$54f2]# vim config.xml
```

```
    <passwordHash>#jbcrypt:$2a$10$KnMSyM3uPS9h69yuDKbx.eZasf7StXdA0k6Twwy
bnCbwAVfsMCAYe</passwordHash>  #此密码是 HMkj8899
    <passwordHash>#jbcrypt:$2a$10$dNekh475/7EU9blh.6Oq5ufFL0owcbQhZ6Usk/.
FRaEVBX3Do/29a</passwordHash>  #此密码是 admin
    [root@jenkins $80e1$660e$54f2]# systemctl restart jenkins.service
```

9.7 部署Java项目

9.7.1 部署 Java 项目之创建私有仓库

前面讲解了用 Jenkins 发布 PHP 代码，但是发布 PHP 代码的需求不多，企业用 Jenkins 部署项目一般以 Java 为主。Java 的项目需要编译、打包。例如，开发的项目上传到 Git 上，然后用 Jenkins 把源代码复制到服务器中，借助一些插件进行编译。编译完成后进行打包，再部署到远程服务器的服务中，如部署到 Tomcat 中。

目前一般使用主流的 GitLab 创建私有库，它是国外的网站，在国内注册时获取不到验证码，但是可以使用 GitHub 登录到 GitLab 上。创建一个私有的项目，如 Jenkins，网址是 https://gitlab.com/humingzhe/Jenkins.git。单击右上角“个人信息”→Settings→SSH Keys，粘贴服务器中的公钥，如图 9-13 所示。

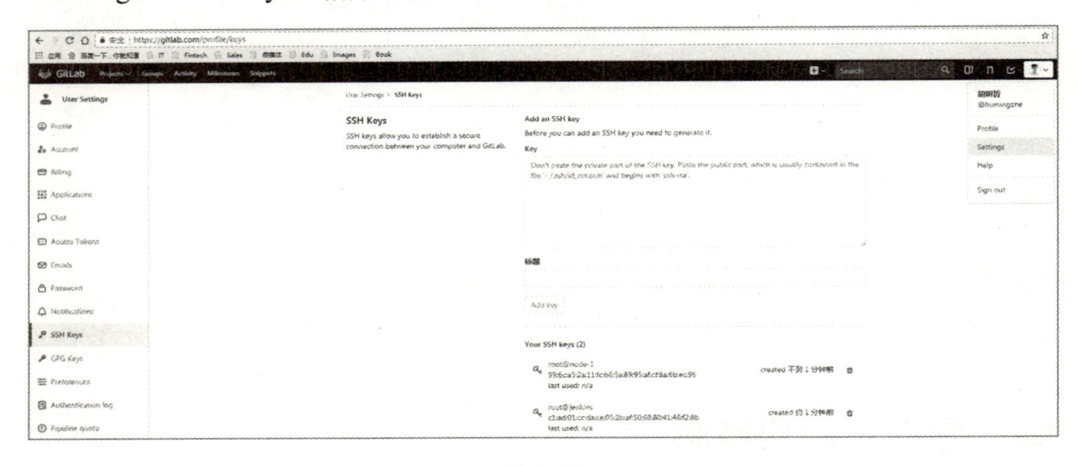

图 9-13

设置好公钥后在服务器中进行全局设置，然后进行项目的复制操作，命令如下。

```
[root@jenkins ~]# mkdir /gitlab
[root@jenkins ~]# cd /gitlab/
[root@jenkins gitlab]# git config --global user.name "胡明哲"
[root@jenkins gitlab]# git config --global user.email
```

```
"admin@humingzhe.com"
    [root@jenkins gitlab]# git clone git@gitlab.com:humingzhe/Jenkins.git
```

复制项目后就可以创建文件或更新代码了，如创建 README.md 文件，在该文件中输入该项目的一些说明，然后上传到 GitLab 中，命令如下。

```
    [root@jenkins gitlab]# cd Jenkins/
    [root@jenkins Jenkins]# vim README.md
    ## 测试 Jenkins 私有仓库
    [root@jenkins Jenkins]# git add README.md
    [root@jenkins Jenkins]# git commit -m "add README"
    [root@jenkins Jenkins]# git push -u origin master
```

9.7.2　部署 Java 项目之下载 Zrlog 源码

下载 Zrlog 源代码到/gitlab/目录下，然后进行解压，解压完毕后移至 Jenkins 目录中，命令如下。

```
    [root@jenkins gitlab]# wget
https://codeload.github.com/94fzb/zrlog/zip/master
    [root@jenkins gitlab]# unzip master
    [root@jenkins gitlab]# ls
    Devops  Jenkins  master  zrlog-master
    [root@jenkins gitlab]# mv zrlog-master/* Jenkins/
```

进入 Jenkins 目录，将当前目录下的 Zrlog 博客源码上传至 GitLab 代码管理平台，命令如下。

```
    [root@jenkins gitlab]# cd Jenkins/
    [root@jenkins Jenkins]# ls
    bin  CHANGELOG.md  common  data  doc  LICENSE  mvnw  mvnw.cmd  pom.xml
README.md  service  web
    [root@jenkins Jenkins]# git add .
    [root@jenkins Jenkins]# git commit -m "add zrlog blog"
    [root@jenkins Jenkins]# git push
```

上传完毕后访问 GitLab，就会看到刚上传的 Zrlog 博客源码，如图 9-14 所示。

add zrlog blog
胡明哲 authored 2 分钟前　　a548bb0a

名称	最后提交	最后更新
bin	add zrlog blog	2 分钟前
common	add zrlog blog	2 分钟前
data	add zrlog blog	2 分钟前
doc	add zrlog blog	2 分钟前
service	add zrlog blog	2 分钟前
web	add zrlog blog	2 分钟前
CHANGELOG.md	add zrlog blog	2 分钟前
LICENSE	add zrlog blog	2 分钟前
README.md	add zrlog blog	2 分钟前
mvnw	add zrlog blog	2 分钟前
mvnw.cmd	add zrlog blog	2 分钟前
pom.xml	add zrlog blog	2 分钟前

图 9-14

9.7.3　安装 Tomcat

在一台新的服务器上安装 JDK 和 Tomcat 服务，然后启动 Tomcat 服务，命令如下。

```
[root@node-1 ~]# yum install -y openjdk
[root@node-1 src]# wget http://apache.fayea.com/tomcat/tomcat-9/
v9.0.5/bin/apache-tomcat-9.0.5.tar.gz
[root@node-1 src]# tar xf apache-tomcat-9.0.5.tar.gz -C /usr/local/
[root@node-1 local]# mv apache-tomcat-9.0.5 tomcat
[root@node-1 local]# /usr/local/tomcat/bin/startup.sh
```

Jenkins 需要用 Tomcat 的 API 接口，该接口是 Tomcat 的用户管理界面，需要设置一个管理员用户，编辑 tomcat-users.xml 配置文件，在该文件</tomcat-users>代码前面添加如下代码。

```
[root@node-1 ~]# cd /usr/local/tomcat/conf/
[root@node-1 conf]# vim tomcat-users.xml
<role rolename="admin"/>
<role rolename="admin-gui"/>
<role rolename="admin-script"/>
<role rolename="manager"/>
<role rolename="manager-gui"/>
<role rolename="manager-script"/>
```

```
    <role rolename="manager-jmx"/>
    <role rolename="manager-status"/>
    <user name="admin" password="humingzhe.com"
roles="admin,manager,admin-gui,admin-script,manager-gui,manager-script,m
anager-jmx,manager-status" />
    </tomcat-users>
```

添加 Tomcat 管理员用户后重启 Tomcat 服务，命令如下。

```
[root@node-1 conf]# /usr/local/tomcat/bin/shutdown.sh
[root@node-1 conf]# /usr/local/tomcat/bin/startup.sh
```

编辑 context.xml 文件，添加服务器 IP 地址，如果不添加 IP 网段，访问后台管理界面会提示错误代码 403，设置代码如下。

```
[root@node-1 conf]# vim ../webapps/manager/META-INF/context.xml
allow="127\.\d+\.\d+\.\d+|::1|0:0:0:0:0:0:0:1|192.158.222.*" />
[root@node-1 conf]# ../bin/shutdown.sh
[root@node-1 conf]# ../bin/startup.sh
```

设置后访问 Tomcat Web 页面，单击 Manager App 按钮会提示输入用户名和密码，输入成功即可访问后台界面，如图 9-15 所示。

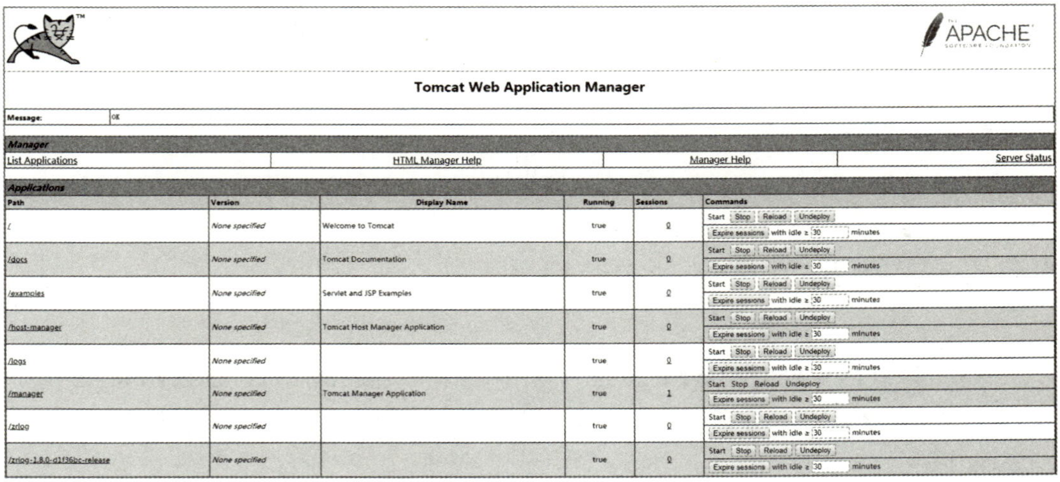

图 9-15

9.7.4　部署 Java 项目之安装 Maven

Maven 工具用于将 Java 项目进行编译和打包，下载 Maven 后进行安装，命令如下。

```
    [root@jenkins local]# wget
http://mirrors.shu.edu.cn/apache/maven/maven-3/3.5.2/binaries/apache-mav
en-3.5.2-bin.tar.gz
    [root@jenkins local]# tar xf apache-maven-3.5.2-bin.tar.gz
    [root@jenkins local]# mv apache-maven-3.5.2 maven
```

在 Jenkins Web 界面设置 Maven，单击“系统管理”→“全局工具配置”，在 Maven Configuration 选项区中单击 Settings file in filesystem 下面的 File Path，如图 9-16 所示。拉到页面底部，在 Maven 选项区中单击“新增 Maven”按钮，如图 9-17 所示。

全局工具配置		
Maven Configuration		
Default settings provider		Settings file in filesystem
	File path	/usr/local/maven/conf/settings.xml
Default global settings provider		Global settings file on filesystem
	File path	/usr/local/maven/conf/settings.xml

图 9-16

Maven		
Maven 安装	Maven	
	Name	Maven-3.5.2
	MAVEN_HOME	/usr/local/maven/
	自动安装	
		删除 Maven
	新增 Maven	
	系统下Maven 安装列表	

图 9-17

9.7.5 部署 Java 项目之安装插件

在 Jenkins 后台安装 Maven Integration plugin 和 Deploy to container Plugin 两个插件，安装位置在“系统管理”→“管理插件”中，检查是否已经安装，安装后重启 Jenkins 服务。

打开 Jenkins Web 界面，单击“新建任务”按钮后就可以看到“构建一个 maven 项目”选项，如图 9-18 所示。

9.7.6 部署 Java 项目之构建 Job

新建任务，选择“构建一个 maven 项目”，任务名称如 Java_code。创建完成后会自动跳转到另外一个页面，在该页面中可以定义任务的细节，源码管理选择 Git，如图 9-19 所示。

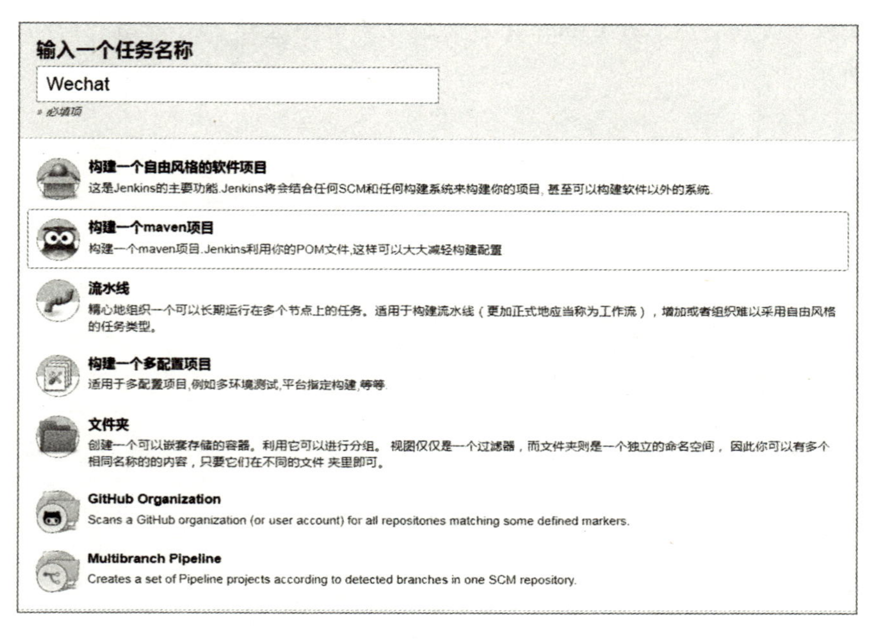

图 9-18

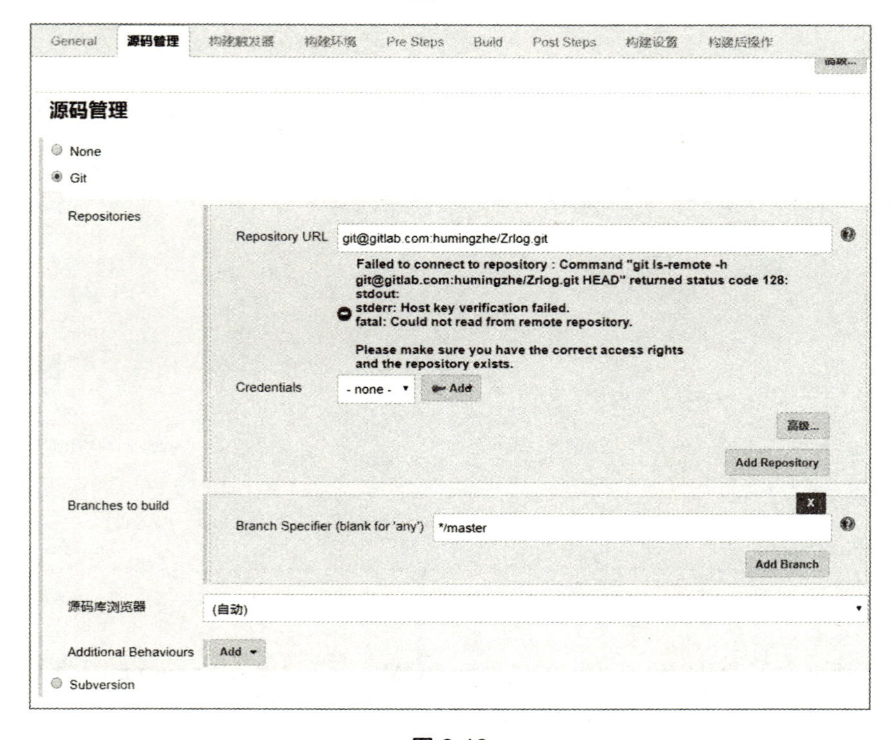

图 9-19

图 9-19 中提示出错，这是因为仓库是私有仓库，需要输入用户名和密码。在 Credentials 处单击 Add 按钮，选择 Jenkins 进行设置，如图 9-20 所示。设置完成后单击 Add 按钮提交，然后在 Credentials 下拉列表框中选择 git，报错会自动消失，如图 9-21 所示。

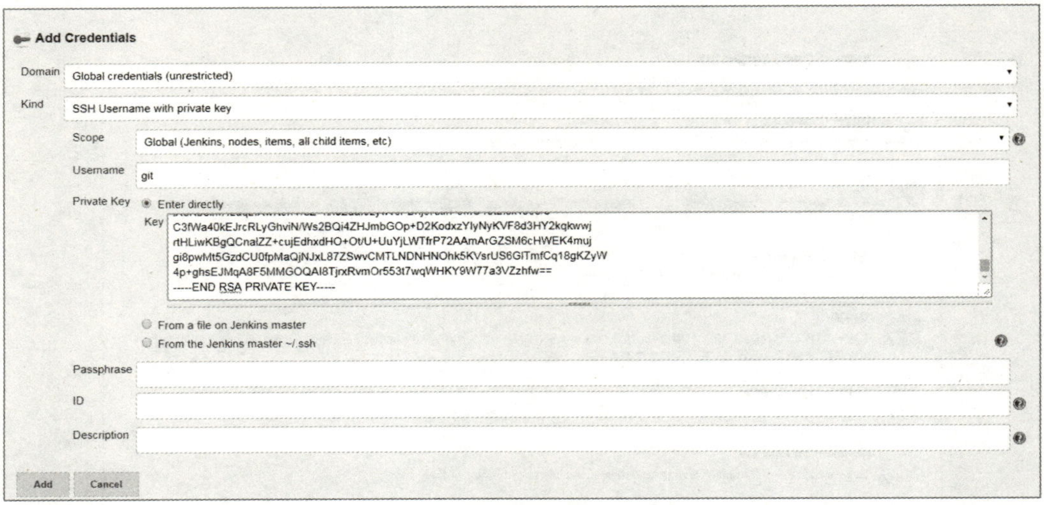

图 9-20

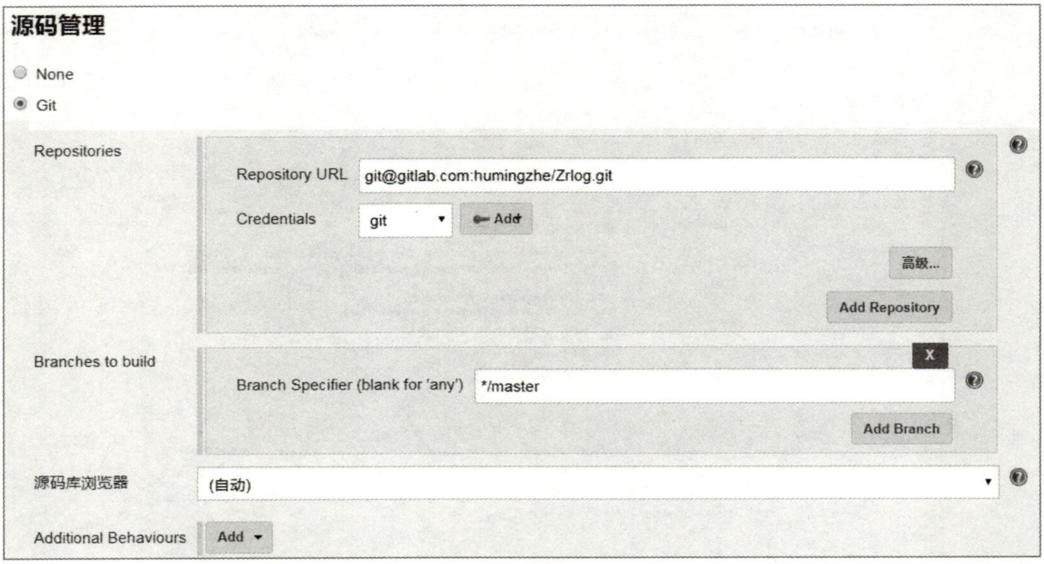

图 9-21

下拉到页面底部的 Build 处，将"Root POM"设置为默认 pom.xml，将 Goals and options 设置为 clean install -D maven.test.skip=true。

9.7.7　部署 Java 项目之手动安装 JDK

在 Jenkins 服务器上安装官方 JDK，由于前面讲解 Tomcat 时用到了官方 JDK，所以此处选择远程复制到当前服务器中，命令如下。

```
    [root@centos7 ~]# scp -r /usr/local/jdk1.9/
root@192.168.222.132:/usr/local/jdk1.9/
    [root@jenkins ~]# ls /usr/local/jdk1.9/
    bin  conf  include  jmods  legal  lib  README.html  release
    [root@jenkins ~]# /usr/local/jdk1.9/bin/java --version
```

进入 Jenkins 后台管理界面，单击“系统管理”→“全局工具设置”→JDK，在“JDK 安装”选项区中单击“新增 JDK”按钮，如图 9-22 所示。

图 9-22

9.7.8　部署 Java 项目之发布 War 包

将 War 包发布到远程的服务器中，先选择 Java_code 任务，进入 Java_code 任务的操作页面，单击“配置”选项，再单击“增加构建后操作步骤”下拉按钮，选择“Deploy war/ear to a container”，然后单击 Credentials 右侧的 Add 按钮，在打开的页面中添加用户名和密码，此处的用户名和密码即 Tomcat 后台登录的用户名和密码，如图 9-23 所示。

图 9-23

单击 Credentials 下拉列表框，选择 admin，在 Tomcat URL 文本框中输入远程 Tomcat 服务器的 IP 地址，如图 9-24 所示。

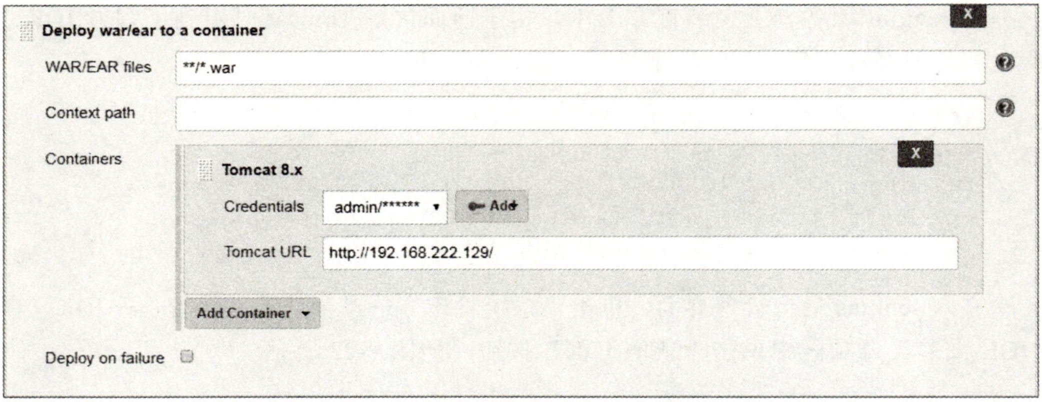

Deploy war/ear to a container

WAR/EAR files **/*.war

Context path

Containers

Tomcat 8.x

Credentials admin/****** Add

Tomcat URL http://192.168.222.129/

Add Container

Deploy on failure

图 9-24

选择 Java_code 任务，单击“立即构建”按钮，构建此任务，构建成功后会在远程 Tomcat 服务器的 webapps 目录下生成 zrlog-1.9.0-SNAPSHOT.war 包，操作如下。

```
[root@centos7 ~]# ls /usr/local/tomcat/webapps/
docs examples host-manager logs manager ROOT zrlog
zrlog-1.8.0-d1f36bc-release  zrlog-1.8.0-d1f36bc-release.war
zrlog-1.9.0-SNAPSHOT  zrlog-1.9.0-SNAPSHOT.war
```

在浏览器的地址栏中输入“IP/zrlog-1.9.0-SNAPSHOT/”即可访问，如图 9-25 所示。

图 9-25

第 10 章 Docker 容器实践

10.1 Docker简介

10.1.1 Docker 主要解决什么问题

Docker 对外宣称的作用是“Build, Ship and Run”，Docker 要解决的核心问题就是快速地做这三件事情。它通过将运行环境和应用程序打包到一起，来解决部署的环境依赖问题，真正做到跨平台的分发和使用。这一点和 DevOps 不谋而合，通过 Docker 可以大大提升开发、测试和运维的效率，在这个移动互联网的时代，如果一个工具能节省人力、提升效率，那么必定会“火”起来。

10.1.2 Docker 的历史

有一家叫 dotCloud 的法国公司，最初也是提供 PaaS 服务的，他们提供了对多种语言的运行环境支持，如 Java、Python、Ruby、NodeJS 等。但是，可能生不逢时，在 PaaS 领域有太多的巨头和大企业了。有一天，Solomon Hykes（Docker 之父）召集了公司员工进行商讨，最后得出的结论是，要和那些大厂商硬干肯定是不行的，那么干脆就把 Docker 项目开源了吧。即使赚不到钱，至少也在开源社区得到一个好名声。因此，在 2013 年 3 月，Docker 正式以开源软件的形式被发布了。正是由于这次开源，让容器领域焕发了第二春，截至 2015 年 11 月，Docker 在 GitHub 上收到了 25 600 个赞，超过 6800 次复制，以及拥有超过 1100 名的贡献者，成为 20 个最具影响力的 GitHub 开源项目。可以说，Docker 是继 Linux 之后，最让人感到兴奋的系统层面的开源项目。据不完全统计，Docker、Red Hat、IBM、Google、Cisco、亚马逊及国内的华为等公司，都在为它贡献代码。在美国，几乎所有的云计算厂商都在“拥抱”Docker 这个生态圈。

10.1.3 Docker 是什么

Docker 其实是容器化技术其中的一种实现，根据之前的介绍可知，容器化技术并不是最近才出现的，那么为什么 Docker 会如此的火爆呢？笔者觉得还是这个时代造就的，因为我们处在一个云计算发展异常迅猛的时代，而云计算又是所有移动互联网、IT 及未来消费行业的基础。从云计算服务的三层架构可以看出，对于传统的 IaaS 层，虚拟机是最基础的组成部分，而虚拟机都是基于 Hyper-V 架构的，也就是说，每个虚拟机都会运行一个完整的操作系统，一个操作系统至少需要占用 5GB 左右的磁盘空间，但是操作系统对我们来说完全无用，我们关心的是虚拟主机所能提供的服务。因此，我们迫切需要轻量级的主机，那就是 Docker 容器。可以看一下如图 10-1 所示的 Docker 的基本结构。

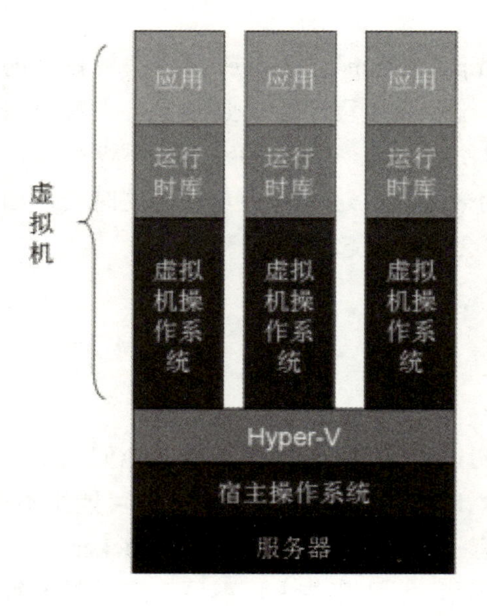

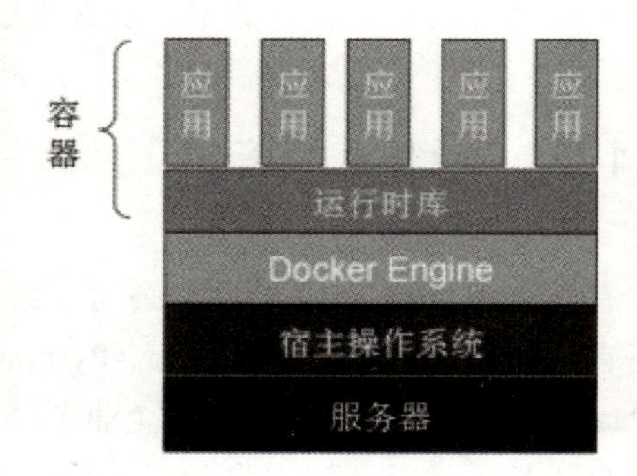

图 10-1

从图 10-1 可以看出，由于容器省去了操作系统，因此整个层级更简化，可以在单台服务器上运行更多的应用，而这正是 IaaS 所需要的，可能腾出 5GB 左右的空间不是什么大事，但是如果需要对外提供成千上万台主机，那么就不得不考虑这个问题了，而这正是容器虚拟化要解决的问题。每到这里，笔者都喜欢举一个例子：波音公司造飞机肯定不会考虑飞机在水上航行的问题，造船厂也绝对不会考虑船在天上飞的问题，汽车制造企业也不会考虑车在水上跑的问题。那么对于众多的移动互联网公司和云计算公司来说，也可以只关注顶层的应用，而不需要考虑操作系统的问题。

10.2　Docker和KVM对比

Docker 容器启动很快，启动和停止都可以秒级实现，而传统的虚拟机则需要数分钟。Docker 容器对系统资源需求很少，一台主机上可以同时运行数百个，甚至上千个 Docker 容器。Docker 通过类似于 Git 设计理念的操作来方便用户获取、分发和更新应用镜像，存储复用，增量更新。Docker 通过 Dockerfile 支持灵活的自动化创建和部署机制，提高工作效率，使流程标准化。

10.3　Docker核心概念

镜像是一个只读的模板，类似于安装系统时用到的 ISO 文件，通过镜像来完成各种应用的部署。容器可以被启动、开始、停止、删除等，每个容器间都是相互隔离的。仓库是一个存放镜像的场所，仓库分为公共仓库和私有仓库。最大的公共仓库是 Docker hub（hub.docker.com），国内公共仓库是 dockerpool.com。

注意

镜像类似于操作系统，容器类似于虚拟机本身。

10.4　安装Docker

Docker 支持在众多平台上安装和运行，如在 Mac、Linux、Windows 和一些云服务器上（亚马逊云、微软云、IBM 云、阿里云、腾讯云、Google 云等）。如图 10-2 所示的界面是 Docker 安装文档页面，文档网址是 https://docs.docker.com/install/。

因为 Docker 是 Linux 的应用，所以后面的章节操作、演示和示例都是在 Linux 环境下运行的。读者需要准备一台安装好 Linux 操作系统的虚拟机。

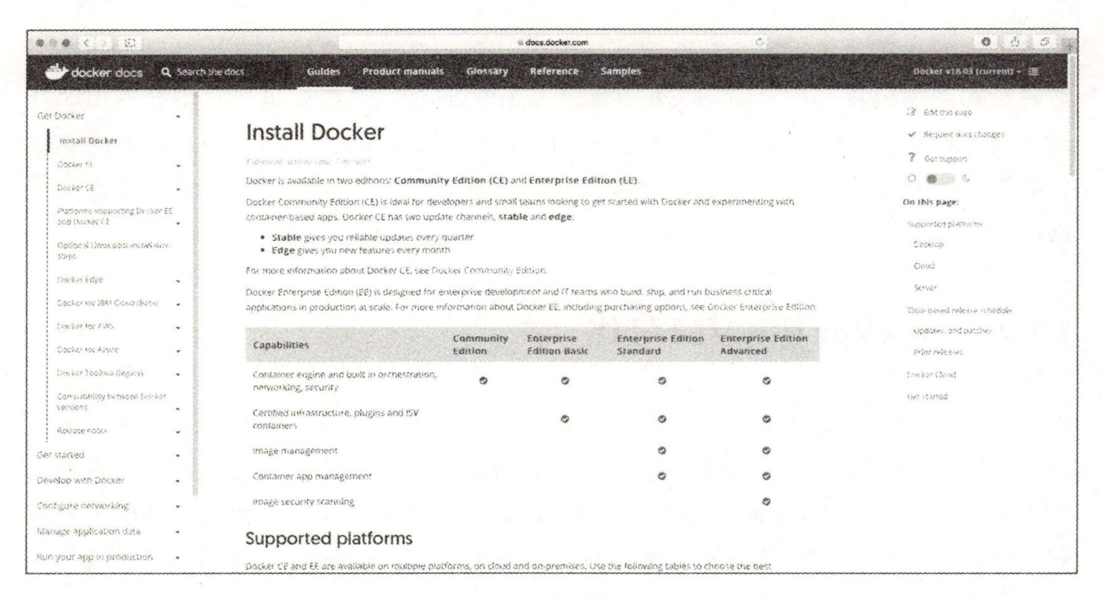

图 10-2

10.4.1 Mac 系统安装 Docker

在 Docker 的安装文档界面（docs.docker.com/install）找到 Desktop，单击 Docker for Mac（Mac OS）进入 Docker 的下载页面，找到 Stable channel，单击下方的 Get Docker for Mac（Stable）进行下载，如图 10-3 所示，会下载一个名为 Docker.dmg 的文件。

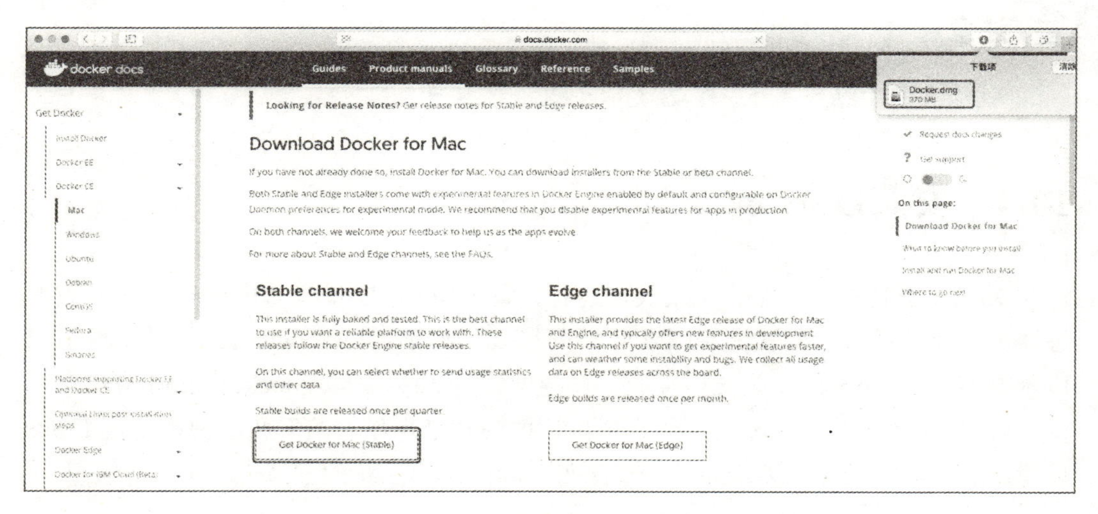

图 10-3

Docker.dmg 文件下载完毕后，双击进行安装。将 Docker 拖动到 Applications 目录中，如图 10-4 所示。

图 10-4

打开 Launchpad（控制台），找到 Docker，单击 Docker 图标即可启动 Docker，会弹出一个提示框，单击“打开”按钮，使用默认选项，最后要求输入 Computer 的密码，验证成功后就会开始启动 Docker，启动成功后会显示 Docker is now up and running。

打开终端窗口，在终端窗口输入 docker －version 命令可查看 Docker 的版本信息，终端窗口会输出 Client 和 Server 的版本号，如图 10-5 所示。

```
humingzhe — -bash — 83×24
Last login: Sun Apr  8 15:13:32 on ttys002
bogon:~ humingzhe$ docker --version
Docker version 18.03.0-ce, build 0520e24
bogon:~ humingzhe$ docker version
Client:
 Version:       18.03.0-ce
 API version:   1.37
 Go version:    go1.9.4
 Git commit:    0520e24
 Built: Wed Mar 21 23:06:22 2018
 OS/Arch:       darwin/amd64
 Experimental:  false
 Orchestrator:  swarm

Server:
 Engine:
  Version:      18.03.0-ce
  API version:  1.37 (minimum version 1.12)
  Go version:   go1.9.4
  Git commit:   0520e24
  Built:        Wed Mar 21 23:14:32 2018
  OS/Arch:      linux/amd64
  Experimental: true
bogon:~ humingzhe$
```

图 10-5

Docker for Mac 用安装包的形式进行安装，里面包含了很多工具，例如 Docker Engine 和 Docker CLI 的 Client，以及 Docker Compose、Docker Machine 和 Kitematic 工具。Kitematic 是图形化界面工具，可以用图形化的界面去运行 Docker，以及下载 Image 和创建 Container。Kitematic 默认没有安装，如果需要安装，可以单击顶部的

Docker 图标，在下拉菜单中选择 Kitematic 选项进行安装，如图 10-6 所示。安装完毕后出现如图 10-7 所示的界面，可以通过这个图形化的界面创建很多 Container。

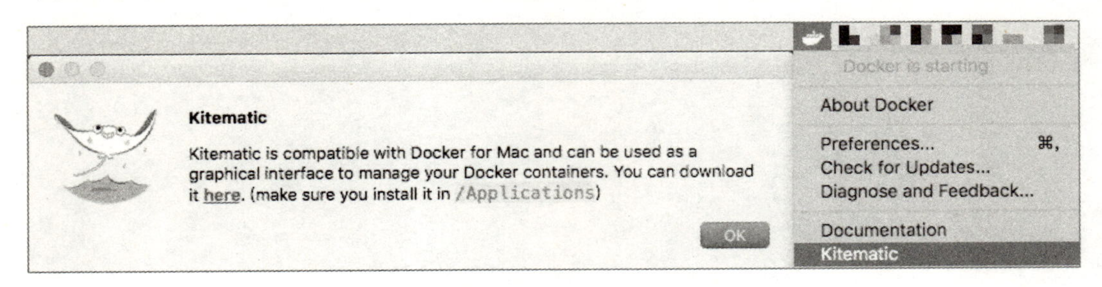

图 10-6

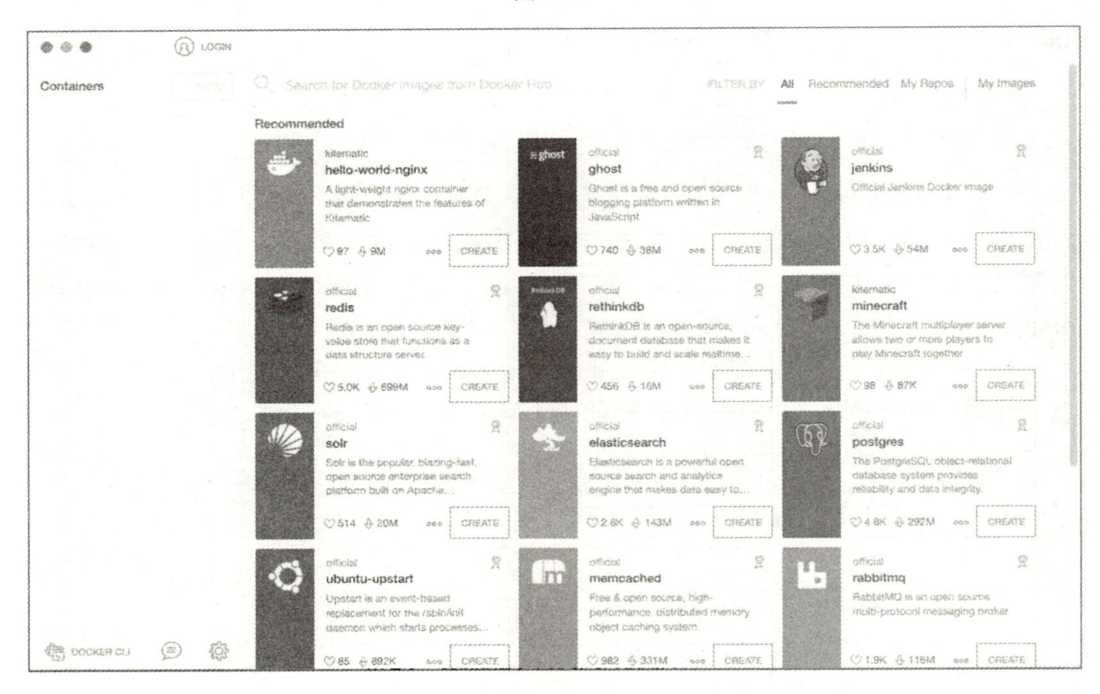

图 10-7

10.4.2 Windows 系统安装 Docker

前面章节提到 Docker 是 Linux 操作系统的一个应用，在 Windows 系统上安装 Linux 系统的应用必然对系统有一些硬性要求。如果想在 Windows 系统上安装 Docker，则操作系统必须是 Windows 10 或 Windows Server 2016。如果操作系统是 Windows 7，那么显然 Docker 是不能安装的。

Docker for Windows 必须是 64 位的 Windows 10 操作系统，而且要支持 Hyper-V。安装过程和在 Mac 系统上的安装过程类似，首先下载 Docker for Windows（Stable）

文件，然后双击 Docker 的安装图标进行安装，完成后会在桌面上创建一个快捷方式，双击快捷方式即可启动 Docker。启动 Docker 时提示未开启 Hyper-V 虚拟化，如图 10-8 所示，需要重启，单击 OK 按钮即可。重启后运行 cmd 命令，在命令行输入 docker --version 和 docker version 命令会出现如图 10-9 所示的界面。

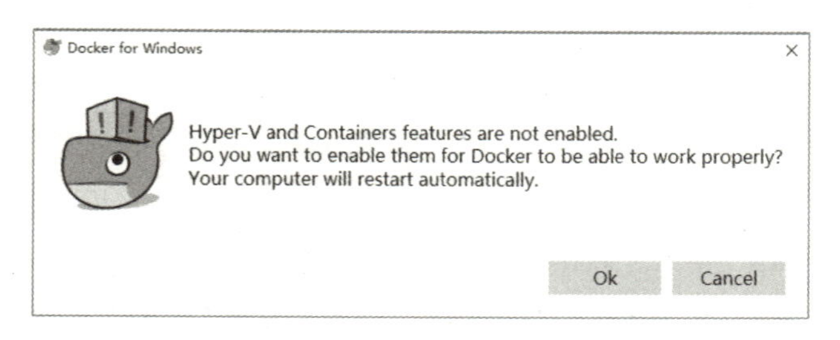

图 10-8

```
管理员: C:\Windows\system32\cmd.exe
Microsoft Windows [版本 10.0.14393]
(c) 2016 Microsoft Corporation。保留所有权利。

C:\Users\Administrator>docker --version
Docker version 18.03.0-ce, build 0520e24

C:\Users\Administrator>docker version
Client:
 Version:       18.03.0-ce
 API version:   1.37
 Go version:    go1.9.4
 Git commit:    0520e24
 Built: Wed Mar 21 23:06:28 2018
 OS/Arch:       windows/amd64
 Experimental:  false
 Orchestrator:  swarm

Server:
 Engine:
  Version:      18.03.0-ce
  API version:  1.37 (minimum version 1.24)
  Go version:   go1.9.4
  Git commit:   0520e24
  Built:        Wed Mar 21 23:21:06 2018
  OS/Arch:      windows/amd64
  Experimental: false

C:\Users\Administrator>_
```

图 10-9

10.4.3　CentOS 7 系统安装 Docker

下载 docker-ce.repo 的 yum 源到/etc/yum.repo.d/目录下，运行命令如下。也可以使用读者 QQ 群文件中的 docker-ce.repo 文件。

```
    [root@docker ~]# curl
https://download.docker.com/linux/centos/docker-ce.repo -o
/etc/yum.repos.d/docker.repo
```

设置好 docker-ce 的 yum 源后，使用 yum 命令安装 docker-ce，运行命令如下。

```
    [root@docker ~]# yum install -y docker-ce
```

注意

使用 yum 命令安装 docker-ce 较慢，可到 GitLab 中下载 docker-ce 的 RPM 包，下载后上传至 Linux 服务器中，再使用 yum 命令安装可自动解决依赖关系。

启动 Docker 服务，添加开机自启动，运行命令如下。

```
[root@docker ~]# systemctl enable docker.service
[root@docker ~]# systemctl start docker.service
```

启动后默认在 iptables 防火墙中生成一些关于 Docker 的规则，如图 10-10 所示。

```
[root@docker ~]# iptables -nvL
Chain INPUT (policy ACCEPT 3987 packets, 237K bytes)
 pkts bytes target     prot opt in     out     source               destination

Chain FORWARD (policy DROP 0 packets, 0 bytes)
 pkts bytes target     prot opt in     out     source               destination
    0     0 DOCKER-USER  all  --  *      *       0.0.0.0/0            0.0.0.0/0
    0     0 DOCKER-ISOLATION  all  --  *      *       0.0.0.0/0            0.0.0.0/0
    0     0 ACCEPT     all  --  *      docker0  0.0.0.0/0            0.0.0.0/0            ctstate RELATED,ESTABLISHED
    0     0 DOCKER     all  --  *      docker0  0.0.0.0/0            0.0.0.0/0
    0     0 ACCEPT     all  --  docker0 !docker0  0.0.0.0/0            0.0.0.0/0
    0     0 ACCEPT     all  --  docker0 docker0  0.0.0.0/0            0.0.0.0/0
```

图 10-10

10.5 Docker镜像管理

10.5.1 下载 Docker 镜像

Docker 镜像可以从 Docker 的官方网站下载，如下载 centos 镜像，运行命令如下。

```
[root@docker ~]# docker pull centos
```

10.5.2 设置阿里云 Docker 加速器

从国外网站下载镜像的速度较慢，还有可能下载失败。为了加快镜像的下载速度，可以使用加速器进行下载，给读者提供两个加速器，分别是阿里云和 Docker Hub Mirror。

1. 下载阿里云 Docker 加速器

登录阿里云网站（https://cr.console.aliyun.com/），输入用户名和密码，登录成功后单击“镜像加速器”→“您的专属加速器地址”，如图 10-11 所示。

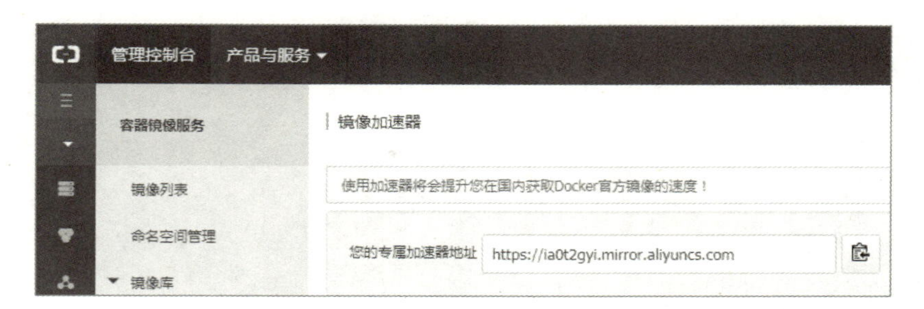

图 10-11

在/etc/docker/daemon.json 配置文件中添加镜像加速器网址链接，然后重启 Docker 服务，运行命令如下。重启后再次下载 centos 镜像会快很多。

```
[root@docker ~]# vim /etc/docker/daemon.json
{"registry-mirrors": ["https://ia0t2gyi.mirror.aliyuncs.com"]}
[root@docker ~]# systemctl daemon-reload
[root@docker ~]# systemctl restart docker.service
```

2．设置 Docker Hub Mirror 加速器

Docker Hub Mirror 加速器获取网址是 https://www.daocloud.io/mirror#accelerator-doc。Docker Hub Mirror 提供了一个 curl 命令来设置镜像加速器，此方法无需手动创建 daemon.json 文件，直接运行 curl 命令会自动创建，运行命令如下。

```
    curl -sSL https://get.daocloud.io/daotools/set_mirror.sh | sh -s
http://ff33ccad.m.daocloud.io
    [root@docker-1 ~]# systemctl daemon-reload
    [root@docker-1 ~]# systemctl restart docker.service
```

10.5.3　Docker 基本命令

使用 docker help 命令可以查看 Docker 所有的子命令，使用 docker help [子命令] 命令可以查看某个子命令如何使用，如查看 cp 命令如何使用，运行命令如下。

```
[root@docker-1 ~]# docker help
[root@docker-1 ~]# docker help cp
```

使用 docker version 命令可查看 Docker 的版本信息，运行命令如下。

```
[root@docker-1 ~]# docker version
```

使用 docker -D info 或 docker info 命令可查看 Docker 的基本信息，运行命令如下。

```
[root@docker-1 ~]# docker -D info
[root@docker-1 ~]# docker info
```

使用 docker images 或 docker image ls 命令可查看本地都有哪些镜像，运行命令如下。

```
[root@docker ~]# docker images
REPOSITORY   TAG     IMAGE ID       CREATED      SIZE
ubuntu       latest  0458a4468cbc   2 weeks ago   112MB
centos       latest  ff426288ea90   4 weeks ago   207MB
[root@docker ~]# docker image ls
REPOSITORY   TAG     IMAGE ID       CREATED      SIZE
ubuntu       latest  0458a4468cbc   2 weeks ago   112MB
centos       latest  ff426288ea90   4 weeks ago   207MB
```

使用 docker search 命令可以搜索某个镜像，如搜索 jumpserver（跳板机）镜像，运行命令如下。

```
[root@docker ~]# docker search jumpserver
```

搜索 jumpserver 会列出很多关于 jumpserver 的镜像，排在第一位的是官方的镜像，官方镜像下面的是个人或组织开发的镜像，也可以自己制作一个 jumpserver 镜像，然后推送到 Docker 官方网站上。

使用 docker tag 命令可以给镜像打标签，如给 centos 打标签，打完标签后会生成一个新的镜像，但是 Image ID 不会发生变化，运行命令如下。

```
[root@docker ~]# docker tag centos humingzhe
[root@docker ~]# docker images
REPOSITORY   TAG      IMAGE ID        CREATED        SIZE
ubuntu       latest   0458a4468cbc    2 weeks ago    112MB
centos       latest   ff426288ea90    4 weeks ago    207MB
humingzhe    latest   ff426288ea90    4 weeks ago    207MB
```

使用 docker run －itd 命令可以把镜像启动为容器，默认在后台启动，不会进入交互命令行。如启动 centos 容器，运行命令如下。

```
[root@docker ~]# docker run -itd centos
32a92fbcead96a2407996e2c57c1b2b44dbf7d1a93f0a9c63046a6fa47f879ba
```

注意

-i 选项表示让容器的标准输入打开，-t 选项表示分配一个伪终端，-d 选项表示后台启动。-i、-t、-d 选项需放在镜像名字前面。

使用 docker ps 或 docker container ls 命令可以查看运行的容器，加上-a 选项可查看所有的容器，包括未运行的容器，运行命令如下。

```
[root@docker ~]# docker ps
[root@docker ~]# docker ps --a
[root@docker ~]# docker container ls
[root@docker ~]# docker container ls -a
```

使用 docker rmi 命令可以删除指定的镜像，如删除 centos 镜像，运行命令如下。

```
[root@docker ~]# docker rmi centos
Untagged: centos:latest
Untagged:
centos@sha256:2671f7a3eea36ce43609e9fe7435ade83094291055f1c96d9d1d1d7c0b
986a5d
```

注意

rmi 命令后面的参数可以是 tag，如果是 tag，则表示删除该 tag。当后面的镜像参数为 Image ID 时，会彻底删除整个镜像，所有的标签也会被一同删除。

10.6　通过容器创建镜像

镜像可以从官网通过 pull 命令直接获取，也可以自定义，如获取 centos 镜像，但 centos 镜像中有一些服务不存在，需要安装，如安装 jumpserver、LNMP 环境，此时可在容器中进行变更操作。使用 yum 安装或编译一些服务，安装后就可以把容器移到镜像中，然后可以将创建的镜像移植到其他服务器上。

使用 docker exec －it 和容器 ID 命令可以进入到某个容器中，如进入到 centos 容器中，使用 df －h 命令可查看系统的磁盘使用情况，容器的磁盘大小和宿主机的磁盘大小是一致的，运行命令如下。

```
[root@docker-1 ~]# docker exec -it f6e0606 bash
[root@f6e060600b33 /]# ls
[root@f6e060600b33 /]# df -h
[root@f6e060600b33 /]# free
```

默认不存在查看 IP 的命令，若想查看 IP 地址，需要安装 net-tools 工具，运行命令如下。

```
[root@f6e060600b33 /]# yum install -y net-tools
```

安装成功后使用 ifconfig 命令可查看容器的 IP，但容器的 IP 和宿主机的 IP 是不同的，按 Ctrl+D 组合键或使用 exit 命令可退出容器到宿主机中。使用 ifconfig 命令查看会发现多了一个 docker0 网卡，如图 10-12 所示。

```
[root@f6e060600b33 /]# ifconfig
eth0: flags=4163<UP,BROADCAST,RUNNING,MULTICAST>  mtu 1500
        inet 172.17.0.2  netmask 255.255.0.0  broadcast 172.17.255.255
        ether 02:42:ac:11:00:02  txqueuelen 0  (Ethernet)
        RX packets 3344  bytes 13198299 (12.5 MiB)
        RX errors 0  dropped 0  overruns 0  frame 0
        TX packets 1711  bytes 97336 (95.0 KiB)
        TX errors 0  dropped 0 overruns 0  carrier 0  collisions 0

lo: flags=73<UP,LOOPBACK,RUNNING>  mtu 65536
        inet 127.0.0.1  netmask 255.0.0.0
        loop  txqueuelen 1  (Local Loopback)
        RX packets 0  bytes 0 (0.0 B)
        RX errors 0  dropped 0  overruns 0  frame 0
        TX packets 0  bytes 0 (0.0 B)
        TX errors 0  dropped 0 overruns 0  carrier 0  collisions 0

[root@f6e060600b33 /]# exit
[root@docker-1 ~]# ifconfig
docker0: flags=4163<UP,BROADCAST,RUNNING,MULTICAST>  mtu 1500
        inet 172.17.0.1  netmask 255.255.0.0  broadcast 172.17.255.255
        inet6 fe80::42:49ff:fe4e:b546  prefixlen 64  scopeid 0x20<link>
        ether 02:42:49:4e:b5:46  txqueuelen 0  (Ethernet)
        RX packets 1711  bytes 73382 (71.6 KiB)
        RX errors 0  dropped 0  overruns 0  frame 0
        TX packets 3336  bytes 13197651 (12.5 MiB)
        TX errors 0  dropped 0 overruns 0  carrier 0  collisions 0

ens33: flags=4163<UP,BROADCAST,RUNNING,MULTICAST>  mtu 1500
        inet 192.168.222.131  netmask 255.255.255.0  broadcast 192.168.222.255
        inet6 fe80::eecb:1a40:94bb:30d7  prefixlen 64  scopeid 0x20<link>
        ether 00:0c:29:8c:e1:8d  txqueuelen 1000  (Ethernet)
        RX packets 358321  bytes 228859321 (218.2 MiB)
        RX errors 0  dropped 0  overruns 0  frame 0
        TX packets 236550  bytes 18576729 (17.7 MiB)
        TX errors 0  dropped 0 overruns 0  carrier 0  collisions 0
```

图 10-12

将修改后的容器做成镜像，然后使用 docker images 命令查看新创建的镜像，新创建的 Image ID 和原始的镜像 ID 不同，运行命令如下。

```
[root@docker-1 ~]# docker commit -m "install net-tools" -a "humingzhe" f6e060600b33 centos_with_nettools
sha256:facebf261028e620effd72fe9a97039d991c36587d5a0281f0949eb9165c8074
[root@docker-1 ~]# docker images
REPOSITORY             TAG       IMAGE ID       CREATED          SIZE
centos_with_nettools   latest    facebf261028   13 seconds ago   304MB
ubuntu                 latest    0458a4468cbc   2 weeks ago      112MB
centos                 latest    ff426288ea90   5 weeks ago      207MB
```

注意

-m 选项后面可加改动信息，-a 选项用于指定与作者相关的信息，f6e060600b33 为容器 ID，ID 后面是新容器的名称，如 centos_with_nettools，类似于在 GitLab 中提交。

10.7　通过模板创建镜像

10.7.1　通过模板导入镜像

模板需要到 openvz.org 网站进行下载，如下载一个 CentOS 6 的 minimal 模板，运行命令如下。国内下载速度较慢，笔者已将该文件上传至 GitLab，读者下载后上传至 Linux 服务器即可。

```
[root@docker-1 ~]# wget http://download.openvz.org/template/
precreated/centos-6-x86-minimal.tar.gz
[root@docker-1 ~]# du -sh centos-6-x86-minimal.tar.gz
201M    centos-6-x86-minimal.tar.gz
```

导入下载完成的 CentOS 6 操作系统模板，导入成功后使用 docker images 命令可查看导入的镜像，运行命令如下。

```
[root@docker-1 ~]# cat centos-6-x86-minimal.tar.gz |docker import -
centos6
sha256:f03e4fc65f14874fb8816647f45d94b39414baac28f3060438f16616951008
4e
[root@docker-1 ~]# docker images
REPOSITORY      TAG       IMAGE ID        CREATED              SIZE
centos6        latest    f03e4fc65f14    About a minute ago   512MB
```

10.7.2　通过镜像导出文件

前面提到可以通过模板导入镜像，既然可以导入，那么也能导出，运行命令如下。

```
[root@docker-1 ~]# docker save -o centos7_with_nettools.tar
centos_with_nettools
[root@docker-1 ~]# du -sh centos7_with_nettools.tar
299M    centos7_with_nettools.tar
```

10.7.3　通过文件恢复镜像

通过镜像导出的文件是可以恢复至本地镜像的，恢复时可使用--input 命令或反向重定向，如图 10-13 所示。

```
[root@docker-1 ~]# docker ps
CONTAINER ID        IMAGE                 COMMAND             CREATED              STATUS              PORTS               NAMES
76823e250d72        centos6               "bash"              12 minutes ago       Up 12 minutes                           unruffled_darwin
f41015bba9c9        centos_with_nettools  "bash"              About an hour ago    Up About an hour                        compassionate_kirch
f6e060600b33        centos                "/bin/bash"         2 hours ago          Up 2 hours                              stoic_shockley
[root@docker-1 ~]# docker rm -f f41015bba9c9
f41015bba9c9
[root@docker-1 ~]# docker rmi facebf261028
Untagged: centos_with_nettools:latest
Deleted: sha256:facebf261028e620effd72fe9a97039d991c36587d5a0281f0949eb9175c8074
Deleted: sha256:61ce1197920b6a6cf9e1e9b65d6c0b7899f5c55b52f64c1ad490eea72e19aab0
[root@docker-1 ~]# docker images
REPOSITORY          TAG                 IMAGE ID            CREATED             SIZE
centos6             latest              f03e4fc65f14        18 minutes ago      512MB
ubuntu              latest              0458a4468cbc        2 weeks ago         112MB
centos              latest              ff426288ea90        5 weeks ago         207MB
[root@docker-1 ~]# docker load --input centos7_with_nettools.tar
9f20aef09b60: Loading layer [==================================================>]  96.87MB/96.87MB
Loaded image: centos_with_nettools:latest
[root@docker-1 ~]# docker images
REPOSITORY             TAG                 IMAGE ID            CREATED              SIZE
centos6                latest              f03e4fc65f14        19 minutes ago       512MB
centos_with_nettools   latest              facebf261028        About an hour ago    304MB
ubuntu                 latest              0458a4468cbc        2 weeks ago          112MB
centos                 latest              ff426288ea90        5 weeks ago          207MB
[root@docker-1 ~]# docker rmi facebf261028
Untagged: centos_with_nettools:latest
Deleted: sha256:facebf261028e620effd72fe9a97039d991c36587d5a0281f0949eb9175c8074
Deleted: sha256:61ce1197920b6a6cf9e1e9b65d6c0b7899f5c55b52f64c1ad490eea72e19aab0
[root@docker-1 ~]# docker load < centos7_with_nettools.tar
9f20aef09b60: Loading layer [==================================================>]  96.87MB/96.87MB
Loaded image: centos_with_nettools:latest
[root@docker-1 ~]# docker images
REPOSITORY             TAG                 IMAGE ID            CREATED              SIZE
centos6                latest              f03e4fc65f14        21 minutes ago       512MB
centos_with_nettools   latest              facebf261028        About an hour ago    304MB
ubuntu                 latest              0458a4468cbc        2 weeks ago          112MB
centos                 latest              ff426288ea90        5 weeks ago          207MB
[root@docker-1 ~]#
```

图 10-13

10.8 Docker的基本管理

10.8.1 Docker 容器管理

使用 docker create 命令可以创建一个容器，如创建 ubuntu 容器，创建的容器默认不会启动，所以用 docker ps 命令查看不到，需要在 ps 命令后加上-a 选项，如图 10-14 所示。

```
[root@docker-1 ~]# docker create -it ubuntu bash
fcf19843b48908e030902064e04e09607772f20c234378dfb8883e971c4cc47c
[root@docker-1 ~]# docker ps
CONTAINER ID        IMAGE               COMMAND             CREATED             STATUS              PORTS               NAMES
76823e250d72        centos6             "bash"              2 hours ago         Up 2 hours                              unruffled_darwin
f6e060600b33        centos              "/bin/bash"         3 hours ago         Up 3 hours                              stoic_shockley
[root@docker-1 ~]# docker ps -a
CONTAINER ID        IMAGE               COMMAND             CREATED             STATUS              PORTS               NAMES
fcf19843b489        ubuntu              "bash"              31 seconds ago      Created                                 vigilant_spence
8b878aed6f69        centos6             "bash"              2 minutes ago       Created                                 eloquent_bartik
76823e250d72        centos6             "bash"              2 hours ago         Up 2 hours                              unruffled_darwin
f6e060600b33        centos              "/bin/bash"         3 hours ago         Up 3 hours                              stoic_shockley
```

图 10-14

使用 docker start 命令启动容器，如启动 ubuntu 容器，启动后即可用 docker ps 命令查看，还可使用 stop 命令停止容器，以及使用 restart 命令重启容器，如图 10-15 所示。

```
[root@docker-1 ~]# docker start fcf19843b489
fcf19843b489
[root@docker-1 ~]# docker ps
CONTAINER ID        IMAGE               COMMAND             CREATED             STATUS              PORTS               NAMES
fcf19843b489        ubuntu              "bash"              11 minutes ago      Up 4 seconds                            vigilant_spence
76823e250d72        centos6             "bash"              2 hours ago         Up 2 hours                              unruffled_darwin
f6e060600b33        centos              "/bin/bash"         4 hours ago         Up 4 hours                              stoic_shockley
[root@docker-1 ~]# docker restart fcf19843b489
fcf19843b489
[root@docker-1 ~]# docker stop fcf19843b489
fcf19843b489
[root@docker-1 ~]# docker ps
CONTAINER ID        IMAGE               COMMAND             CREATED             STATUS              PORTS               NAMES
76823e250d72        centos6             "bash"              2 hours ago         Up 2 hours                              unruffled_darwin
f6e060600b33        centos              "/bin/bash"         4 hours ago         Up 4 hours                              stoic_shockley
[root@docker-1 ~]# docker ps -a
CONTAINER ID        IMAGE               COMMAND             CREATED             STATUS                      PORTS               NAMES
fcf19843b489        ubuntu              "bash"              12 minutes ago      Exited (0) 9 seconds ago                        vigilant_spence
8b878aed6f69        centos6             "bash"              13 minutes ago      Created                                         eloquent_bartik
76823e250d72        centos6             "bash"              2 hours ago         Up 2 hours                                      unruffled_darwin
f6e060600b33        centos              "/bin/bash"         4 hours ago         Up 4 hours                                      stoic_shockley
```

图 10-15

前面讲的 docker run 命令相当于先创建再启动。

使用 docker run –d 命令可以使容器在后台运行，运行命令如下。

```
[root@docker-1 ~]# docker run -d centos bash
```

使用 docker run –name 命令可以给容器自定义名字，如给 centos 容器自定义名字为 humingzhhe，如图 10-16 所示。

```
[root@docker-1 ~]# docker run --name humingzhe -itd centos bash
6685beef35c5979fb6012f28ec1be46237a32207fa5e2fca33d499a076f13d39
[root@docker-1 ~]# docker ps
CONTAINER ID        IMAGE               COMMAND             CREATED             STATUS              PORTS               NAMES
6685beef35c5        centos              "bash"              5 seconds ago       Up 4 seconds                            humingzhe
76823e250d72        centos6             "bash"              2 hours ago         Up 2 hours                              unruffled_darwin
f6e060600b33        centos              "/bin/bash"         4 hours ago         Up 4 hours                              stoic_shockley
[root@docker-1 ~]# docker exec -it humingzhe bash
```

图 10-16

使用 docker run –rm 命令可以让容器退出并删除它，运行命令如下。

```
[root@docker-1 ~]# docker run --rm -it centos bash -c "sleep 30"
```

使用 docker logs 命令可以获取容器运行的历史信息，运行命令如下。

```
[root@docker-1 ~]# docker run  -itd centos bash -c "echo humingzhe;"
f75f73b16f139314aa72e97c3c5a36502e7bd3d6534a1be461d22063b1677e50
[root@docker-1 ~]# docker logs f75f73b16f13931
humingzhe
```

使用 docker attach container_id 命令可以进入一个后台运行的容器，但是该命令不是很好用，要想退出终端需要输入 exit 命令，而输入 exit 命令后容器也跟着退出了。使用 docker exec -it container_id bash 命令可以临时打开一个终端窗口，并且使用 exit 命令退出后容器依然运行。

使用 docker rm container_id 命令可以删除一个容器，如果容器处于运行状态，则需要加-f 选项进行删除，如图 10-17 所示。

```
[root@docker-1 ~]# docker ps -a
CONTAINER ID        IMAGE               COMMAND                  CREATED             STATUS                       PORTS               NAMES
f75f73b17f13        centos              "bash -c 'echo humin…"   6 minutes ago       Exited (0) 6 minutes ago                         unruffled_minsky
b0b2e1cb72df        centos              "bash -c 'echo humin…"   6 minutes ago       Exited (0) 6 minutes ago                         hardcore_johnson
6685beef35c5        centos              "bash"                   14 minutes ago      Up 14 minutes                                    humingzhe
c23a31beb440        centos              "bash"                   18 minutes ago      Exited (0) 18 minutes ago                        nervous_benz
fcf19843b489        ubuntu              "bash"                   34 minutes ago      Exited (0) 22 minutes ago                        vigilant_spence
8b878aed6f69        centos6             "bash"                   35 minutes ago      Created                                          eloquent_bartik
76823e250d72        centos6             "bash"                   3 hours ago         Up 3 hours                                       unruffled_darwin
f6e060600b33        centos              "/bin/bash"              4 hours ago         Up 4 hours                                       stoic_shockley
[root@docker-1 ~]# docker rm f75f73b17f13
f75f73b17f13
[root@docker-1 ~]# docker rm b0b2e1cb72df
b0b2e1cb72df
[root@docker-1 ~]# docker rm c23a31beb440
c23a31beb440
[root@docker-1 ~]# docker ps -a
CONTAINER ID        IMAGE               COMMAND             CREATED             STATUS                       PORTS               NAMES
6685beef35c5        centos              "bash"              14 minutes ago      Up 14 minutes                                    humingzhe
fcf19843b489        ubuntu              "bash"              35 minutes ago      Exited (0) 23 minutes ago                        vigilant_spence
8b878aed6f69        centos6             "bash"              36 minutes ago      Created                                          eloquent_bartik
76823e250d72        centos6             "bash"              3 hours ago         Up 3 hours                                       unruffled_darwin
f6e060600b33        centos              "/bin/bash"         4 hours ago         Up 4 hours                                       stoic_shockley
[root@docker-1 ~]# docker rm -f 76823e250d72
76823e250d72
[root@docker-1 ~]#
```

图 10-17

使用 docker export container_id 命令可以导出容器，并且可以迁移到其他服务器中，迁移到其他服务器中需要使用 cat filename | docker import －[镜像名]命令，如图 10-18 所示。

```
[root@docker-1 ~]# docker ps
CONTAINER ID        IMAGE               COMMAND             CREATED             STATUS              PORTS               NAMES
6685beef35c5        centos              "bash"              19 minutes ago      Up 19 minutes                           humingzhe
f6e060600b33        centos              "/bin/bash"         4 hours ago         Up 4 hours                              stoic_shockley
[root@docker-1 ~]# docker export f6e060600b33 > centos.tar
[root@docker-1 ~]# cat centos.tar |docker import - humingzhe_test
sha256:d2bd89d4dfc37a80bff1f421ae602cc850e80f96419f3fd68b3a52ecf734db07
[root@docker-1 ~]# docker images
REPOSITORY            TAG                 IMAGE ID            CREATED             SIZE
humingzhe_test        latest              d2bd89d4dfc3        2 minutes ago       284MB
centos6               latest              f03e4fc65f14        3 hours ago         512MB
centos_with_nettools  latest              facebf261028        4 hours ago         304MB
ubuntu                latest              0458a4468cbc        2 weeks ago         112MB
centos                latest              ff426288ea90        5 weeks ago         207MB
[root@docker-1 ~]#
```

图 10-18

10.8.2 Docker 仓库管理

镜像是从官网的公共仓库中直接拉取的，在推送时也是推送到公共仓库中，但是很多企业做的项目不想公开发布，这时可以用 registry 镜像在本地搭建私有 Docker 仓库（类似于 GitLab 私有库），运行命令如下。

```
[root@docker-1 ~]# docker pull registry
```

启动 registry 镜像，如图 10-19 所示。使用-p 选项可以把容器的端口映射到宿主机上，冒号左边是宿主机监听端口，冒号右边是容器监听端口。

```
[root@docker-1 ~]# docker run -d -p 5000:5000 registry
709fa1e244a02d332d56a35e991af3ff1de4eba5bfee542acf8a67b21d09be43
[root@docker-1 ~]# docker ps
CONTAINER ID        IMAGE               COMMAND                  CREATED              STATUS              PORTS                    NAMES
709fa1e244a0        registry            "/entrypoint.sh /etc…"   7 seconds ago        Up 5 seconds        0.0.0.0:5000->5000/tcp   pensive_saha
6685beef35c5        centos              "bash"                   About an hour ago    Up About an hour                             humingzhe
f6e060600b33        centos              "/bin/bash"              5 hours ago          Up 5 hours                                   stoic_shockley
```

图 10-19

将 CentOS 6 镜像传至私有仓库中，上传前要给 CentOS 6 镜像打标签，如图 10-20 所示。

```
[root@docker-1 ~]# docker tag centos6 192.168.222.131:5000/centos6
[root@docker-1 ~]# docker images
REPOSITORY                      TAG          IMAGE ID          CREATED            SIZE
humingzhe_test                  latest       d2bd89d4dfc3      37 minutes ago     284MB
192.168.222.131:5000/centos6    latest       f03e4fc65f14      3 hours ago        512MB
centos6                         latest       f03e4fc65f14      3 hours ago        512MB
centos_with_nettools            latest       facebf261028      4 hours ago        304MB
ubuntu                          latest       0458a4468cbc      2 weeks ago        112MB
registry                        latest       d1fd7d86a825      4 weeks ago        33.3MB
centos                          latest       ff426288ea90      5 weeks ago        207MB
```

图 10-20

使用 docker push 命令可以把标记后的镜像推送至私有仓库，但是推送后报错，提示需要输入 HTTPS，运行命令如下。

```
[root@docker-1 ~]# docker push 192.168.222.131:5000/centos6
The push refers to repository [192.168.222.131:5000/centos6]
Get https://192.168.222.131:5000/v2/: http: server gave HTTP response to
HTTPS client
```

若不想使用 HTTPS，需编辑/etc/docker/daemon.json 配置文件指定一个私有仓库地址，运行命令如下。

```
[root@docker-1 ~]# vim /etc/docker/daemon.json
{"insecure-registries":["192.168.222.131:5000"]}
[root@docker-1 ~]# systemctl restart docker.service
```

重启 Docker 服务后，registry 处于停止状态，需要启动后再进行推送，如图 10-21 所示。

```
[root@docker-1 ~]# docker ps -a
CONTAINER ID    IMAGE       COMMAND                  CREATED              STATUS                          PORTS    NAMES
b197c244b7db    registry    "/entrypoint.sh /etc…"   3 minutes ago        Exited (2) 7 seconds ago                 vibrant_brahmagupta
709fa1e244a0    registry    "/entrypoint.sh /etc…"   28 minutes ago       Exited (2) 2 minutes ago                 pensive_saha
6685beef35c5    centos      "bash"                   About an hour ago    Exited (137) 13 minutes ago              humingzhe
fcf19843b489    ubuntu      "bash"                   2 hours ago          Exited (0) About an hour ago             vigilant_spence
8b878aed6f69    centos6     "bash"                   2 hours ago          Created                                  eloquent_bartik
f6e060600b33    centos      "/bin/bash"              5 hours ago          Exited (137) 13 minutes ago              stoic_shockley
[root@docker-1 ~]# docker start 709fa1e244a0
709fa1e244a0
[root@docker-1 ~]# docker push 192.168.222.131:5000/centos6
The push refers to repository [192.168.222.131:5000/centos6]
0a2f11f7b1ef: Pushed
latest: digest: sha256:221c0ab1a46401467075f3f6b6cf2aa989ab7baa59e7a2bb1414f137f666f0f8 size: 529
[root@docker-1 ~]#
```

图 10-21

使用 curl 127.0.0.1:5000/v2/_catalog 命令可以查看推送到私有库中的镜像，运行命令如下。

```
[root@docker-1 ~]# curl 127.0.0.1:5000/v2/_catalog
{"repositories":["centos6"]}
```

使用 docker pull+私有仓库 IP+Port 命令可下载私有库中的镜像，运行命令如下。

```
[root@docker-1 ~]# docker pull 192.168.222.131:5000/ubuntu
```

搭建的私有仓库肯定不是当前一台服务器使用，其他服务器也要使用，在使用的服务器的/etc/docker/daemon.json 配置文件中设置私有库 IP 和端口，然后重启 Docker 服务并进行拉取，如图 10-22 所示。

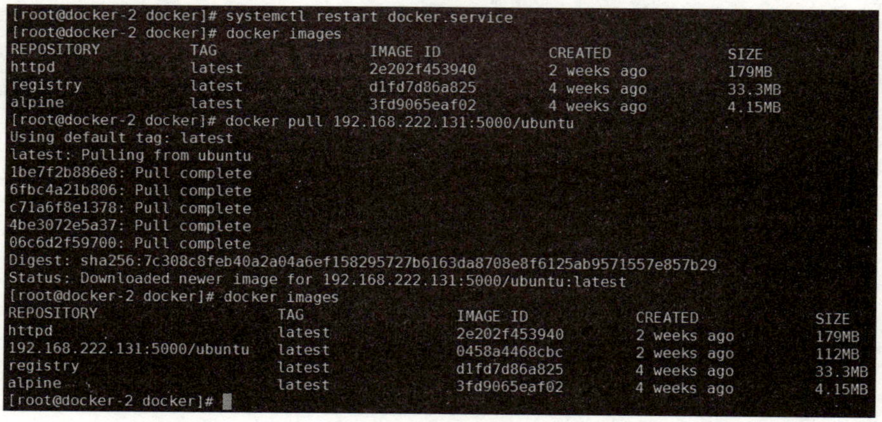

```
[root@docker-2 docker]# systemctl restart docker.service
[root@docker-2 docker]# docker images
REPOSITORY          TAG                 IMAGE ID            CREATED             SIZE
httpd               latest              2e202f453940        2 weeks ago         179MB
registry            latest              d1fd7d86a825        4 weeks ago         33.3MB
alpine              latest              3fd9065eaf02        4 weeks ago         4.15MB
[root@docker-2 docker]# docker pull 192.168.222.131:5000/ubuntu
Using default tag: latest
latest: Pulling from ubuntu
1be7f2b886e8: Pull complete
6fbc4a21b806: Pull complete
c71a6f8e1378: Pull complete
4be3072e5a37: Pull complete
06c6d2f59700: Pull complete
Digest: sha256:7c308c8feb40a2a04a6ef158295727b6163da8708e8f6125ab9571557e857b29
Status: Downloaded newer image for 192.168.222.131:5000/ubuntu:latest
[root@docker-2 docker]# docker images
REPOSITORY                    TAG                 IMAGE ID            CREATED             SIZE
httpd                         latest              2e202f453940        2 weeks ago         179MB
192.168.222.131:5000/ubuntu   latest              0458a4468cbc        2 weeks ago         112MB
registry                      latest              d1fd7d86a825        4 weeks ago         33.3MB
alpine                        latest              3fd9065eaf02        4 weeks ago         4.15MB
[root@docker-2 docker]#
```

图 10-22

10.8.3 Docker 数据管理

容器是由镜像启动的，无论是关闭容器，还是删除容器，容器中存储的数据和更改的数据都会消失，这就意味着数据可能会有一定的风险。但是，可以把宿主机的某个目录挂载到容器中，如创建/data/目录，将容器生成的新数据全部写到/data/目录下，也就是写到宿主机的磁盘中，这样即使删除了容器，数据也不会消失，挂载命令如下。

```
[root@docker-1 ~]# ls /data/
humingzhe.cn  humingzhe.com  mariadb  mysql  redis  www
[root@docker-1 ~]# docker run -itd -v /data/:/data/ centos bash
93a1f4963fad66e8b48b0f3c4b9d9abc552f1eff60396f3371520dc3d410f3de
[root@docker-1 ~]# docker exec -it 93a1f496 bash
[root@93a1f4963fad /]# ls /data/
humingzhe.cn  humingzhe.com  mariadb  mysql  redis  www
```

在容器的/data/目录下创建一个 mongodb 的新目录，退出容器，使用宿主机查看/data/目录，还会看到 mongodb 目录，运行命令如下。

```
[root@93a1f4963fad /]# mkdir /data/mongodb
[root@93a1f4963fad /]# exit
[root@docker-1 ~]# ls /data/
```

```
humingzhe.cn  humingzhe.com  mariadb  mongodb  mysql  redis  www
```

注意

冒号前面的/data/是宿主机的本地目录，冒号后面的/data/是容器的目录，该目录如果不存在，则会在容器中自动创建。

挂载目录时可指定容器名称，如果不指定，则会随机定义。上面没有指定容器名称，系统生成了一个 unruffled_euclid 的名称。

当需要定义多个容器共享数据时，可以搭建一个专门的数据卷容器，其他容器直接挂载该数据卷即可，类似于 Linux 中的 NFS。创建新的容器，把刚创建的 unruffled_euclid 容器作为数据卷，运行命令如下。

```
[root@docker-1 ~]# docker run -itd --volumes-from unruffled_euclid
centos6 bash
7e0622ed8697bc5476f40f8d09a854cfa88416e776d03c63731c730111a72dfb
[root@docker-1 ~]# docker exec -it 7e0622e bash
[root@7e0622ed8697 /]# ls /data/
humingzhe.cn  humingzhe.com  mariadb  mongodb  mysql  redis  www
```

docker run -tid -v 命令支持另外一种写法，可以不加冒号。如果不加冒号，就不需要把宿主机的目录做映射。-v 选项的意思是把该目录作为一个公共的目录，这就意味着要做一个数据卷容器，创建数据卷容器运行命令如下。

```
[root@docker-1 ~]# docker run -itd -v /data/ --name unruffled_euclid
centos bash
2f9d0f227244f6011fa3b75152d169e49a93ee106a0e68bcf077163ccacb6861
```

在创建新容器时，可加上--volumes-from 选项，该选项的意思是挂载该数据卷，运行命令如下。

```
[root@docker-1 ~]# docker run -itd --volumes-from unruffled_euclid centos
bash
a19f5db7d9749deb91891ffab958e8792ae7160641655fb3f5701a591fce9928
[root@docker-1 ~]# docker exec -it a19f5db7 bash
```

10.9　Docker数据卷备份与恢复

10.9.1　Docker 数据卷备份

数据卷容器给本地宿主机目录做映射，根本不需要备份，只要定期复制硬盘数据即可。

如果没有做映射，直接执行 docker run -v /data/需要借助中间容器进行备份。例如，在宿主机磁盘上创建一个/data/backup 目录，再用 unruffled_euclid 数据卷新创建一个容器，同时把本地创建的/data/backup 目录挂载到该容器的/bak 目录下，这样在容器的/bak 目录下新建文件，就可以直接在/data/backup/目录中看到，最后把数据卷容器中的/data/目录下的文件打包成 data.tar 文件并放到/bak 目录下，运行命令如下。

```
[root@docker-1 ~]# docker run --volumes-from unruffled_euclid -v /data/backup/:/bak centos tar cvf /bak/data.tar /data/
[root@docker-1 ~]# ls /data/backup/
data.tar
```

10.9.2 Docker 数据卷恢复

先创建一个数据卷容器，再创建一个新的容器，并挂载该数据卷容器，最后将 data.tar 包进行解包，运行命令如下。

```
[root@docker-1 ~]# docker run -itd -v /data/ --name unruffled_euclid2 centos bash
f03b71c8761630e57783f7909a203904d27ef64125cf83e5189903b661e9116c
[root@docker-1 ~]# docker exec -it f03b71c8 bash
[root@f03b71c87616 /]# ls /data/
[root@f03b71c87616 /]# exit
[root@docker-1 ~]# docker run --volumes-from unruffled_euclid2 -v /data/backup/:/bak centos tar xf /bak/data.tar
[root@docker-1 ~]# docker exec -it f03b71c8 bash
[root@f03b71c87616 /]# ls /data/
backup  humingzhe.cn  humingzhe.com  mariadb  mongodb  mysql  redis  www
```

10.10 Docker网络模式与外部访问容器

10.10.1 Docker 网络模式

Docker 共有 4 种网络模式，下面分别介绍。

- host 模式：在执行 docker run 命令时用--net=host 选项指定，该模式下 Docker 使用的网络和宿主机使用的网络是一样的，在容器内看到的网卡 IP 就是宿主机的 IP。

- container 模式：用选项--net=container：container_id/container_name 指定，该模式是多个容器共同使用网络，看到的 IP 都是一样的。
- none 模式：用--net=none 选项指定，该模式下不会设置任何网络。
- bridge 模式：用--net=bridge 选项指定，是 Docker 默认的网络模式。该模式会为每个容器分配一个独立的 Network Namespace，类似于 VMWare 的 NAT 网络模式。同一个宿主机上的所有容器会在同一个网段下，相互之间可以通信。

10.10.2　外部访问容器

Docker 默认使用 bridge 网络模式，Docker 宿主机启动了一个容器，容器本来监听的是 80 端口，但是容器本身只在局域网内生效，宿主机之外的服务器是访问不到的，要想其他的服务器也能访问，可以做端口映射。

进入一个容器，在容器中安装 Nginx 服务，运行命令如下。

```
[root@docker-1 ~]# docker exec -it e9d0ab bash
[root@e9d0abd69350 /]# yum install -y epel-release
[root@e9d0abd69350 /]# yum install -y nginx
[root@e9d0abd69350 /]# rpm -qa nginx
nginx-1.12.2-1.el7.x86_64
```

将容器导出为镜像，再使用新镜像创建容器并指定端口映射，运行命令如下。

```
    [root@docker-1 ~]# docker commit -m "install nginx" -a "humingzhe"
e9d0abd69350 centos_with_nginx
    sha256:c48f60c0a1542b72e5e93776979977c75de68d278937a505c7c838708646c9
9f
    [root@docker-1 ~]# docker images
    REPOSITORY          TAG       IMAGE ID      CREATED          SIZE
    centos_with_nginx  latest    c48f60c0a154  13 seconds ago   389MB
    [root@docker-1 ~]# docker run -itd -p 8888:80 centos_with_nginx bash
    aac9c66a0d0bac503140bef03de0b04aad2d1dcba541a64020449397c7e9e0a5
    [root@docker-1 ~]# docker ps
```

10.10.3　Operation Not Permitted 解决方案

新建的容器启动 Nginx 或 httpd 服务时会报 Failed to get D-Bus connection: Operation not permitted 错误，这是因为 dbus-daemon 服务没有启动，解决该问题的方法是，在创建容器时加上--privileged -e "container=docker"选项，进入到容器中，

再启动 Nginx 服务时就不会报错了。使用 ps aux 命令可查看 Nginx 的进程，如图 10-23 所示。

```
[root@docker-1 ~]# docker rm -f aac9c66a0d0b
aac9c66a0d0b
[root@docker-1 ~]# docker run -itd --privileged -e "container=docker" -p 8888:80 centos_with_nginx /usr/sbin/init
7fba51a16336d4bf1ca6be497e59550ecb5630b46058a76852ef18554619a8b8
[root@docker-1 ~]# docker exec -it 7fba51a1 bash
[root@7fba51a16336 /]# systemctl start nginx
[root@7fba51a16336 /]# ps aux |grep nginx
root        84  0.0  0.2 122924  2104 ?        Ss   14:05   0:00 nginx: master process /usr/sbin/nginx
nginx       85  0.0  0.3 123388  3132 ?        S    14:05   0:00 nginx: worker process
root        87  0.0  0.0   9044   660 pts/1    S+   14:06   0:00 grep --color=auto nginx
[root@7fba51a16336 /]#
```

图 10-23

容器中的 Nginx 默认是 80 端口，在外部访问需要用 8888 端口，运行命令如下。同理，使用外部浏览器时输入 IP 地址+端口号也可以访问，如图 10-24 所示。

```
[root@docker-1 ~]# curl localhost:8888
[root@docker ~]# curl 192.168.222.131:8888
[root@docker-2 ~]# curl 192.168.222.131:8888
```

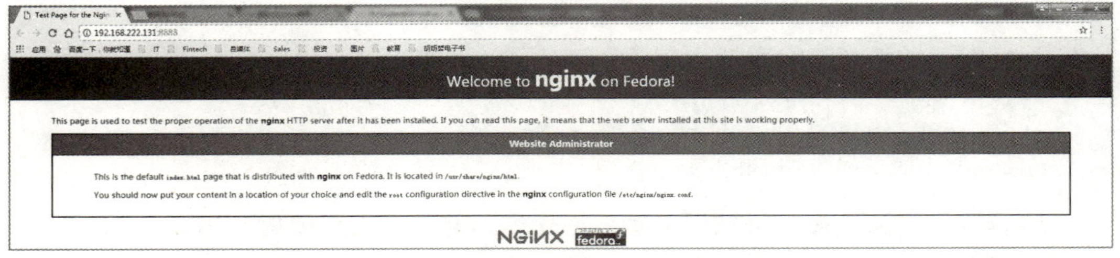

图 10-24

10.11 设置桥接网络

pipework 属于第三方开源网络工具，它能够实现 Docker 容器和宿主机使用同一个交换机、在同一个网段中，这样就可以和外部的其他服务器进行通信了，还可以把 Docker 容器看作独立的服务器，如安装 LNMP 服务，运行后可以直接让外部用户去访问，同样也支持安装 sshd 服务，使用户远程登录。

进入宿主机网卡设置目录，通过 ens33 网卡复制一个新网卡，如 br0 网卡。复制后编辑 br0 网卡的配置文件，将 Type 代码改为 TYPE=Bridge 即可，运行命令如下。

```
[root@docker-1 ~]# cd /etc/sysconfig/network-scripts/
[root@docker-1 network-scripts]# cp ifcfg-ens33 ifcfg-br0
[root@docker-1 network-scripts]# vim ifcfg-br0
TYPE=Bridge
NAME=br0
```

```
DEVICE=br0
```

编辑 ens33 网卡，注释掉 UUID、IPADDR、GATEWAY、NETMASK 和 DNS，新增一行代码，运行命令如下，这样做的目的是将 ens33 网卡的 IP 设置到 br0 网卡上。

```
[root@docker-1 network-scripts]# vim ifcfg-ens33
#UUID=7cde1071-823a-4d1f-a7a5-82dd46e969f4
#IPADDR=192.168.222.131
#GATEWAY=192.168.222.2
#NETMASK=255.255.255.0
#DNS1=114.114.114.114
BRIDGE=br0
```

网卡设置后重启网络服务（设置成功可 ping 通，否则无法 ping 通，终端也会断开），运行 ifconfig 命令即可看到 br0 网卡，如图 10-25 所示。

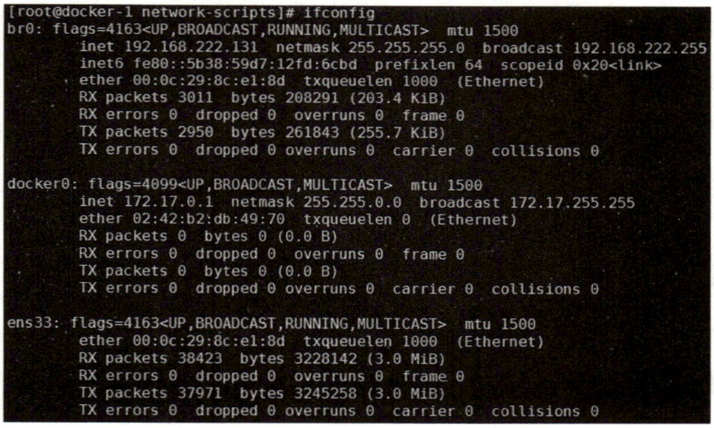

```
[root@docker-1 network-scripts]# ifconfig
br0: flags=4163<UP,BROADCAST,RUNNING,MULTICAST>  mtu 1500
        inet 192.168.222.131  netmask 255.255.255.0  broadcast 192.168.222.255
        inet6 fe80::5b38:59d7:12fd:6cbd  prefixlen 64  scopeid 0x20<link>
        ether 00:0c:29:8c:e1:8d  txqueuelen 1000  (Ethernet)
        RX packets 3011  bytes 208291 (203.4 KiB)
        RX errors 0  dropped 0  overruns 0  frame 0
        TX packets 2950  bytes 261843 (255.7 KiB)
        TX errors 0  dropped 0 overruns 0  carrier 0  collisions 0

docker0: flags=4099<UP,BROADCAST,MULTICAST>  mtu 1500
        inet 172.17.0.1  netmask 255.255.0.0  broadcast 172.17.255.255
        ether 02:42:b2:db:49:70  txqueuelen 0  (Ethernet)
        RX packets 0  bytes 0 (0.0 B)
        RX errors 0  dropped 0  overruns 0  frame 0
        TX packets 0  bytes 0 (0.0 B)
        TX errors 0  dropped 0 overruns 0  carrier 0  collisions 0

ens33: flags=4163<UP,BROADCAST,RUNNING,MULTICAST>  mtu 1500
        ether 00:0c:29:8c:e1:8d  txqueuelen 1000  (Ethernet)
        RX packets 38423  bytes 3228142 (3.0 MiB)
        RX errors 0  dropped 0  overruns 0  frame 0
        TX packets 37971  bytes 3245258 (3.0 MiB)
        TX errors 0  dropped 0 overruns 0  carrier 0  collisions 0
```

图 10-25

安装 pipework 工具，使用 git 命令可直接复制，复制成功后进入 pipework 目录，把 pipework 可执行文件复制到/usr/local/bin/目录下，然后使用 pipework 命令直接运行，运行命令如下。

```
[root@docker-1 ~]# git clone https://github.com/jpetazzo/pipework
[root@docker-1 ~]# cd pipework/
[root@docker-1 pipework]# ls
docker-compose.yml  doctoc  LICENSE  pipework  pipework.spec
README.md
[root@docker-1 pipework]# cp pipework /usr/local/bin/
[root@docker-1 pipework]# pipework
```

开启一个容器，并给刚开启的容器设置 IP，然后进入容器，使用 ifconfig 命令查看，会出现刚设置的 IP 192.168.222.136，ping humingzhe.com 可以 ping 通，说明是可以连接外网的，运行命令如下。

```
    [root@docker-1 ~]# docker run -itd --net=none centos_with_nginx bash
    8caaf8537deb5dbda94ce0ed7b53c016830c2b0bb3b280e9d8f3729850266562
    [root@docker-1 ~]# docker exec -it 8caaf853 bash
    [root@8caaf8537deb /]# exit
    [root@docker-1 ~]# pipework br0 8caaf8537
192.168.222.136/24@192.168.222.2
    [root@docker-1 ~]# docker exec -it 8caaf853 bash
    [root@8caaf8537deb /]# ifconfig
    eth1: flags=4163<UP,BROADCAST,RUNNING,MULTICAST>  mtu 1500
        inet 192.168.222.136  netmask 255.255.255.0  broadcast
192.168.222.255
    [root@8caaf8537deb /]# ping humingzhe.com
    PING humingzhe.com (120.79.160.60) 56(84) bytes of data.
    64 bytes from 120.79.160.60 (120.79.160.60): icmp_seq=1 ttl=128 time=42.6
ms
```

注意

192.168.222.136 是容器的 IP，@后面的 IP 是网关 IP。若想使用 LNMP 或 SSHD 等一些服务，则需要安装相应软件包，如安装 Nginx 服务，安装后启动即可在浏览器中进行访问。

10.12 DockerFile创建镜像

创建镜像可以用现有的镜像制作成容器，在容器中进行一些操作，最后导出为镜像；还有一种方法是在 OpenVZ 官网下载模板制作镜像。本节讲解通过 DockerFile 创建镜像。

10.12.1 DockerFile 格式

在 DockerFile 中可以写一些语句。FROM 用于指定基于哪个镜像，格式是 FROM<image>或 FROM<image>:<tag>，示例代码如下。

```
FROM centos
FROM centos:latest
```

MAINTAINER 用于指定作者信息，格式是 MAINTAIN <name>，示例如下。

```
MAINTAINER  humingzhe  admin@humingzhe.com
```

RUN 是镜像操作指令，格式是 RUN <command> 或 RUN ["executable", "param1", "param2"]，示例如下。

```
RUN  yum install  httpd
RUN ["/bin/bash", "-c", "echo hello"]
```

CMD 用于执行一些命令，它有 3 种语法格式，与 RUN 的语法格式类似，CMD 指定容器启动时用到的命令只能有一条，示例格式如下。

```
CMD ["executable", "param1", "param2"]
CMD command param1 param2
CMD ["param1", "param2"]
CMD ["/bin/bash", "/usr/local/nginx/sbin/nginx", "-c",
"/usr/local/nginx/conf/nginx.conf"]
```

EXPOSE 是用来暴露端口的命令，比如把 22 端口、80 端口、8443 端口暴露出来并赋值，示例格式如下。

```
EXPOSE 22 80 8443
```

注意

在启动容器时，可以加上-P（大写）选项让系统自动分配端口。如果想指定具体的端口，可使用-p（小写）选项来指定。

ENV 用来定义环境变量，格式是<key><value>，示例格式如下。

```
ENV PATH /usr/local/mysql/bin:$PATH
```

注意

ENV 主要为后续的 RUN 指令提供一个环境变量，也可以自定义一些环境变量，如 ENV MYSQL_version5.7。

ADD 命令是将本地的文件或目录复制到容器的某个目录下。其中，源为 DockerFile 所在目录的相对路径，也可以是 URL，示例格式如下。

```
ADD <conf/vhosts> </usr/local/nginx/conf>
```

COPY 命令和 ADD 命令类似，语法格式也一致，不同的是 COPY 命令不支持 URL 远程下载。

ENTRYPOINT 命令的格式类似于 CMD 命令的格式，容器启动时要执行的命令

也类似于 CMD 命令，只有一条生效，如果写多条语句，则只有最后一条语句会生效。不同的是，CMD 可以被 docker run 命令指定覆盖，而 ENTRYPOINT 不能被覆盖。比如，容器名是 humingzhe，在 dockerfile 中指定 CMD 格式是 CMD ["/bin/echo", "test"]，启动容器命令是 docker run humingzhe，这样会输出 test。如果启动容器的命令是 docker run -it humingzhe /bin/bash，则什么都不会输出。

ENTRYPOINT 不会被覆盖，而且比 CMD 或 docker run 指定的命令要靠前执行 ENTRYPOINT ["echo", "test"]。执行 docker run -it humingzhe 123 命令会先输出 test，再输出 123，这相当于执行命令 echo test 123。

VOLUME 是用来指定挂载点的，可以创建一个从本地或其他容器挂载的挂载点，格式是 VOLUME ["/data"]。

USER 指定运行容器的用户，格式是 USER daemon。

WORKDIR 用于指定一个目录，指定目录后，在目录下进行一些操作，如运行一些命令时，先进入到路径下，或者在启动容器时，使用 CMD、ENTRYPOINT 指定工作目录。

10.12.2 DockerFile 示例安装 Nginx

在读者 QQ 群文件中找到 nginx_conf 文件，将其下载到 root 目录。创建 Dockerfile 配置文件，在该配置文件中编辑并安装 Nginx 的内容，示例代码如下。

```
    [root@docker-1 ~]# vim Dockerfile
    ## Set the base image to CentOS
    FROM centos
    # File Author / Maintainer
    MAINTAINER humingzhe admin@humingzhe.com
    # Install necessary tools
    RUN yum install -y pcre-devel wget net-tools gcc zlib zlib-devel make
openssl-devel
    # Install Nginx
    ADD http://nginx.org/download/nginx-1.8.0.tar.gz .
    RUN tar zxvf nginx-1.8.0.tar.gz
    RUN mkdir -p /usr/local/nginx
    RUN cd nginx-1.8.0 && ./configure --prefix=/usr/local/nginx && make &&
make install
    RUN rm -fv /usr/local/nginx/conf/nginx.conf
    COPY .nginx_conf /usr/local/nginx/conf/nginx.conf
    # Expose ports
    EXPOSE 80
    # Set the default command to execute when creating a new container
```

```
ENTRYPOINT /usr/local/nginx/sbin/nginx && tail -f /etc/passwd
```

创建镜像，使用-t 选项指定镜像名称为 centos_nginx，使用 docker images 命令可看到新建的镜像，运行命令如下。

```
[root@docker-1 ~]# docker build -t centos_nginx .
[root@docker-1 ~]# docker images
REPOSITORY        TAG         IMAGE ID        CREATED           SIZE
centos_nginx      latest      6dba380c03e5    5 minutes ago     374MB
```

指定端口进入该容器，使用 ps aux 命令可查看 Nginx 进程，如图 10-26 所示。

```
[root@docker-1 ~]# docker run -itd -p 81:80 centos_nginx bash
8b3b0c97ce38b68a737d353db16c06f5051f0ed1d9c9f94cc2886ae035bdafa3
[root@docker-1 ~]# docker exec -it 8b3b0c97 bash
[root@8b3b0c97ce38 /]# ps aux |grep nginx
root         1  0.1  0.1  11636  1328 pts/0    Ss+  04:31   0:00 /bin/sh -c /usr/local/nginx/sbin/nginx && tail -f /etc/passwd bash
root         6  0.0  0.0  24836   788 ?        Ss   04:31   0:00 nginx: master process /usr/local/nginx/sbin/nginx
nobody       8  0.0  0.3  27280  3372 ?        S    04:31   0:00 nginx: worker process
nobody       9  0.0  0.3  27280  3372 ?        S    04:31   0:00 nginx: worker process
root        23  0.0  0.0   9044   664 pts/1    S+   04:32   0:00 grep --color=auto nginx
[root@8b3b0c97ce38 /]#
```

图 10-26

使用 curl 命令访问 81 端口或使用外部浏览器访问 81 端口，都可以访问 Nginx 欢迎界面，如图 10-27 所示。

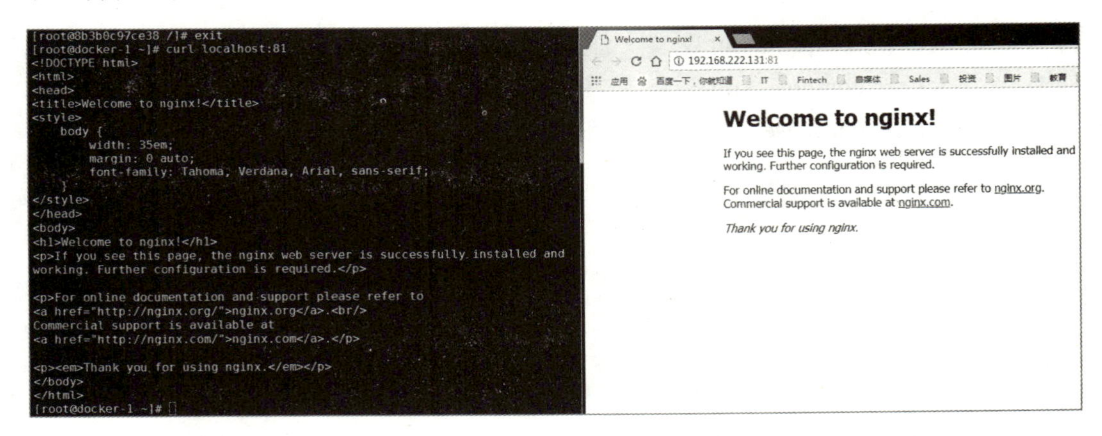

图 10-27

10.13　docker compose部署服务与示例

使用 docker-compose 命令可以快捷、高效地管理容器的启动、停止、重启等操作，类似于 Linux 中的 shell 脚本，基于 yaml 语法，在 YAML 文件中可以描述应用的架构，比如使用的镜像、数据卷、网络模式、监听端口等信息。也可以在一个 compose 文件中定义一个多容器的应用（比如 jumpserver），然后通过该 compose 启动这个应用。

使用 curl 命令到官网下载最新版本的 docker compose，下载完毕后赋予 755 权限，使用 docker-compose version 命令可查看版本信息，运行命令如下。如果下载速度非常慢，可到读者 QQ 群文件中下载，然后上传到相应目录即可。

```
[root@docker-1 ~]# curl -L
https://github.com/docker/compose/releases/download/1.10.0-rc1/docker-compose-`uname -s`-`uname -m` > /usr/local/bin/docker-compose
[root@docker-1 ~]# du -sh /usr/local/bin/docker-compose
8.5M    /usr/local/bin/docker-compose
[root@docker-1 ~]# chmod 755 !$
chmod 755 /usr/local/bin/docker-compose
[root@docker-1 ~]# docker-compose version
docker-compose version 1.10.0-rc1, build a0f95af
```

注意

Compose 分为 Version1 和 Version2（Compose 1.6.0+，Docker Engine 1.10.0+）。Version2 支持更多的指令。没有声明版本的，默认是 Version1，声明版本的则是 Version2，Version1 将来会被弃用。

编辑 docker-compose.yml 配置文件，添加代码，代码文件在读者 QQ 群文件中，文件名是 docker-compose.yml。

使用 docker-compose up 命令启动配置文件中定义的容器，运行命令如下。

```
[root@docker-1 ~]# docker-compose up -d
Creating root_app1_1 ...
Creating root_app2_1 ...
Creating root_app1_1
Creating root_app2_1 ... done
```

使用 docker-compose stop 命令可以停止配置文件中定义的容器，使用 docker-compose –help 命令可查看 docker-compose 支持的指令，运行命令如下。

```
[root@docker-1 ~]# docker-compose stop
[root@docker-1 ~]# docker-compose --help
```

docker-compose 语法的参考文档网址是 http://www.web3.xin/index/article/182.html。可以在 Docker 官方的 GitHub 中学习 Docker 官方提供的一些 Dockerfile 优秀案例。Dockerfile 优秀案例的网址是 https://github.com/docker-library，里面有许多优秀的 Dockerfile 案例，比如 MySQL 的 Dockerfile 语法，如图 10-28 和图 10-29 所示。这些优秀案例可以帮助我们更好地理解和学习 Dockerfile。

```
76 lines (64 sloc) | 3.43 KB                                  Raw  Blame  History
1   FROM debian:stretch-slim
2
3   # add our user and group first to make sure their IDs get assigned consistently, regardless of whatever dependencies get added
4   RUN groupadd -r mysql && useradd -r -g mysql mysql
5
6   RUN apt-get update && apt-get install -y --no-install-recommends gnupg dirmngr && rm -rf /var/lib/apt/lists/*
7
8   # add gosu for easy step-down from root
9   ENV GOSU_VERSION 1.7
10  RUN set -x \
11          && apt-get update && apt-get install -y --no-install-recommends ca-certificates wget && rm -rf /var/lib/apt/lists/* \
12          && wget -O /usr/local/bin/gosu "https://github.com/tianon/gosu/releases/download/$GOSU_VERSION/gosu-$(dpkg --print-architecture)" \
13          && wget -O /usr/local/bin/gosu.asc "https://github.com/tianon/gosu/releases/download/$GOSU_VERSION/gosu-$(dpkg --print-architecture
14          && export GNUPGHOME="$(mktemp -d)" \
15          && gpg --keyserver ha.pool.sks-keyservers.net --recv-keys B42F6819007F00F88E364FD4036A9C25BF357DD4 \
16          && gpg --batch --verify /usr/local/bin/gosu.asc /usr/local/bin/gosu \
17          && gpgconf --kill all \
18          && rm -rf "$GNUPGHOME" /usr/local/bin/gosu.asc \
19          && chmod +x /usr/local/bin/gosu \
20          && gosu nobody true \
21          && apt-get purge -y --auto-remove ca-certificates wget
22
23  RUN mkdir /docker-entrypoint-initdb.d
24
25  RUN apt-get update && apt-get install -y --no-install-recommends \
26  # for MYSQL_RANDOM_ROOT_PASSWORD
27                  pwgen \
28  # for mysql_ssl_rsa_setup
29                  openssl \
30  # FATAL ERROR: please install the following Perl modules before executing /usr/local/mysql/scripts/mysql_install_db:
31  # File::Basename
32  # File::Copy
33  # Sys::Hostname
34  # Data::Dumper
```

图 10-28

```
51  RUN echo "deb http://repo.mysql.com/apt/debian/ stretch mysql-${MYSQL_MAJOR}" > /etc/apt/sources.list.d/mysql.list
52
53  # the "/var/lib/mysql" stuff here is because the mysql-server postinst doesn't have an explicit way to disable the mysql_install_db codepa
54  # also, we set debconf keys to make APT a little quieter
55  RUN { \
56                  echo mysql-community-server mysql-community-server/data-dir select ''; \
57                  echo mysql-community-server mysql-community-server/root-pass password ''; \
58                  echo mysql-community-server mysql-community-server/re-root-pass password ''; \
59                  echo mysql-community-server mysql-community-server/remove-test-db select false; \
60          } | debconf-set-selections \
61          && apt-get update && apt-get install -y mysql-community-client="${MYSQL_VERSION}" mysql-community-server-core="${MYSQL_VERSION}" &&
62          && rm -rf /var/lib/mysql && mkdir -p /var/lib/mysql /var/run/mysqld \
63          && chown -R mysql:mysql /var/lib/mysql /var/run/mysqld \
64  # ensure that /var/run/mysqld (used for socket and lock files) is writable regardless of the UID our mysqld instance ends up having at run
65          && chmod 777 /var/run/mysqld
66
67  VOLUME /var/lib/mysql
68  # Config files
69  COPY config/ /etc/mysql/
70  COPY docker-entrypoint.sh /usr/local/bin/
71  RUN ln -s usr/local/bin/docker-entrypoint.sh /entrypoint.sh # backwards compat
72  ENTRYPOINT ["docker-entrypoint.sh"]
73
74  EXPOSE 3306 33060
75  CMD ["mysqld"]
```

图 10-29

第 11 章 搭建 Kubernetes 集群

11.1 Kubernetes（K8S）简介

目前，Kubernetes 已经成为容器领域当之无愧的事实标准。Google、Microsoft、IBM 等巨头们已经在容器领域博弈了多年，国内的 Baidu、Alibaba、Tencent、360 等企业也都将容器技术列入集团重点发展的战略技术之一，而且大量的中小企业也正在走容器化的道路。

想要掌握 Kubernetes 并不是那么容易的，无论是初学者，还是有一定基础的读者，都会遇到诸多的难题。下面列举几个学习 Kubernetes 时常见的问题。

（1）Kubernetes 过于复杂，内容较多，根本无从下手。

（2）文档看完还是搞不清 Kubernetes 的脉络。

（3）Kubernetes 版本更新太快，内容非常多，每次都要花很多的时间去钻研，完全跟不上节奏。

（4）支持 Kubernetes 的项目越来越多，例如 Linkerd 2.0 如何在 Kubernetes 上部署、如何基于 Kubernetes 构建 AI 平台等。如此之多的开源项目，如此之多的代码，学习起来非常头疼。

为什么会这样呢？因为 Kubernetes 并不是我们过去所认识的典型的开源项目，学习 Kubernetes 不仅要深耕文档和代码，更重要的是要深度理解 Kubernetes 的设计思想和初衷。

本章用通俗易懂的语言深度剖析 Kubernetes 的各个层并讲解其特性，使读者理解 Kubernetes 所蕴含的思想。

Kubernetes 开源后许多巨头公司对 Kubernetes 社区做了很多贡献，如 Microsoft、

Red Hat、IBM、Docker 等。

Kubernetes 官方网址是 www.kubernetes.io。Kubernetes 是 Google 公司在 2014 年 6 月开源的一个容器集群管理系统，使用 Golang 语言开发，简称 K8S（以 K 开头，以 S 结尾，中间包含 8 个字母）。K8S 是从 Google 公司内部一个叫作 Borg 的容器集群管理系统衍生出来的，Borg 已经在 Google 公司大规模生产运行十年之久。

K8S 主要用于自动化部署、扩展和管理容器应用，提供了资源调度、部署管理、服务发现、扩容缩容、监控等一系列功能，这些功能是将容器推送到生产环境很关键的一步。2015 年 7 月，Kubernetes v1.0 正式发布，截至 2018 年 9 月 29 日，最新稳定版本是 v1.12。Kubernetes 的目标是让部署容器化应用更加简单、高效。

Kubernetes 的主要功能有：数据卷、应用程序健康检查、复制应用程序实例、弹性伸缩、服务发现、负载均衡、滚动更新、服务编排、资源监控、提供认证和授权等，具体如下。

- 数据卷：Pod（Pod 是 Kubernetes 创建或部署的最小、最简单的基本单位，一个 Pod 代表集群上正在运行的一个进程）中容器之间共享数据，可以使用数据卷。一个 Pod 中可能存在多个容器，多个容器可以通过数据卷的形式共享数据。
- 应用程序健康检查：容器内的服务可能因进程堵塞而无法处理请求，可以设置监控来检查策略，保证应用的健壮性。
- 复制应用程序实例：控制器维护着 Pod 副本数量，保证一个 Pod 或一组同类的 Pod 数量始终可用。
- 弹性伸缩：根据设定的目标（CPU 利用率）自动缩放 Pod 副本数。
- 服务发现：使用环境变量或 DNS 服务插件保证容器中的程序能够发现 Pod 入口访问地址。
- 负载均衡：一组 Pod 副本分配一个私有的集群 IP 地址，负载均衡转发请求到后端容器。在集群内部，其他 Pod 可通过 ClusterIP 访问应用。
- 滚动更新：更新服务不中断，一次更新一个 Pod，而不是同时删除整个服务，类似于灰度发布。
- 服务编排：通过文件描述可以部署服务，使得应用程序部署变得更加高效。
- 资源监控：Node（节点）组件集成 cAdvisor 资源收集工具，可通过 Heapster 汇总整个集群节点资源数据，然后存储到 InflunDB 时序数据库中，再由 Grafana 展示。
- 提供认证和授权：支持角色访问控制（RBAC）认证授权等策略。

11.2 Kubernetes基本概念

在 Kubernetes 中，对象的概念是非常重要的，在后面的章节中会不断遇到，希望读者能够把这些概念牢记于心。

Kubernetes 基本对象有：Pod、Service、Volume、Namespace 和 Lable，下面分别介绍。

- Pod：最小部署单元，一个 Pod 由一个或多个容器组成，Pod 中容器共享存储和网络，在同一台 Docker 主机上运行。
- Service：是一个应用服务抽象，定义了 Pod 逻辑集合和访问这个 Pod 集合的策略。Service 代理 Pod 集合的对外表现是一个访问入口，被分配一个集群 IP 地址，来自这个 IP 的请求将负载均衡转发给后端 Pod 中的容器。Service 通过 Lable Selector 选择一组 Pod 提供服务。
- Volume：数据卷，共享 Pod 中容器使用的数据。
- Namespace：命名空间，在逻辑上将对象分配到不同的命名空间，可以是不同的项目、用户等，可设置控制策略，从而实现多租户。命名空间也称为虚拟集群。
- Lable：标签，用于区分对象（比如 Pod、Service），以键/值对存在；每个对象可以有多个标签，通过标签关联对象。

上述所讲的 Pod、Service 等偏系统底部，对用户层操作较少。对用户层操作较多的是基于基本对象更高层次的抽象，包括 ReplicaSet、Deployment、StatefulSet、DaemonSet 和 Job，下面分别介绍。

- ReplicaSet（RS）：下一代 Replication Controller（RC）。确保在任何给定的时间指定 Pod 副本数量，并提供声明式更新等功能。RC 与 RS 唯一的区别就是 Lable selector 的支持不同，RS 支持新的基于集合的标签，而 RC 仅支持基于等式的标签。
- Deployment：是一个更高层次的 API 对象，它管理的是 ReplicaSets 和 Pod，并提供声明式更新等功能。官方建议使用 Deployment 管理 ReplicaSets，而不是直接使用 ReplicaSets，这就意味着可能永远不需要直接操作 ReplicaSet 对象。
- StatefulSet：适合持久性的应用程序，有唯一的网络标识符（IP），持久存储，能够有序地部署、扩展、删除和滚动更新。
- DaemonSet：确保所有（一些）节点运行同一个 Pod。当节点加入到 Kubernetes 集群中时，Pod 会被调度到该节点上运行，当节点从集群中移除时，DaemonSet 的 Pod 会被删除。删除 DaemonSet 时会清理它创建的所有 Pod。
- Job：一次性任务，运行完成后用 Pod 销毁，不再重新启动新容器，还可以让任务定时运行。

11.3　Kubernetes架构和组件功能

Kubernetes 的架构如图 11-1 所示，最上层的 kubectl 是用户层面，用户可以使用 kubectl 工具去管理整个集群。kubectl 节点有三个组件，分别是 kube-scheduler、kube-apiserver 和 kube-controller-manager。这三个组件是在 Master 节点中运行的。

图 11-1 右侧的虚线框中是共享存储，Etcd 是一个分布式的键值存储，主要保存集群中的一些状态。比如，创建的 Pod 和 Service 相关信息都会存储在分布式存储中。

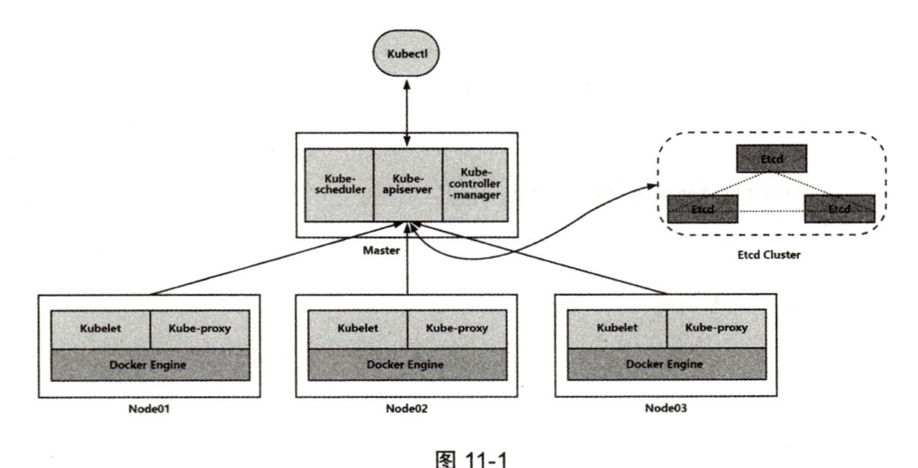

图 11-1

图 11-1 的下方是具体的集群节点，有 kubelet 和 kube-proxy 两个组件。kubelet 主要是对 Pod 和服务器的状态进行管理，kube-proxy 主要负责网络策略管理和负载均衡。节点上必须安装 Docker 服务，使用 kubelet 管理 Docker-Engine 并进行容器创建。

11.3.1　Master 组件功能介绍

kube-apiserver：Kubernetes API，集群的统一入口，各组件的协调者，以 HTTP API 提供接口服务，所有对象资源的增、删、改、查和监听操作都交给 API Server 处理后，再提交给 Etcd 存储。

kube-controller-manager：处理集群中常规的后台任务，一个资源对应一个控制器，ControllerManager 就是负责管理这些控制器的。

kube-scheduler：根据调度算法为新创建的 Pod 选择一个 Node（节点）。

11.3.2　Node 组件功能介绍

kubelet：是 Master 在 Node 上的 Agent，管理本机运行容器的生命周期，比如创建容器、Pod 挂载数据卷、下载 secret、获取容器和节点状态等。kubelet 将每个 Pod 转换成一组容器。

kube-proxy：在 Node 上实现 Pod 网络代理，维护网络规则和四层负载均衡工作。

Docker 或 rocket/rkt：运行容器。

11.3.3 三方组件 Etcd 介绍

Etcd 是第三方服务，是分布式键值存储系统，用于保存集群状态，比如保存 Pod、Service 等对象信息。

11.4 Kubernetes Cluster部署

11.4.1 集群环境规划

Kubernetes 集群搭建的内容较多，而且非常容易出现问题，因为涉及很多组件之间的关系，还涉及 TLS 证书，该证书是数字加密证书，用于集群组件之间的加密通信，保证数据的安全性。

集群部署环境采用当前最新的 Linux 操作系统的 CentOS 7.5 版本，Kubernetes 采用最新 1.10 版本，Docker 采用最新的 18.03-ce 版本，Etcd 采用最新的 3.0 版本。

准备 5 台服务器或在本地搭建 5 台虚拟机，Master 节点的 IP 是 192.168.10.5，Node-1 节点的 IP 是 192.168.10.6，Node-2 节点的 IP 是 192.168.10.7，Node-3 节点的 IP 是 192.168.10.8，Node-4 节点的 IP 是 192.168.10.9，Node-5 节点的 IP 是 192.168.10.10，5 台服务器部署组件情况如表 11-1 所示。

表 11-1

<table>
<tr><th>角　色</th><th>IP</th><th>组　件</th><th>节点推荐配置</th></tr>
<tr><td>Master</td><td>192.168.10.5</td><td>kube-apiserver
kube-controller-manager
kube-scheduler
etcd</td><td rowspan="6">CPU 4 核 8GB</td></tr>
<tr><td>Node-1</td><td>192.168.10.6</td><td rowspan="5">kubelet
kube-proxy
docker
flannel
etcd</td></tr>
<tr><td>Node-2</td><td>192.168.10.7</td></tr>
<tr><td>Node-3</td><td>192.168.10.8</td></tr>
<tr><td>Node-4</td><td>192.168.10.9</td></tr>
<tr><td>Node-5</td><td>192.168.10.10</td></tr>
</table>

注意

关闭 SELinux 和 Firewalld（防火墙），否则会给后续的实例操作造成极大的影响。

安装系统自动补全工具，命令如下。

```
[root@master ~]# yum install -y bash-completion-extras.noarch
bash-completion.noarch
```

11.4.2　安装 Docker 服务

Docker 只需在 Node 上进行安装，无需在 Maser 上进行安装，安装后启动 Docker 服务，命令如下。

```
[root@node-1 ~]# yum install -y yum-utils device-mapper-persistent-data
lvm2
[root@node-1 ~]# yum-config-manager --add-repo
https://download.docker.com/linux/centos/docker-ce.repo
[root@node-1 ~]# yum install -y docker-ce
[root@node-5 ~]# systemctl enable docker && systemctl start docker && ps
aux|grep docker && docker version
```

给 Node 服务器配置 Docker 官方国内镜像源和私有镜像仓库，运行命令如下。

```
[root@node-1 ~]# cat << EOF > /etc/docker/daemon.json
{
   "registry-mirrors": [ "https://registry.docker-cn.com"],
   "insecure-registries":["192.168.10.11:5000"]
}
EOF
[root@node-2 ~]# cat << EOF > /etc/docker/daemon.json
{
   "registry-mirrors": [ "https://registry.docker-cn.com"],
   "insecure-registries":["192.168.10.18:5000"]
}
EOF
```

11.4.3　自签 TLS 证书

自签 TLS 证书的组件和使用的证书如表 11-2 所示，在 etcd、kube-apiserver、kubelet、kube-proxy 和 kebectl 组件中会用到。

表 11-2

组　　件	使用的证书
etcd	ca.pem、server.pem、server-key.pem
kube-apiserver	ca.pem、server.pem、server-key.pem
kubelet	ca.pem、ca-key.pem
kube-proxy	ca.pem、kube-proxy.pem、kube-proxy-key.pem
kubectl	ca.pem、admin.pem、admin-key.pem

在自签 TLS 证书前需安装生成证书的工具 cfssl，该工具极大地简化了证书的生成过程。OpenSSL 工具也可以生成证书，但是步骤较多、较烦琐，容易出现问题。cfssl 工具能够很方便地生成证书，可生成证书的配置模板，只要稍作修改即可。

创建存储证书的目录，如在/root/目录下创建 ssl 目录作为存储证书的目录，命令如下。

```
[root@master ~]# mkdir ssl
```

进入 ssl 目录，下载 cfssl 证书安装工具，如果该工具的下载速度非常慢，则可打开网址 https://gitlab.com/humingzhe/Linux 进行下载，或者直接去官网（https://pkg.cfssl.org/）下载。

```
[root@master ~]# cd ssl/
[root@master ssl]# wget https://pkg.cfssl.org/R1.2/cfssl_linux-amd64
[root@master ssl]# wget
https://pkg.cfssl.org/R1.2/cfssljson_linux-amd64
[root@master ssl]# wget
https://pkg.cfssl.org/R1.2/cfssl-certinfo_linux-amd64
[root@master ssl]# ls
cfssl-certinfo_linux-amd64  cfssljson_linux-amd64  cfssl_linux-amd64
```

注意

生产环境中可以不使用证书，但要确保整个集群不会在外部暴露。不使用证书可以不加密传输，减少传输的时间；加密则会进行双向认证，认证会增加时间开销。但是在生成环境中还是建议使用证书，安全才是第一位的！

赋予刚下载的 3 个 cfssl 文件执行权限，然后将 3 个文件移至/usr/local/bin/目录下，运行命令如下。

```
[root@master ssl]# chmod +x cfssl-certinfo_linux-amd64
cfssljson_linux-amd64 cfssl_linux-amd64
[root@master ssl]# mv cfssl-certinfo_linux-amd64
/usr/local/bin/cfssl-certinfo
```

```
[root@master ssl]# mv cfssljson_linux-amd64 /usr/local/bin/cfssljson
[root@master ssl]# mv cfssl_linux-amd64 /usr/local/bin/cfssl
[root@master ssl]# ls /usr/local/bin/
cfssl  cfssl-certinfo  cfssljson
```

使用 cfssl －help 命令可查看生成证书的指令，使用 print-defaults 命令可以打印生成的证书模板文件，在生成的证书模板文件中进行相应的修改，运行命令如下。

```
[root@master ssl]# cfssl print-defaults config > config.json
[root@master ssl]# cfssl print-defaults csr > csr.json
```

笔者提供了一个已经修改好的证书模板文件，可以直接使用，文件在 GitLab 中，文件名是 certificate.sh，下载后上传到服务器中并赋予执行权限，运行命令如下。

```
[root@master ssl]# chmod +x certificate.sh
[root@master ssl]# ls -l certificate.sh
-rwxr-xr-x 1 root root 1763 Feb 15 16:58 certificate.sh
```

证书全部生成完毕后，将*.pem 的证书文件留下。其余后面用不到，可全部删除，命令如下。

```
[root@master ssl]# ls *pem
admin-key.pem  admin.pem  ca-key.pem  ca.pem  kube-proxy-key.pem
kube-proxy.pem  server-key.pem  server.pem
[root@master ssl]# ls | grep -v pem
admin.csr
admin-csr.json
ca-config.json
ca.csr
ca-csr.json
certificate.sh
kube-proxy.csr
kube-proxy-csr.json
server.csr
server-csr.json
[root@master ssl]# ls | grep -v pem | xargs -i rm {}
[root@master ssl]# ls
admin-key.pem  admin.pem  ca-key.pem  ca.pem  kube-proxy-key.pem
kube-proxy.pem  server-key.pem  server.pem
```

11.4.4　部署 Etcd 集群

Etcd 集群是作为存储用的，必须在集群部署之前设置好。可在 https://gitlab.com/

humingzhe/Linux 中或直接访问 Etcd 的 GitLab 地址下载 Etcd 二进制软件包，然后上传至 Linux 服务器，解压该软件包，运行命令如下。

```
[root@master ~]# rz -E
rz waiting to receive.
[root@master ~]# tar xf etcd-v3.3.10-linux-amd64.tar.gz
```

为了方便后期统一维护和管理，一定要把 K8S 所有的集群组件都放在统一的目录中，如放在/opt/kubernetes/目录中，该目录若不存在，则需要创建。在该目录下再创建 bin、cfg、ssl 目录，运行命令如下。bin 目录中放二进制软件包，cfg 目录中放配置文件，ssl 目录中放证书文件。

```
[root@master ~]# mkdir /opt/kubernetes
[root@master ~]# mkdir /opt/kubernetes/{bin,cfg,ssl}
```

将解压后的 Etcd 软件包目录中的 etcd 和 etcdctl 目录移至/opt/kubernetes/bin 目录下，运行命令如下。

```
[root@master ~]# mv etcd-v3.3.10-linux-amd64/etcd /opt/kubernetes/bin/
[root@master ~]# mv etcd-v3.3.10-linux-amd64/etcdctl
/opt/kubernetes/bin/
[root@master ~]# chmod +x /opt/kubernetes/bin/etcd*
[root@master ~]# ls /opt/kubernetes/bin/
etcd  etcdctl
```

在/opt/kubernetes/cfg 目录下创建 etcd 配置文件，添加配置代码，示例代码如下。

```
[root@master ~]# vim /opt/kubernetes/cfg/etcd
#[Member]
ETCD_NAME="etcd01"
ETCD_DATA_DIR="/var/lib/etcd/default.etcd"
ETCD_LISTEN_PEER_URLS="https://192.168.10.5:2380"
ETCD_LISTEN_CLIENT_URLS="https://192.168.10.5:2379"

#[Clustering]
ETCD_INITIAL_ADVERTISE_PEER_URLS="https://192.168.10.5:2380"
ETCD_ADVERTISE_CLIENT_URLS="https://192.168.10.5:2379"
ETCD_INITIAL_CLUSTER="etcd01=https://192.168.10.5:2380,etcd02=https:/
/192.168.10.6:2380,etcd03=https://192.168.10.7:2380,etcd04=https://192.1
68.10.8:2380,etcd05=https://192.168.10.9:2380,etcd06=https://192.168.10.
10:2380"
ETCD_INITIAL_CLUSTER_TOKEN="etcd-cluster"
ETCD_INITIAL_CLUSTER_STATE="new"
```

将/root/ssl 目录下的全部 server*.pem 和 ca*.pem 文件复制到/opt/kubernetes/ssl/

目录下，运行命令如下。

```
[root@master ~]# cp ssl/server*pem ssl/ca*.pem /opt/kubernetes/ssl/
[root@master ~]# ls /opt/kubernetes/ssl/
ca-key.pem  ca.pem  server-key.pem  server.pe
```

创建 etcd.service 配置文件，通过 systemd 管理 Etcd 服务，etcd.service 配置文件代码可到 GitLab 中进行下载，下载后上传至 Linux 服务器的/usr/lib/systemd/system/目录下。启动 etcd.service 服务并设置开机自启动，启动时 systemd 会有一点问题，但是不影响运行，按 Ctrl+C 组合键退出，使用 ps aux 命令查看进程是否处于运行状态，如图 11-2 所示。

```
[root@master ~]# vim /usr/lib/systemd/system/etcd.service
[root@master ~]# systemctl enable etcd.service && systemctl start etcd.service
Created symlink from /etc/systemd/system/multi-user.target.wants/etcd.service to /usr/lib/systemd/system/etcd.service.
^C
[root@master ~]# ps aux|grep etcd
root      1805  104  0.5 10567144 44052 ?      Ssl  16:56   0:32 /opt/kubernetes/bin/etcd --name=etcd01 --data-dir=/v
ar/lib/etcd/default.etcd --listen-peer-urls=https://192.168.10.5:2380 --listen-client-urls=https://192.168.10.5:2379,h
ttp://127.0.0.1:2379 --advertise-client-urls=https://192.168.10.5:2379 --initial-advertise-peer-urls=https://192.168.1
0.5:2380 --initial-cluster=etcd01=https://192.168.10.5:2380,etcd02=https://192.168.10.6:2380,etcd03=https://192.168.10
.7:2380,etcd04=https://192.168.10.8:2380,etcd05=https://192.168.10.9:2380,etcd06=https://192.168.10.10:2380 --initial-
cluster-token=etcd01=https://192.168.10.5:2380,etcd02=https://192.168.10.6:2380,etcd03=https://192.168.10.7:2380,etcd0
4=https://192.168.10.8:2380,etcd05=https://192.168.10.9:2380,etcd06=https://192.168.10.10:2380 --initial-cluster-state
=new --cert-file=/opt/kubernetes/ssl/server.pem --key-file=/opt/kubernetes/ssl/server-key.pem --peer-cert-file=/opt/ku
bernetes/ssl/server.pem --peer-key-file=/opt/kubernetes/ssl/server-key.pem --trusted-ca-file=/opt/kubernetes/ssl/ca.pe
m --peer-trusted-ca-file=/opt/kubernetes/ssl/ca.pem
root      1814  0.0  0.0 112660   968 pts/1    S+   16:56   0:00 grep --color=auto etcd
[root@master ~]#
```

图 11-2

为了方便后期管理日常文件，需要为集群做互信，生成 SSH 证书，执行 ssh-keygen 命令并一直按 Enter 键，会生成私钥和公钥，运行命令如下。

```
[root@master ~]# ssh-keygen
[root@master ~]# ls /root/.ssh/
id_rsa  id_rsa.pub  known_hosts
```

将公钥复制到 Node 上，然后就可以免秘钥登录 Node，运行命令如下。

```
[root@master ~]# ssh-copy-id root@192.168.10.6
[root@master ~]# ssh-copy-id root@192.168.10.7
[root@master ~]# ssh-copy-id root@192.168.10.8
[root@master ~]# ssh-copy-id root@192.168.10.9
[root@master ~]# ssh-copy-id root@192.168.10.10
[root@master ~]# ssh root@192.168.10.10
Last login: Tue May  1 14:44:54 2018 from 192.168.10.1
[root@node-5 ~]# logout
Connection to 192.168.10.10 closed.
```

将/opt/kubernetes/bin/和/opt/kubernetes/cfg/目录下的文件远程复制到其他 Node 中，命令如下。

```
[root@node-1 ~]# mkdir -p /opt/kubernetes/{bin,cfg,ssl}
[root@master ~]# scp -r /opt/kubernetes/bin/
root@192.168.10.6:/opt/kubernetes/
[root@node-1 ~]# ls /opt/kubernetes/bin/
etcd  etcdctl
[root@master ~]# scp -r /opt/kubernetes/cfg/
root@192.168.10.6:/opt/kubernetes/
```

将/opt/kuber/netes/ssl/目录下的数字证书和/usr/lib/system/中的 etcd.service 配置文件同样远程复制到其他 Node 中，如图 11-3 和图 11-4 所示。

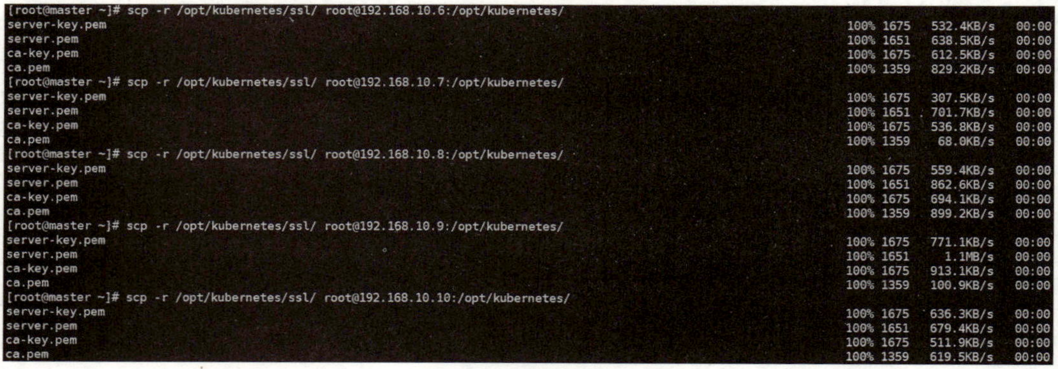

```
[root@master ~]# scp -r /opt/kubernetes/ssl/ root@192.168.10.6:/opt/kubernetes/
server-key.pem                                   100% 1675   532.4KB/s   00:00
server.pem                                       100% 1651   638.5KB/s   00:00
ca-key.pem                                       100% 1675   612.5KB/s   00:00
ca.pem                                           100% 1359   829.2KB/s   00:00
[root@master ~]# scp -r /opt/kubernetes/ssl/ root@192.168.10.7:/opt/kubernetes/
server-key.pem                                   100% 1675   307.5KB/s   00:00
server.pem                                       100% 1651   701.7KB/s   00:00
ca-key.pem                                       100% 1675   536.8KB/s   00:00
ca.pem                                           100% 1359    68.0KB/s   00:00
[root@master ~]# scp -r /opt/kubernetes/ssl/ root@192.168.10.8:/opt/kubernetes/
server-key.pem                                   100% 1675   559.4KB/s   00:00
server.pem                                       100% 1651   862.6KB/s   00:00
ca-key.pem                                       100% 1675   694.1KB/s   00:00
ca.pem                                           100% 1359   899.2KB/s   00:00
[root@master ~]# scp -r /opt/kubernetes/ssl/ root@192.168.10.9:/opt/kubernetes/
server-key.pem                                   100% 1675   771.1KB/s   00:00
server.pem                                       100% 1651     1.1MB/s   00:00
ca-key.pem                                       100% 1675   913.1KB/s   00:00
ca.pem                                           100% 1359   100.9KB/s   00:00
[root@master ~]# scp -r /opt/kubernetes/ssl/ root@192.168.10.10:/opt/kubernetes/
server-key.pem                                   100% 1675   636.3KB/s   00:00
server.pem                                       100% 1651   679.4KB/s   00:00
ca-key.pem                                       100% 1675   511.9KB/s   00:00
ca.pem                                           100% 1359   619.5KB/s   00:00
```

图 11-3

```
[root@master ~]# scp -r /usr/lib/systemd/system/etcd.service root@192.168.10.6:/usr/lib/systemd/system/
etcd.service                                     100% 1049   300.7KB/s   00:00
[root@master ~]# scp -r /usr/lib/systemd/system/etcd.service root@192.168.10.7:/usr/lib/systemd/system/
etcd.service                                     100% 1049   333.6KB/s   00:00
[root@master ~]# scp -r /usr/lib/systemd/system/etcd.service root@192.168.10.8:/usr/lib/systemd/system/
etcd.service                                     100% 1049   445.8KB/s   00:00
[root@master ~]# scp -r /usr/lib/systemd/system/etcd.service root@192.168.10.9:/usr/lib/systemd/system/
etcd.service                                     100% 1049   297.8KB/s   00:00
[root@master ~]# scp -r /usr/lib/systemd/system/etcd.service root@192.168.10.10:/usr/lib/systemd/system/
etcd.service                                     100% 1049   347.5KB/s   00:00
[root@master ~]#
```

图 11-4

复制后修改其他 Node 上/opt/kubernetes/cfg 目录中的 Etcd 配置文件，将 etcd01 分别改为 etcd02、etcd03、etcd04、etcd05、etcd06，同时对 IP 也进行更改，更改完毕后启动 Etcd 服务并设置开机自启动，运行命令如下。

```
[root@node-5 ~]# systemctl enable etcd.service && systemctl start
etcd.service
[root@node-1 ~]# systemctl enable etcd.service && systemctl start
etcd.service
```

在以后的日常工作中会经常用到 etcdctl 工具，为了提高工作效率，在/etc/profile 配置文件中为 etcdctl 工具设置环境变量，命令如下。

```
[root@master ~]# vim /etc/profile
PATH=$PATH:/opt/kubernetes/bin
```

```
[root@master ~]# source !$
source /etc/profile
```

使用 etcdctl 命令测试 Etcd 集群部署是否正常，输入测试命令后终端会输出 cluster is healthy 的信息，如图 11-5 所示。

```
[root@master ~]# cd ssl/
[root@master ssl]# etcdctl \
> --ca-file=ca.pem --cert-file=server.pem --key-file=server-key.pem \
> --endpoints="https://192.168.10.5:2379,https://192.168.10.6:2379,https://192.168.10.7:2379,https://192.168.10.8:2379,https://192.168.10.9:2379,https://
192.168.10.10:2379" \
> cluster-health
member 4b970befb437ab02 is healthy: got healthy result from https://192.168.10.5:2379
member 4e68a846c8a9a1e3 is healthy: got healthy result from https://192.168.10.9:2379
member 81e355655aac87c2 is healthy: got healthy result from https://192.168.10.10:2379
member 85a4097fdfa0ef6e is healthy: got healthy result from https://192.168.10.6:2379
member b576a7fee38735d7 is healthy: got healthy result from https://192.168.10.8:2379
member fb5ef4fb3f0ac88d is healthy: got healthy result from https://192.168.10.7:2379
cluster is healthy
[root@master ssl]#
```

图 11-5

11.4.5　Flannel 集群网络工作原理

Overlay Network 即覆盖网络，是在基础的网络上叠加的一种虚拟网络技术模式，该网络中的主机通过虚拟链路实现连接。

VXLAN 是 Overlay Network 技术的一种实现。VXLAN 是将源数据包封装到 UDP 中，并使用基础网络的 IP/MAC 作为外层报文头进行封装，然后在以太网上传输，到达目的地后由隧道端点解封装并将数据发送给目标地址。

Flannel 是 CoreOS 团队针对 Kubernetes 设计的一个网络规划服务，简单来说，它的功能是让集群中的不同节点主机创建的 Docker 容器都具有全集群唯一的虚拟 IP 地址。Flannel 实质上是 Overlay Network 的一种，也是将源数据包封装在另一种网络包里面进行路由转发和通信，目前支持 UDP、VXLAN、AWS VPC 和 GCE 路由等数据转发方式。

多主机容器网络通信的其他主流方案有：隧道方案（Weave、OpenvSwitch）和路由方案（Calico）等。

在 Kubernetes 网络模型中，假设每个物理节点都具备一个“属于同一个内网 IP 段内”的“专用的子网 IP”。例如，节点 A 为 10.0.1.0/24，节点 B 为 10.0.2.0/24，节点 C 为 10.0.3.0/24。在默认的 Docker 配置中，每个节点上的 Docker 服务会分别负责所在节点容器的 IP 分配。这样导致的一个问题是，不同节点上的容器可能获得相同的内外 IP 地址。这些容器之间能够通过 IP 地址相互 ping 通。

Flannel 设计的目的就是为集群中的所有节点重新规划 IP 地址的使用规则，从而促使不同节点上的容器能够获得“同属一个内网”且“不重复的”IP 地址，让属于不同节点上的容器能够直接通过内网 IP 通信。默认的节点间数据通信方式是 UDP 转发。数据从源容器中发出后，经由所在主机的 Docker0 虚拟网卡转发到 flannel0

虚拟网卡，这是一个 P2P 的虚拟网卡，flanneld 服务监听在网卡的另外一端，原理如图 11-6 所示。

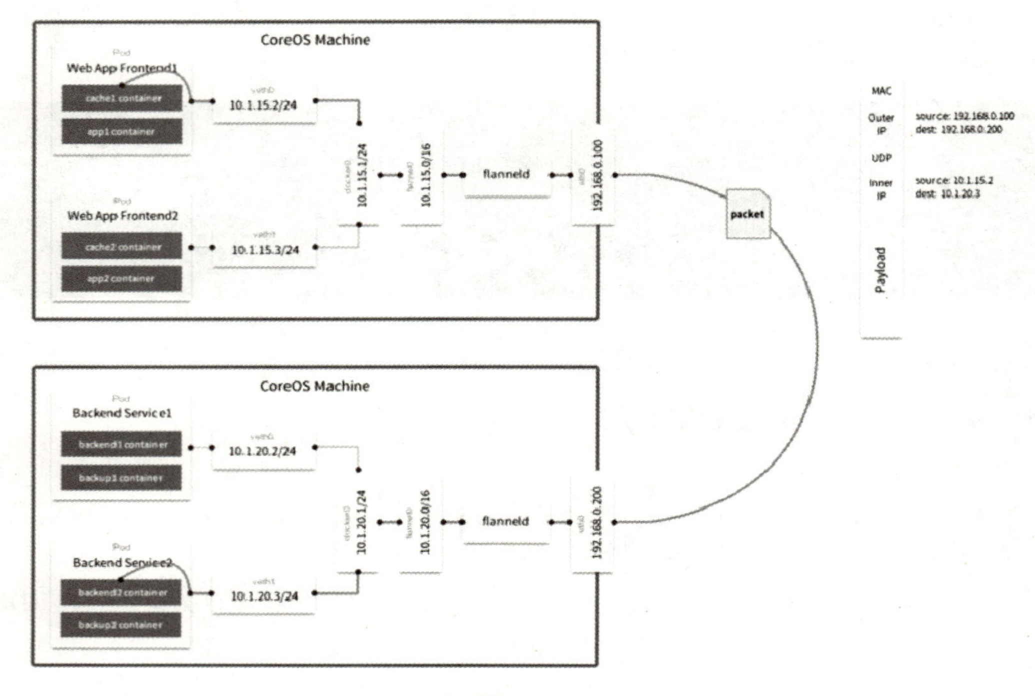

图 11-6

Flannel 通过 Etcd 服务维护了节点间的路由。源主机的 Flanneld 服务将原本的数据内容进行 UDP 封装后，根据自己的路由表投递给目的节点的 Flanneld 服务，数据到达后被解包，接着直接进入目的节点的 flannel0 虚拟网卡，然后被转发到目的主机的 Docker0 虚拟网卡，最后就像本机容器通信一样由 Docker0 路由到达目标容器。这样整个数据包的传递就结束了。

11.4.6 部署 Flannel 集群网络

下载 Flannel 二进制软件包至 Linux 服务器中，然后进行解压，解压后可以看到两个文件，即 mk-docker-opts.sh 脚本和 flannel 可执行程序，运行命令如下。

```
    [root@master flannel]# wget https://github.com/coreos/flannel/
releases/download/v0.10.0/flannel-v0.10.0-linux-amd64.tar.gz
    [root@master flannel]# tar xf flannel-v0.10.0-linux-amd64.tar.gz
    [root@master flannel]# ls
    flanneld  flannel-v0.10.0-linux-amd64.tar.gz  mk-docker-opts.sh
README.md
```

将 mk-docker-opts.sh 和 flannel 两个文件移至 Node 的/opt/kubernetes/bin/目录下，命令如下。

```
    [root@master flannel]# scp flanneld mk-docker-opts.sh
root@192.168.10.6:/opt/kubernetes/bin/
    [root@master flannel]# scp flanneld mk-docker-opts.sh
root@192.168.10.7:/opt/kubernetes/bin/
```

在 Node 中创建 flannel 配置文件，代码如下。

```
    [root@node-1 ~]# vim /opt/kubernetes/cfg/flanneld
    FLANNEL_OPTIONS="--etcd-endpoints=https://192.168.10.5:2379,https://1
92.168.10.6:2379,https://192.168.10.7:2379,https://192.168.10.8:2379,htt
ps://192.168.10.9:2379,https://192.168.10.10:2379
-etcd-cafile=/opt/kubernetes/ssl/ca.pem
-etcd-certfile=/opt/kubernetes/ssl/server.pem
-etcd-keyfile=/opt/kubernetes/ssl/server-key.pem
```

在 systemd 中创建 flanneld.service 脚本来管理 Flannel 网络，flanneld.service 启动脚本代码如下。

```
    [root@node-1 ~]# cat <<EOF >/usr/lib/systemd/system/flanneld.service
    [Unit]
    Description=Flanneld overlay address etcd agent
    After=network-online.target network.target
    Before=docker.service

    [Service]
    Type=notify
    EnvironmentFile=/opt/kubernetes/cfg/flanneld
    ExecStart=/opt/kubernetes/bin/flanneld --ip-masq \$FLANNEL_OPTIONS
    ExecStartPost=/opt/kubernetes/bin/mk-docker-opts.sh -k
DOCKER_NETWORK_OPTIONS -d /run/flannel/subnet.env
    Restart=on-failure

    [Install]
    WantedBy=multi-user.target

    EOF
```

创建 flanneld.service 脚本后启动 Flanneld 服务，并设置开机自启动，命令如下。

```
    [root@node-3 ~]# systemctl daemon-reload && systemctl enable
flanneld.service && systemctl start flanneld.service
    Job for flanneld.service failed because a timeout was exceeded. See
```

```
"systemctl status flanneld.service" and "journalctl -xe" for details.
```

启动 Flanneld 网络时提示报错，查看系统日志信息，提示没有获取到 Key，如图 11-7 所示，表示获取不到网段。

```
[root@node-1 ~]# tail /var/log/messages
Apr 28 19:20:08 node-1 flanneld: timed out
Apr 28 19:20:08 node-1 flanneld: E0428 19:20:08.358871    1969 main.go:349] Couldn't fetch network config: 100: Key not found (/coreos.com) [8]
Apr 28 19:20:09 node-1 flanneld: timed out
Apr 28 19:20:09 node-1 flanneld: E0428 19:20:09.368735    1969 main.go:349] Couldn't fetch network config: 100: Key not found (/coreos.com) [8]
Apr 28 19:20:10 node-1 flanneld: timed out
Apr 28 19:20:10 node-1 flanneld: E0428 19:20:10.378664    1969 main.go:349] Couldn't fetch network config: 100: Key not found (/coreos.com) [8]
Apr 28 19:20:11 node-1 flanneld: timed out
Apr 28 19:20:11 node-1 flanneld: E0428 19:20:11.386066    1969 main.go:349] Couldn't fetch network config: 100: Key not found (/coreos.com) [8]
Apr 28 19:20:12 node-1 flanneld: timed out
Apr 28 19:20:12 node-1 flanneld: E0428 19:20:12.394744    1969 main.go:349] Couldn't fetch network config: 100: Key not found (/coreos.com) [8]
[root@node-1 ~]#
```

图 11-7

写入分配的子网段到 Etcd 中供 Flanneld 网络使用。在 Master 中执行如图 11-8 所示的命令，执行后再去 Node 中启动 Flanneld 服务。

```
[root@master ssl]# /opt/kubernetes/bin/etcdctl --ca-file=ca.pem --cert-file=server.pem --key-file=serv
er-key.pem --endpoints="https://192.168.10.5:2379,https://192.168.10.6:2379,https://192.168.10.7:2379,
https://192.168.10.8:2379,https://192.168.10.9:2379,https://192.168.10.10:2379" set /coreos.com/networ
k/config '{ "Network": "172.17.0.0/16", "Backend": {"Type": "vxlan"}}'
{ "Network": "172.17.0.0/16", "Backend": {"Type": "vxlan"}}
[root@master ssl]#
```

图 11-8

Flannel 从 Etcd 中获取到分配的网段后会生成一个 Flannel 的虚拟网卡，如图 11-9 所示。

```
[root@node-1 ~]# ifconfig
docker0: flags=4099<UP,BROADCAST,MULTICAST>  mtu 1500
        inet 172.17.0.1  netmask 255.255.0.0  broadcast 172.17.255.255
        ether 02:42:7b:a2:1c:91  txqueuelen 0  (Ethernet)
        RX packets 0  bytes 0 (0.0 B)
        RX errors 0  dropped 0  overruns 0  frame 0
        TX packets 0  bytes 0 (0.0 B)
        TX errors 0  dropped 0 overruns 0  carrier 0  collisions 0

ens33: flags=4163<UP,BROADCAST,RUNNING,MULTICAST>  mtu 1500
        inet 192.168.10.6  netmask 255.255.255.0  broadcast 192.168.10.255
        inet6 fe80::177:f654:f3de:66c4  prefixlen 64  scopeid 0x20<link>
        ether 00:0c:29:c8:31:72  txqueuelen 1000  (Ethernet)
        RX packets 453677  bytes 119415444 (113.8 MiB)
        RX errors 0  dropped 0  overruns 0  frame 0
        TX packets 395032  bytes 43457689 (41.4 MiB)
        TX errors 0  dropped 0 overruns 0  carrier 0  collisions 0

flannel.1: flags=4163<UP,BROADCAST,RUNNING,MULTICAST>  mtu 1450
        inet 172.11.76.0  netmask 255.255.255.255  broadcast 0.0.0.0
        inet6 fe80::6806:54ff:fefc:de15  prefixlen 64  scopeid 0x20<link>
        ether 6a:06:54:fc:de:15  txqueuelen 0  (Ethernet)
        RX packets 0  bytes 0 (0.0 B)
        RX errors 0  dropped 0  overruns 0  frame 0
        TX packets 0  bytes 0 (0.0 B)
        TX errors 0  dropped 8 overruns 0  carrier 0  collisions 0
```

图 11-9

在/run/flannel/subnet.env 文件中会生成子网变量文件，代码如下。子网在所有的集群中都是唯一的。

```
[root@node-1 ~]# cat /run/flannel/subnet.env
```

```
    DOCKER_OPT_BIP="--bip=172.11.76.1/24"
    DOCKER_OPT_IPMASQ="--ip-masq=false"
    DOCKER_OPT_MTU="--mtu=1450"
    DOCKER_NETWORK_OPTIONS=" --bip=172.11.76.1/24 --ip-masq=false
--mtu=1450"
```

在 Docker 服务的启动脚本中指定该子网 IP，使 Docker 服务应用 Flannel 网络。修改后重启 Docker 服务，命令如下。

```
[root@node-1 ~]# vim /usr/lib/systemd/system/docker.service
[Service]
Type=notify
EnvironmentFile=/run/flannel/subnet.env
ExecStart=/usr/bin/dockerd  $DOCKER_NETWORK_OPTIONS
[root@node-1 ~]# systemctl daemon-reload
[root@node-1 ~]# systemctl restart docker.service
```

使用 ifconfig 命令查看网卡会发现 Docker 虚拟网卡和 Flannel 虚拟网卡的 IP 在同一个网段，如图 11-10 所示。

在 Node-2 中重复上述操作步骤，即可完成 Node 的 Flannel 网络部署工作。

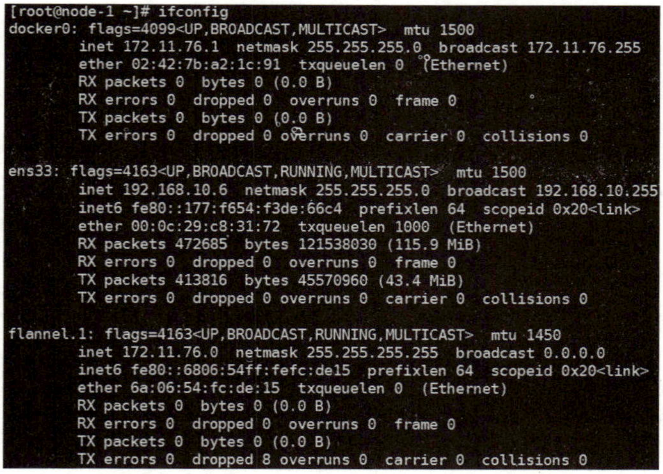

```
[root@node-1 ~]# ifconfig
docker0: flags=4099<UP,BROADCAST,MULTICAST>  mtu 1500
        inet 172.11.76.1  netmask 255.255.255.0  broadcast 172.11.76.255
        ether 02:42:7b:a2:1c:91  txqueuelen 0  (Ethernet)
        RX packets 0  bytes 0 (0.0 B)
        RX errors 0  dropped 0  overruns 0  frame 0
        TX packets 0  bytes 0 (0.0 B)
        TX errors 0  dropped 0 overruns 0  carrier 0  collisions 0

ens33: flags=4163<UP,BROADCAST,RUNNING,MULTICAST>  mtu 1500
        inet 192.168.10.6  netmask 255.255.255.0  broadcast 192.168.10.255
        inet6 fe80::177:f654:f3de:66c4  prefixlen 64  scopeid 0x20<link>
        ether 00:0c:29:c8:31:72  txqueuelen 1000  (Ethernet)
        RX packets 472685  bytes 121538030 (115.9 MiB)
        RX errors 0  dropped 0  overruns 0  frame 0
        TX packets 413816  bytes 45570960 (43.4 MiB)
        TX errors 0  dropped 0 overruns 0  carrier 0  collisions 0

flannel.1: flags=4163<UP,BROADCAST,RUNNING,MULTICAST>  mtu 1450
        inet 172.11.76.0  netmask 255.255.255.255  broadcast 0.0.0.0
        inet6 fe80::6806:54ff:fefc:de15  prefixlen 64  scopeid 0x20<link>
        ether 6a:06:54:fc:de:15  txqueuelen 0  (Ethernet)
        RX packets 0  bytes 0 (0.0 B)
        RX errors 0  dropped 0  overruns 0  frame 0
        TX packets 0  bytes 0 (0.0 B)
        TX errors 0  dropped 8 overruns 0  carrier 0  collisions 0
```

图 11-10

测试 overlay 网络是否有问题，可以使用两个 Node 互 ping 对方网关，命令如下。

```
[root@node-1 ~]# ping 172.11.25.1
PING 172.11.25.1 (172.11.25.1) 56(84) bytes of data.
64 bytes from 172.11.25.1: icmp_seq=1 ttl=64 time=2.10 ms
64 bytes from 172.11.25.1: icmp_seq=2 ttl=64 time=0.888 ms
[root@node-2 ~]# ping 172.11.76.1
PING 172.11.76.1 (172.11.76.1) 56(84) bytes of data.
```

```
64 bytes from 172.11.76.1: icmp_seq=1 ttl=64 time=1.91 ms
64 bytes from 172.11.76.1: icmp_seq=2 ttl=64 time=0.917 ms
```

注意

部署 Flannel 网络需要注意两点：第一，需要将分配 Node 子网的网段写入 Etcd 中，默认前缀是/coreos.com/network/config，前缀必须要写，如果不写，则会导致 Flanneld 网络启动不了；第二，保证节点之间可以互信，如果启用防火墙，则可以添加 iptables -I INPUT -s 192.168.10.0/24 -j ACCEPT 规则。

11.4.7 创建 Node 的 Kubeconfig 文件

将 kubectl 客户端文件上传到/opt/kubernetes/bin/目录下，然后添加可执行权限，命令如下。

```
[root@master bin]# ls
etcd  etcdctl  kubectl
[root@master bin]# chmod +x kubectl
```

Kubeconfig 文件用于 Node 上的 kube-proxy 和 kubelet 与集群进行通信时做认证。将 Kubeconfig 文件上传到服务器中，并赋予可执行权限，命令如下。

```
[root@master ssl]# chmod +x kubeconfig.sh
[root@master ssl]# ls -l kubeconfig.sh
-rwxr-xr-x 1 root root 1517 Feb 16 21:01 kubeconfig.sh
```

创建 Kubeconfig 文件有三个步骤。第一步创建 TLS Bootstrapping Token 文件，执行如下命令会生成一个 Token 文件，包括 token 的随机字符、用户名和角色分组，内容如下。

```
[root@master ~]# export BOOTSTRAP_TOKEN=$(head -c 16 /dev/urandom | od -An -t x | tr -d ' ')
[root@master ~]# cat > token.csv <<EOF
${BOOTSTRAP_TOKEN},kubelet-bootstrap,10001,"system:kubelet-bootstrap"
EOF
[root@master ~]# cat token.csv
3935c0f885e5fe12c109a08f54261cd7,kubelet-bootstrap,10001,"system:kubelet-bootstrap"
```

第二步创建 kubelet bootstrapping kubeconfig 文件，指定 K8S 集群的 HTTPS 访问入口，并设置集群相关参数，设置后的信息可在 bootstrap.kubeconfig 文件中查看，命令如下。

```
[root@master ~]# export KUBE_APISERVER="https://192.168.10.5:6443"
[root@master ssl]# kubectl config set-cluster kubernetes \
--certificate-authority=./ca.pem \
--embed-certs=true \
--server=${KUBE_APISERVER} \
--kubeconfig=bootstrap.kubeconfig
Cluster "kubernetes" set.
[root@master ssl]# kubectl config set-credentials kubelet-bootstrap \
--token=${BOOTSTRAP_TOKEN} \
--kubeconfig=bootstrap.kubeconfig
User "kubelet-bootstrap" set.
[root@master ssl]# kubectl config set-context default \
--cluster=kubernetes \
--user=kubelet-bootstrap \
--kubeconfig=bootstrap.kubeconfig
Context "default" created.
[root@master ssl]# kubectl config use-context default
--kubeconfig=bootstrap.kubeconfig
Switched to context "default".
[root@master ssl]# cat bootstrap.kubeconfig
```

第三步创建 kube-proxy kubeconfig 文件，该文件的创建方法和第二步文件的创建方法类似，设置后的信息可在 kube-proxy.kubeconfig 文件中查看，命令如下。

```
[root@master ssl]# kubectl config set-cluster kubernetes \
--certificate-authority=./ca.pem \
--embed-certs=true \
--server=${KUBE_APISERVER} \
--kubeconfig=kube-proxy.kubeconfig
Cluster "kubernetes" set.
[root@master ssl]# kubectl config set-credentials kube-proxy \
--client-certificate=./kube-proxy.pem \
--client-key=./kube-proxy-key.pem \
--embed-certs=true \
--kubeconfig=kube-proxy.kubeconfig
User "kube-proxy" set.
[root@master ssl]# kubectl config set-context default \
--cluster=kubernetes \
--user=kube-proxy \
--kubeconfig=kube-proxy.kubeconfig
Context "default" created.
[root@master ssl]# kubectl config use-context default
--kubeconfig=kube-proxy.kubeconfig
```

```
Switched to context "default".
[root@master ssl]# cat kube-proxy.kubeconfig
```

11.4.8 部署 Master 节点组件

目前 GitHub 上 K8S 二进制软件包的最新版本是 v.1.12，选择下载 Server Binaries 软件包，因为该软件包包含了 Client Binaries 包。Server Binaries 软件包下载网址是 https://github.com/kubernetes/kubernetes/blob/master /CHANGELOG-1.12.md#v1122。打开 Server Binaries 软件包下载页面，下载 server-linux-amd64 压缩包即可，其网址是 https://dl.k8s.io/v1.12.2/ kubernetes-server- linux-amd64.tar.gz。

下载完 K8S 二进制软件包后，需要部署到 Master 节点上，此处为读者提供了笔者事先修改好的 Master 节点组件，该组件可在 GitLab 中下载，网址是 https://gitlab.com/humingzhe/Linux/tree/master/Kubernetes/Kubernetes_Cluster。

在 Master 节点需要部署 3 个组件：apiserver、controller-manager 和 scheduler。

将 master.zip 压缩包上传到/root/目录下，然后进行解压操作，将 kube-controller-manager、kube-apiserver 和 kube-scheduler 移动到/opt/kubernetes/bin/目录下，移动后赋予可执行权限，命令如下。

```
[root@master ~]# unzip master.zip
[root@master ~]# cd master/
[root@master master]# mv kube-controller-manager kube-scheduler
kube-apiserver /opt/kubernetes/bin/
[root@master master]# chmod +x /opt/kubernetes/bin/*
```

给 master 目录中的脚本赋予可执行权限，然后执行该脚本，命令如下。

```
[root@master master]# chmod +x *.sh
[root@master master]# ls
apiserver.sh  controller-manager.sh  kubectl  scheduler.sh
[root@master master]# ./apiserver.sh 192.168.10.5
https://192.168.10.5:2379,https://192.168.10.6:2379,https://192.168.10.7
:2379,https://192.168.10.8:2379,https://192.168.10.9:2379,https://192.16
8.10.10:2379
Created syamlink from
/etc/systemd/system/multi-user.target.wants/kube-apiserver.service to
/usr/lib/systemd/system/kube-apiserver.service.
```

将生成的 token.csv 文件复制到/opt/kubernetes/cfg/目录下，命令如下。

```
[root@master master]# cp /root/token.csv /opt/kubernetes/cfg/
```

```
[root@master master]# ls /opt/kubernetes/cfg/
etcd  flanneld  kube-apiserver  token.csv
```

启动 kube-apiserver 服务，如图 11-11 所示。

```
[root@master ~]# systemctl enable kube-apiserver.service && systemctl start kube-apiserver.service
[root@master ~]# ps aux|grep kube-apiserver
root      2759  109  2.4 340968 198572 ?      Ssl  18:21   0:08 /opt/kubernetes/bin/kube-apiserver --logtostderr=true --v=4 --etcd-servers=https://192.
168.10.5:2379,https://192.168.10.6:2379,https://192.168.10.7:2379,https://192.168.10.8:2379,https://192.168.10.9:2379,https://192.168.10.10:2379 --insecu
re-bind-address=127.0.0.1 --bind-address=192.168.10.5 --insecure-port=8080 --secure-port=6443 --advertise-address=192.168.10.5 --allow-privileged=true --
service-cluster-ip-range=10.10.10.0/24 --admission-control=NamespaceLifecycle,LimitRanger,SecurityContextDeny,ServiceAccount,ResourceQuota,NodeRestrictio
n --authorization-mode=RBAC,Node --kubelet-https=true --enable-bootstrap-token-auth --token-auth-file=/opt/kubernetes/cfg/token.csv --service-node-port-r
ange=30000-50000 --tls-cert-file=/opt/kubernetes/ssl/server.pem --tls-private-key-file=/opt/kubernetes/ssl/server-key.pem --client-ca-file=/opt/kubernete
s/ssl/ca.pem --service-account-key-file=/opt/kubernetes/ssl/ca-key.pem --etcd-cafile=/opt/kubernetes/ssl/ca.pem --etcd-certfile=/opt/kubernetes/ssl/serve
r.pem --etcd-keyfile=/opt/kubernetes/ssl/server-key.pem
root      2768  0.0  0.0 112660   976 pts/1    R+   18:21   0:00 grep --color=auto kube-apiserver
[root@master ~]#
```

图 11-11

执行 controller-manager.sh 脚本文件，需要指定 apiserver IP 地址，因为上述配置文件中指定了两个 IP 地址，一个是 HTTPS 安全 IP 地址，另一个是本地 IP 地址。指定安全 IP 地址需要指定证书文件，因为在本地执行，所以指定本地 IP（127.0.0.1）即可，命令如下。

```
[root@master master]# ./controller-manager.sh 127.0.0.1
Created syamlink from /etc/systemd/system/multi-user.target.wants/
kube-controller-manager.service to /usr/lib/systemd/system/kube-
controller-manager.service.
```

kube-controller-manager 进程启动成功，如图 11-12 所示。

执行 scheduler.sh 脚本文件，指定 apiserver IP 地址 127.0.0.1，执行后查看进程是否存在，命令如下。

```
[root@master master_pkg]# ./scheduler.sh 127.0.0.1
Created syamlink from
/etc/systemd/system/multi-user.target.wants/kube-scheduler.service to
/usr/lib/systemd/system/kube-scheduler.service.
```

```
[root@master master]# ps aux|grep kube-controller-manager
root      2815  3.2  0.5 123284 41492 ?       Ssl  18:22   0:00 /opt/kubernetes/bin/kube-controller-
manager --logtostderr=true --v=4 --master=127.0.0.1:8080 --leader-elect=true --address=127.0.0.1 --ser
vice-cluster-ip-range=10.10.10.0/24 --cluster-name=kubernetes --cluster-signing-cert-file=/opt/kuberne
tes/ssl/ca.pem --cluster-signing-key-file=/opt/kubernetes/ssl/ca-key.pem --service-account-private-key
-file=/opt/kubernetes/ssl/ca-key.pem --root-ca-file=/opt/kubernetes/ssl/ca.pem
root      2823  0.0  0.0 112660   976 pts/1    R+   18:22   0:00 grep --color=auto kube-controller-ma
nager
[root@master master]#
```

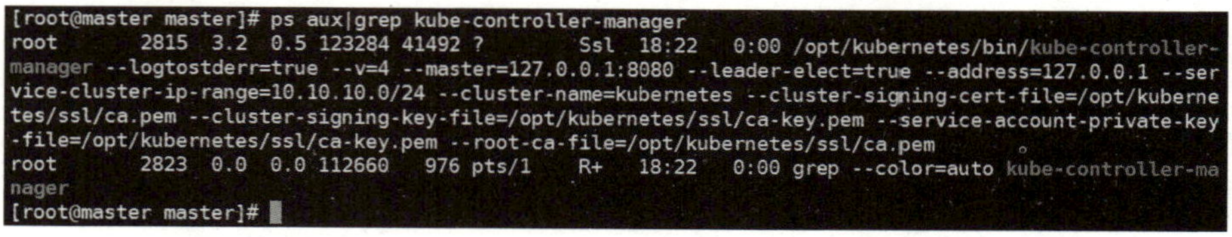

图 11-12

kube-scheduler 进程启动成功，如图 11-13 所示。

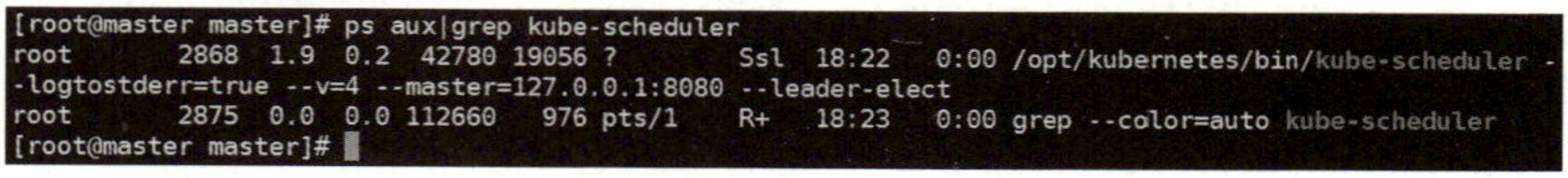

```
[root@master master]# ps aux|grep kube-scheduler
root      2868  1.9  0.2  42780 19056 ?       Ssl  18:22   0:00 /opt/kubernetes/bin/kube-scheduler -
-logtostderr=true --v=4 --master=127.0.0.1:8080 --leader-elect
root      2875  0.0  0.0 112660   976 pts/1    R+   18:23   0:00 grep --color=auto kube-scheduler
[root@master master]#
```

图 11-13

上述组件全部启动成功后可以使用 kubectl get cs 命令查看集群的状态，如图 11-14 所示。

```
[root@master master]# kubectl get cs
NAME                 STATUS    MESSAGE             ERROR
scheduler            Healthy   ok
controller-manager   Healthy   ok
etcd-0               Healthy   {"health":"true"}
etcd-2               Healthy   {"health":"true"}
etcd-1               Healthy   {"health":"true"}
etcd-5               Healthy   {"health":"true"}
etcd-3               Healthy   {"health":"true"}
etcd-4               Healthy   {"health":"true"}
[root@master master]#
```

图 11-14

11.4.9 部署 Node 组件

将前面章节生成的 Kubeconfig 文件复制到其他 Node 中，命令如下。

```
[root@master ssl]# scp *kubeconfig
root@192.168.10.6:/opt/kubernetes/cfg/
[root@master ssl]# scp *kubeconfig
root@192.168.10.7:/opt/kubernetes/cfg/
```

上传 Node 组件到 Node 服务器中，然后进行解压；将 kubelet、kube-proxy 移动到/opt/kubernetes/bin/目录下，赋予执行权限，命令如下。

```
[root@node-1 ~]# unzip node.zip
[root@node-1 ~]# cd node/
[root@node-1 node]# mv kubelet kube-proxy /opt/kubernetes/bin/
[root@node-1 node]# chmod +x /opt/kubernetes/bin/*
[root@node-1 node]# ls /opt/kubernetes/bin/
etcd  etcdctl  flanneld  kubelet  kube-proxy  mk-docker-opts.sh
```

赋予 kubelet.sh 和 proxy.sh 脚本执行权限，并执行该脚本，命令如下。

```
[root@node-1 ~]# chmod +x kubelet.sh proxy.sh
[root@node-1 ~]# ./kubelet.sh 192.168.10.6 10.10.10.2
Created syamlink from /etc/systemd/system/multi-user.target.
wants/kubelet.service to /usr/lib/systemd/system/kubelet.service.
```

使用 ps aux|grep kube 命令查看进程时发现没有启动，这时应该查看系统日志，因为系统日志中显示的信息是最全面的，系统日志中显示 error: failed to run Kubelet: cannot create certificate signing request: certificatesigningrequests.certificates.k8s.io is forbidden: User "kubelet-bootstrap"，这句话的意思是无法创建证书请求，因为是使用 kubelet-bootstrap 账户去请求的，没有权限，需要在 Master 节点中创建该权限并绑定

角色，命令如下。

```
[root@master ssl]# kubectl create clusterrolebinding kubelet-bootstrap
--clusterrole=system:node-bootstrapper --user=kubelet-bootstrap

clusterrolebinding "kubelet-bootstrap" created
```

绑定角色后重启 kubelet 服务，再去查看进程，发现已经有了，如图 11-15 所示。

```
[root@node-1 node]# ./kubelet.sh 192.168.10.6 10.10.10.2
Created symlink from /etc/systemd/system/multi-user.target.wants/kubelet.service to /usr/lib/systemd/system/kubelet.service.
[root@node-1 node]# ps aux|grep kubelet
root      8634  3.6  1.8 278248 34676 ?        Ssl  18:45   0:01 /opt/kubernetes/bin/kubelet --logtostderr=true --v=4 --add
ress=192.168.10.6 --hostname-override=192.168.10.6 --kubeconfig=/opt/kubernetes/cfg/kubelet.kubeconfig --experimental-bootst
rap-kubeconfig=/opt/kubernetes/cfg/bootstrap.kubeconfig --cert-dir=/opt/kubernetes/ssl --allow-privileged=true --cluster-dns
=10.10.10.2 --cluster-domain=cluster.local --fail-swap-on=false --pod-infra-container-image=registry.cn-hangzhou.aliyuncs.co
m/google-containers/pause-amd64:3.0
root      8707  0.0  0.0 112660   972 pts/0    R+   18:46   0:00 grep --color=auto kubelet
[root@node-1 node]#
```

图 11-15

执行 proxy.sh 脚本，同样查看进程，命令如下。

```
[root@node-1 node]# ./proxy.sh 192.168.10.6
Created syamlink from
/etc/systemd/system/multi-user.target.wants/kube-proxy.service to
/usr/lib/systemd/system/kube-proxy.service.
[root@node-1 node]# ps aux|grep kube-proxy
root        8783  4.3  1.1  43792 20672 ?        Ssl  18:47   0:00
/opt/kubernetes/bin/kube-proxy --logtostderr=true --v=4
--hostname-override=192.168.10.6
--kubeconfig=/opt/kubernetes/cfg/kube-proxy.kubeconfig
```

在 Master 节点中使用 kubectl get csr 命令可以查看证书的请求信息，如图 11-16 所示。状态是 Pending 等待，要将其修改为允许状态，如图 11-17 所示。

```
[root@master ~]# kubectl get csr
NAME                                                   AGE       REQUESTOR           CONDITION
node-csr-M_TRS_zcV7qwjYQk4UMmjllA0y2XcedHsMjPuW9a6-M   1m        kubelet-bootstrap   Pending
node-csr-0zuXHPSH56z0AF9wD4gtSr9gybQntje0R0l71kG0Azw   2m        kubelet-bootstrap   Pending
node-csr-fkGNxAvDS6wQ0r-L9z7vVJP5Nw4bPCNmUgw-WKtoxqA   4m        kubelet-bootstrap   Pending
node-csr-ob1szavcvC6A00sRBaw0S1ojkDbFbR5A9_gnUsvSIGQ   22s       kubelet-bootstrap   Pending
node-csr-xzJz2zQJF5siWzLzE6kxLU3-iLZWS9NV67M31xEUrCA   10m       kubelet-bootstrap   Approved,Issued
```

图 11-16

```
[root@master ~]# kubectl certificate approve node-csr-ob1szavcvC6A00sRBaw0S1ojkDbFbR5A9_gnUsvSIGQ
certificatesigningrequest.certificates.k8s.io "node-csr-ob1szavcvC6A00sRBaw0S1ojkDbFbR5A9_gnUsvSIGQ" approved
[root@master ~]# kubectl certificate approve node-csr-xzJz2zQJF5siWzLzE6kxLU3-iLZWS9NV67M31xEUrCA
certificatesigningrequest.certificates.k8s.io "node-csr-xzJz2zQJF5siWzLzE6kxLU3-iLZWS9NV67M31xEUrCA" approved
[root@master ~]# kubectl get csr
NAME                                                   AGE       REQUESTOR           CONDITION
node-csr-M_TRS_zcV7qwjYQk4UMmjllA0y2XcedHsMjPuW9a6-M   3m        kubelet-bootstrap   Approved,Issued
node-csr-0zuXHPSH56z0AF9wD4gtSr9gybQntje0R0l71kG0Azw   4m        kubelet-bootstrap   Approved,Issued
node-csr-fkGNxAvDS6wQ0r-L9z7vVJP5Nw4bPCNmUgw-WKtoxqA   6m        kubelet-bootstrap   Approved,Issued
node-csr-ob1szavcvC6A00sRBaw0S1ojkDbFbR5A9_gnUsvSIGQ   2m        kubelet-bootstrap   Approved,Issued
node-csr-xzJz2zQJF5siWzLzE6kxLU3-iLZWS9NV67M31xEUrCA   12m       kubelet-bootstrap   Approved,Issued
[root@master ~]#
```

图 11-17

在 Node 的/opt/kubernetes/ssl/目录下会自动生成 kubelet-client.crt 和 kubelet-client.key 两个证书，证书如下。

```
    [root@node-1 ~]# ls /opt/kubernetes/ssl/
    ca-key.pem  ca.pem  kubelet-client.crt  kubelet-client.key
kubelet.crt  kubelet.key  server-key.pem  server.pem
```

同理，按照上述操作步骤对其他 Node 进行部署，部署完毕后通过 kubectl get node 命令可以看到所有 Node 的信息，如图 11-18 所示。

```
[root@master ~]# kubectl get node
NAME           STATUS   ROLES    AGE   VERSION
192.168.10.10  Ready    <none>   1m    v1.12.0
192.168.10.6   Ready    <none>   9m    v1.12.0
192.168.10.7   Ready    <none>   1m    v1.12.0
192.168.10.8   Ready    <none>   2m    v1.12.0
192.168.10.9   Ready    <none>   2m    v1.12.0
[root@master ~]#
```

图 11-18

设置自动补全命令，命令如下。

```
    [root@master ~]# source <(kubectl completion bash)
```

11.4.10 集群部署 Nginx 服务

前面已经将集群成功地部署好，使用 kubectl get node 命令可以查看集群有哪些节点，使用 kubectl get cs 命令可以查看集群的状态。kubectl get cs 命令是 kubectl get componentstatus 命令的简写。

创建示例，查看镜像是否处于运行状态，然后使用 kubectl get pod 命令查看创建的容器，使用 kubectl get all 命令查看所有的资源对象，如图 11-19 所示。

```
[root@master ~]# kubectl run nginx --image=nginx --replicas=5
deployment.apps "nginx" created
[root@master ~]# kubectl get pod
NAME                     READY   STATUS    RESTARTS   AGE
nginx-65899c769f-27k6n   1/1     Running   0          1m
nginx-65899c769f-9lh87   1/1     Running   0          1m
nginx-65899c769f-bp9z4   1/1     Running   0          1m
nginx-65899c769f-nnbps   1/1     Running   0          1m
nginx-65899c769f-vzjvt   1/1     Running   0          1m
[root@master ~]# kubectl get all
NAME                         READY   STATUS    RESTARTS   AGE
pod/nginx-65899c769f-27k6n   1/1     Running   0          1m
pod/nginx-65899c769f-9lh87   1/1     Running   0          1m
pod/nginx-65899c769f-bp9z4   1/1     Running   0          1m
pod/nginx-65899c769f-nnbps   1/1     Running   0          1m
pod/nginx-65899c769f-vzjvt   1/1     Running   0          1m

NAME                 TYPE        CLUSTER-IP   EXTERNAL-IP   PORT(S)   AGE
service/kubernetes   ClusterIP   10.10.10.1   <none>        443/TCP   2m

NAME                    DESIRED   CURRENT   UP-TO-DATE   AVAILABLE   AGE
deployment.apps/nginx   5         5         5            5           1m

NAME                               DESIRED   CURRENT   READY   AGE
replicaset.apps/nginx-65899c769f   5         5         5       1m
[root@master ~]#
```

图 11-19

使用 kubectl get pod -o wide 命令可以查看运行的容器具体在哪个节点上，如图 11-20 所示。

```
[root@master ~]# kubectl get pod -o wide
NAME                      READY   STATUS    RESTARTS   AGE   IP            NODE
nginx-65899c769f-27k6n    1/1     Running   0          3m    172.17.66.2   192.168.10.7
nginx-65899c769f-9lh87    1/1     Running   0          3m    172.17.86.2   192.168.10.8
nginx-65899c769f-bp9z4    1/1     Running   0          3m    172.17.20.2   192.168.10.9
nginx-65899c769f-nnbps    1/1     Running   0          3m    172.17.83.2   192.168.10.6
nginx-65899c769f-vzjvt    1/1     Running   0          3m    172.17.25.2   192.168.10.10
[root@master ~]#
```

图 11-20

使用 kubectl expose 命令发布 Nginx 服务，让其在外部可以进行访问，命令如下。

```
[root@master ~]# kubectl expose deployment nginx --port=88
--target-port=80 --type=NodePort
service "nginx" exposed
```

使用 kubectl get svc 命令可以查看 Nginx 服务生成的集群 IP，如图 11-21 所示。该 IP 地址是在/opt/kubernetes/cfg/kube-apiserver 文件中定义的集群网段。

```
[root@master ~]# kubectl get svc
NAME         TYPE        CLUSTER-IP    EXTERNAL-IP   PORT(S)        AGE
kubernetes   ClusterIP   10.10.10.1    <none>        443/TCP        5m
nginx        NodePort    10.10.10.56   <none>        88:46152/TCP   30s
[root@master ~]#
```

图 11-21

在任意节点中输入 10.10.10.56:88 即可访问 Nginx 的欢迎界面，如图 11-22 所示。

```
[root@node-1 ~]# elinks 10.10.10.56:88
                                                              Welcome to nginx!
                          Welcome to nginx!

  If you see this page, the nginx web server is successfully installed and working. Further configuration is required.

  For online documentation and support please refer to nginx.org.
  Commercial support is available at nginx.com.

  Thank you for using nginx.
```

图 11-22

在 curl 命令后面或在外部浏览器中输入任意节点的 IP 地址加 NodePort 端口号，都可以访问 Nginx 欢迎页面，如图 11-23 和图 11-24 所示。

```
[root@node-1 ~]# curl 10.10.10.56:88
<!DOCTYPE html>
<html>
<head>
<title>Welcome to nginx!</title>
<style>
    body {
        width: 35em;
        margin: 0 auto;
        font-family: Tahoma, Verdana, Arial, sans-serif;
    }
</style>
</head>
<body>
<h1>Welcome to nginx!</h1>
<p>If you see this page, the nginx web server is successfully installed and
working. Further configuration is required.</p>
```

图 11-23

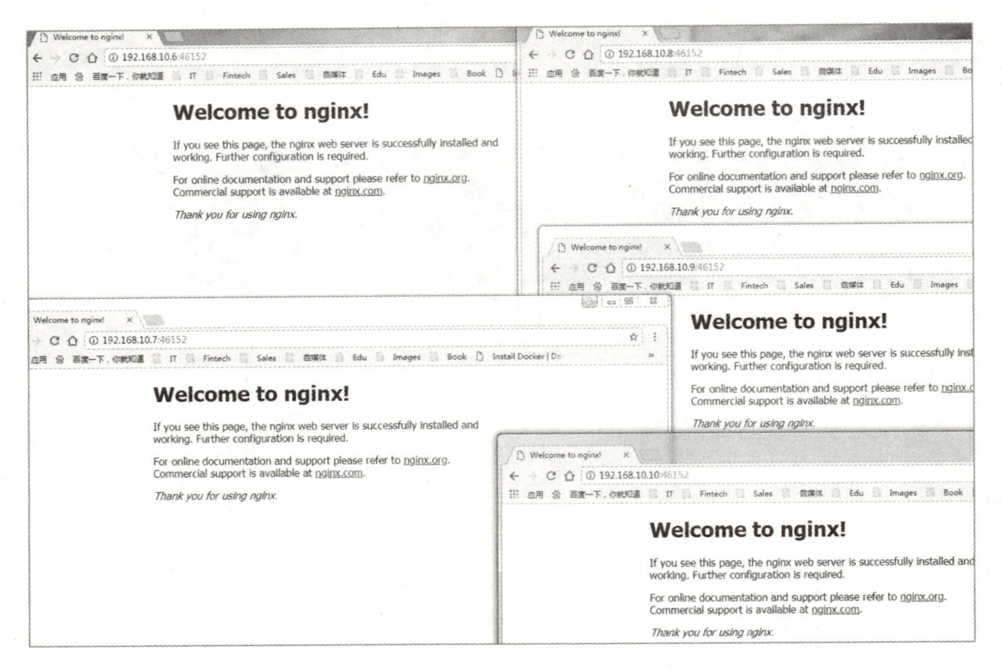

图 11-24

注意

使用 elinks 命令、curl 命令和浏览器访问外部 IP 无法打开 Nginx 欢迎页面，可能是因为 Docker 服务、Kubelet 服务或 Flanneld 网络没有启动，此时可查看节点的这些服务进程是否被启动，如果没有启动，启动即可，稍等片刻后再使用 elinks 命令、curl 命令和外部浏览器访问，即可打开 Nginx 欢迎页面。如果想让 Master 节点也可以访问，只要在 Master 节点上部署好 Flanneld 网络服务即可。

使用 kubectl logs 命令可以查看 Nginx 的日志信息，如图 11-25 所示，可以看到刚刚用浏览器请求 Nginx 的信息。

```
[root@master ~]# kubectl get pod
NAME                     READY     STATUS    RESTARTS   AGE
nginx-65899c769f-27k6n   1/1       Running   1          20m
nginx-65899c769f-9lh87   1/1       Running   1          20m
nginx-65899c769f-bp9z4   1/1       Running   1          20m
nginx-65899c769f-nnbps   1/1       Running   1          20m
nginx-65899c769f-vzjvt   1/1       Running   1          20m
[root@master ~]# kubectl logs nginx-65899c769f-27k6n
172.17.66.1 - - [02/May/2018:07:42:13 +0000] "GET / HTTP/1.1" 200 612 "-" "Mozilla/5.0 (Windows NT 6.1; Win64; x64) AppleWebKit/53
7.36 (KHTML, like Gecko) Chrome/67.0.3371.0 Safari/537.36" "-"
2018/05/02 07:42:14 [error] 5#5: *3 open() "/usr/share/nginx/html/favicon.ico" failed (2: No such file or directory), client: 172.
17.66.1, server: localhost, request: "GET /favicon.ico HTTP/1.1", host: "192.168.10.7:46152", referrer: "http://192.168.10.7:46152
/"
172.17.66.1 - - [02/May/2018:07:42:14 +0000] "GET /favicon.ico HTTP/1.1" 404 572 "http://192.168.10.7:46152/" "Mozilla/5.0 (Window
s NT 6.1; Win64; x64) AppleWebKit/537.36 (KHTML, like Gecko) Chrome/67.0.3371.0 Safari/537.36" "-"
2018/05/02 07:45:33 [error] 5#5: *7 open() "/usr/share/nginx/html/favicon.ico" failed (2: No such file or directory), client: 172.
17.66.1, server: localhost, request: "GET /favicon.ico HTTP/1.1", host: "192.168.10.7:46152", referrer: "http://192.168.10.7:46152
/"
172.17.66.1 - - [02/May/2018:07:45:33 +0000] "GET /favicon.ico HTTP/1.1" 404 572 "http://192.168.10.7:46152/" "Mozilla/5.0 (Window
s NT 6.1; Win64; x64) AppleWebKit/537.36 (KHTML, like Gecko) Chrome/67.0.3371.0 Safari/537.36" "-"
[root@master ~]#
```

图 11-25

11.5　Kubectl管理工具

11.5.1　Kubectl 管理工具远程连接集群

Kubectl 客户端工具的主要功能是管理 Kubernetes 集群中的资源。使用 kubectl 工具可以对资源进行创建、删除和更改等操作。Kubectl 工具默认连接本地 apiserver127.0.0.1:8080。通过-s 选项可以指定集群 HTTP 非安全 IP 地址和端口进行访问，命令如下。

```
[root@master ~]# kubectl -s 127.0.0.1:8080 get node
NAME            STATUS    ROLES     AGE       VERSION
192.168.10.10    Ready     <none>    3h         v1.10.2
192.168.10.6    Ready     <none>    3h         v1.10.2
192.168.10.7    Ready     <none>    3h         v1.10.2
192.168.10.8    Ready     <none>    3h         v1.10.2
192.168.10.9    Ready     <none>    3h         v1.10.2
```

Kubectl 默认通过 HTTP 访问 API 集群接口，若想在集群其他节点通过 HTTPS 访问集群接口，需要将 Kubectl 工具复制到集群其他的节点中，命令如下。

```
[root@master ~]# scp /opt/kubernetes/bin/kubectl
root@192.168.10.6:/usr/local/bin/
[root@node-1 ~]# ls /usr/local/bin/
kubectl
```

创建 ca 证书和 admin 证书。admin 证书用于在客户端管理集群，所以需要将 admin 证书复制到客户端访问集群的节点中，命令如下。

```
[root@master ssl]# scp admin*pem root@192.168.10.6:/root/
[root@node-1 ~]# cp /opt/kubernetes/ssl/ca.pem ./
[root@node-1 ~]# ls
admin-key.pem  admin.pem  anaconda-ks.cfg  ca.pem  node  node.zip
```

使用 kubectl config 命令调用 config 文件中的配置信息去访问集群，默认会在.kube/目录下创建缓存目录，config 文件也要放在此处才会被引用。使用 kubectl 命令可以快速生成 config 配置文件，设置集群项中名为 Kubernetes 的 apiserver 地址与根证书即可，命令如下。

```
[root@node-1 ~]# kubectl config set-cluster kubernetes
--server=https://192.168.10.5:6443 --certificate-authority=ca.pem
Cluster "kubernetes" set.
[root@node-1 ~]# ls .kube/
```

```
config
```

设置用户项中 cluster-admin 用户证书认证字段，命令如下。

```
[root@node-1 ~]# kubectl config set-credentials cluster-admin
--certificate-authority=ca.pem --client-key=admin-key.pem
--client-certificate=admin.pem
User "cluster-admin" set.
```

设置环境项中名为 default 的默认集群和用户，命令如下。

```
[root@node-1 ~]# kubectl config set-context default --cluster=kubernetes
--user=cluster-admin
Context "default" created.
```

使用 cat/root/.kube/config 命令查看配置信息，命令如下。

```
[root@node-1 ~]# kubectl config use-context default
Switched to context "default".
[root@node-1 ~]# cat /root/.kube/config
```

若能够使用 kubectl get node 命令正常获取到集群节点信息，则说明没有问题。如果将/root/.kube 隐藏目录下的 config 配置文件移走，则会提示没有指定要连接集群的 IP 和端口，如图 11-26 所示。

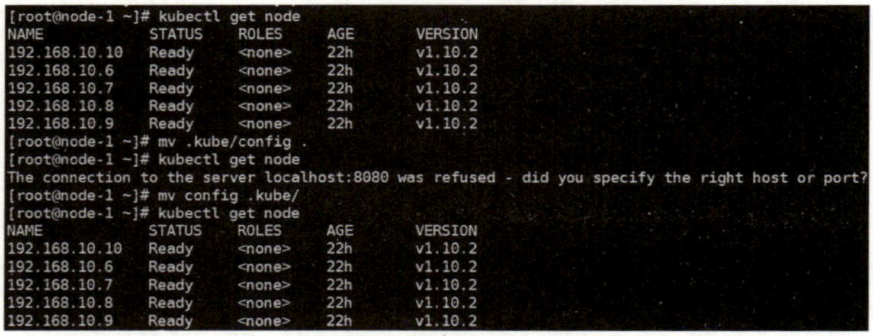

```
[root@node-1 ~]# kubectl get node
NAME            STATUS    ROLES     AGE       VERSION
192.168.10.10   Ready     <none>    22h       v1.10.2
192.168.10.6    Ready     <none>    22h       v1.10.2
192.168.10.7    Ready     <none>    22h       v1.10.2
192.168.10.8    Ready     <none>    22h       v1.10.2
192.168.10.9    Ready     <none>    22h       v1.10.2
[root@node-1 ~]# mv .kube/config .
[root@node-1 ~]# kubectl get node
The connection to the server localhost:8080 was refused - did you specify the right host or port?
[root@node-1 ~]# mv config .kube/
[root@node-1 ~]# kubectl get node
NAME            STATUS    ROLES     AGE       VERSION
192.168.10.10   Ready     <none>    22h       v1.10.2
192.168.10.6    Ready     <none>    22h       v1.10.2
192.168.10.7    Ready     <none>    22h       v1.10.2
192.168.10.8    Ready     <none>    22h       v1.10.2
192.168.10.9    Ready     <none>    22h       v1.10.2
```

图 11-26

使用 kubectl get all 命令可以查看默认命名空间中所有的资源，如图 11-27 所示。

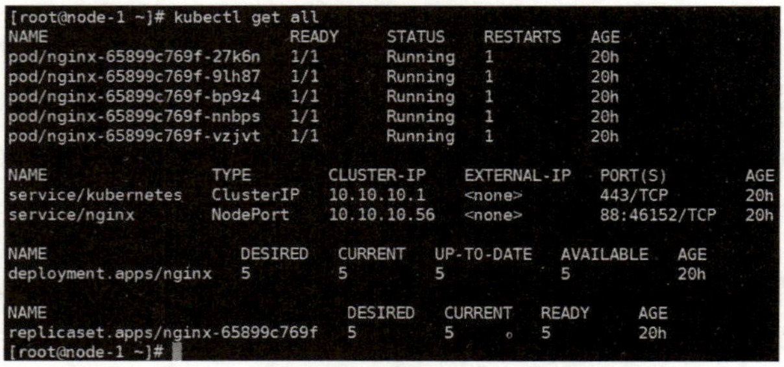

```
[root@node-1 ~]# kubectl get all
NAME                         READY     STATUS    RESTARTS   AGE
pod/nginx-65899c769f-27k6n   1/1       Running   1          20h
pod/nginx-65899c769f-9lh87   1/1       Running   1          20h
pod/nginx-65899c769f-bp9z4   1/1       Running   1          20h
pod/nginx-65899c769f-nnbps   1/1       Running   1          20h
pod/nginx-65899c769f-vzjvt   1/1       Running   1          20h

NAME                 TYPE        CLUSTER-IP    EXTERNAL-IP   PORT(S)        AGE
service/kubernetes   ClusterIP   10.10.10.1    <none>        443/TCP        20h
service/nginx        NodePort    10.10.10.56   <none>        88:46152/TCP   20h

NAME                    DESIRED   CURRENT   UP-TO-DATE   AVAILABLE   AGE
deployment.apps/nginx   5         5         5            5           20h

NAME                               DESIRED   CURRENT   READY     AGE
replicaset.apps/nginx-65899c769f   5         5         5         20h
[root@node-1 ~]#
```

图 11-27

将上述配置和文件打包移植到其他服务器集群的节点中也可以实现访问。

11.5.2　Kubectl 管理命令

如表 11-3 所示是 Kubectl 常用的管理命令及其功能描述。

表 11-3 中的命令是使用 kubectl --help 命令列出来的，在每个列出的命令后面也可加--help 选项查看命令的使用状态，如查看 create 命令如何使用，命令如下。

```
[root@node-1 ~]# kubectl create --help
```

表 11-3

类　型	命　令	描　述
基础命令	create	通过文件名或标准输入创建资源
	expose	将一个资源公开为一个新的 Service
	run	在集群中运行一个特定的镜像
	set	在对象上设置特定的功能
	get	显示一个或多个资源
	explain	文档参考资料
	edit	使用默认的编辑器编辑一个资源
	delete	通过文件名、标准输入、资源名称或标签选择器来删除资源
部署命令	rollout	管理资源的发布
	rolling-update	对给定的复制控制器进行滚动更新
	scale	扩容或缩容 Pod 数量，重新设定 Deployment（在继承 Pod 和 Replicaset 的所有特性的同时，可以实现对模板进行实时滚动更新并具备线上的应用程序生命循环的特性）、ReplicaSet（在继承 Pod 的所有特性的同时，可以利用预先创建好的模板定义副本数量并自动控制，通过改变 Pod 副本数量实现 Pod 的扩容和缩容）、RC（ReplicationController，Pod 的复制抽象，用于解决 Pod 的扩容和缩容问题）和 Job（一次性任务，运行完成后 Pod 被销毁，不再重新启动新容器，也可以让任务定时运行）的 size
	autoscale	创建一个自动选择的扩容或缩容，并设置 Pod 数量
集群管理命令	certificate	修改证书资源
	cluster-info	显示集群信息
	top	显示资源（CPU/Memory/Storage）使用情况，需要 Heapster 运行
	cordon	标记节点不可调度
	uncordon	标记节点可调度
	drain	维护期间排除节点
	taint	更新一个节点或多个节点的污点

续表

类　型	命　令	描　述
故障诊断和调试命令	describe	显示特定资源或资源组的详细信息
	logs	在一个 Pod 中打印一个容器日志。如果 Pod 只有一个容器，则容器名称是可选的
	attach	附加一个运行的容器
	exec	执行命令到容器
	port-forward	转发一个或多个本地端口到一个 Pod
	proxy	运行一个 proxy 到 Kubernetes API Server
	cp	复制文件或目录到容器中
	auth	检查授权
高级命令	apply	通过文件名或标准输入对资源应用配置
	patch	使用补丁修改、更新资源的字段
	replace	通过文件名或标准输入替换一个资源
	convert	不同的 API 版本之间转换配置文件
设置命令	label	更新资源上的标签
	annotate	更新资源上的注释
	completion	用于实现 Kubectl 工具自动补全
其他命令	api-versions	打印受支持的 API 版本
	config	修改 kubeconfig 文件（用于访问 API，比如配置认证信息）
	help	所有命令的帮助
	plugin	运行一个命令行插件
	version	打印客户端和服务版本信息

11.5.3　Kubectl 工具管理集群应用

本小节将根据 Kubernetes 在集群中部署应用管理周期的需求来讲解如何使用 Kubectl 管理工具。

1. 使用 Kubectl 工具创建资源

为了避免混淆，删除前面创建的资源，删除资源需要先删除 Deployment 和 Service，删除命令如下。

```
[root@master ~]# kubectl delete deploy/nginx
deployment.extensions "nginx" deleted
[root@master ~]# kubectl delete svc/nginx
service "nginx" deleted
[root@master ~]# kubectl get all
```

在集群中创建一个新资源，如创建目前企业主流的 Nginx 服务，该服务是从

Docker 上获取的，命令如下。

```
    [root@master ~]# kubectl run nginx --replicas=5
--labels="app=nginx-example" --image=nginx:1.10 --port=80
    deployment.apps "nginx" created
```

创建后使用 kubectl get all 命令查看资源的状态，如图 11-28 所示。

```
[root@master ~]# kubectl get all
NAME                          READY     STATUS    RESTARTS   AGE
pod/nginx-5b8c7bbc4c-2qwbh    1/1       Running   0          3m
pod/nginx-5b8c7bbc4c-grb2f    1/1       Running   0          3m
pod/nginx-5b8c7bbc4c-kmhjj    1/1       Running   0          3m
pod/nginx-5b8c7bbc4c-n7b4s    1/1       Running   0          3m
pod/nginx-5b8c7bbc4c-zz6t8    1/1       Running   0          3m

NAME                 TYPE        CLUSTER-IP   EXTERNAL-IP   PORT(S)   AGE
service/kubernetes   ClusterIP   10.10.10.1   <none>        443/TCP   3h

NAME                     DESIRED   CURRENT   UP-TO-DATE   AVAILABLE   AGE
deployment.apps/nginx    5         5         5            5           3m

NAME                                DESIRED   CURRENT   READY     AGE
replicaset.apps/nginx-5b8c7bbc4c    5         5         5         3m
[root@master ~]#
```

图 11-28

注意

通过 run 命令创建的对象默认都是 Deployment。使用 kubectl run--help 命令可以查看 run 命令的帮助信息。

2. 使用 Kubectl 工具查看资源信息

使用 kubectl describe 命令可以查看 Pod 的详细信息，重点查看最后几行信息，如图 11-29 所示。

```
Events:
  Type    Reason                 Age        From                  Message
  ----    ------                 ----       ----                  -------
  Normal  Scheduled              4m         default-scheduler     Successfully assigned nginx-5b8c7bbc4c-2qwbh to 192.168.10.8
  Normal  SuccessfulMountVolume  <invalid>  kubelet, 192.168.10.8  MountVolume.SetUp succeeded for volume "default-token-lpg9d"
  Normal  Pulling                <invalid>  kubelet, 192.168.10.8  pulling image "nginx:1.10"
  Normal  Pulled                 <invalid>  kubelet, 192.168.10.8  Successfully pulled image "nginx:1.10"
  Normal  Created                <invalid>  kubelet, 192.168.10.8  Created container
  Normal  Started                <invalid>  kubelet, 192.168.10.8  Started container
[root@master ~]#
```

图 11-29

使用 kubectl get svc 命令可以查看 Service。如果想列出多个资源，可以用逗号进行分隔，如列出 svc 和 Deployment 资源，如图 11-30 所示。

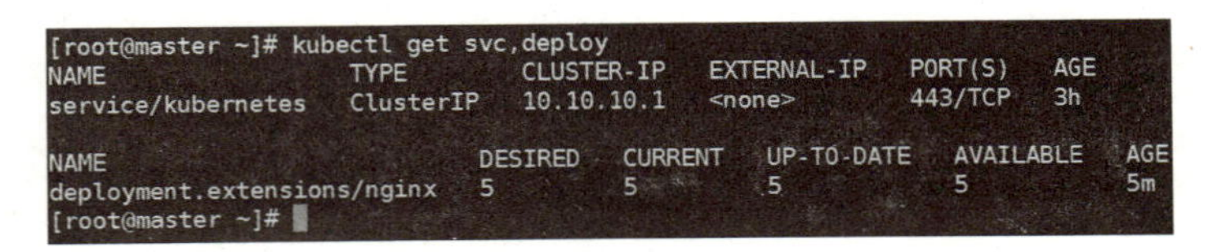

```
[root@master ~]# kubectl get svc,deploy
NAME                 TYPE        CLUSTER-IP   EXTERNAL-IP   PORT(S)   AGE
service/kubernetes   ClusterIP   10.10.10.1   <none>        443/TCP   3h

NAME                           DESIRED   CURRENT   UP-TO-DATE   AVAILABLE   AGE
deployment.extensions/nginx    5         5         5            5           5m
[root@master ~]#
```

图 11-30

在创建 Nginx deployment 时指定了一个标签（Labels），可以通过标签去匹配具体的 Pod，如图 11-31 所示。

```
[root@master ~]# kubectl get pods --show-labels
NAME                     READY     STATUS    RESTARTS   AGE       LABELS
nginx-5b8c7bbc4c-2qwbh   1/1       Running   0          7m        app=nginx-example,pod-template-hash=1647366707
nginx-5b8c7bbc4c-grb2f   1/1       Running   0          7m        app=nginx-example,pod-template-hash=1647366707
nginx-5b8c7bbc4c-kmhjj   1/1       Running   0          7m        app=nginx-example,pod-template-hash=1647366707
nginx-5b8c7bbc4c-n7b4s   1/1       Running   0          7m        app=nginx-example,pod-template-hash=1647366707
nginx-5b8c7bbc4c-zz6t8   1/1       Running   0          7m        app=nginx-example,pod-template-hash=1647366707
[root@master ~]#
```

图 11-31

使用 kubectl get pods -o wide 命令可以查看资源的详细信息，如查看 Pod 被分配到哪个节点上和哪个 IP 上。使用 kubectl get deploy -o wide 命令可以具体地查看对象使用了哪些镜像，如图 11-32 所示。

```
[root@master ~]# kubectl get pods -o wide
NAME                     READY     STATUS    RESTARTS   AGE       IP            NODE
nginx-5b8c7bbc4c-2qwbh   1/1       Running   0          7m        172.17.86.2   192.168.10.8
nginx-5b8c7bbc4c-grb2f   1/1       Running   0          7m        172.17.66.2   192.168.10.7
nginx-5b8c7bbc4c-kmhjj   1/1       Running   0          7m        172.17.83.2   192.168.10.6
nginx-5b8c7bbc4c-n7b4s   1/1       Running   0          7m        172.17.20.2   192.168.10.9
nginx-5b8c7bbc4c-zz6t8   1/1       Running   0          7m        172.17.25.2   192.168.10.10
[root@master ~]# kubectl get deploy -o wide
NAME      DESIRED   CURRENT   UP-TO-DATE   AVAILABLE   AGE       CONTAINERS   IMAGES       SELECTOR
nginx     5         5         5            5           7m        nginx        nginx:1.10   app=nginx-example
[root@master ~]#
```

图 11-32

3．使用 Kubectl 工具发布应用

部署的应用如果正常就可以发布了。如果部署完应用后不进行发布，那么用户是无法访问的。通过 expose 命令可以使应用在外部暴露一个端口，此方法相当于给应用加负载均衡，因为 Pod 被分配在不同的节点上，对用户只有一个入口，最后负载均衡到里面。该需求是由节点上的 kube-proxy 实现的，发布命令如下。

```
[root@master ~]# kubectl expose deployment nginx --port=88
--type=NodePort --target-port=80 --name=nginx-service
service "nginx-service" exposed
```

Nginx 服务发布后，可以使用 kubectl get svc 命令查看创建的 Service，如图 11-33 所示。在图 11-33 中可以看到服务的名称、服务的类型和集群的 IP，创建的 Service 都会为集群分配 Cluster-IP，Cluster-IP 是在启动 kube-apiserver 组件时指定的网段，如图 11-34 所示。

```
[root@master ~]# kubectl get svc
NAME            TYPE        CLUSTER-IP    EXTERNAL-IP   PORT(S)        AGE
kubernetes      ClusterIP   10.10.10.1    <none>        443/TCP        3h
nginx-service   NodePort    10.10.10.87   <none>        88:38157/TCP   12s
[root@master ~]#
```

图 11-33

```
[root@master ~]# ps aux|grep kube-apiserver
root       2759  2.9  3.5 520152 286776 ?      Ssl  May01  10:17 /opt/kubernetes/bin/kube-apiserver --logtostderr=true --v=4 --et
cd-servers=https://192.168.10.5:2379,https://192.168.10.6:2379,https://192.168.10.7:2379,https://192.168.10.8:2379,https://192.168
.10.9:2379,https://192.168.10.10:2379 --insecure-bind-address=127.0.0.1 --bind-address=192.168.10.5 --insecure-port=8080 --secure-
port=6443 --advertise-address=192.168.10.5 --allow-privileged=true --service-cluster-ip-range=10.10.10.0/24 --admission-control=Na
mespaceLifecycle,LimitRanger,SecurityContextDeny,ServiceAccount,ResourceQuota,NodeRestriction --authorization-mode=RBAC,Node --kub
elet-https=true --enable-bootstrap-token-auth --token-auth-file=/opt/kubernetes/cfg/token.csv --service-node-port-range=30000-5000
0 --tls-cert-file=/opt/kubernetes/ssl/server.pem --tls-private-key-file=/opt/kubernetes/ssl/server-key.pem --client-ca-file=/opt/k
ubernetes/ssl/ca.pem --service-account-key-file=/opt/kubernetes/ssl/ca-key.pem --etcd-cafile=/opt/kubernetes/ssl/ca.pem --etcd-cer
tfile=/opt/kubernetes/ssl/server.pem --etcd-keyfile=/opt/kubernetes/ssl/server-key.pem
root       4029  0.0  0.0 112660   976 pts/0    S+   00:13   0:00 grep --color=auto kube-apiserver
[root@master ~]#
```

图 11-34

通过外部浏览器，输入节点 IP 地址加暴露在外部的 NodePort 端口号即可访问 Nginx 页面，如图 11-35 所示。也可以使用 curl 命令或 elinks 命令加内部 IP 和端口号进行访问，如图 11-36 所示。

图 11-35

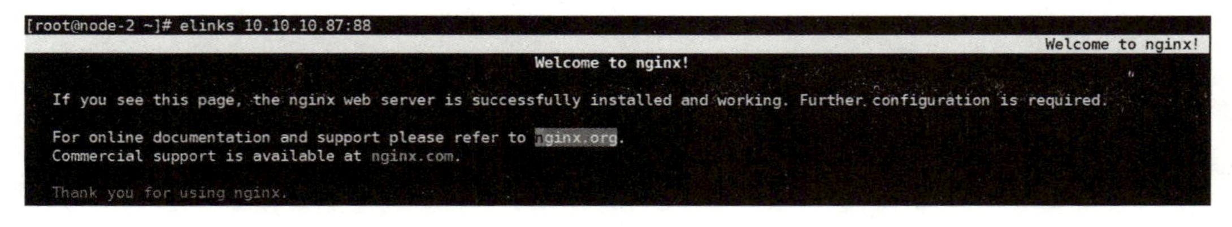

图 11-36

4．使用 Kubectl 工具排查故障

在部署应用时是无法做到每次都部署成功的，会出现一些不可避免的情况。当然，出现问题对我们来说是一件好事，可以提高我们排查错误的水平和能力。

应用出现问题时可以使用 describe 命令查看资源事件的类型，类型可以是 deploy、rs 和 po。在类型后面加上资源名称，还可查看 svc 的具体配置信息，命令如下。

```
[root@master ~]# kubectl describe po/nginx-f95d765f9-dss2v
[root@master ~]# kubectl describe deploy/nginx
[root@master ~]# kubectl describe rs/nginx-f95d765f9
[root@master ~]# kubectl describe svc
[root@master ~]# kubectl describe svc nginx-service
```

使用 kubectl logs 命令可以查看 Pod 运行时输出的日志信息，如图 11-37 所示。

```
[root@node-1 ~]# kubectl get pods
NAME                   READY     STATUS    RESTARTS   AGE
nginx-f95d765f9-4298m   1/1       Running   0          7m
nginx-f95d765f9-b8d7g   1/1       Running   0          7m
nginx-f95d765f9-qpx4l   1/1       Running   0          7m
[root@node-1 ~]# kubectl logs nginx-f95d765f9-4298m
[root@node-1 ~]# kubectl logs nginx-f95d765f9-qpx4l
192.168.10.6 - - [21/Apr/2018:13:34:02 +0000] "GET / HTTP/1.1" 200 612 "-" "ELinks/0.12pre6 (textmode; Linux; 113x28-2)" "-"
[root@node-1 ~]#
```

图 11-37

使用 kubectl exec 命令可以进入 Pod 查看具体应用的一些状态信息，如修改 index.html 文件中的内容为 HuMingZhe，命令如下。

```
    [root@node-1 ~]# kubectl exec -it nginx-f95d765f9-qpx41 bash
    root@nginx-f95d765f9-qpx4l:/# cd /usr/share/nginx/html/
    root@nginx-f95d765f9-qpx4l:/usr/share/nginx/html# echo "HuMingZhe" >
index.html
    root@nginx-f95d765f9-qpx4l:/usr/share/nginx/html# cat index.html
    HuMingZh
```

5. 使用 Kubectl 工具更新应用

应用部署好之后，后期会做一些升级，或者更换一些镜像。更新镜像可以使用 kubectl set 命令，如前面小节中 Nginx 镜像版本是 1.10，现在更新为 1.13.8，查询是否更新成功可以使用 kubectl describe 命令，命令如下。

```
    [root@master ~]# kubectl set image deployment/nginx nginx=nginx:1.13.8
    deployment "nginx" image updated
    [root@master ~]# kubectl describe pod nginx-f56ffcb44-41pg5
    Containers:
      nginx:
        Container ID:   docker://4bac9958cbd72c64a23f874af4c6a9b82b80ffaa
fdde7c0840319395929f5d1a
        Image:          nginx:1.13.8
        Image ID:
docker-pullable://nginx@sha256:0ffc09487404ea43807a1fd9e33d9e924d2c8b48a
7b7897e4d1231a396052ff9
        Port:           80/TCP
        State:          Running
          Started:      Fri, 02 Mar 2018 21:45:27 +0800
```

使用 kubectl edit 命令可以编辑 Nginx 的资源，找到 Nginx 版本所在位置，将 nginx:1.13.8 改为 nginx:1.12，命令如下。

```
    [root@master ~]# kubectl edit deploy/nginx
        spec:
```

```
      containers:
      - image: nginx:1.12
deployment "nginx" edited
[root@master ~]# kubectl describe pod nginx-667655d949-8gtrz
nginx:
    Container ID:   docker://8ddb2c2326542bf08c856c3a1f713c3c0616ad
fa06a53ed279e780a019d9c062
    Image:          nginx:1.12
    Image ID:
docker-pullable://nginx@sha256:813be5ee155f914d063e8ed39878267995d31d704
6a9010001f0a9e8192412a5
```

使用 curl -I 命令加上内部 IP 地址和端口号也可以查看 Nginx 的版本信息，命令如下。

```
[root@node-1 ~]# curl -I 10.10.10.155:88
HTTP/1.1 200 OK
Server: nginx/1.12.2
Date: Sat, 21 Apr 2018 13:41:57 GMT
Content-Type: text/html
Content-Length: 10
Last-Modified: Sat, 21 Apr 2018 13:40:02 GMT
Connection: keep-alive
ETag: "5adb3f32-a"
Accept-Ranges: bytes
```

项目升级到最新版本后，使用 kubectl rollout status 命令可以查看发布的状态信息，命令如下。

```
[root@master ~]# kubectl rollout status deploy/nginx
deployment "nginx" successfully rolled out
```

使用 kubectl rollout history 命令可以查看发布的版本，共显示了 3 个版本，命令如下。

```
[root@master ~]# kubectl rollout history deploy/nginx
deployments "nginx"
REVISION  CHANGE-CAUSE
1         <none>
2         <none>
3         <none>
```

查看代码时发现版本信息默认全部为 none，这是因为系统没有记录版本信息，可以在修改 Nginx 版本信息的后面加上--record 选项，让系统记录版本信息，命令如下。

```
[root@master ~]# kubectl set image deployment/nginx nginx=nginx:1.13.8
--record
deployment "nginx" image updated
[root@master ~]# kubectl rollout history deploy/nginx
deployments "nginx"
REVISION  CHANGE-CAUSE
1         <none>
3         <none>
5         <none>
6         <none>
7         kubectl set image deployment/nginx nginx=nginx:1.13.8
--record=true
```

现在有 3 个 Nginx 副本数提供服务，在特殊的购物节日，系统访问量会倍增，此时就需要更多的 Nginx 服务来支撑更大的并发，可以进行扩容来增加 Pod 的数量。使用 kubectl scale 命令可以针对 deploy 进行伸缩处理，如将 Nginx 副本数改为 10，再查看 Pod 数量就会变为 10 个，当购物节日过去后，不再需要这么多 Nginx 副本数，可以改回 3 个，如图 11-38 所示。

```
[root@master ~]# kubectl scale deployment nginx --replicas=10
deployment.extensions "nginx" scaled
[root@master ~]# kubectl get pods
NAME                     READY     STATUS    RESTARTS   AGE
nginx-5b8c7bbc4c-2qwbh   1/1       Running   1          28m
nginx-5b8c7bbc4c-8gwrs   1/1       Running   0          2m
nginx-5b8c7bbc4c-9tr9v   1/1       Running   0          2m
nginx-5b8c7bbc4c-grb2f   1/1       Running   0          28m
nginx-5b8c7bbc4c-kmhjj   1/1       Running   2          28m
nginx-5b8c7bbc4c-n7b4s   1/1       Running   0          28m
nginx-5b8c7bbc4c-pn5hk   1/1       Running   0          2m
nginx-5b8c7bbc4c-wdx7h   1/1       Running   0          2m
nginx-5b8c7bbc4c-xxbhg   1/1       Running   0          2m
nginx-5b8c7bbc4c-zz6t8   1/1       Running   0          28m
[root@master ~]# kubectl scale deployment nginx --replicas=3
deployment.extensions "nginx" scaled
[root@master ~]# kubectl get pods -o wide
NAME                     READY     STATUS    RESTARTS   AGE       IP            NODE
nginx-5b8c7bbc4c-grb2f   1/1       Running   0          29m       172.17.66.2   192.168.10.7
nginx-5b8c7bbc4c-n7b4s   1/1       Running   0          29m       172.17.20.2   192.168.10.9
nginx-5b8c7bbc4c-zz6t8   1/1       Running   0          29m       172.17.25.2   192.168.10.10
[root@master ~]#
```

图 11-38

6. 使用 Kubectl 工具资源回滚

更新镜像到最新版本后发现镜像存在问题，可以使用 kubectl rollout undo 命令实现回滚，默认回滚到上一个版本，命令如下。

```
[root@master ~]# kubectl rollout undo deployment/nginx
deployment "nginx"
```

7. 使用 Kubectl 工具删除资源

当不再需要这些应用时，可以进行删除，先删除 Deployment，再删除 Service，命令如下。

```
[root@master ~]# kubectl get all
[root@master ~]# kubectl delete deploy/nginx
deployment "nginx" deleted
[root@master ~]# kubectl delete svc/nginx-service
service "nginx-service" deleted
[root@master ~]# kubectl get all
NAME           TYPE        CLUSTER-IP   EXTERNAL-IP   PORT(S)   AGE
svc/kubernetes   ClusterIP   10.10.10.1   <none>        443/TCP   2h
```

第 12 章

Kubernetes 管理维护与运用

12.1 YAML配置文件管理资源

YAML 是配置文件的格式。YAML 文件是由一些易读的字段和指令组成的。K8S 使用 YAML 配置文件需要注意如下事项。

- 定义配置文件时，指定最新稳定版 API（当前最新稳定版是 v1 版本）。最新版本的 API 可以通过 kubectl api-versions 命令进行查看。前面创建 kubernetes-dashboard 时的 YAML 文件中指定的就是最新版本的 API 接口。
- 配置文件应该存储在集群之外的版本控制仓库中。如果需要，可以快速回滚配置、重新创建和恢复。
- 应该使用 YAML 格式编写配置文件，而不是使用 JSON 格式编写配置文件。尽管 YAML 和 JSON 格式都可以使用，但 YAML 格式的文件对用户更加友好。
- 可以将相关对象组合成单个文件，这样更容易管理。部署 UI 时笔者提供了三个 YAML 文件，这三个 YAML 文件是对一个文件进行拆分后生成的。
- 不要指定没必要的默认值，简单和最小配置可以减少错误的发生。
- 在注释中描述对象有利于后期的管理与维护。

在前面的章节中讲解部署 K8S 时用到了 YAML 文件，但是 YAML 文件中语法格式的作用没有进行详细的说明，下面以 nginx-deployment 的 YAML 文件为例进行讲解，文字说明如图 12-1 所示。

创建好 nginx-deployment.yaml 文件后，使用 kubectl create -f 命令指定 YAML 配置文件，命令如下。

```
[root@master ~]# kubectl create -f nginx-deployment.yaml
```

```
deployment "nginx-deployment" created
[root@master ~]# kubectl get all
[root@master ~]# kubectl describe deploy nginx-deployment
```

```
[root@master ~]# vim nginx-deployment.yaml

apiVersion: apps/v1beta2 # API版本，通过kubectl api-versions列出
kind: Deployment          # 指定对象的名称
metadata:                 # 元数据
  name: nginx-deployment # 指定创建deployment对象的名称，还可增加命令空间，如namespace: default，还可以写labels，格式如下所示
  namespace: default
  labels:
    web: nginx
spec:     #具体控制器选项
  replicas: 3     #指定控制器deployment保证副本的数量
  selector:       # 选择器
    matchLabels:  #匹配具体pod标签
      app: nginx
  template:       # 创建具体的pod
    metadata:     # pod创建的元数据
      labels:     # pod指定的标签的app nginx 通过控制器匹配app: nginx来管理下面的pod
        app: nginx
    spec:         # 具体容器的选项
      containers:
      - name: nginx # 容器名称 nginx
        image: nginx:1.10 # 镜像版本
        ports: # 容器端口是80
        - containerPort: 80
```

图 12-1

创建 nginx-service.yaml 文件，代码如下。

```
[root@master ~]# vim nginx-service.yaml
apiVersion: v1
kind: Service
metadata:
  name: nginx-service
  labels:
    app: nginx
spec:
  ports:
  - port: 88 # 集群 IP88 端口
    targetPort: 80 # 容器 IP80 端口
  selector:
    app: nginx
```

发布 nginx-service 服务，然后使用 kubectl describe svc 命令并指定 nginx-service 可查看 IP 地址，命令如下。

```
[root@master ~]# kubectl create -f nginx-service.yaml
service "nginx-service" created
[root@master ~]# kubectl get svc
NAME            TYPE        CLUSTER-IP     EXTERNAL-IP   PORT(S)   AGE
kubernetes       ClusterIP   10.10.10.1      <none>         443/TCP   4h
nginx-service    ClusterIP   10.10.10.221    <none>          88/TCP    1m
[root@master ~]# kubectl describe svc nginx-service
```

使用 curl 命令并指定发布的 IP 地址可查看 Nginx 的访问页面和 Nginx 版本信息，

命令如下。

```
[root@node-2 ~]# curl -I 10.10.10.221:88
HTTP/1.1 200 OK
Server: nginx/1.10.3
Date: Fri, 02 Mar 2018 15:58:57 GMT
Content-Type: text/html
Content-Length: 612
Last-Modified: Tue, 31 Jan 2017 15:01:11 GMT
Connection: keep-alive
ETag: "5890a6b7-264"
Accept-Ranges: bytes
```

同样，可以通过日志查看刚才具体传送的请求信息，命令如下。

```
[root@master ~]# kubectl get pods
NAME                     READY     STATUS    RESTARTS   AGE
nginx-f95d765f9-4298m    1/1       Running   0          14m
nginx-f95d765f9-b8d7g    1/1       Running   0          14m
nginx-f95d765f9-qpx4l    1/1       Running   0          14m
[root@master ~]# kubectl logs nginx-f95d765f9-qpx4l
192.168.10.6 - - [21/Apr/2018:13:34:02 +0000] "GET / HTTP/1.1" 200 612
"-" "ELinks/0.12pre6 (textmode; Linux; 113x28-2)" "-"
192.168.10.6 - - [21/Apr/2018:13:41:57 +0000] "HEAD / HTTP/1.1" 200 0 "-"
"curl/7.29.0" "-"
```

12.2 Pod管理

Pod 是集群中部署应用最小的单元，可以理解为一个 Pod 就是一个容器，但实际上，一个 Pod 可以是多个容器。学习 Pod 是为了后期编写应用时更方便、更清晰地进行配置。

在生产环境中，很少对 Pod 进行直接创建和管理。通常使用 Deployment 控制器去管理 Pod，这样就具备了调度、弹性伸缩、滚动更新等一系列特性。

12.2.1 Pod 基本管理

创建 Pod 对象需要先创建一个 pod.yaml 文件，在 YAML 文件中写入如下代码。

```
[root@master ~]# vim pod.yaml
apiVersion: v1 #指定 API 版本
```

```
kind: Pod        #创建对象
metadata:        #对象元数据，标识对象资源，方便后期进行匹配查询
  name: nginx-pod # 对象名称
  labels:
    app: nginx
spec:   #描述对象具体关联容器等资源
  containers:          #具体管理容器的配置情况
  - name: nginx      #Nginx 的名称
    image: nginx#使用 Nginx 的镜像
```

pod.yaml 文件编辑完成后，可以使用 kubectl 命令进行 Pod 资源的创建，创建命令如下。

```
[root@master ~]# kubectl create -f pod.yaml
pod "nginx-pod" created
```

Pod 资源创建成功后，使用 kubectl get pods 命令可以查看刚刚创建的名称为 nginx-pod 的 Pod，命令如下。

```
[root@master ~]# kubectl get pods
NAME                        READY     STATUS     RESTARTS   AGE
nginx-5b8c7bbc4c-grb2f   1/1        Running    0           34m
nginx-5b8c7bbc4c-n7b4s   1/1        Running    0           34m
nginx-5b8c7bbc4c-zz6t8   1/1        Running    0           34m
nginx-pod                 1/1       Running    0          7s
```

nginx-pod 在 default 命名空间中，使用 kubectl get pods 命令可以查看。如果在别的命名空间中，则需要加上-n 选项指定命名空间，如指定 kube-system 命名空间，列出 kube-system 命名空间中的 Pod，运行命令如下。

```
[root@master ~]# kubectl get pods -n kube-system
NAME                                    READY     STATUS     RESTARTS   AGE
kubernetes-dashboard-7f8c9bff96-f6llr   1/1        Running    3
3h
```

要想查看某个 Pod 的具体情况，可以通过 describe 命令实现。使用 describe 命令可以查看所有资源的类型描述，在 describe 命令后面指定资源类型为 Pod 就可以查看 nginx-pod 的具体信息，命令如下。

```
[root@master ~]# kubectl describe pod nginx-pod
Name:          nginx-pod
Namespace:     default
Node:          192.168.10.8/192.168.10.8
Start Time:    Thu, 03 May 2018 14:07:36 +0800
Labels:        app=nginx
```

```
Annotations:  <none>
Status:       Running
IP:           172.17.37.2
```

在生产环境中 Pod 并不是一成不变的，多数情况下会对 Pod 的资源进行替换。其实，替换就是直接删除后再进行创建，因为 Pod 不具备控制器的管理功能，若想更改 Pod 的信息，就要先删除 Pod，再去创建。例如，将 Nginx 镜像版本 1.10 改为 1.13，运行命令如下。

```
[root@master ~]# kubectl delete -f pod.yaml
pod "nginx-pod" deleted
[root@master ~]# vim pod.yaml
apiVersion: v1
kind: Pod
metadata:
  name: nginx-pod
  labels:
    app: nginx
spec:
  containers:
  - name: nginx
    image: nginx:1.13
[root@master ~]# kubectl create -f pod.yaml
pod "nginx-pod" created
[root@master ~]# kubectl describe pod nginx-pod
Containers:
  nginx:
    Image:          nginx:1.13
```

还可以使用 replace 命令删除镜像之后再进行重建，但是应用该 Pod 会出现一些问题，需要在命令后面加上--force 选项，命令如下。

```
[root@master ~]# kubectl replace -f pod.yaml --force
pod "nginx-pod" deleted
pod "nginx-pod" replaced
```

使用 kubectl apply -f pod.yaml 命令也可以达到替换的效果，运行命令如下。

```
[root@master ~]# kubectl apply -f pod.yaml
Warning: kubectl apply should be used on resource created by either kubectl
create --save-config or kubectl apply
pod "nginx-pod" configured
```

使用 kubectl delete 命令并指定 Pod 类型加 Pod 名称可以删除 Pod，如删除

nginx-pod，运行命令如下。

```
[root@master ~]# kubectl delete pod nginx-pod
pod "nginx-pod" deleted
```

12.2.2　Pod 资源限制

在日常工作中，有些容器占用的资源多，有些容器占用的资源少。当容器占用的资源异常多时，就会导致整个宿主机处于响应慢、超时等状态，影响部分用户的访问。

限制 Pod 使用资源是为了防止在异常情况下资源使用的溢出，从而影响整个宿主机。Pod 资源限制示例代码如下。

```
[root@master ~]# vim pod.yaml
apiVersion: v1
kind: Pod
metadata:
  name: nginx-pod
  labels:
    app: nginx
spec:
  containers:
  - name: nginx
    image: nginx
    resources:
      requests:
        memory: "64Mi"
        cpu: "250m"
      limits:
        memory: "128Mi"
        cpu: "500m"
```

使用 kubectl 命令创建该资源，运行命令如下。

```
[root@master ~]# kubectl create -f pod.yaml
pod "nginx-pod" created
```

创建完成后可以通过 describe 命令查看 nginx-pod 具体的信息，在 Limits 处可以看到对资源做的一些限制，如图 12-2 所示。

查看 nginx-pod 具体分配的服务器 IP，可以看到是如何限制的，如图 12-3～图 12-5 所示。

```
[root@master ~]# kubectl describe pod nginx-pod
Name:           nginx-pod
Namespace:      default
Node:           192.168.10.9/192.168.10.9
Start Time:     Thu, 03 May 2018 14:15:10 +0800
Labels:         app=nginx
Annotations:    <none>
Status:         Running
IP:             172.17.20.3
Containers:
  nginx:
    Container ID:   docker://5ab757413ad240f151ea972e4f2096c6f003ec5e7d72f403491135433e69d372
    Image:          nginx
    Image ID:       docker-pullable://nginx@sha256:0fb320e2a1b1620b4905facb3447e3d84ad36da0b2c8aa8fe3a5a81d1187b884
    Port:           <none>
    Host Port:      <none>
    State:          Running
      Started:      Thu, 03 May 2018 14:15:15 +0800
    Ready:          True
    Restart Count:  0
    Limits:
      cpu:     500m
      memory:  128Mi
    Requests:
      cpu:        250m
      memory:     64Mi
```

图 12-2

```
[root@master ~]# kubectl get pods -o wide
NAME                     READY     STATUS    RESTARTS   AGE       IP            NODE
nginx-5b8c7bbc4c-grb2f   1/1       Running   0          44m       172.17.66.2   192.168.10.7
nginx-5b8c7bbc4c-n7b4s   1/1       Running   0          44m       172.17.20.2   192.168.10.9
nginx-5b8c7bbc4c-zz6t8   1/1       Running   0          44m       172.17.25.2   192.168.10.10
nginx-pod                1/1       Running   0          2m        172.17.20.3   192.168.10.9
[root@master ~]#
```

图 12-3

```
[root@node-4 ~]# docker ps
CONTAINER ID        IMAGE                                                               COMMAND                  CREATED             STATUS              PORTS
     NAMES
5ab757413ad2        nginx                                                               "nginx -g 'daemon of…"   5 minutes ago       Up 5 minutes
     k8s_nginx_nginx-pod_default_eacf3d98-4d5e-11e8-908f-000c292cf20b_0
fa8176059d31        registry.cn-hangzhou.aliyuncs.com/google-containers/pause-amd64:3.0   "/pause"                 5 minutes ago       Up 5 minutes
     k8s_POD_nginx-pod_default_eacf3d98-4d5e-11e8-908f-000c292cf20b_0
e897650a34b1        nginx                                                               "nginx -g 'daemon of…"   42 minutes ago      Up 42 minutes
     k8s_nginx_nginx-5b8c7bbc4c-n7b4s_default_024ece18-4d59-11e8-908f-000c292cf20b_0
518f42771d38        registry.cn-hangzhou.aliyuncs.com/google-containers/pause-amd64:3.0   "/pause"                 44 minutes ago      Up 44 minutes
     k8s_POD_nginx-5b8c7bbc4c-n7b4s_default_024ece18-4d59-11e8-908f-000c292cf20b_0
[root@node-4 ~]#
```

图 12-4

```
[root@node-4 ~]# docker inspect 5ab757413ad2
[
    {
        "Id": "5ab757413ad240f151ea972e4f2096c6f003ec5e7d72f403491135433e69d372",
        "Created": "2018-05-03T06:15:14.752112757Z",
        "Path": "nginx",
        "Args": [
            "-g",
            "daemon off;"
        ],
            "Isolation": "",
            "CpuShares": 256,
            "Memory": 134217728,
            "NanoCpus": 0,
            "CgroupParent": "/kubepods/burstable/podeacf3d98-4d5e-11e8-908f-000c292cf20b",
            "BlkioWeight": 0,
            "BlkioWeightDevice": null,
            "BlkioDeviceReadBps": null,
            "BlkioDeviceWriteBps": null,
            "BlkioDeviceReadIOps": null,
            "BlkioDeviceWriteIOps": null,
            "CpuPeriod": 100000,
            "CpuQuota": 50000,
            "CpuRealtimePeriod": 0,
            "CpuRealtimeRuntime": 0,
            "CpusetCpus": "",
            "CpusetMems": "",
            "Devices": [],
            "DeviceCgroupRules": null,
            "DiskQuota": 0,
            "KernelMemory": 0,
            "MemoryReservation": 0,
            "MemorySwap": 134217728,
            "MemorySwappiness": null,
            "OomKillDisable": false,
            "PidsLimit": 0,
            "Ulimits": null,
            "CpuCount": 0,
            "CpuPercent": 0,
            "IOMaximumIOps": 0,
            "IOMaximumBandwidth": 0
        },
```

图 12-5

12.2.3　Pod 调度约束

在集群中可能运行着多个环境，如开发测试环境和生产环境，它们用的是同一个数据库，只不过项目不同。要想创建一个应用，先将 Pod 运行到具体的节点上，如运行到测试环境中或生产环境中，然后通过字段进行约束。

Pod.spec.nodeName 命令强制约束 Pod 调度到指定节点上，Pod.spec.nodeSelector 命令通过 lable-selector 机制选择节点，示例代码如下。

```
[root@master ~]# kubectl delete -f pod.yaml
pod "nginx-pod" deleted
[root@master ~]# vim pod.yaml
apiVersion: v1
kind: Pod
metadata:
  name: nginx-pod
  labels:
       app: nginx
spec:
  nodeName: 192.168.10.7 #指定 IP 地址
  #nodeSelector:
  #  env_role: dev
  containers:
  - name: nginx
    image: nginx
[root@master ~]# kubectl create -f pod.yaml
pod "nginx-pod" created
```

使用 kubectl get pods -o wide 命令可以查看分配的 IP 地址和上述指定地址是否一致，命令如下。

```
[root@master ~]# kubectl get pods -o wide
NAME          EADY   STATUS    RESTARTS  AGE    IP           NODE
nginx-pod      1/1    Running   0         1m     172.11.84.4  192.168.10.7
```

将一个 Pod 分配到 192.168.222.12 服务器上，在 pod.yaml 文件中注释掉 nodeName: 192.168.222.12，打开 nodeSelector 和 env_role: dev，命令如下。

```
[root@master ~]# vim pod.yaml
apiVersion: v1
kind: Pod
metadata:
  name: nginx-pod2
```

```
    labels:
         app: nginx
  spec:
    #nodeName: 192.168.10.7
    nodeSelector:
      env_role: dev
    containers:
    - name: nginx
      image: nginx
```

使用 nodeSelector 选择器匹配 dev 标签，默认只有如下所示的标签，dev 标签不存在。

```
[root@master ~]# kubectl describe node 192.168.10.7
Name:                192.168.10.7
Roles:               <none>
Labels:              beta.kubernetes.io/arch=amd64
                   beta.kubernetes.io/os=linux
                   kubernetes.io/hostname=192.168.10.7
```

使用 kubectl label 命令可以设置标签，然后分配到 192.168.10.7 服务器上，如果不约束创建这两个 Pod，有 90%的概率会分配到两个节点上。指定完节点名称后设置 dev 标签，运行命令如下。

```
[root@master ~]# kubectl label nodes 192.168.10.7 env_role=dev
node "192.168.222.12" labeled
```

创建完成后，查看 192.168.10.7 服务器会发现多出一个 dev 标签，命令如下。

```
[root@master ~]# kubectl describe node 192.168.10.7
Name:                192.168.10.7
Roles:               <none>
Labels:              beta.kubernetes.io/arch=amd64
                   beta.kubernetes.io/os=linux
                   env_role=dev
                   kubernetes.io/hostname=192.168.10.7
```

创建 Pod 资源，nginx-pod2 还会被分配到 192.168.10.7 服务器上，这是因为哪个节点拥有 dev 标签就会被分配到哪个节点上，运行命令如下。

```
[root@master ~]# kubectl create -f pod.yaml
pod "nginx-pod2" created
[root@master ~]# kubectl get pods -o wide
NAME         READY     STATUS     RESTARTS   AGE      IP            NODE
nginx-pod     1/1       Running    0          21m      172.11.84.4
```

```
192.168.10.7
    nginx-pod2    1/1     Running   0          18s       172.11.84.5
192.168.10.7
```

综上，最终 Pod 被分配到 192.168.10.7 服务器上，说明约束起了作用。

12.2.4　Pod 重启策略

Pod 有三种重启策略：Always，当容器停止时，总是重建容器，是默认策略；OnFailure，当容器异常退出（退出状态码非 0）时，会重启容器，如果容器是正常关闭的，则不会做任何操作；Never，当容器退出时，不会重启容器。设置重启策略只需在容器的下面定义三种策略中的任意一种，示例代码如下。

```
[root@master ~]# vim pod.yaml
apiVersion: v1
kind: Pod
metadata:
  name: nginx-pod
  labels:
    app: nginx
spec:
  containers:
  - name: nginx
    image: nginx
  restartPolicy: OnFailure
[root@master ~]# kubectl create -f pod.yaml
pod "nginx-pod" created
```

12.2.5　Pod 健康检查

在一个节点中会运行多个容器，Kubernetes 会对容器的状态做判断，如果容器宕掉了，会重新建立容器或重启容器。假设容器的状态是正常的，但是容器内部的应用出现了异常，那么对于 Kubernetes 来说，它会认为容器是正常的，默认还会将请求转发到容器中。为了防止这样的情况发生，可以采用 Probe 机制去检测容器中的应用状态。容器中应用状态返回的数值是非 0 或非指定状态码，就会退出容器，从而让 Kubernetes 知道容器是异常的。

Probe 机制有两种类型：livenessProbe，如果检查失败，就把容器“杀”死，根据 Pod 的 restartPolicy 进行操作；readinessProbe，如果检查失败，就把 Pod 从 Service endpoints 中删除。

Probe 支持以下三种检查方法。

- httpGet：发送 HTTP 请求，若返回 200～400 状态码，则表示成功。比如，请求访问网站的首页，如果网页状态码是 200，则表示访问正常；网页状态码是 300 也表示正常，可能是重定向，因为服务是正常的。如果返回的是非 200～400 的状态码，则会认为应用有问题，会将容器“杀”死。
- exec：执行 shell 命令，返回状态码为 0 表示成功。容器内运行的应用服务如果不是 HTTP 服务，则可以通过 exec 执行脚本判断某个文件是否存在。比如，运行某个应用时会生成一个 pid，将应用的 pid 放在某个文件中，通过 shell 命令来判断该文件是否存在，如果存在就说明应用是正常的，返回 0。
- tcpSocket：发起 TCP Socket 并建立成功。

健康检查 Pod 文件的示例代码如下。

```
[root@master ~]# vim pod.yaml
apiVersion: v1
kind: Pod
metadata:
  name: nginx-pod
  labels:
    app: nginx
spec:
  containers:
  - name: nginx
    image: nginx
    ports:
    - containerPort: 80        # 指定容器端口
    livenessProbe:             # 检查 livenessProbe 机制
      httpGet:                 # 通过 httpGet 请求去判断本机的 80 端口
        path: /index.html      # 查看 index.html 页面是否存在，如果存在表示应用正常
        port: 80               # 如果不存在，则对该容器做相应处理
```

创建 Pod，然后查看 Pod 是否在运行，命令如下。

```
[root@master ~]# kubectl create -f pod.yaml
pod "nginx-pod" created
[root@master ~]# kubectl get pods
NAME         READY     STATUS    RESTARTS   AGE
nginx-pod    1/1       Running   0          15s
```

查看 Pod 描述信息，在描述信息中可以看到关于健康检查的描述，如图 12-6 所示。

```
[root@master ~]# kubectl describe pod nginx-pod
Name:           nginx-pod
Namespace:      default
Node:           192.168.10.7/192.168.10.7
Start Time:     Sat, 21 Apr 2018 22:02:41 +0800
Labels:         app=nginx
Annotations:    <none>
Status:         Running
IP:             172.11.57.3
Containers:
  nginx:
    Container ID:   docker://a3c5445092f375333b5d7c08d8e3fc8db0b638bdc34655584d0d3590b35bb633
    Image:          nginx
    Image ID:       docker-pullable://nginx@sha256:18156dcd747677b03968621b2729d46021ce83a5bc15118e5bcced925fb4ebb9
    Port:           80/TCP
    State:          Running
      Started:      Sat, 21 Apr 2018 22:02:44 +0800
    Ready:          True
    Restart Count:  0
    Liveness:       http-get http://:80/index.html delay=0s timeout=1s period=10s #success=1 #failure=3
    Environment:    <none>
    Mounts:
      /var/run/secrets/kubernetes.io/serviceaccount from default-token-jw8qc (ro)
```

图 12-6

由于是 Nginx，因此查看日志会发现每隔 10s 就会请求访问 index.html 页面，如图 12-7 所示。

```
[root@master ~]# kubectl logs nginx-pod -f
172.11.57.1 - - [21/Apr/2018:14:02:48 +0000] "GET /index.html HTTP/1.1" 200 612 "-" "kube-probe/1.9" "-"
172.11.57.1 - - [21/Apr/2018:14:02:58 +0000] "GET /index.html HTTP/1.1" 200 612 "-" "kube-probe/1.9" "-"
172.11.57.1 - - [21/Apr/2018:14:03:08 +0000] "GET /index.html HTTP/1.1" 200 612 "-" "kube-probe/1.9" "-"
172.11.57.1 - - [21/Apr/2018:14:03:18 +0000] "GET /index.html HTTP/1.1" 200 612 "-" "kube-probe/1.9" "-"
172.11.57.1 - - [21/Apr/2018:14:03:28 +0000] "GET /index.html HTTP/1.1" 200 612 "-" "kube-probe/1.9" "-"
172.11.57.1 - - [21/Apr/2018:14:03:38 +0000] "GET /index.html HTTP/1.1" 200 612 "-" "kube-probe/1.9" "-"
172.11.57.1 - - [21/Apr/2018:14:03:48 +0000] "GET /index.html HTTP/1.1" 200 612 "-" "kube-probe/1.9" "-"
172.11.57.1 - - [21/Apr/2018:14:03:58 +0000] "GET /index.html HTTP/1.1" 200 612 "-" "kube-probe/1.9" "-"
172.11.57.1 - - [21/Apr/2018:14:04:08 +0000] "GET /index.html HTTP/1.1" 200 612 "-" "kube-probe/1.9" "-"
172.11.57.1 - - [21/Apr/2018:14:04:18 +0000] "GET /index.html HTTP/1.1" 200 612 "-" "kube-probe/1.9" "-"
```

图 12-7

打开一个新的 Master 节点的终端窗口，使用 kubectl logs nginx-pod -f 命令查看请求信息，在另外一个 Master 节点的终端窗口进入 nginx-pod 容器中，将 index.html 页面删除，在新打开的终端窗口界面中会输出如图 12-8 所示的信息。

```
[root@master ~]# kubectl exec -it nginx-pod bash
root@nginx-pod:/# cd /usr/share/nginx/html/
root@nginx-pod:/usr/share/nginx/html# ls
50x.html  index.html
root@nginx-pod:/usr/share/nginx/html# rm -rf index.html
root@nginx-pod:/usr/share/nginx/html#
```

```
2018/04/14 08:53:46 [error] 5#5: *21 open() "/usr/share/nginx/html/index.html" failed (2: No such file o
r directory), client: 172.17.90.1, server: localhost, request: "GET /index.html HTTP/1.1", host: "172.17
.90.3:80"
172.17.90.1 - - [14/Apr/2018:08:53:46 +0000] "GET /index.html HTTP/1.1" 404 170 "-" "kube-probe/1.9" "-"
2018/04/14 08:53:56 [error] 5#5: *22 open() "/usr/share/nginx/html/index.html" failed (2: No such file o
r directory), client: 172.17.90.1, server: localhost, request: "GET /index.html HTTP/1.1", host: "172.17
.90.3:80"
172.17.90.1 - - [14/Apr/2018:08:53:56 +0000] "GET /index.html HTTP/1.1" 404 170 "-" "kube-probe/1.9" "-"
2018/04/14 08:54:06 [error] 5#5: *23 open() "/usr/share/nginx/html/index.html" failed (2: No such file o
r directory), client: 172.17.90.1, server: localhost, request: "GET /index.html HTTP/1.1", host: "172.17
.90.3:80"
```

图 12-8

使用 describe 命令可以查看 nginx-pod 的描述信息，在描述信息的最下面会提示进行健康检查时没有检测到 index.html 页面，返回 404 状态码，然后会重新创建容器并启动容器，如图 12-9 所示。

```
Events:
  Type    Reason                 Age   From                      Message
  ----    ------                 ----  ----                      -------
  Normal  Scheduled              19h   default-scheduler         Successfully assigned nginx-pod to 192.168.10.7
  Normal  SuccessfulMountVolume  19h   kubelet, 192.168.10.7     MountVolume.SetUp succeeded for volume "default-token-jw8qc"
  Normal  Pulling                19h   kubelet, 192.168.10.7     pulling image "nginx"
  Normal  Pulled                 19h   kubelet, 192.168.10.7     Successfully pulled image "nginx"
  Normal  Created                19h   kubelet, 192.168.10.7     Created container
  Normal  Started                19h   kubelet, 192.168.10.7     Started container
[root@master ~]#
```

图 12-9

回到进入容器的终端窗口会发现，Nginx 自动退出容器了。重新进入容器后 index.html 页面又存在了，这就是 Pod 的健康检查机制，如图 12-10 所示。

```
[root@master ~]# kubectl exec -it nginx-pod bash
root@nginx-pod:/# cd /usr/share/nginx/html/
root@nginx-pod:/usr/share/nginx/html# ls
50x.html  index.html
root@nginx-pod:/usr/share/nginx/html# rm -rf index.html
root@nginx-pod:/usr/share/nginx/html# command terminated with exit code 137
[root@master ~]# kubectl exec -it nginx-pod bash
root@nginx-pod:/# cd /usr/share/nginx/html/
root@nginx-pod:/usr/share/nginx/html# ls
50x.html  index.html
```

图 12-10

12.2.6 Pod 问题定位

使用 describe 命令可以查看 Pod 的具体信息，需要关注的是 Events（事件），通过 Events 可以查看 nginx-pod 执行到了哪一步。

使用 logs 命令可以查看具体输出的日志信息。

使用 exec 命令可以进入容器进一步排查应用层的问题。

12.3 Service

Service 在 Kubernetes 集群中是非常重要的，它将请求转发到后端具体的 Pod 中。Service 对外也有负载均衡功能，提供一个统一的入口，以此代理所有应用的 Pod。

12.3.1 网络代理模式

Service 有三种网络代理模式：Userspace、Iptables 和 Ipvs。

在 Service 早期的版本中，它是通过如图 12-11 所示的一个用户空间实现网络代理模式的。客户端具体请求的是 Service IP，Service IP 就是使用 kubectl get svc 命令得到的 Cluster-IP，命令如下。

```
[root@master ~]# kubectl get svc
NAME           TYPE        CLUSTER-IP    EXTERNAL-IP  PORT(S)        AGE
kubernetes     ClusterIP   10.10.10.1    <none>       443/TCP        3h
nginx-service  NodePort    10.10.10.87   <none>       88:38157/TCP   48m
```

Service 会将请求交给 kube-proxy，kube-proxy 在节点中是以一个组件的方式运行的，该组件主要负责网络，命令如下。

```
[root@node-1 ~]# ps aux|grep kube-proxy
root      8783  0.1  1.1  44848 21220 ?      Ssl  08:34   0:39
/opt/kubernetes/bin/kube-proxy --logtostderr=true --v=4
--hostname-override=192.168.10.6
--kubeconfig=/opt/kubernetes/cfg/kube-proxy.kubeconfig
```

用户发起 HTTP 请求后，先去请求 Service IP，查看 Service IP 会发现在节点上创建了一个 iptables 规则，命令如下。

```
[root@node-1 ~]# iptables-save |grep 88
-A KUBE-SERVICES -d 10.10.10.87/32 -p tcp -m comment --comment
"default/nginx-service: cluster IP" -m tcp --dport 88 -j
KUBE-SVC-GKN7Y2BSGW4NJTYL
```

注意

在节点上访问虚拟 IP，会在本机上转发到 kube-proxy 组件，kube-proxy 组件再转发到 Pod。这种模式是早期的 Kubernetes Service 网络代理模式。

在现在的版本中，默认使用 iptables 代理模式，如图 12-12 所示。访问 Cluster-IP 后会直接转发到 Pod 中，这种工作方式直接应用 Iptables 规则，该规则每个节点上都有。Pod 是负载均衡之后关联的节点，iptables 会直接转发到容器的 IP 上，再去具体处理请求。与 userspace 模式相比，iptables 模式的性能要好很多，它不经过 kube-proxy 用户态，而且是内核态的、集成的四层转发，所以性能非常好。

kube-proxy 起到的作用是从 kube-apiserver 中拉取与 Service 相关联的 endpoints 并生成规则，在 Master 节点中用 kubectl get ep 命令获取 nginx-service 的 Cluster-IP 中的 172.11.52.2，在节点上是可以搜索到的，命令如下。

```
[root@master ~]# kubectl get ep
NAME           ENDPOINTS                                             AGE
kubernetes     192.168.10.5:6443                                     4h
nginx-service  172.17.20.2:80,172.17.25.2:80,172.17.66.2:80          53m
[root@node-1 ~]# iptables-save |grep 172.17.20.2
-A KUBE-SEP-SEBITSZIC7CZOJ5G -s 172.17.20.2/32 -m comment --comment
```

```
"default/nginx-service:" -j KUBE-MARK-MASQ
    -A KUBE-SEP-SEBITSZIC7CZOJ5G -p tcp -m comment --comment
"default/nginx-service:" -m tcp -j DNAT --to-destination 172.17.20.2:80
```

lvs 是基于 ipvs 这种内核级的负载均衡模块实现的负载均衡。如图 12-13 所示，用户请求的是 Service IP，并且 Service IP 在宿主机上，但是 Service IP 后面代理的是 Pod。ipvs 模式和 iptables 模式类似，不同的是，ipvs 模式直接通过内核模块的 ipvs 实现网络转发，而 iptables 模式通过 iptables 规则实现转发。两者相比，ipvs 的性能更好，因为它是一个专业级开源的负载均衡解决方案。

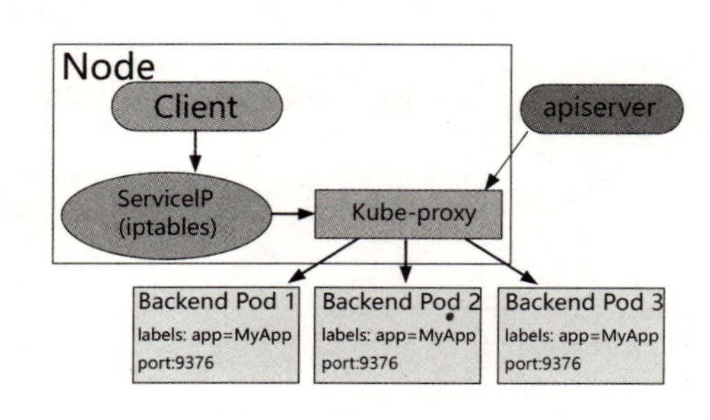

图 12-11

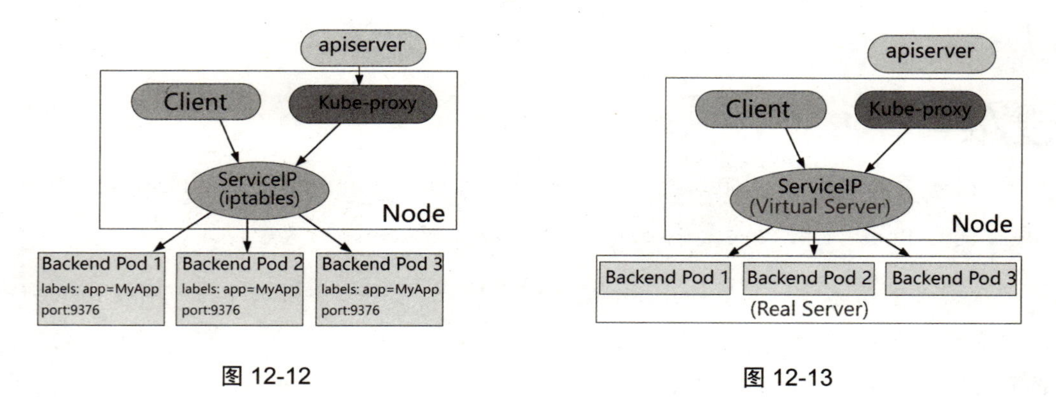

图 12-12

图 12-13

Service 官方文档的网址是 https://kubernetes.io/docs/concepts/services-networking/service。

注意

ipvs 网络模式目前还在公测阶段，不建议用到生产环境中。

12.3.2 服务代理

创建 Service 时需要创建 Service 对象，代码如下。

```
[root@master ~]# vim service.yaml
apiVersion: v1  # API 版本
kind: Service   # Service 对象
metadata:       # Service 元数据
  name: my-service  # 名称
spec:
  selector:       # 选择器
    app: MyApp
  ports:     # 暴露 Service 端口
  - name: http  # 如果是两个端口映射,则需指定两个 name 进行区分,如 HTTP 和 HTTPS
    prot ocol: TCP    # 协议类型 TCP
    port: 80     # 暴露端口
    targetPort: 80 # 目标端口（容器端口）
  - name: https
    protocol: TCP
    port: 443
    targetPort: 443
```

创建 Service.yaml 文件，运行命令如下。

```
[root@master ~]# kubectl create -f service.yaml
service "my-service" created
```

查看刚刚创建的 Service，如图 12-14 所示。

```
[root@master ~]# kubectl get svc
NAME            TYPE        CLUSTER-IP     EXTERNAL-IP   PORT(S)          AGE
kubernetes      ClusterIP   10.10.10.1     <none>        443/TCP          4h
my-service      ClusterIP   10.10.10.59    <none>        80/TCP,443/TCP   26s
nginx-service   NodePort    10.10.10.87    <none>        88:38157/TCP     1h
[root@master ~]#
```

图 12-14

my-service 是刚创建的，默认类型是 Cluster-IP，也会被分配一个 Cluster-IP，该网段是在创建 kube-apiserver 网段时指定的 service-cluster-ip。同样也会暴露端口，可以通过 IP+Port 去访问具体代理的应用。

使用 kubectl get endpoints 命令可以查看具体的 Service 后端的 Pod 的 IP，命令如下。可以看到，my-service 并没有 Pod 的 IP，这是因为在 Service.yaml 文件中并没有匹配某个 Deployment 可以为它代理，它唯一的匹配就是标签选择器。所以，一般创建一个项目并部署到集群中，首先需要创建一个 Deployment，再创建一个 Service，它们之间通过标签进行关联。

```
[root@master ~]# kubectl get endpoints my-service
NAME          ENDPOINTS    AGE
my-service    <none>       1m
```

查看 nginx-pod 的标签，让 my-service 代理 nginx-pod，只需指定标签 app=nginx 即可，然后查看 nginx-pod IP，如图 12-15 所示。将 IP 192.168.10.9 加入到 my-service 中。使用 kubectl edit 命令编辑 svc/my-service，使 my-service 直接应用修改后的字段，将 selector 中的 app 改为 nginx，如图 12-16 所示。修改后使用 kubectl get svc 命令查看，发生没有任何变化，当使用 kubectl get ep（endpoint 缩写）查看时，发现 IP 映射在了 my-service 中，如图 12-17 所示。有了 endpoint 后，使用 Cluster-IP（10.10.10.59）+Port 命令就可以访问代理的具体 Pod，如图 12-18 所示。

```
[root@master ~]# kubectl get pods --show-labels
NAME                     READY   STATUS    RESTARTS   AGE    LABELS
nginx-5b8c7bbc4c-grb2f   1/1     Running   1          1h     app=nginx-example,pod-template-hash=1647366707
nginx-5b8c7bbc4c-n7b4s   1/1     Running   1          1h     app=nginx-example,pod-template-hash=1647366707
nginx-5b8c7bbc4c-zz6t8   1/1     Running   1          1h     app=nginx-example,pod-template-hash=1647366707
nginx-pod                1/1     Running   1          29m    app=nginx
[root@master ~]# kubectl get pods --show-labels -o wide
NAME                     READY   STATUS    RESTARTS   AGE    IP            NODE           LABELS
nginx-5b8c7bbc4c-grb2f   1/1     Running   1          1h     172.17.2.2    192.168.10.7   app=nginx-example,pod-template-hash=1647366707
nginx-5b8c7bbc4c-n7b4s   1/1     Running   1          1h     172.17.84.2   192.168.10.9   app=nginx-example,pod-template-hash=1647366707
nginx-5b8c7bbc4c-zz6t8   1/1     Running   1          1h     172.17.70.2   192.168.10.10  app=nginx-example,pod-template-hash=1647366707
nginx-pod                1/1     Running   1          29m    172.17.84.3   192.168.10.9   app=nginx
[root@master ~]#
```

图 12-15

```
[root@master ~]# kubectl edit svc/my-service

# Please edit the object below. Lines beginning with a '#' will be ignored,
# and an empty file will abort the edit. If an error occurs while saving this file will be
# reopened with the relevant failures.
#
apiVersion: v1
kind: Service
metadata:
  creationTimestamp: 2018-05-01T17:11:44Z
  name: my-service
  namespace: default
  resourceVersion: "35145"
  selfLink: /api/v1/namespaces/default/services/my-service
  uid: b9193e04-4d62-11e8-908f-000c292cf20b
spec:
  clusterIP: 10.10.10.59
  ports:
  - name: http
    port: 80
    protocol: TCP
    targetPort: 80
  - name: https
    port: 443
    protocol: TCP
    targetPort: 443
  selector:
    app: nginx
  sessionAffinity: None
  type: ClusterIP
status:
  loadBalancer: {}
```

图 12-16

```
[root@master ~]# kubectl get svc
NAME            TYPE        CLUSTER-IP    EXTERNAL-IP   PORT(S)           AGE
kubernetes      ClusterIP   10.10.10.1    <none>        443/TCP           4h
my-service      ClusterIP   10.10.10.59   <none>        80/TCP,443/TCP    6m
nginx-service   NodePort    10.10.10.87   <none>        88:38157/TCP      1h
[root@master ~]# kubectl get ep
NAME            ENDPOINTS                                               AGE
kubernetes      192.168.10.5:6443                                       4h
my-service      172.17.84.3:80,172.17.84.3:443                          6m
nginx-service   172.17.2.2:80,172.17.70.2:80,172.17.84.2:80             1h
[root@master ~]#
```

图 12-17

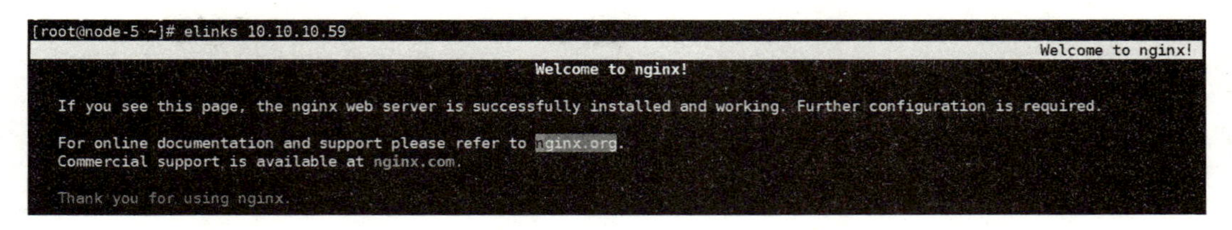

图 12-18

注意

在任意节点都可以通过 Cluster-IP 访问后端负载均衡的 Pod。

12.3.3　服务发现

Service 服务发现分为两种，一种是环境变量，另一种是 DNS。

在 Kubernetes 集群中，假设部署 Tomcat 集群，而 Tomcat 集群可能有多个副本，这些副本分配在不同的节点中。创建 MySQL 数据库，让 Tomcat 副本都可以访问到 MySQL 数据库，此时就会用到 Service。Service 为副本进行服务，创建一个统一的入口，因为多个 Tomcat 或多个其他服务要想访问，只能访问一个 IP+Port。所以，多个副本就是 Service 提供的一个负载均衡的功能。

企业生产环境中主流的是微服务。微服务是由多个组件和模块组成的。它们之间为了实现弹性伸缩、提高性能，就会在集群中部署很多的副本。它们之间可能会相互调用，如果相互调用，就会用 Service 去关联 deployment 副本，通过内部 Cluster-IP+Port 进行访问，从而实现两者之间的通信。

比如，第一组组件是 10 个副本，第二组组件也是 10 个副本。第二组组件要访问第一组的 Service IP，就需要用服务发现去获取 IP。这样就可以让第二组组件中的 Pod 都能知道第一组的组件虚拟 IP 具体是哪个 IP，通过该 IP 就可以访问第一组组件。

1. 环境变量

当一个 Pod 运行到节点时，kubelet 会为每个容器添加一组环境变量，Pod 容器中的程序就可以使用这些环境变量发现 Service。环境变量名格式如下：{SVCNAME}_SERVICE_HOST、{SVCNAME}_SERVICE_PORT。其中，服务名和端口名用大写字母，连字符转换为下画线。

创建一个 busybox 测试镜像，代码如下。

```
[root@master ~]# vim busybox.yaml
apiVersion: v1
```

```
kind: Pod
metadata:
  name: busybox
  namespace: default
spec:
  containers:
  - image: busybox
    command:
      - sleep
      - "3600"
    imagePullPolicy: IfNotPresent
    name: busybox
  restartPolicy: Always
[root@master ~]# kubectl create -f busybox.yaml
pod "busybox" created
```

进入 busybox 镜像，输入 env 命令查看当前系统的变量，如图 12-19 所示。输入 env 后会输出很多大写字母的变量，这些变量会为今后创建的每个 Pod 都生成一组变量信息，其中最关键的是{SVCNAME}_SERVICE_HOST 和{SVCNAME}_SERVICE_PORT。比如，前面小节中创建的 my-service 的变量信息如图 12-20 所示。变量名以服务名开头，以 SERVICE_HOST 结尾。

```
[root@master ~]# kubectl exec -it busybox sh
/ # env
KUBERNETES_PORT=tcp://10.10.10.1:443
KUBERNETES_SERVICE_PORT=443
MY_SERVICE_PORT_443_TCP_ADDR=10.10.10.59
NGINX_SERVICE_PORT_88_TCP_ADDR=10.10.10.87
MY_SERVICE_PORT_80_TCP=tcp://10.10.10.59:80
HOSTNAME=busybox
```

图 12-19

```
[root@master ~]# kubectl get ep
NAME            ENDPOINTS                                           AGE
kubernetes      192.168.10.5:6443                                   4h
my-service      172.17.84.3:80,172.17.84.3:443                      17m
nginx-service   172.17.2.2:80,172.17.70.2:80,172.17.84.2:80        1h
[root@master ~]#
```

图 12-20

在终端输出变量就可以获取具体 Service 的虚拟 IP，命令如下，从而实现了利用变量与另外一组组件通信。

```
[root@master ~]# kubectl exec -it busybox sh
/ # echo ${MY_SERVICE_SERVICE_HOST}
10.10.10.59
/ # echo ${MY_SERVICE_SERVICE_PORT}
80
```

```
[root@master ~]# kubectl get svc
NAME            TYPE        CLUSTER-IP    EXTERNAL-IP  PORT(S)          AGE
kubernetes      ClusterIP   10.10.10.1    <none>       443/TCP          4h
my-service      ClusterIP   10.10.10.59   <none>       80/TCP,443/TCP   20m
nginx-service   NodePort    10.10.10.87   <none>       88:38157/TCP     1h
```

注意

环境变量存在一定的限制。(1)Pod 和 Service 的创建顺序是有要求的，Service 必须在 Pod 创建之前被创建，因为每次创建 Pod 时会把当前所有的 Service 的环境变量都注入后面创建的 Pod 中，所以要先创建 Service，声明 Service 是为哪个 Pod 做代理的，再去创建 Pod，这样后面的 Pod 就具备了创建的 Service 的环境变量，否则环境变量不会被设置到 Pod 中。(2) Pod 只能获取同一个 NameSpace 中的 Service 环境变量。如果 Service 和 Pod 跨命名空间，则无法获取环境变量。

2. KubeDNS

在集群中运行 DNS 服务时，创建的每个 Service 都有 DNS 记录，所有的 Pod 都可以通过 DNS 名称解析相应的 Service IP。所以，在实际的生产环境中，一般都会采用 DNS。

Kube-dns 的工作原理如图 12-21 所示。在 Kubernetes 集群中使用 kube-dns 服务，可以在 DNS 域中进行记录的增加和查找，可通过集群的 API 获取相关的 Service。Dnsmasq 组件用于进行 DNS 缓存，当 Pod 有查询操作时，就会在 Dnsmasq 中做缓存，提高查询的速度。

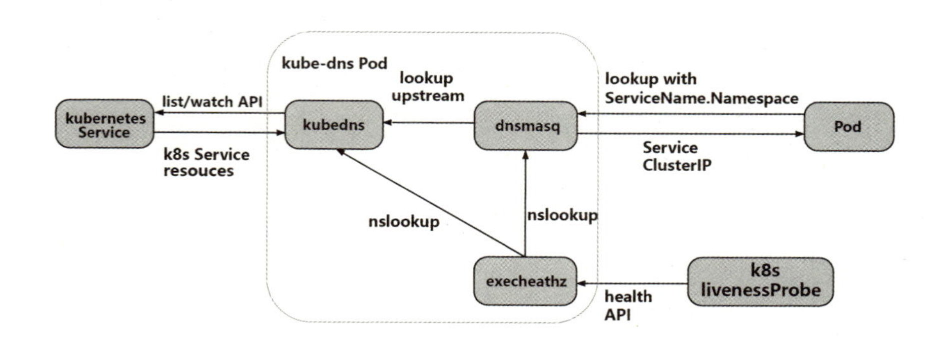

图 12-21

使用 DNS 需要安装插件，插件可在 GitLab 中下载，然后上传至 Master 节点服务器，最后使用 kubectl create 命令创建 DNS 服务，命令如下。

```
[root@master ~]# vim kube-dns.yaml
[root@master ~]# kubectl create -f kube-dns.yaml
service "kube-dns" created
serviceaccount "kube-dns" created
configmap "kube-dns" created
deployment "kube-dns" created
```

查看 kube-dns 服务，因为将 DNS 服务放在了 kube-system 命名空间下，所以还需要加-n 选项指定命名空间，如图 12-22 所示。如果不指定命名空间，则在默认的 default 中查看不到。图 12-22 中的 Cluster-IP 是指定的 IP，若想设置 IP，可在 YAML 文件中加上“clusterIP：字段”，如图 12-23 所示。

```
[root@master ~]# kubectl get all -n kube-system
NAME                                           READY     STATUS    RESTARTS   AGE
pod/kube-dns-774db77856-wjzd7                  3/3       Running   0          1m
pod/kubernetes-dashboard-7f8c9bff96-f6llr      1/1       Running   3          4h

NAME                           TYPE        CLUSTER-IP    EXTERNAL-IP   PORT(S)         AGE
service/kube-dns               ClusterIP   10.10.10.2    <none>        53/UDP,53/TCP   1m
service/kubernetes-dashboard   NodePort    10.10.10.63   <none>        80:35765/TCP    4h

NAME                                   DESIRED   CURRENT   UP-TO-DATE   AVAILABLE   AGE
deployment.apps/kube-dns               1         1         1            1           1m
deployment.apps/kubernetes-dashboard   1         1         1            1           4h

NAME                                              DESIRED   CURRENT   READY     AGE
replicaset.apps/kube-dns-774db77856               1         1         1         1m
replicaset.apps/kubernetes-dashboard-7f8c9bff96   1         1         1         4h
[root@master ~]#
```

图 12-22

```
spec:
  selector:
    k8s-app: kube-dns
  clusterIP: 10.10.10.2
  ports:
  - name: dns
    port: 53
    protocol: UDP
  - name: dns-tcp
    port: 53
    protocol: TCP
```

图 12-23

创建一个容器，做 DNS 解析测试，使用 kubectl get all 命令查看所有资源对象，然后用 busybox 做解析，命令如下。

```
[root@master ~]# kubectl exec -ti busybox -- nslookup kubernetes.default
Server:    10.10.10.2
Address 1: 10.10.10.2 kube-dns.kube-system.svc.cluster.local

Name:      kubernetes.default
Address 1: 10.10.10.1 kubernetes.default.svc.cluster.local
```

在容器中解析其他的 Service 服务也是可以的，命令如下。

```
[root@master ~]# kubectl exec -ti busybox -- nslookup
```

```
nginx-service.default
    Server:    10.10.10.2
    Address 1: 10.10.10.2 kube-dns.kube-system.svc.cluster.local

    Name:      nginx-service.default
    Address 1: 10.10.10.87 nginx-service.default.svc.cluster.local
```

12.3.4　发布服务

发布服务的类型有：ClusterIP、NodePort 和 LoadBalancer。

- ClusterIP 是分配的一个内部集群的 IP 地址，只能在集群内部访问，同 NameSpace 内的 Pod，默认是 ServiceType。
- NodePort 是分配的一个内部集群的 IP 地址，在每个节点上启用一个端口来暴露服务，可以在集群外部进行访问。访问地址是“<NodeIP>:<NodePort>”。
- LoadBalancer 是分配的一个内部集群的 IP 地址，在每个节点上启用一个端口来暴露服务，除此之外，Kubernetes 会请求底层云平台上的负载均衡器，将每个节点（[NodeIP]:[NodePort]）作为后端添加进去。

下面通过 NodePort 来实现外部 Service 的访问，删除原有的 nginx-service，命令如下。

```
[root@master ~]# kubectl get svc
NAME           TYPE        CLUSTER-IP    EXTERNAL-IP   PORT(S)          AGE
kubernetes     ClusterIP   10.10.10.1    <none>        443/TCP          4h
my-service     ClusterIP   10.10.10.59   <none>        80/TCP,443/TCP   28m
nginx-service  NodePort    10.10.10.87   <none>        88:38157/TCP     1h
[root@master ~]# kubectl delete svc/nginx-service
service "nginx-service" deleted
```

创建一个 nginx-deployment 文件，代码如下。

```
[root@master ~]# vim nginx-deployment.yaml
apiVersion: apps/v1beta2
kind: Deployment
metadata:
  name: nginx-deployment
  namespace: default
  labels:
    web: nginx
spec:
  replicas: 3
```

```
  selector:
    matchLabels:
      app: nginx
  template:
    metadata:
      labels:
        app: nginx
    spec:
      containers:
      - name: nginx
        image: nginx:1.10
        ports:
        - containerPort: 80
[root@master ~]# kubectl create -f nginx-deployment.yaml
deployment "nginx-deployment" created
```

创建 nginx-service，代码如下。

```
[root@master ~]# vim nginx-service.yaml
apiVersion: v1
kind: Service
metadata:
  name: nginx-service
  labels:
    app: nginx
spec:
  ports:
  - port: 80
    targetPort: 80
  selector:
    app: nginx
  type: NodePort # 指定类型为NodePort，在外部暴露一个随机的端口。
[root@master ~]# kubectl create -f nginx-service.yaml
service "nginx-service" created
```

使用 kubectl get svc 命令可以查看刚才创建的 nginx-service 的类型是 NodePort，并分配一个 Cluster-IP，代码如下。

```
[root@master ~]# kubectl get svc
NAME            TYPE        CLUSTER-IP     EXTERNAL-IP  PORT(S)         AGE
kubernetes      ClusterIP   10.10.10.1     <none>       443/TCP         4h
my-service      ClusterIP   10.10.10.59    <none>       80/TCP,443/TCP  42m
nginx-service   NodePort    10.10.10.186   <none>       80:44722/TCP    1m
```

使用 curl 命令或 elinks 命令可以通过内部的 10.10.10.186:80 访问 Nginx 欢迎页面，也可以通过外部浏览器，以及 curl 和 elinks 命令加任意节点 IP 和 NodePort 端口号访问 Nginx 欢迎页面，如图 12-24 所示。

图 12-24

NodePort 端口范围是可以自定义的，在 apiserver 中可以看到，笔者自定义的端口范围是 30000～50000，如图 12-25 所示。当然，也可以指定 1000～65535 端口的范围，只要在/opt/kubernetes/cfg/kube-apiserver 文件中定义即可。

```
[root@master ~]# ps aux|grep kube-apiserver
root       2759  2.7  3.5 520152 287272 ?      Ssl  May01  12:47 /opt/kubernetes/bin/kube-apiserver --logtostderr=true --v=4 --et
cd-servers=https://192.168.10.5:2379,https://192.168.10.6:2379,https://192.168.10.7:2379,https://192.168.10.8:2379,https://192.168
.10.9:2379,https://192.168.10.10:2379 --insecure-bind-address=127.0.0.1 --bind-address=192.168.10.5 --insecure-port=8080 --secure-
port=6443 --advertise-address=192.168.10.5 --allow-privileged=true --service-cluster-ip-range=10.10.10.0/24 --admission-control=Na
mespaceLifecycle,LimitRanger,SecurityContextDeny,ServiceAccount,ResourceQuota,NodeRestriction --authorization-mode=RBAC,Node --kub
elet-https=true --enable-bootstrap-token-auth --token-auth-file=/opt/kubernetes/cfg/token.csv --service-node-port-range=30000-5000
0 --tls-cert-file=/opt/kubernetes/ssl/server.pem --tls-private-key-file=/opt/kubernetes/ssl/server-key.pem --client-ca-file=/opt/k
ubernetes/ssl/ca.pem --service-account-key-file=/opt/kubernetes/ssl/ca-key.pem --etcd-cafile=/opt/kubernetes/ssl/ca.pem --etcd-cer
tfile=/opt/kubernetes/ssl/server.pem --etcd-keyfile=/opt/kubernetes/ssl/server-key.pem
root       4700  0.0  0.0 112660   976 pts/1    S+   01:59   0:00 grep --color=auto kube-apiserver
[root@master ~]#
```

图 12-25

12.4　Ingress

Ingress 在企业中使用广泛，主要用于发布集群应用。如果用 Internet Service 发布服务，那么用户访问时直接到 Service，再到具体的 Pod 中；如果用 Internet Ingress 发布服务，那么用户访问时先到 Ingress，再从 Ingress 到 Service 具体关联的 Pod 中，如图 12-26 所示。

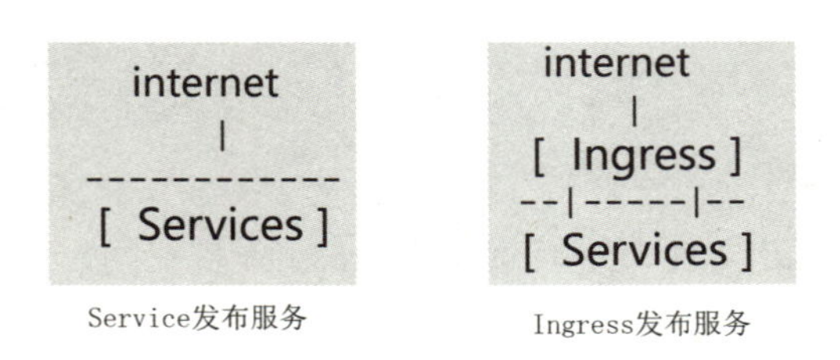

图 12-26

Ingress 在每个节点上创建一个负载均衡，直接的负载均衡代理服务所有的 Pod，这样效率明显更高，如图 12-27 所示。

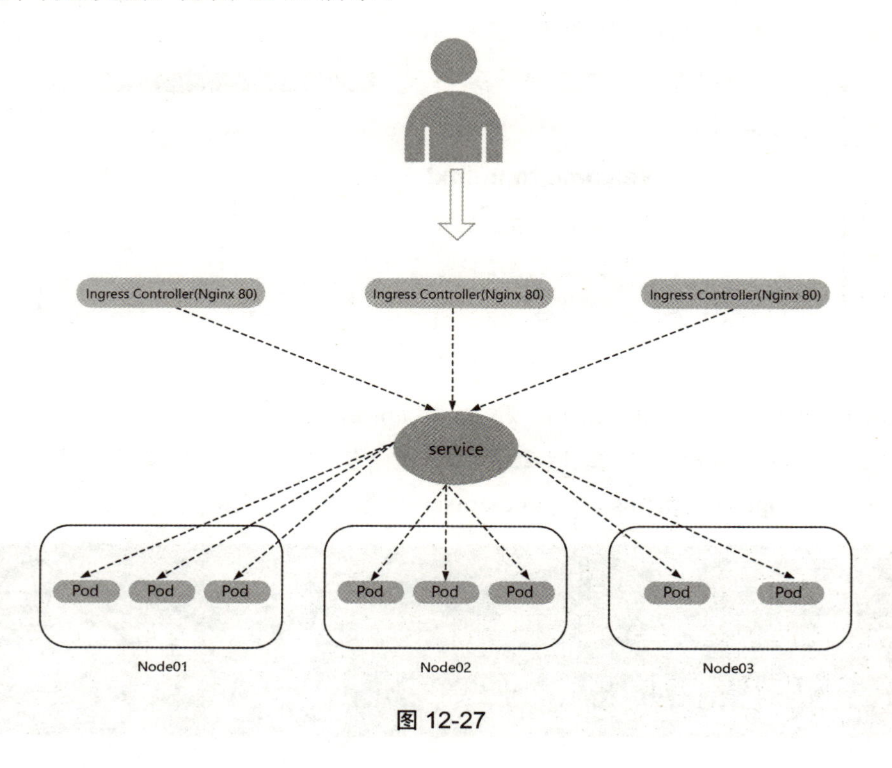

图 12-27

12.4.1 部署 Ingress

创建 Ingress 目录，从 GitLab 上下载 YAML 文件并上传至服务器，然后创建命名空间，命令如下。

```
[root@master ~]# mkdir ingress
[root@master ~]# cd ingress/
[root@master ingress]# ls
default-backend.yaml  deployment.yaml  namespace.yaml  rbac.yaml
tcp-services-configmap.yaml  udp-services-configmap.yaml
[root@master ingress]# kubectl create -f namespace.yaml
namespace "ingress-nginx" created
```

创建 default-backend 文件，然后查看 Ingress 命名空间，发现 default-backend 文件在 Ingress 命名空间中，命令如下。

```
[root@master ingress]# kubectl create -f default-backend.yaml
deployment "default-http-backend" created
```

```
service "default-http-backend" created
[root@master ingress]# kubectl get pods -n ingress-nginx
NAME                                    READY   STATUS    RESTARTS   AGE
default-http-backend-757754bc6b-pv5gv   1/1     Running   0          15s
```

创建 TCP 和 UDP 文件，命令如下。

```
[root@master ingress]# kubectl create -f tcp-services-configmap.yaml
configmap "tcp-services" created
[root@master ingress]# kubectl create -f udp-services-configmap.yaml
configmap "udp-services" created
```

注意

TCP 和 UDP 文件为空。

创建 RBAC 集群角色，该角色会创建一个 Server Account，命令如下。

```
[root@master ingress]# kubectl create -f rbac.yaml
serviceaccount "nginx-ingress-serviceaccount" created
clusterrole "nginx-ingress-clusterrole" created
role "nginx-ingress-role" created
rolebinding "nginx-ingress-role-nisa-binding" created
clusterrolebinding "nginx-ingress-clusterrole-nisa-binding" created
```

创建 deployment 文件来具体部署 Ingress 控制器，命令如下。

```
[root@master ingress]# kubectl create -f deployment.yaml
deployment "nginx-ingress-controller" created
```

全部创建完毕后，再次查看 Ingress 命名空间中的文件是否全部正常运行，如图 12-28 所示。

```
[root@master ingress]# kubectl get pods -n ingress-nginx
NAME                                       READY     STATUS    RESTARTS   AGE
default-http-backend-757754bc6b-pv5gv      1/1       Running   0          8m
nginx-ingress-controller-746b4d9b7-rl29b   0/1       Running   3          3m
[root@master ingress]#
```

图 12-28

12.4.2　HTTP 与 HTTPS 测试

创建 nginx 和 httpd 的 Deployment，命令如下。

```
[root@master ingress]# kubectl run --image=nginx nginx
deployment "nginx" created
```

```
[root@master ingress]# kubectl run --image=httpd httpd
deployment "httpd" created
```

查看 Pod 发现它们已经处于运行状态，如图 12-29 所示。

```
[root@master ingress]# kubectl get pods
NAME                     READY     STATUS    RESTARTS   AGE
busybox                  1/1       Running   0          57m
httpd-5db7c66c79-vlnrr   1/1       Running   0          1m
nginx-65899c769f-2wdrw   1/1       Running   0          1m
nginx-pod                1/1       Running   1          1h
[root@master ingress]# kubectl get pods -o wide
NAME                     READY     STATUS    RESTARTS   AGE       IP            NODE
busybox                  1/1       Running   0          57m       172.17.75.3   192.168.10.6
httpd-5db7c66c79-vlnrr   1/1       Running   0          1m        172.17.75.4   192.168.10.6
nginx-65899c769f-2wdrw   1/1       Running   0          1m        172.17.2.2    192.168.10.7
nginx-pod                1/1       Running   1          1h        172.17.84.3   192.168.10.9
[root@master ingress]#
```

图 12-29

给刚创建的 nginx 和 httpd 创建 Service 以暴露统一入口，让 Ingress 控制器获取 nginx 和 httpd 的 Pod，命令如下。

```
[root@master ingress]# kubectl expose deployment nginx --port=80
service "nginx" exposed
[root@master ingress]# kubectl expose deployment httpd --port=80
service "httpd" exposed
```

查看 httpd 和 nginx 的 Cluster-IP 地址，如图 12-30 所示。访问该地址，查看是否可以输出 Ingress Httpd 和 Ingress Nginx，如果可以输出，表示可以正常访问，如下所示。

```
[root@node-1 ~]# curl 10.10.10.154
Ingress Nginx
[root@node-1 ~]# curl 10.10.10.239
Ingress Httpd!
```

```
[root@master ingress]# kubectl get svc
NAME         TYPE        CLUSTER-IP     EXTERNAL-IP   PORT(S)          AGE
httpd        ClusterIP   10.10.10.239   <none>        80/TCP           18s
kubernetes   ClusterIP   10.10.10.1     <none>        443/TCP          5h
my-service   ClusterIP   10.10.10.59    <none>        80/TCP,443/TCP   1h
nginx        ClusterIP   10.10.10.154   <none>        80/TCP           26s
[root@master ingress]#
```

图 12-30

创建 Ingress 规则，让规则生成一段虚拟主机，代码如下。

```
[root@master ingress]# vim ingress-httpd.yaml
apiVersion: extensions/v1beta1
kind: Ingress
metadata:
  name: ingress-httpd
```

```
spec:
  rules:
  - host: nginx.linux.com   # 测试域名
    http:
      paths:
      - backend:
          serviceName: nginx
          servicePort: 80
  - host: httpd.linux.com   # 测试域名
    http:
      paths:
      - backend:
          serviceName: httpd
          servicePort: 80
[root@master ingress]# kubectl create -f ingress-httpd.yaml
ingress "ingress-httpd" created
```

查看刚创建的 Ingress 规则，命令如下。

```
[root@master ingress]# kubectl get ingress
NAME            HOSTS                            ADDRESS   PORTS    AGE
ingress-httpd   nginx.linux.com,httpd.linux.com            80       6s
```

在浏览器的地址栏中分别输入 nginx.linux.com 和 httpd.linux.com 两个域名，如果返回刚才自定义的信息，则表示设置成功，如图 12-31 所示。

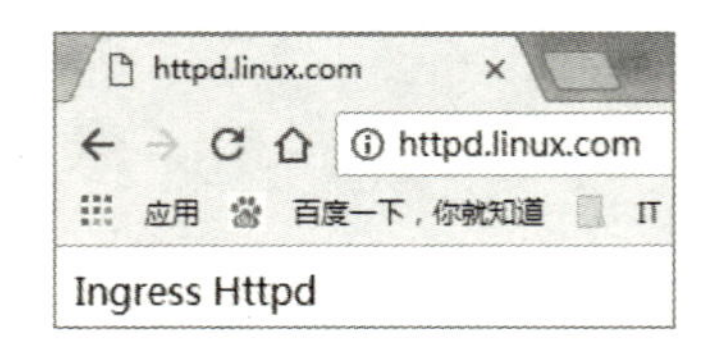

图 12-31

httpd.linux.com 和 nginx.linux.com 都只能在 192.168.10.7 节点上访问，若想在其他节点上访问，需要增加 Ingress 控制器到其他节点上的连接，命令如下。

```
    [root@master ingress]# kubectl scale deploy/nginx-ingress-controller
--replicas=5 -n ingress-nginx
    deployment.extensions "nginx-ingress-controller" scaled
    [root@master ingress]# kubectl get pods -n ingress-nginx
    [root@master ingress]# kubectl get pods -n ingress-nginx -o wide
```

进入 Nginx 容器中，通过 ps 命令查看进程，发现实际上运行的就是 nginx，如图 12-32 所示。

```
[root@master ingress]# kubectl exec -it nginx-ingress-controller-58dc5c46b9-dgb7w  bash -n ingress-nginx
root@node-2:/# ls
Dockerfile  etc                    lib64                     proc  sys
bin         home                   media                     root  tmp
boot        ingress-controller     mnt                       run   usr
build.sh    install_lua_resty_waf.sh  nginx-ingress-controller  sbin  var
dev         lib                    opt                       srv
root@node-2:/# ps aux
USER       PID %CPU %MEM    VSZ   RSS TTY      STAT START   TIME COMMAND
root         1  0.0  0.0   4040   352 ?        Ss   12:12   0:00 /usr/bin/dumb-
root         5  0.4  0.9  38384 18152 ?        Ssl  12:12   0:01 /nginx-ingress
root        17  0.0  1.7 110744 32756 ?        S    12:12   0:00 nginx: master
nobody      29  0.0  1.9 377116 36584 ?        Sl   12:12   0:00 nginx: worker
root        62  0.0  0.1  18188  2072 pts/0    Ss   12:18   0:00 bash
root        66  1.0  0.0  36624  1564 pts/0    R+   12:19   0:00 ps aux
root@node-2:/#
```

图 12-32

12.4.3 部署 Ingress TLS

创建 ca-csr 证书，运行命令如下。

```
[root@master ~]# mkdir https
[root@master ~]# cd https/
[root@master https]# cfssl print-defaults csr > ca-csr.json
[root@master https]# vim ca-csr.json
{
    "CN": "humingzhe",
    "key": {
        "algo": "rsa",
        "size": 2048
    },
    "names": [
        {
            "C": "CN",
            "L": "BeiJing",
            "ST": "BeiJing"
        }
    ]
}
```

创建 ca-config 证书，命令如下。

```
[root@master https]# cfssl print-defaults config > ca-config.json
```

生成证书，命令如下。

```
[root@master https]# cfssl gencert --initca ca-csr.json | cfssljson -bare
ca -
[root@master https]# ls
ca-config.json  ca.csr  ca-csr.json  ca-key.pem  ca.pem
```

为网站生成证书，命令如下。

```
[root@master https]# cfssl print-defaults csr > server-csr.json
[root@master https]# vim server-csr.json
{
    "CN": "www.humingzhe.cn",
    "key": {
        "algo": "rsa",
        "size": 2048
    },
    "names": [
        {
            "C": "CN",
            "L": "BeiJing",
            "ST": "BeiJing"
        }
    ]
}
```

生成 service-csr 证书，命令如下。

```
[root@master https]# cfssl gencert -ca=ca.pem -ca-key=ca-key.pem
--config=ca-config.json --profile=www server-csr.json | cfssljson -bare
server
[root@master https]# ls server*
server.csr  server-csr.json  server-key.pem  server.pem
```

将 ca 证书和 server 证书导入集群管理中，方便日常工作使用，命令如下。

```
[root@master https]# kubectl create secret tls humingzhe-https --key
server-key.pem --cert server.pem
secret "humingzhe-https" created
```

使用 kubectl get secret 命令进行查看，如图 12-33 所示。

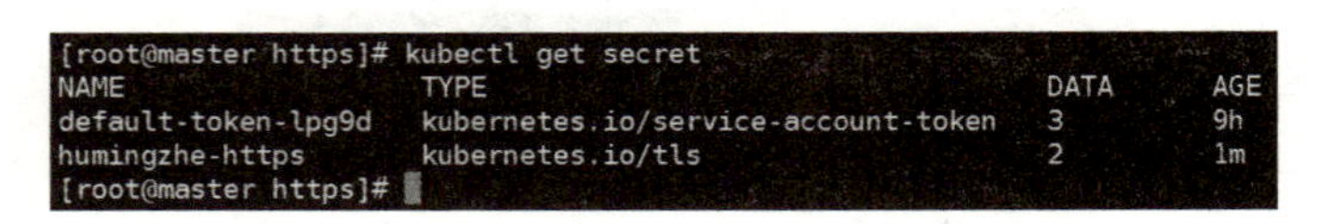

图 12-33

创建 https.yaml 文件，在文件中添加如下代码。

```
[root@master ingress]# vim https.yaml
apiVersion: extensions/v1beta1
kind: Ingress
metadata:
```

```
  name: https-test
spec:
  tls:
  - hosts:
    - www.humingzhe.cn
    secretName: humingzhe-https
  rules:
  - host: www.humingzhe.cn
    http:
      paths:
      - backend:
          serviceName: nginx
          servicePort: 80
```

创建 HTTPS Pod，命令如下。

```
[root@master ingress]# kubectl create -f https.yaml
ingress.extensions "https-test" created
```

创建好之后查看 Ingress，如图 12-34 所示。

```
[root@master ingress]# kubectl get ingress
NAME         HOSTS                                ADDRESS   PORTS     AGE
http-test    nginx.linux.com,httpd.linux.com                80        5s
https-test   www.humingzhe.cn                               80, 443   3m
[root@master ingress]#
```

图 12-34

在本地 hosts 文件中设置 IP 地址和需要访问的域名，命令如下。

```
C:\Windows\System32\drivers\etc\hosts
192.168.10.7    www.humingzhe.cn
```

打开浏览器，在地址栏中输入 www.humingzhe.cn，会访问 Ingress Nginx 页面，如图 12-35 所示。

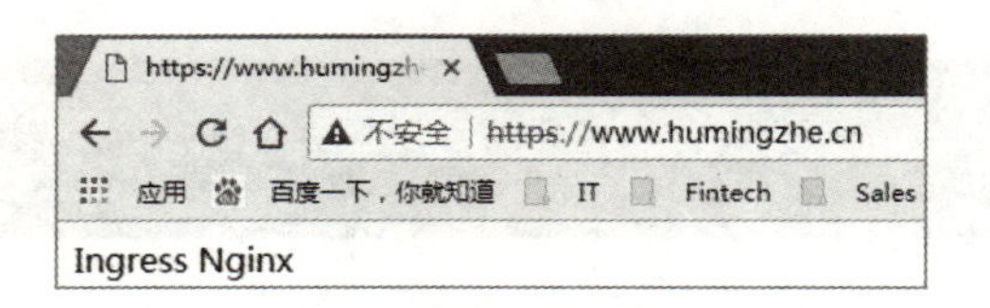

图 12-35

单击浏览器地址栏中显示的“不安全”这 3 个字，选择“证书”选项后即可在打开的对话框中看到“颁发给：www.humingzhe.cn”，如图 12-36 所示。

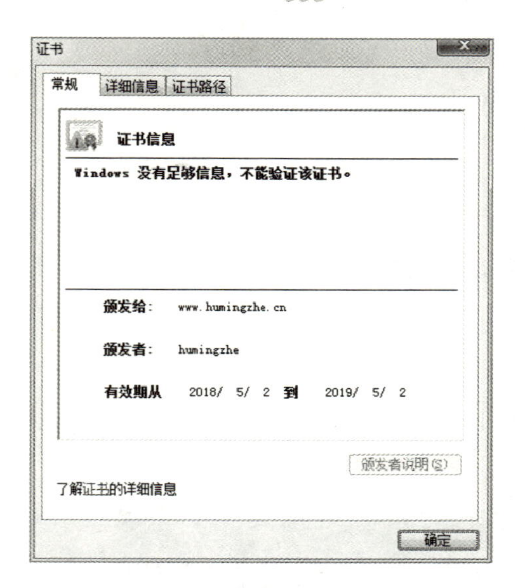

图 12-36

12.5　数据管理

Volume 对 Kubernetes 集群应用数据进行管理，有几个常见的类型，分别是 emptyDir、hostPath、NFS 和 GlusterFS。

12.5.1　emptyDir

emptyDir 表示空目录数据卷，相当于 Docker 自身的数据卷，会在宿主机上创建宿主机目录，然后挂载到容器中。

创建 emptyDir.yaml 文件，在文件中添加如下代码。

```
[root@master ~]# mkdir Volume
[root@master ~]# cd Volume/
[root@master Volume]# vim emptyDir.yaml
apiVersion: v1
kind: Pod
metadata:
  name: redis-pod
spec:
  containers:
  - image: redis
    name: redis
    volumeMounts:
```

```
    - mountPath: /cache
      name: cache-volume
  volumes:
  - name: cache-volume
    emptyDir: {}
```

创建 redis-pod 资源，命令如下。

```
[root@master Volume]# kubectl create -f emptyDir.yaml
pod "redis-pod" created
```

使用 kubectl get pods 命令可以查看刚创建的 redis-pod 是否处于运行状态，如图 12-37 所示。

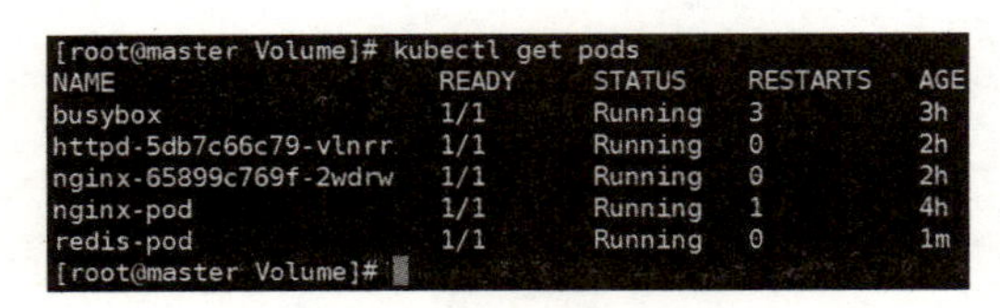

图 12-37

进入到 redis 中，命令如下。

```
[root@master Volume]# kubectl exec -it redis-pod bash
root@redis-pod:/data#
```

刚才挂载到了容器中的/cache 目录下，使用 mount 命令可以过滤出挂载的目录，命令如下。

```
root@redis-pod:/cache# mount |grep /cache
/dev/sda3 on /cache type xfs (rw,relatime,attr2,inode64,noquota)
```

使用 kubectl describe 命令可以查看分配到哪个节点上，如下代码表示分配到了 192.168.10.7 节点上。

```
[root@master Volume]# kubectl describe pods redis-pod
Name:         redis-pod
Namespace:    default
Node:         192.168.10.7/192.168.10.7
Start Time:   Fri, 04 May 2018 14:45:45 +0800
Labels:       <none>
Annotations:  <none>
Status:       Running
IP:           172.17.2.4
```

在 192.168.10.7 节点上执行 docker ps 命令查看 redis 的进程，如图 12-38 所示。执行 docker inspect dc4dd934edbb 命令会发现容器挂载了/cache 目录，如图 12-39 所示。

```
[root@node-2 ~]# docker ps
CONTAINER ID        IMAGE                                                    COMMAND                  CREATED             STATUS
      PORTS               NAMES
dc4dd934edbb        redis                                                    "docker-entrypoint.s…"   7 minutes ago       Up 7 minutes
                          k8s_redis_redis-pod_default_997fd1d0-4d8c-11e8-908f-000c292cf20b_0
```

图 12-38

```
{
    "Type": "bind",
    "Source": "/var/lib/kubelet/pods/997fd1d0-4d8c-11e8-908f-000c292cf20b/volumes/kubernetes.io~empty-dir/cache-volume",
    "Destination": "/cache",
    "Mode": "rslave",
    "RW": true,
    "Propagation": "rslave"
},
```

图 12-39

图 12-39 中的“/var/lib/kubelet/pods/997fd1d0-4d8c-11e8-908f-000c292cf20b/volumes/kubernetes.io~empty-dir/cache-volume”就是 emptyDir 在 host 上的真实路径。

emptyDir 是在主机上创建的临时目录，其优点是能够方便地为 Pod 中的容器提供共享存储，而不需要进行额外的配置。但它不具备持久性，如果 Pod 不存在了，emptyDir 也会随之删除。所以，emptyDir 适合 Pod 中的容器需要临时共享存储空间的场景。

12.5.2　hostPath

删除 emptyDir Pod，命令如下。

```
[root@master Volume]# kubectl delete -f emptyDir.yaml
pod "redis-pod" deleted
```

hostPath 是将宿主机上的目录文件挂载到容器中，创建 hostPath.yaml 文件，代码如下。

```
[root@master Volume]# vim hostPath.yaml
apiVersion: v1
kind: Pod
metadata:
  name: redis
spec:
  containers:
  - image: redis
    name: redis
    volumeMounts:
    - mountPath: /tmp-test #挂载到容器的哪个目录下
      name: test-volume
  volumes:
```

```
  - name: test-volume
    hostPath:
      path: /tmp
      type: Directory
```

创建 hostPath Pod，命令如下。

```
[root@master Volume]# kubectl create -f hostPath.yaml
pod "redis-pod" created
```

进入 redis-pod 容器中，命令如下。

```
[root@master Volume]# kubectl exec -it redis-pod bash
```

使用 ls 命令列出根目录下的所有目录，发现有一个 tmp-test 目录，这个目录就是刚刚在 YAML 文件中指定的目录，进入该目录，用 ls 命令列出该目录下的所有目录和文件，如图 12-40 所示。

```
root@redis-pod:/data# ls /
bin   data  etc   lib    media  opt   root  sbin  sys  tmp-test  var
boot  dev   home  lib64  mnt    proc  run   srv   tmp  usr
root@redis-pod:/data# cd /tmp-test/
root@redis-pod:/tmp-test# ls
ks-script-UJKyty
systemd-private-de07477238e845a18c7a22a4eaaa2805-chronyd.service-0xwup0
systemd-private-de07477238e845a18c7a22a4eaaa2805-vgauthd.service-1j1lh5
systemd-private-de07477238e845a18c7a22a4eaaa2805-vmtoolsd.service-FRdNjn
yum.log
```

图 12-40

查看 redis-pod 被分配到哪个节点上，如图 12-41 所示。可以看到，还是被分配到了 192.168.10.7 节点上。

```
[root@master Volume]# kubectl get pods -o wide
NAME                      READY   STATUS    RESTARTS   AGE   IP            NODE
busybox                   1/1     Running   3          3h    172.17.75.3   192.168.10.6
httpd-5db7c66c79-vlnrr    1/1     Running   0          2h    172.17.75.4   192.168.10.6
nginx-65899c769f-2wdrw    1/1     Running   0          2h    172.17.2.2    192.168.10.7
nginx-pod                 1/1     Running   1          4h    172.17.84.3   192.168.10.9
redis-pod                 1/1     Running   0          6m    172.17.2.4    192.168.10.7
[root@master Volume]#
```

图 12-41

查看 192.168.10.7 节点的/tmp 目录会发现，该目录和 redis-pod 容器中/tmp-test 目录所显示的内容一致，如图 12-42 所示。

```
[root@node-2 ~]# ll /tmp/
total 4
-rwx------. 1 root root 836 Apr 30 21:00 ks-script-UJKyty
drwx------  3 root root  17 May  1 14:15 systemd-private-de07477238e845a18c7a22a4eaaa2805-chronyd.service-0xwup0
drwx------  3 root root  17 May  1 14:15 systemd-private-de07477238e845a18c7a22a4eaaa2805-vgauthd.service-1j1lh5
drwx------  3 root root  17 May  1 14:15 systemd-private-de07477238e845a18c7a22a4eaaa2805-vmtoolsd.service-FRdNjn
-rw-------. 1 root root   0 Apr 30 20:50 yum.log
[root@node-2 ~]#
```

图 12-42

日常工作中可能将宿主机上的目录或文件都挂载到容器中，而这些文件和目录在每个节点上都要有，所以容器就起到了搜集信息的作用，这也是 hostPath 的主要应用场景。即使 Pod 被销毁了，hostPath 对应的目录依然存在，从这一点来看，hostPath 持久性要比 emptyDir 持久性好很多。但是，一旦主机崩溃，hostPath 自然也就无法进行访问了。

12.5.3 NFS

NFS 是网络存储，通过挂载去访问里面的资源。在 Lanmp 节点上安装 NFS，然后启动它，命令如下。

```
[root@lanmp ~]# yum install -y nfs-utils
[root@lanmp ~]# systemctl enable rpcbind.service && systemctl start rpcbind.service
[root@lanmp ~]# systemctl enable nfs-server.service && systemctl start nfs-server.service
[root@lanmp ~]# ps aux|grep nfs
```

编辑 NFS 主配置文件/etc/exports 并设置共享目录，命令如下。

```
[root@lanmp ~]# vim /etc/exports
/opt/nfs/data *(insecure,rw,no_root_squash)
```

创建/opt/nfs/data 目录，在该目录下创建几个文件，命令如下。

```
[root@lanmp ~]# mkdir /opt/nfs/data -p
[root@lanmp ~]# cd /opt/nfs/data/
[root@lanmp data]# ls
data.html  index.html
```

在 Master 节点测试是否可以挂载成功，如果可以挂载成功，则卸载，如图 12-43 所示。

```
[root@master Volume]# mount 192.168.10.10:/opt/nfs/data /mnt
[root@master Volume]# df -h
Filesystem                    Size  Used Avail Use% Mounted on
/dev/sda3                      44G  2.6G   42G   6% /
devtmpfs                      3.9G     0  3.9G   0% /dev
tmpfs                         3.9G     0  3.9G   0% /dev/shm
tmpfs                         3.9G  425M  3.4G  11% /run
tmpfs                         3.9G     0  3.9G   0% /sys/fs/cgroup
/dev/sda1                    1014M  135M  880M  14% /boot
tmpfs                         781M     0  781M   0% /run/user/0
192.168.10.10:/opt/nfs/data    47G  2.6G   45G   6% /mnt
[root@master Volume]# ls /mnt/
data.html  index.html
[root@master Volume]#
```

图 12-43

在 Master 节点上传 Nginx Deployment.yaml 文件，添加如下代码。

```
[root@master Volume]# vim nginx-deployment.yaml
apiVersion: extensions/v1beta1
kind: Deployment
metadata:
  name: nginx-deployment
spec:
  replicas: 5
  template:
    metadata:
      labels:
        app: nginx
    spec:
      containers:
      - name: nginx
        image: nginx
        volumeMounts:
        - name: www
          mountPath: /usr/share/nginx/html
        ports:
        - containerPort: 80
      volumes:
      - name: www
        nfs:
          server: 192.168.10.10
          path: /opt/nfs/data
```

创建 nginx-deployment 文件，命令如下。

```
[root@master Volume]# kubectl create -f nginx-deployment.yaml
deployment.extensions "nginx-deployment" created
```

创建成功后使用 kubectl get pods 命令查看副本是否处于运行状态，如图 12-44 所示，处于运行状态。

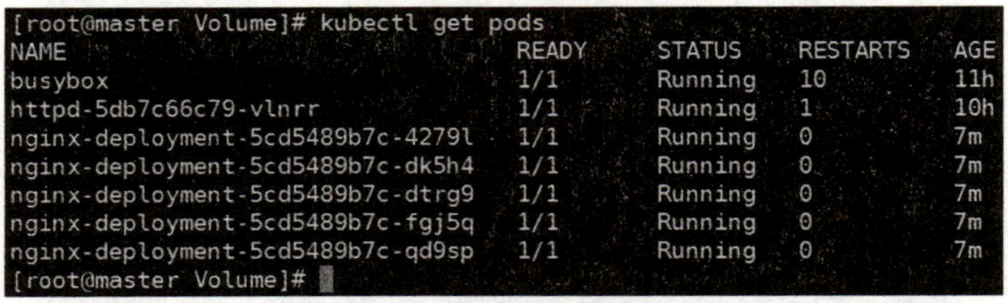

```
[root@master Volume]# kubectl get pods
NAME                                READY   STATUS    RESTARTS   AGE
busybox                             1/1     Running   10         11h
httpd-5db7c66c79-vlnrr              1/1     Running   1          10h
nginx-deployment-5cd5489b7c-4279l   1/1     Running   0          7m
nginx-deployment-5cd5489b7c-dk5h4   1/1     Running   0          7m
nginx-deployment-5cd5489b7c-dtrg9   1/1     Running   0          7m
nginx-deployment-5cd5489b7c-fgj5q   1/1     Running   0          7m
nginx-deployment-5cd5489b7c-qd9sp   1/1     Running   0          7m
[root@master Volume]#
```

图 12-44

创建 Service，使其可以进行访问，命令如下。

```
[root@master Volume]# cp ../nginx-service.yaml .
[root@master Volume]# vim nginx-service.yaml
[root@master Volume]# kubectl create -f nginx-service.yaml
service "nginx-service" created
```

查看 nginx-deployment Pod 的 IP 地址是否被 endpoint 代理，如图 12-45 所示。

```
[root@master Volume]# kubectl get pods -o wide
NAME                                READY     STATUS    RESTARTS   AGE       IP            NODE
busybox                             1/1       Running   10         11h       172.17.75.3   192.168.10.5
httpd-5db7c66c79-vlnrr              1/1       Running   1          10h       172.17.75.4   192.168.10.6
nginx-deployment-5cd5489b7c-4279l   1/1       Running   0          14m       172.17.75.5   192.168.10.6
nginx-deployment-5cd5489b7c-dk5h4   1/1       Running   0          14m       172.17.2.4    192.168.10.7
nginx-deployment-5cd5489b7c-dtrg9   1/1       Running   0          14m       172.17.37.3   192.168.10.8
nginx-deployment-5cd5489b7c-fgj5q   1/1       Running   0          14m       172.17.84.2   192.168.10.9
nginx-deployment-5cd5489b7c-qd9sp   1/1       Running   0          14m       172.17.2.2    192.168.10.7
[root@master Volume]# kubectl get ep
NAME            ENDPOINTS                                                AGE
httpd           172.17.75.4:80                                           10h
kubernetes      192.168.10.5:6443                                        15h
my-service      172.17.2.2:80,172.17.2.4:80,172.17.37.3:80 + 7 more...   11h
nginx-service   172.17.2.2:80,172.17.2.4:80,172.17.37.3:80 + 2 more...   59s
[root@master Volume]#
```

图 12-45

使用 kubectl get svc 命令查看 nginx-service 的 Cluster-IP 地址和内外部端口号，如图 12-46 所示。

```
[root@master Volume]# kubectl get svc
NAME            TYPE        CLUSTER-IP      EXTERNAL-IP   PORT(S)          AGE
httpd           ClusterIP   10.10.10.239    <none>        80/TCP           10h
kubernetes      ClusterIP   10.10.10.1      <none>        443/TCP          15h
my-service      ClusterIP   10.10.10.59     <none>        80/TCP,443/TCP   11h
nginx-service   NodePort    10.10.10.162    <none>        80:49594/TCP     3m
[root@master Volume]#
```

图 12-46

在任意节点使用 curl 命令加 Cluster-IP 地址就可以访问 index.html 页面中的内容，如图 12-47 所示。也可以使用外部浏览器进行访问，如图 12-48 所示。

```
[root@node-2 ~]# curl 10.10.10.162
Hello NFS Volume
[root@node-2 ~]# curl 10.10.10.162
Hello NFS Volume
```

图 12-47

图 12-48

12.5.4　GlusterFS

GlusterFS 是分布式存储，可以保证数据的可靠性，提高处理性能，而 NFS 很难满足这样的需求。所以，在企业生产环境中都使用分布式存储。GlusterFS 就是企业中使用的主流的分布式存储，在使用 GlusterFS 前需要进行部署。

部署准备环境如表 12-1 所示。

表 12-1

节　　点	备　　注
192.168.10.11 （lanmp）	关闭 Selinux 和 Firewalld 防火墙
192.168.10.12 （mysql）	

GlusterFS 部署参考文档的网址是 https://docs.gluster.org/en/latest/Quick-Start-Guide/Quickstart/。

GlusterFS 官方文档的描述是：至少有两个虚拟磁盘，一个用于操作系统安装，另一个用于服务 GlusterFS 存储（SDB）。

综上所述，在两个节点上需要关闭虚拟机并增加一块磁盘，然后打开虚拟机，使用 mkfs.xfs 命令进行磁盘分区格式化，如图 12-49 所示。

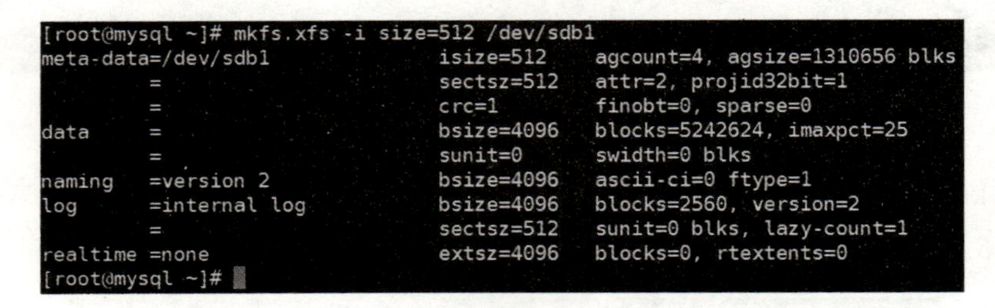

图 12-49

创建/data/brick1 目录，将/dev/sdb1 挂载到/data/brick1 目录中，命令如下。

```
[root@lanmp ~]# mkdir -p /data/brick1
[root@lanmp ~]# echo '/dev/sdb1 /data/brick1 xfs defaults 1 2' >>
/etc/fstab
[root@lanmp ~]# mount -a && mount
[root@lanmp ~]# df -h
Filesystem      Size  Used Avail Use% Mounted on
/dev/sda3        36G  4.1G   32G  12% /
devtmpfs        3.9G     0  3.9G   0% /dev
tmpfs           3.9G     0  3.9G   0% /dev/shm
tmpfs           3.9G  8.7M  3.9G   1% /run
tmpfs           3.9G     0  3.9G   0% /sys/fs/cgroup
/dev/sda1       497M  128M  370M  26% /boot
tmpfs           781M     0  781M   0% /run/user/0
/dev/sdb1        20G   33M   20G   1% /data/brick1
```

分别在两个节点上安装 GlusterFS 分布式存储，命令如下。

```
[root@lanmp ~]# yum install centos-release-gluster -y
[root@lanmp ~]# yum install glusterfs-server -y
```

安装完成后启动 Glusterd 服务，命令如下。

```
[root@lanmp ~]# systemctl enable glusterd.service && systemctl start glusterd.service
[root@lanmp ~]# ps aux|grep gluster
```

注意

启动成功后节点之间需要添加 iptables 规则，否则不能进行互信。当然，也可以选择清空 iptables 规则，或者将节点之间的 IP 添加到 iptables 规则中，使其允许双方访问。

编辑所有节点的/etc/hosts 配置文件，使节点之间可以互信，命令如下。

```
[root@mysql ~]# vim /etc/hosts
192.168.10.11 lanmp
192.168.10.12 mysql
```

设置可信池，使 lanmp 节点和 mysql 节点关联。简而言之，就是在 lanmp 节点关联 mysql 节点，并且在 mysql 节点关联 lanmp 节点，命令如下。

```
[root@lanmp ~]# gluster peer probe mysql
peer probe: success.
[root@lanmp ~]# gluster peer probe lanmp
peer probe: success.
```

注意

一旦建立了可信池，只有受信任的成员才可以探测到池中的新服务器。新服务器不能探测池，必须从池中进行探测。

使用 gluster peer status 命令可以查看节点的状态，如图 12-50 所示。

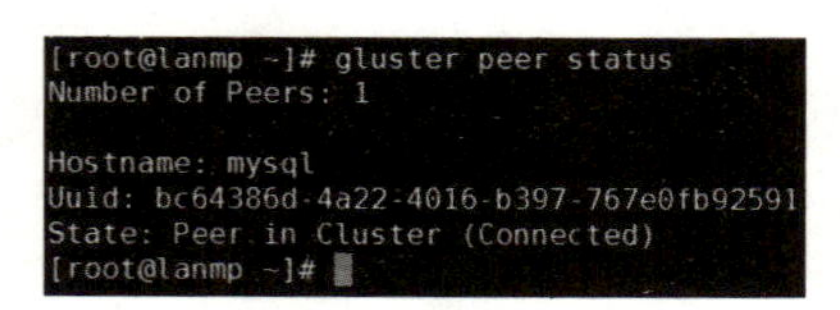

图 12-50

加入到集群中以后，就需要在 lanmp 节点和 mysql 节点上创建一个数据卷目录，命令如下。

```
[root@lanmp ~]# mkdir -p /data/brick1/gv0
[root@mysql ~]# mkdir -p /data/brick1/gv0
```

创建完数据卷目录后，可以在任意节点创建数据卷，然后启动，命令如下。

```
[root@lanmp ~]# gluster volume create gv0 replica 2
```

```
lanmp:/data/brick1/gv0 mysql:/data/brick1/gv0
    Do you still want to continue?
     (y/n) y
    [root@lanmp ~]# gluster volume start gv0
    volume start: gv0: success
```

数据卷启动成功后可以使用 gluster volume info 命令查看 Volume 的信息，如图 12-51 所示。

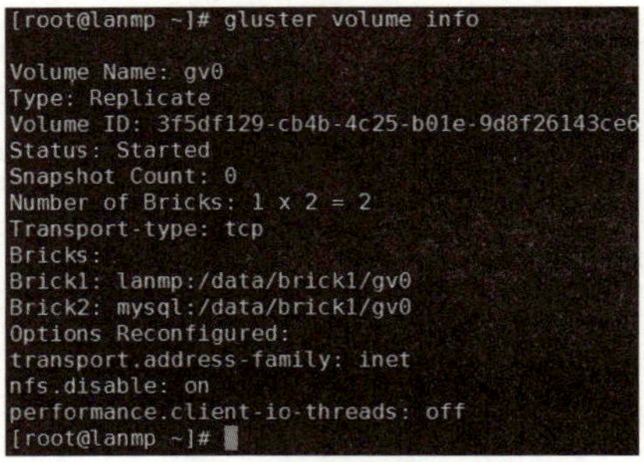

```
[root@lanmp ~]# gluster volume info

Volume Name: gv0
Type: Replicate
Volume ID: 3f5df129-cb4b-4c25-b01e-9d8f26143ce6
Status: Started
Snapshot Count: 0
Number of Bricks: 1 x 2 = 2
Transport-type: tcp
Bricks:
Brick1: lanmp:/data/brick1/gv0
Brick2: mysql:/data/brick1/gv0
Options Reconfigured:
transport.address-family: inet
nfs.disable: on
performance.client-io-threads: off
[root@lanmp ~]#
```

图 12-51

注意

如果数据卷启动失败，则可以通过服务器上的/var/log/glusterfs/glusterd.log 日志文件查看启动失败信息。

以上操作全部完成后就可以测试 GlusterFS 数据卷了，可以测试任意节点，但是必须要确保节点上存在 glusterfs-fuse。默认是不存在 glusterfs-fuse 的，需要使用 yum 进行安装，命令如下。

```
[root@node-3 ~]# yum install -y glusterfs-fuse
[root@node-3 ~]# mount -t glusterfs lanmp:/gv0 /mnt/
```

挂载成功后在/mnt 目录下新建一个 index.html 文件，在文件中写入 Hello GlusterFS，命令如下。

```
[root@node-3 ~]# vim /mnt/index.html
Hello GlusterFS
```

查看 glusterfs 节点上的/data/brick1/gv0/目录，会多出来一个 index.html 文件，文件中的内容和刚创建的 index.html 文件中的内容一致，命令如下。

```
[root@lanmp ~]# ls /data/brick1/gv0/
.glusterfs/ index.html
[root@lanmp ~]# cat /data/brick1/gv0/index.html
Hello GlusterFS
```

注意

在另一个节点上也有 index.html 文件，这是因为有两个副本，即使某个节点宕掉，还有一个节点在运行，供用户访问，这就是分布式存储的好处。

Kubernetes 集群有多个节点，可以利用 Kubernetes 的 Service 将多个节点整合到一起。创建 GlusterFS 目录，将事先准备好的配置文件上传至该目录，命令如下。

```
[root@master ~]# mkdir GlusterFS
[root@master ~]# cd GlusterFS/
[root@master GlusterFS]# rz -E
rz waiting to receive.
[root@master GlusterFS]# ls
glusterfs-endpoints.json  glusterfs-service.json
nginx-deployment.yaml
```

注意

配置文件可从 GitHub 上获取，网址是 https://github.com/kubernetes/kubernetes/tree/8fd414537b5143ab039cb910590237cabf4af783/examples/volumes/glusterfs。

修改 glusterfs-endpoints.json 配置文件中的 IP 地址为自己的 GlusterFS 节点的 IP 地址，然后使用 kubectl 命令进行创建，命令如下。

```
[root@master GlusterFS]# vim glusterfs-endpoints.json
[root@master GlusterFS]# kubectl create -f glusterfs-endpoints.json
endpoints "glusterfs-cluster" created
```

创建成功后使用 kubectl get ep 命令可以查看创建的 glusterfs-cluster 关联的 IP，如图 12-52 所示。

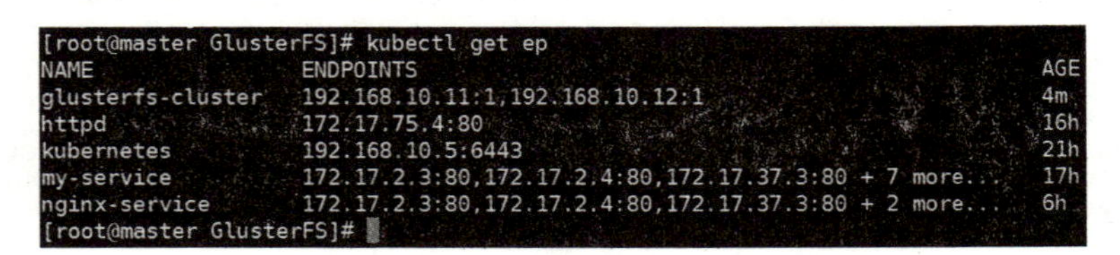

图 12-52

创建 glusterfs-service，命令如下。

```
[root@master GlusterFS]# kubectl create -f glusterfs-service.json
service "glusterfs-cluster" created
```

使用 kubectl get svc 命令可以查看 glusterfs-service 的 IP 地址，如图 12-53 所示。

```
[root@master GlusterFS]# kubectl get svc
NAME                TYPE        CLUSTER-IP     EXTERNAL-IP   PORT(S)          AGE
glusterfs-cluster   ClusterIP   10.10.10.250   <none>        1/TCP            20s
httpd               ClusterIP   10.10.10.239   <none>        80/TCP           16h
kubernetes          ClusterIP   10.10.10.1     <none>        443/TCP          21h
my-service          ClusterIP   10.10.10.59    <none>        80/TCP,443/TCP   17h
nginx-service       NodePort    10.10.10.162   <none>        80:49594/TCP     6h
[root@master GlusterFS]#
```

图 12-53

创建 nginx-deployment Pod，命令如下。

```
[root@master GlusterFS]# kubectl create -f nginx-deployment.yaml
deployment.extensions "nginx-deployment" created
service "nginx-service" created
```

创建完成后使用 kubectl get pods 命令查看 Pod 的运行状态，显示 Running，表示启动成功，如图 12-54 所示。

```
[root@master GlusterFS]# kubectl get pods
NAME                               READY     STATUS    RESTARTS   AGE
busybox                            1/1       Running   16         17h
httpd-5db7c66c79-vlnrr             1/1       Running   1          16h
nginx-deployment-d66f6c7fc-5wfbg   1/1       Running   0          16s
nginx-deployment-d66f6c7fc-8xdc6   1/1       Running   0          15s
nginx-deployment-d66f6c7fc-jlhcs   1/1       Running   0          16s
nginx-deployment-d66f6c7fc-mjsvc   1/1       Running   0          16s
nginx-deployment-d66f6c7fc-rr4nq   1/1       Running   0          15s
[root@master GlusterFS]#
```

图 12-54

使用 kubectl get svc 命令可以查看 nginx-service 暴露在外部的端口号，如图 12-55 所示。

```
[root@master GlusterFS]# kubectl get svc
NAME                TYPE        CLUSTER-IP     EXTERNAL-IP   PORT(S)          AGE
glusterfs-cluster   ClusterIP   10.10.10.250   <none>        1/TCP            25m
httpd               ClusterIP   10.10.10.239   <none>        80/TCP           16h
kubernetes          ClusterIP   10.10.10.1     <none>        443/TCP          22h
my-service          ClusterIP   10.10.10.59    <none>        80/TCP,443/TCP   18h
nginx-service       NodePort    10.10.10.88    <none>        80:37445/TCP     4m
[root@master GlusterFS]#
```

图 12-55

打开外部浏览器，在地址栏中输入任意节点的 IP 地址和 nginx-service 暴露在外部的端口号，就可以访问 Hello GlusterFS，如图 12-56 所示。

图 12-56

进入到 Nginx 的 Pod 中也可以查看挂载的 GlusterFS，如图 12-57 所示。

```
[root@master GlusterFS]# kubectl exec -it nginx-deployment-d66f6c7fc-mjsvc bash
root@nginx-deployment-d66f6c7fc-mjsvc:/# mount |grep glusterfs
192.168.10.11:gv0 on /usr/share/nginx/html type fuse.glusterfs (rw,relatime,user_id=0,group_id=0,default_permissions,allow_other,max_read=131072)
root@nginx-deployment-d66f6c7fc-mjsvc:/# cat /usr/share/nginx/html/index.html
Hello GlusterFS
root@nginx-deployment-d66f6c7fc-mjsvc:/#
```

图 12-57

前面讲过 GlusterFS 是分布式存储，即使其中一个节点宕掉，用户也不会受到任何影响。图 12-57 中显示挂载的是 192.168.10.11 节点，现在不妨把 192.168.10.11 节点的 GlusterFS 服务关闭，命令如下。

```
[root@lanmp ~]# systemctl stop glusterd.service
[root@lanmp ~]# ps aux|grep glusterd
[root@lanmp ~]# gluster peer status
Connection failed. Please check if gluster daemon is operational.
```

再次打开外部浏览器，访问任意节点，依然可以访问如图 12-58 所示的页面。

图 12-58

12.5.5　PersistentVolume

PersistentVolume 即持久化存储数据卷，是在企业生产环境中使用广泛的一种存储方式。PersistentVolume 和数据卷的区别在于，PersistentVolume 会在后端存储上做一定的抽象管理，这种抽象管理归属于集群调用，会将抽象管理作为集群的资源进行分配。

PersistentVolume 有两个概念，一个是 PersistentVolume（PV），另一个是 PersistentVolumeClaim（PVC）。PV 是对后端存储的一种抽象，后端可以是 NFS，也可以是 GlusterFS。PVC 会消费 PV，也就是消费后端的存储。将存储进行抽象作为集群的资源进行管理，那么就要创建 PVC 去消费 PV。有了这种抽象的概念，在使用过程中就不需要考虑后端是什么类型的存储，只要考虑如何使用 PVC 去消费 PV 的资源就可以了。

PersistentVolume 工作流程是：Pod 申请 PVC 作为卷来使用，集群通过 PVC 查找对应的 PV，最终挂载给 Pod。

PersistentVolume 支持的 PV 类型如表 12-2 所示。

表 12-2

PV 类型	PV 类型
GCEPersistentDisk	AWSElasticBlockStore
AzureFile	AzureDisk
FC（Fibre Channel）	FlexVolume
Flocker	NFS
iSCSI	RBD（Ceph Block Device）
CephFS	Cinder（OpenStack block storage）
Glusterfs	VsphereVolume
Quobyte Volumes	HostPath
VMware Photon	Portworx Volumes
ScaleIO Volumes	StorageOS

使用 NFS 存储时，需要在 YAML 文件中指定 NFS 的路径，而 GlusterFS 也需要指定路径。

1．创建 NFS-PV 存储

创建一个 NFS 的 PV 存储，来看一下它的格式与 NFS 和 GlusterFS 有何不同，代码如下。

```
[root@master PV]# vim nfs-pv.yaml
apiVersion: v1
kind: PersistentVolume
metadata:
 name: nfs-pv
spec:
 capacity:
  storage: 5Gi
 accessModes:
  - ReadWriteMany
 persistentVolumeReclaimPolicy: Recycle
 nfs:
  path: /opt/nfs/data
  server: 192.168.10.11
```

创建 nfs-pv 资源，创建完成后使用 kubectl get pv,pvc 命令查看资源的状态，如图 12-59 所示。

```
[root@master PV]# vim nfs-pv.yaml
[root@master PV]# kubectl create -f nfs-pv.yaml
persistentvolume "nfs-pv" created
[root@master PV]# kubectl get pv,pvc
NAME                     CAPACITY   ACCESS MODES   RECLAIM POLICY   STATUS      CLAIM    STORAGECLASS   REASON   AGE
persistentvolume/nfs-pv  5Gi        RWX            Recycle          Available                                  8s
[root@master PV]#
```

图 12-59

在图 12-59 中，名称是 nfs-pv，容量是 5GB，访问模式是 RWX，回收策略是 Recycle（自动回收），状态是 Available（可用，表示可以被新的 PVC 进行挂载）。如表 12-3 所示是 NFS-PV 存储的访问模式、回收策略和状态的类型。

表 12-3

类型名称	类　型
访问模式	ReadWriteOnce（RWO）、ReadOnlyMany（ROX）、ReadWriteMany（RWX）
回收策略	Retain、Recycle、Delete
状态	Available、Bound、Released、Failed

单纯地创建一个 PV 是不能直接使用的，因为需要通过 PVC 去消费 PV。创建 PVC，代码如下。

```
[root@master PV]# vim pvc.yaml
apiVersion: v1
kind: PersistentVolumeClaim
metadata:
  name: pvc-one
spec:
  accessModes:
    - ReadWriteMany
  resources:
    requests:
      storage: 5Gi
```

注意

PVC 是统一的，无需考虑后端是 NFS 还是 GlusterFS。PVC 与 PV 之间的绑定是通过存储容量和访问模式实现的。此处申请的是 5GB，会优先匹配 PV 容量刚好是 5GB 的。如果申请的是 3GB，PV 中分别有 5GB、10GB 和 20GB，同样也会优先匹配 5GB 的。也就是说，它会优先匹配与它数值最接近的一个。另外一个是访问模式，若 PVC 中的访问模式与 PV 中的创建模式一致，则会进行匹配。

创建 PVC 资源，创建后查看 PV 和 PVC 资源，如图 12-60 所示。从图中可以看到，状态变成了 Bound（绑定）。由此可见，它们二者之间已经进行了关联。

```
[root@master PV]# kubectl create -f pvc.yaml
persistentvolumeclaim "pvc-one" created
[root@master PV]# kubectl get pv,pvc
NAME                     CAPACITY   ACCESS MODES   RECLAIM POLICY   STATUS   CLAIM             STORAGECLASS   REASON    AGE
persistentvolume/nfs-pv  5Gi        RWX            Recycle          Bound    default/pvc-one                            23m

NAME                               STATUS   VOLUME   CAPACITY   ACCESS MODES   STORAGECLASS   AGE
persistentvolumeclaim/pvc-one      Bound    nfs-pv   5Gi        RWX                           55s
[root@master PV]#
```

图 12-60

此时，就可以创建应用去使用 PVC 了，代码如下。

```
[root@master PV]# vim pvc-app.yaml
apiVersion: v1
kind: Pod
metadata:
  name: pvc-pod
spec:
  containers:
    - name: nginx
      image: nginx
      volumeMounts:
      - mountPath: "/usr/share/nginx/html"
        name: wwwroot    # 名称是 wwwroot，挂载到 Nginx 下面
  volumes:
    - name: wwwroot
      persistentVolumeClaim:
        claimName: pvc-one # 明确指定要使用 pvc-one 这个 PV，使用 pvc-one 的 PV
就会使用 nfs-pv 作为存储
```

创建 pvc-app 资源，创建完成后使用 kubectl get pods 命令查看 pvc-pod 的运行状态，如图 12-61 所示。

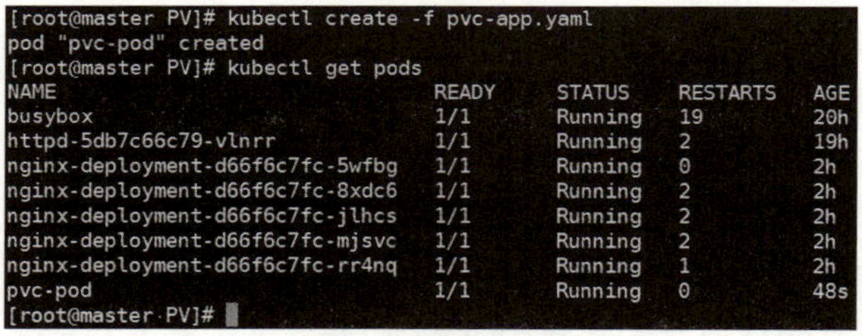

```
[root@master PV]# kubectl create -f pvc-app.yaml
pod "pvc-pod" created
[root@master PV]# kubectl get pods
NAME                                READY     STATUS    RESTARTS   AGE
busybox                             1/1       Running   19         20h
httpd-5db7c66c79-vlnrr              1/1       Running   2          19h
nginx-deployment-d66f6c7fc-5wfbg    1/1       Running   0          2h
nginx-deployment-d66f6c7fc-8xdc6    1/1       Running   2          2h
nginx-deployment-d66f6c7fc-jlhcs    1/1       Running   2          2h
nginx-deployment-d66f6c7fc-mjsvc    1/1       Running   2          2h
nginx-deployment-d66f6c7fc-rr4nq    1/1       Running   1          2h
pvc-pod                             1/1       Running   0          48s
[root@master PV]#
```

图 12-61

使用-o wide 命令查看 pvc-pod 的内网 IP 地址，获得 IP 地址后，在任意节点访问 pvc-app，如图 12-62 所示。图 12-62 显示的内容正是 NFS 存储上的内容，可以到 NFS 存储的节点上查看，如图 12-63 所示。也可以将 NFS 存储节点的 index.html 文件中的内容进行更改，更改后在任意节点进行访问，会发生变化，如图 12-64 所示。

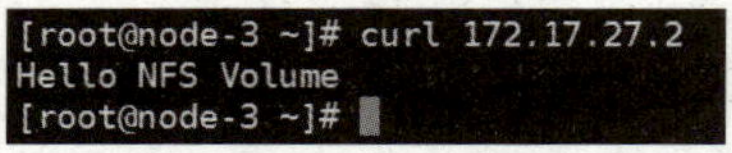

```
[root@node-3 ~]# curl 172.17.27.2
Hello NFS Volume
[root@node-3 ~]#
```

图 12-62

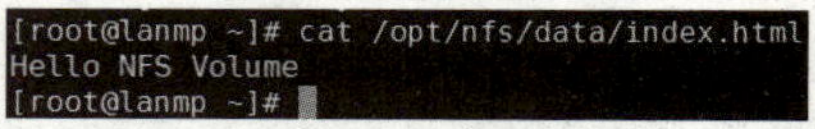

```
[root@lanmp ~]# cat /opt/nfs/data/index.html
Hello NFS Volume
[root@lanmp ~]#
```

图 12-63

```
[root@node-4 ~]# curl 172.17.27.2
Hello NFS Volume
Hello NFS Volume
Hello NFS Volume
Hello NFS Volume
[root@node-4 ~]#
```

图 12-64

2. 创建 GlusterFS-PV 存储

前面创建了 NFS 的 PV 存储，现在创建一个 GlusterFS 的 PV 存储，代码如下。

```
[root@master PV]# vim glusterfs-pv.yaml
apiVersion: v1
kind: PersistentVolume
metadata:
  name: gluster-pv
spec:
  capacity:
    storage: 10Gi
  accessModes:
    - ReadWriteMany
  glusterfs:
    endpoints: "glusterfs-cluster"
    path: "gv0"
    readOnly: false
```

创建 GlusterFS-PV 存储资源，命令如下。

```
[root@master PV]# kubectl create -f glusterfs-pv.yaml
persistentvolume "gluster-pv" created
```

创建完成后查看 PV 和 PVC，默认回收策略是删除，状态是可用，如图 12-65 所示。

```
[root@master ~]# kubectl get pv,pvc
NAME                          CAPACITY   ACCESS MODES   RECLAIM POLICY   STATUS      CLAIM             STORAGECLASS   REASON    AGE
persistentvolume/gluster-pv   10Gi       RWX            Retain           Available                                              1m
persistentvolume/nfs-pv       5Gi        RWX            Recycle          Bound       default/pvc-one                            1h

NAME                              STATUS    VOLUME    CAPACITY   ACCESS MODES   STORAGECLASS   AGE
persistentvolumeclaim/pvc-one     Bound     nfs-pv    5Gi        RWX                           1h
[root@master ~]#
```

图 12-65

删除前面创建的 pvc-app 应用，如图 12-66 所示。

```
[root@master PV]# kubectl delete persistentvolumeclaim/pvc-one
persistentvolumeclaim "pvc-one" deleted
[root@master PV]# vim pvc-app.yaml
[root@master PV]# kubectl get pv,pvc
NAME                          CAPACITY   ACCESS MODES   RECLAIM POLICY   STATUS      CLAIM     STORAGECLASS   REASON    AGE
persistentvolume/gluster-pv   10Gi       RWX            Retain           Available                                      4m
persistentvolume/nfs-pv       5Gi        RWX            Recycle          Available                                      11m
[root@master PV]#
```

图 12-66

创建一个 GlusterFS-PVC 和应用，代码如下。

```
[root@master PV]# vim pvc-app.yaml
apiVersion: v1
kind: PersistentVolumeClaim
metadata:
  name: pvc-one
spec:
  accessModes:
    - ReadWriteMany
  resources:
    requests:
      storage: 11Gi
---
apiVersion: v1
kind: Pod
metadata:
  name: pvc-pod
spec:
  containers:
    - name: nginx
      image: nginx
      volumeMounts:
      - mountPath: "/usr/share/nginx/html"
        name: wwwroot
  volumes:
    - name: wwwroot
      persistentVolumeClaim:
        claimName: pvc-one
```

创建一个 20GB 的 glusterfs-pv，命令如下。

```
[root@master PV]# vim glusterfs-pv.yaml
apiVersion: v1
kind: PersistentVolume
metadata:
  name: gluster-pv-two
spec:
  capacity:
    storage: 20Gi
  accessModes:
    - ReadWriteMany
  glusterfs:
    endpoints: "glusterfs-cluster"
```

```
    path: "gv0"
    readOnly: false
```

创建 gluster-pv-two 资源，命令如下。

```
[root@master PV]# kubectl create -f glusterfs-pv.yaml
persistentvolume "gluster-pv-two" created
```

查看 PV 和 PVC，如图 12-67 所示。

```
[root@master PV]# kubectl get pv,pvc
NAME                              CAPACITY   ACCESS MODES   RECLAIM POLICY   STATUS      CLAIM     STORAGECLASS   REASON    AGE
persistentvolume/gluster-pv       10Gi       RWX            Retain           Available                                      10m
persistentvolume/gluster-pv-two   20Gi       RWX            Retain           Available                                      56s
persistentvolume/nfs-pv           5Gi        RWX            Recycle          Available                                      2s
[root@master PV]#
```

图 12-67

创建 pvc-app 应用，命令如下。

```
[root@master PV]# kubectl create -f pvc-app.yaml
persistentvolumeclaim "pvc-one" created
pod "pvc-pod" created
```

创建完成后，查看 PV 和 PVC 资源，会默认绑定 20GB 的资源，如图 12-68 所示。

```
[root@master PV]# kubectl get pv,pvc
NAME                              CAPACITY   ACCESS MODES   RECLAIM POLICY   STATUS      CLAIM             STORAGECLASS   REASON    AGE
persistentvolume/gluster-pv       10Gi       RWX            Retain           Available                                              13m
persistentvolume/gluster-pv-two   20Gi       RWX            Retain           Bound       default/pvc-one                                3m
persistentvolume/nfs-pv           5Gi        RWX            Recycle          Available                                              2m

NAME                            STATUS    VOLUME           CAPACITY   ACCESS MODES   STORAGECLASS   AGE
persistentvolumeclaim/pvc-one   Bound     gluster-pv-two   20Gi       RWX                           11s
[root@master PV]#
```

图 12-68

前面申请的是 11GB，但是绑定的却是 20GB。为什么不选择绑定 10GB？这是因为遵循了尽可能往大容量方向去绑定的规则，不会往小容量方向去绑定。如果创建的是 9GB 或 8GB，则默认绑定 10GB，而不是 20GB。

绑定后查看 pvc-pod 的 IP 地址，如图 12-69 所示。

```
[root@master PV]# kubectl get pods -o wide
NAME                                READY     STATUS    RESTARTS   AGE       IP            NODE
busybox                             1/1       Running   33         21d       172.17.82.4   192.168.10.6
httpd-5db7c66c79-vlnrr              1/1       Running   8          21d       172.17.82.3   192.168.10.6
nginx-deployment-d66f6c7fc-56htg    1/1       Running   2          20d       172.17.82.6   192.168.10.6
nginx-deployment-d66f6c7fc-5wfbg    1/1       Running   3          20d       172.17.99.4   192.168.10.8
nginx-deployment-d66f6c7fc-6c9cz    1/1       Running   1          20d       172.17.81.2   192.168.10.10
nginx-deployment-d66f6c7fc-86h59    1/1       Running   1          20d       172.17.99.5   192.168.10.8
nginx-deployment-d66f6c7fc-rr4nq    1/1       Running   7          20d       172.17.82.5   192.168.10.6
pvc-pod                             1/1       Running   1          19m       172.17.27.2   192.168.10.7
[root@master PV]#
```

图 12-69

在任意节点访问 172.17.27.2，就可以输出如图 12-70 所示的结果。输出该结果说明是正常的。

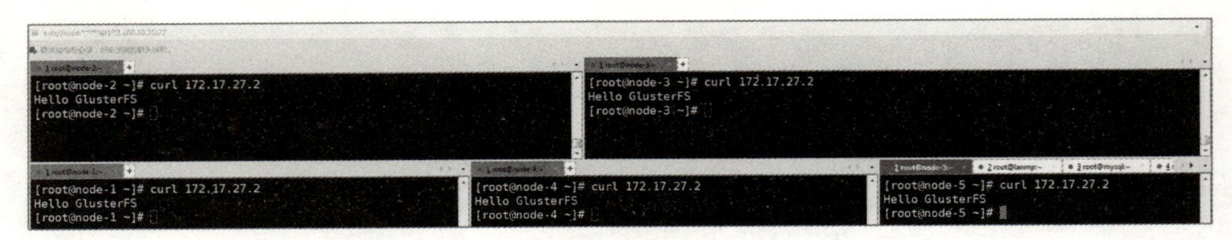
[root@node-2 ~]# curl 172.17.27.2
Hello GlusterFS
[root@node-2 ~]#
[root@node-3 ~]# curl 172.17.27.2
Hello GlusterFS
[root@node-3 ~]#
[root@node-1 ~]# curl 172.17.27.2
Hello GlusterFS
[root@node-1 ~]#
[root@node-4 ~]# curl 172.17.27.2
Hello GlusterFS
[root@node-4 ~]#
[root@node-5 ~]# curl 172.17.27.2
Hello GlusterFS
[root@node-5 ~]#

图 12-70

第 13 章 Kubernetes高可用架构和项目案例

13.1 Kubernetes Dashboard

为了方便后期维护与管理，可以在 Kubernetes 上安装 Web UI 插件来实现用 Web 界面管理集群资源。

在 Master 节点中创建 UI 目录，将模板文件下载到该目录中，命令如下。

```
[root@master ~]# mkdir UI
[root@master ~]# cd UI/
[root@master UI]# ls
dashboard-deployment.yaml  dashboard-rbac.yaml
dashboard-service.yaml
```

使用 kubectl 命令创建 dashboard-rbac.ymal 文件，命令如下。

```
[root@master UI]# kubectl create -f dashboard-rbac.yaml
serviceaccount "kubernetes-dashboard" created
clusterrolebinding "kubernetes-dashboard-minimal" created
```

dashboard-deployment.yaml 是用来创建 UI 应用的控制器，可以管理 Pod 副本。dashboard-service.yaml 可以将 UI 进行发布，达到用户访问的目的。使用 kubectl 命令分别创建 dashboard-deployment.yaml 和 dashboard-service.yaml，命令如下。

```
[root@master UI]# kubectl create -f dashboard-deployment.yaml
deployment "kubernetes-dashboard" created
[root@master UI]# kubectl create -f dashboard-service.yaml
service "kubernetes-dashboard" created
```

创建完毕后可以在客户端使用 kubectl get all -n kube-system 命令查看部署情况，如图 13-1 所示，类型是 NodePort，Cluster-IP 是 10.10.10.63，内部端口是 80，外部端口是 35765。

```
[root@master UI]# kubectl get all -n kube-system
NAME                                        READY     STATUS    RESTARTS   AGE
pod/kubernetes-dashboard-7f8c9bff96-f6llr   1/1       Running   0          45s

NAME                           TYPE       CLUSTER-IP    EXTERNAL-IP   PORT(S)        AGE
service/kubernetes-dashboard   NodePort   10.10.10.63   <none>        80:35765/TCP   38s

NAME                                   DESIRED   CURRENT   UP-TO-DATE   AVAILABLE   AGE
deployment.apps/kubernetes-dashboard   1         1         1            1           45s

NAME                                              DESIRED   CURRENT   READY     AGE
replicaset.apps/kubernetes-dashboard-7f8c9bff96   1         1         1         45s
[root@master UI]#
```

图 13-1

使用 kubectl get ns 命令可以查看有多少个命名空间，如图 13-2 所示。

```
[root@master PV]# kubectl get ns
NAME            STATUS    AGE
default         Active    9d
ingress-nginx   Active    8d
kube-public     Active    9d
kube-system     Active    9d
[root@master PV]#
```

图 13-2

在浏览器的地址栏中输入节点 IP 和暴露在外部的随机端口 35765，即可访问 Kubernetes Web 管理界面，如图 13-3 所示。当访问 35765 端口时，会将请求自动转发到具体提供服务的容器中。Kubernetes UI 界面非常简陋，在日常工作中使用 Kubernetes UI 的概率较小。

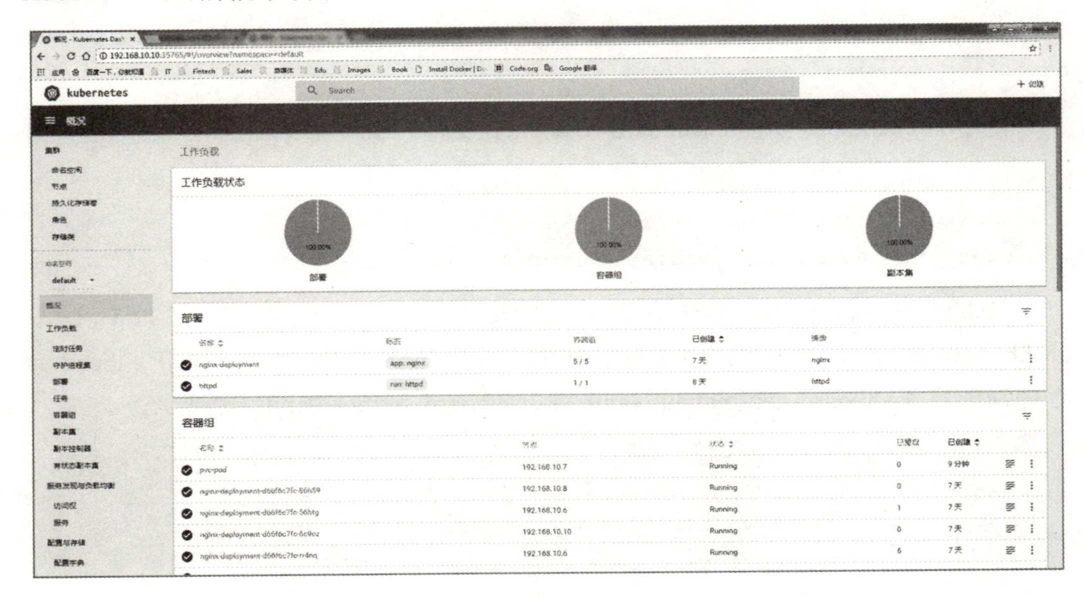

图 13-3

13.2　部署集群应用

创建一个 lnmp 目录，然后进入到该目录中，将 lnmp 环境放到 lnmp 目录下，以便部署集群 Web 应用，命令如下。

```
[root@master ~]# mkdir lnmp
[root@master ~]# cd lnmp/
[root@master lnmp]# ls
mysql-deployment.yaml  nginx-deployment.yaml  php-deployment.yaml
pv.yaml
```

创建三个 PV，分别是 mysql-pv、wp-pv-one 和 wp-pv-two。mysql-pv 是 20GB，用来存放 MySQL 数据库；wp-pv-one 和 wp-pv-two 均是 5GB，用来存放 WordPress 程序。mysql-pv 使用的是 GlusterFS 存储，wp-pv-one 和 wp-pv-two 使用的是 NFS 存储。具体详见 pv.yaml 文件。

将前面 5GB 的 nfs-pv 删除以避免混淆，如图 13-4 所示。

```
[root@master lnmp]# kubectl get pv,pvc
NAME                                  CAPACITY   ACCESS MODES   RECLAIM POLICY   STATUS      CLAIM             STORAGECLASS   REASON    AGE
persistentvolume/gluster-pv           10Gi       RWX            Retain           Available                                              19m
persistentvolume/gluster-pv-two       20Gi       RWX            Retain           Bound       default/pvc-one                            9m
persistentvolume/nfs-pv               5Gi        RWX            Recycle          Available                                              8m

NAME                                STATUS    VOLUME           CAPACITY   ACCESS MODES   STORAGECLASS   AGE
persistentvolumeclaim/pvc-one       Bound     gluster-pv-two   20Gi       RWX                           6m
[root@master lnmp]# kubectl delete persistentvolume/nfs-pv
persistentvolume "nfs-pv" deleted
[root@master lnmp]# kubectl get pv,pvc
NAME                                  CAPACITY   ACCESS MODES   RECLAIM POLICY   STATUS      CLAIM             STORAGECLASS   REASON    AGE
persistentvolume/gluster-pv           10Gi       RWX            Retain           Available                                              19m
persistentvolume/gluster-pv-two       20Gi       RWX            Retain           Bound       default/pvc-one                            9m

NAME                                STATUS    VOLUME           CAPACITY   ACCESS MODES   STORAGECLASS   AGE
persistentvolumeclaim/pvc-one       Bound     gluster-pv-two   20Gi       RWX                           6m
[root@master lnmp]#
```

图 13-4

创建 pv.yaml 文件中的三个 PV，命令如下。

```
[root@master lnmp]# kubectl create -f pv.yaml
persistentvolume "mysql-pv" created
persistentvolume "wp-pv-one" created
persistentvolume "wp-pv-two" created
```

创建完成后查看 PV 的状态是否为可用，如果为可用状态，则表示正常，如图 13-5 所示。

```
[root@master lnmp]# kubectl get pv,pvc
NAME                                  CAPACITY   ACCESS MODES   RECLAIM POLICY   STATUS      CLAIM             STORAGECLASS   REASON    AGE
persistentvolume/gluster-pv           10Gi       RWX            Retain           Available                                              22m
persistentvolume/gluster-pv-two       20Gi       RWX            Retain           Bound       default/pvc-one                            12m
persistentvolume/mysql-pv             20Gi       RWX            Retain           Available                                              9s
persistentvolume/wp-pv-one            5Gi        RWX            Retain           Available                                              9s
persistentvolume/wp-pv-two            5Gi        RWX            Retain           Available                                              9s

NAME                                STATUS    VOLUME           CAPACITY   ACCESS MODES   STORAGECLASS   AGE
persistentvolumeclaim/pvc-one       Bound     gluster-pv-two   20Gi       RWX                           9m
[root@master lnmp]#
```

图 13-5

上面创建的三个 PV 分配如下：MySQL 使用 20GB 的 mysql-pv，PHP 使用 5GB 的 wp-pv-one，Nginx 使用 5GB 的 wp-pv-two。

在 mysql-deployment.yaml 文件中先创建一个 Service，主要是让其他服务连接 WordPress，如图 13-6 所示。下面还要创建 PVC，因为 PV 已经创建好了，如图 13-7 所示。接着创建一个 Deployment，在 Deployment 中传入一个变量给 MySQL，使用 MySQL 的官方镜像，传入一个 MySQL 的 ROOT 密码，ROOT 密码存储在 secret 中，创建好之后直接引用该密码即可。下面处理 MySQL 数据库的端口和数据卷，挂载的是 mysql-pv-claim 的 PVC，如图 13-8 所示。

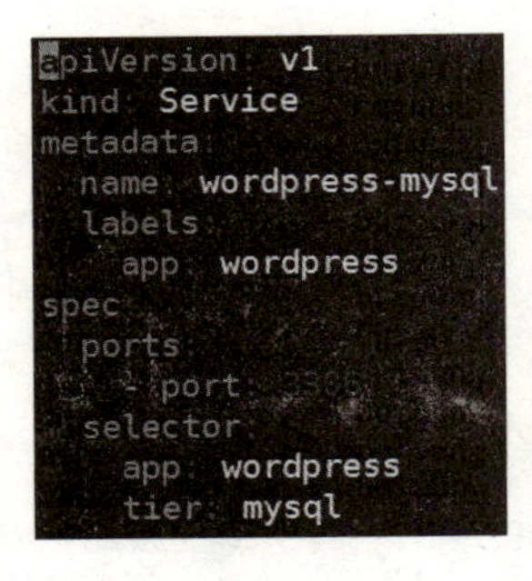

```
apiVersion: v1
kind: Service
metadata:
  name: wordpress-mysql
  labels:
    app: wordpress
spec:
  ports:
    - port: 3306
  selector:
    app: wordpress
    tier: mysql
```

图 13-6

```
apiVersion: v1
kind: PersistentVolumeClaim
metadata:
  name: mysql-pv-claim
  labels:
    app: wordpress
spec:
  accessModes:
    - ReadWriteMany
  resources:
    requests:
      storage: 20Gi
```

图 13-7

```
spec:
  containers:
  - name: mysql
    image: mysql:5.6
    env:
    - name: MYSQL_ROOT_PASSWORD
      valueFrom:
        secretKeyRef:
          name: mysql-pass
          key: password
    ports:
    - containerPort: 3306
      name: mysql
    volumeMounts:
    - name: mysql-persistent-storage
      mountPath: /var/lib/mysql
  volumes:
  - name: mysql-persistent-storage
    persistentVolumeClaim:
      claimName: mysql-pv-claim
```

图 13-8

创建 secret 的 MySQL 数据库密码，用来引用 MySQL 数据库，命令如下。

```
[root@master lnmp]# kubectl create secret generic mysql-pass
--from-literal=password=123456
secret "mysql-pass" created
```

密码创建完成后部署 MySQL 数据库，命令如下。

```
[root@master lnmp]# kubectl create -f mysql-deployment.yaml
```

```
service "wordpress-mysql" created
persistentvolumeclaim "mysql-pv-claim" created
deployment.apps "wordpress-mysql" created
```

部署完 MySQL 数据库后，查看 MySQL 的运行状态，如图 13-9 所示。

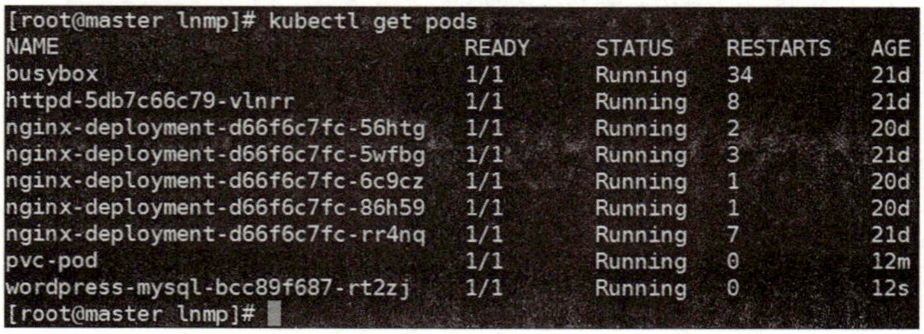

```
[root@master lnmp]# kubectl get pods
NAME                                 READY     STATUS    RESTARTS   AGE
busybox                              1/1       Running   34         21d
httpd-5db7c66c79-vlnrr               1/1       Running   8          21d
nginx-deployment-d66f6c7fc-56htg     1/1       Running   2          20d
nginx-deployment-d66f6c7fc-5wfbg     1/1       Running   3          21d
nginx-deployment-d66f6c7fc-6c9cz     1/1       Running   1          20d
nginx-deployment-d66f6c7fc-86h59     1/1       Running   1          20d
nginx-deployment-d66f6c7fc-rr4nq     1/1       Running   7          21d
pvc-pod                              1/1       Running   0          12m
wordpress-mysql-bcc89f687-rt2zj      1/1       Running   0          12s
[root@master lnmp]#
```

图 13-9

部署 PHP 服务，命令如下。

```
[root@master lnmp]# kubectl create -f php-deployment.yaml
service "wordpress-php" created
persistentvolumeclaim "wp-pvc-two" created
deployment.apps "wordpress-php" created
```

部署完 PHP 服务后，查看 PHP 服务 Pod 的运行状态，如图 13-10 所示。

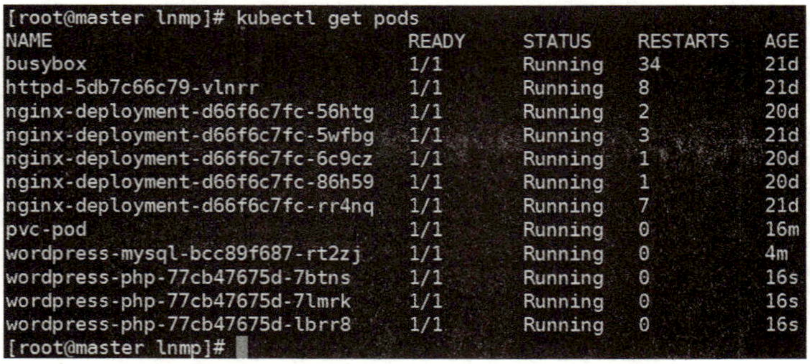

```
[root@master lnmp]# kubectl get pods
NAME                                 READY     STATUS    RESTARTS   AGE
busybox                              1/1       Running   34         21d
httpd-5db7c66c79-vlnrr               1/1       Running   8          21d
nginx-deployment-d66f6c7fc-56htg     1/1       Running   2          20d
nginx-deployment-d66f6c7fc-5wfbg     1/1       Running   3          21d
nginx-deployment-d66f6c7fc-6c9cz     1/1       Running   1          20d
nginx-deployment-d66f6c7fc-86h59     1/1       Running   1          20d
nginx-deployment-d66f6c7fc-rr4nq     1/1       Running   7          21d
pvc-pod                              1/1       Running   0          16m
wordpress-mysql-bcc89f687-rt2zj      1/1       Running   0          4m
wordpress-php-77cb47675d-7btns       1/1       Running   0          16s
wordpress-php-77cb47675d-7lmrk       1/1       Running   0          16s
wordpress-php-77cb47675d-lbrr8       1/1       Running   0          16s
[root@master lnmp]#
```

图 13-10

部署 Nginx 服务，命令如下。

```
[root@master lnmp]# kubectl create -f nginx-deployment.yaml
configmap "nginx-wp-config" created
service "wordpress-nginx" created
persistentvolumeclaim "wp-pv-one" created
deployment.apps "wordpress-nginx" created
```

查看 Nginx 服务 Pod 的运行状态，如图 13-11 所示。

```
[root@master lnmp]# kubectl get pods
NAME                                READY     STATUS    RESTARTS   AGE
busybox                             1/1       Running   35         21d
httpd-5db7c66c79-vlnrr              1/1       Running   8          21d
nginx-deployment-d66f6c7fc-56htg    1/1       Running   2          20d
nginx-deployment-d66f6c7fc-5wfbg    1/1       Running   3          21d
nginx-deployment-d66f6c7fc-6c9cz    1/1       Running   1          20d
nginx-deployment-d66f6c7fc-86h59    1/1       Running   1          20d
nginx-deployment-d66f6c7fc-rr4nq    1/1       Running   7          21d
pvc-pod                             1/1       Running   0          1h
wordpress-mysql-bcc89f687-rt2zj     1/1       Running   0          1h
wordpress-nginx-cb9fc4765-chrbt     1/1       Running   0          1m
wordpress-nginx-cb9fc4765-klbrg     1/1       Running   0          1m
wordpress-nginx-cb9fc4765-wbswx     1/1       Running   0          1m
wordpress-php-5d57ccf845-6vjkl      1/1       Running   0          25m
wordpress-php-5d57ccf845-7ck2c      1/1       Running   0          25m
wordpress-php-5d57ccf845-kwfgm      1/1       Running   0          25m
[root@master lnmp]#
```

图 13-11

一共创建了三个 SVC，使用 kubectl get svc 命令可以查看创建的 wordpress-mysql、wordpress-nginx 和 wordpress-php 这三个 Service，如图 13-12 所示。

```
[root@master lnmp]# kubectl get svc
NAME                TYPE        CLUSTER-IP      EXTERNAL-IP   PORT(S)        AGE
glusterfs-cluster   ClusterIP   10.10.10.250    <none>        1/TCP          21d
httpd               ClusterIP   10.10.10.239    <none>        80/TCP         21d
kubernetes          ClusterIP   10.10.10.1      <none>        443/TCP        22d
nginx-service       NodePort    10.10.10.88     <none>        80:37445/TCP   21d
wordpress-mysql     ClusterIP   10.10.10.208    <none>        3306/TCP       2h
wordpress-nginx     NodePort    10.10.10.104    <none>        80:36206/TCP   1m
wordpress-php       ClusterIP   10.10.10.236    <none>        9000/TCP       1h
[root@master lnmp]#
```

图 13-12

查看 PV 和 PVC 是否全部绑定成功，如图 13-13 所示。

```
[root@master lnmp]# kubectl get pv,pvc
NAME                                 CAPACITY   ACCESS MODES   RECLAIM POLICY   STATUS     CLAIM                          STORAGECLASS   REASON    AGE
persistentvolume/gluster-pv          10Gi       RWX            Retain           Released   default/wp-pv-one                                       2h
persistentvolume/gluster-pv-two      20Gi       RWX            Retain           Bound      default/pvc-one                                         2h
persistentvolume/mysql-pv            20Gi       RWX            Retain           Bound      default/mysql-pv-claim                                  1h
persistentvolume/wp-pv-one           5Gi        RWX            Retain           Bound      default/wp-pv-one                                       1h
persistentvolume/wp-pv-two           5Gi        RWX            Retain           Bound      default/wp-pv-two                                       43m

NAME                                       STATUS    VOLUME           CAPACITY   ACCESS MODES   STORAGECLASS   AGE
persistentvolumeclaim/mysql-pv-claim       Bound     mysql-pv         20Gi       RWX                           1h
persistentvolumeclaim/pvc-one              Bound     gluster-pv-two   20Gi       RWX                           1h
persistentvolumeclaim/wp-pv-one            Bound     wp-pv-one        5Gi        RWX                           1h
persistentvolumeclaim/wp-pv-two            Bound     wp-pv-two        5Gi        RWX                           43m
[root@master lnmp]#
```

图 13-13

Nginx 使用的是 NFS 存储，可以让其在外部暴露一个随机端口号进行访问，但是 NFS 存储的路径下没有文件，所以访问 36206 端口时会显示“403 Forbidden”，如图 13-14 所示。

图 13-14

在 NFS 存储路径/opt/nfs/data 目录下创建一个 HTML 页面，在 HTML 页面中输入一些内容，再次刷新浏览器即可显示刚才在 HTML 文件中输入的内容，如图 13-15 所示。

图 13-15

在 NFS 存储路径下创建 index.php 文件，在文件中输入 phpinfo()函数，打开浏览器，查看是否会输出服务器的配置信息，如图 13-16 所示。

PHP Version 5.6.31

System	Linux wordpress-php-5d57ccf845-7ck2c 3.10.0-693.el7.x86_64 #1 SMP Tue Aug 22 21:09:27 UTC 2017 x86_64
Build Date	May 22 2018 13:10:47
Configure Command	'./configure' '--prefix=/usr/local/php' '--with-config-file-path=/usr/local/php/etc' '--with-mysql' '--with-mysqli' '--with-openssl' '--with-zlib' '--with-curl' '--with-gd' '--with-jpeg-dir' '--with-png-dir' '--with-iconv' '--enable-fpm' '--enable-zip' '--enable-mbstring'
Server API	FPM/FastCGI
Virtual Directory Support	disabled
Configuration File (php.ini) Path	/usr/local/php/etc
Loaded Configuration File	/usr/local/php/etc/php-fpm.conf
Scan this dir for additional .ini files	(none)
Additional .ini files parsed	(none)
PHP API	20131106
PHP Extension	20131226
Zend Extension	220131226
Zend Extension Build	API220131226,NTS
PHP Extension Build	API20131226,NTS
Debug Build	no
Thread Safety	disabled
Zend Signal Handling	disabled
Zend Memory Manager	enabled
Zend Multibyte Support	provided by mbstring
IPv6 Support	enabled
DTrace Support	disabled
Registered PHP Streams	https, ftps, compress.zlib, php, file, glob, data, http, ftp, phar, zip
Registered Stream Socket Transports	tcp, udp, unix, udg, ssl, sslv3, tls, tlsv1.0, tlsv1.1, tlsv1.2
Registered Stream Filters	zlib.*, convert.iconv.*, string.rot13, string.toupper, string.tolower, string.strip_tags, convert.*, consumed, dechunk

This program makes use of the Zend Scripting Language Engine:
Zend Engine v2.6.0, Copyright (c) 1998-2016 Zend Technologies

Configuration

图 13-16

不断地按 Ctrl+F5 组合键强制刷新，会看到第一行的 System 容器的主机名不断地变化，如图 13-17 所示。

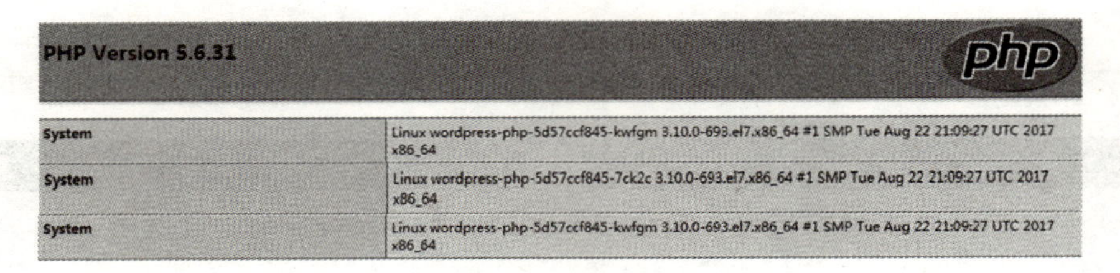

PHP Version 5.6.31

System	Linux wordpress-php-5d57ccf845-kwfgm 3.10.0-693.el7.x86_64 #1 SMP Tue Aug 22 21:09:27 UTC 2017 x86_64
System	Linux wordpress-php-5d57ccf845-7ck2c 3.10.0-693.el7.x86_64 #1 SMP Tue Aug 22 21:09:27 UTC 2017 x86_64
System	Linux wordpress-php-5d57ccf845-kwfgm 3.10.0-693.el7.x86_64 #1 SMP Tue Aug 22 21:09:27 UTC 2017 x86_64

图 13-17

部署 WordPress 博客程序，从 WordPress 官网下载程序源码包并上传到服务器中，命令如下。

```
[root@lanmp data]# wget
https://cn.wordpress.org/wordpress-4.9.4-zh_CN.tar.gz
```

下载完毕后解压 WordPress 博客程序，命令如下。

```
[root@lanmp data]# tar xf wordpress-4.9.4-zh_CN.tar.gz
[root@lanmp data]# du -sh wordpress
31M wordpr2ss
```

在浏览器的地址栏中依次输入任意节点的 IP 地址、暴露在外部的 NodePort 随机端口号和 wordpress 后缀，即可访问 WordPress 博客程序的安装界面，如图 13-18 所示。

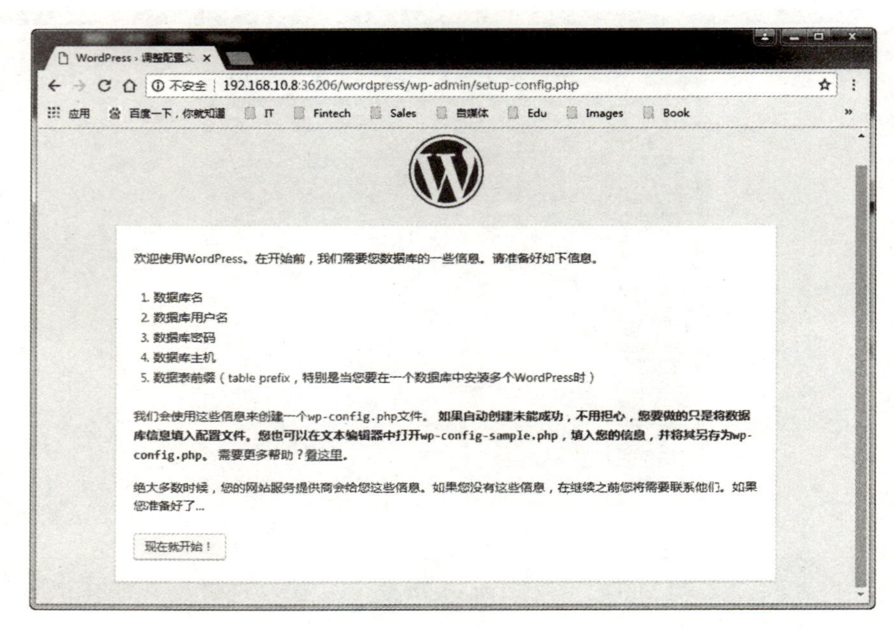

图 13-18

因为在前面没有创建数据库，所以还需要进入 MySQL 容器创建 MySQL 数据库，命令如下。

```
[root@master lnmp]# kubectl exec -it wordpress-mysql-bcc89f687-rt2zj bash
root@wordpress-mysql-bcc89f687-rt2zj:/# mysql -uroot -p123456
mysql> create database wordpress;
Query OK, 1 row affected (0.07 sec)
```

创建完数据库后单击“现在就开始！”按钮，在打开的页面中输入对应的信息，如图 13-19 所示。

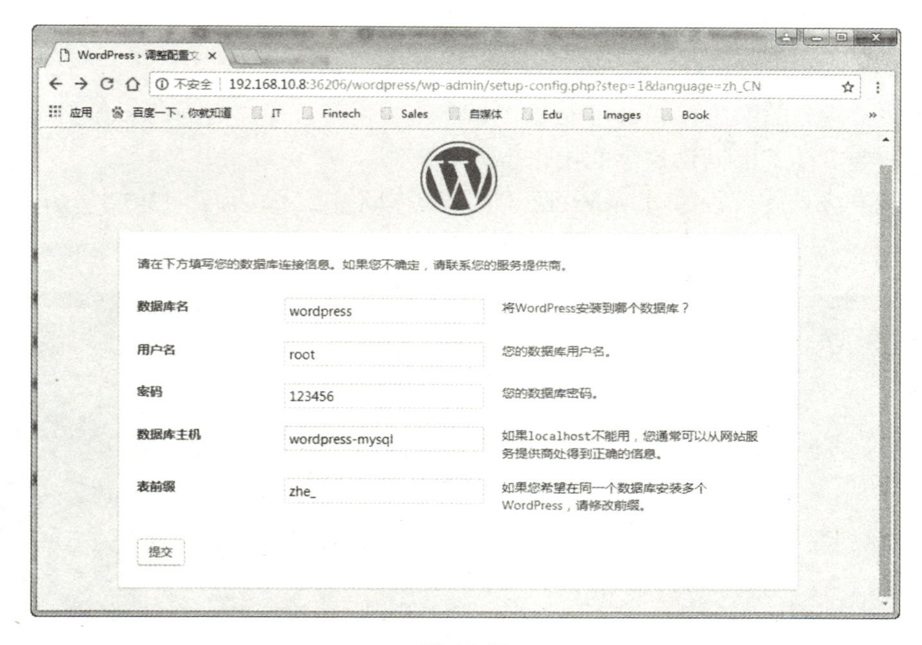

图 13-19

在图 13-19 中，数据库主机是通过 kubectl get svc 命令查看 Service 名称的，如图 13-20 所示。

```
[root@master lnmp]# kubectl get svc
NAME                TYPE        CLUSTER-IP     EXTERNAL-IP   PORT(S)          AGE
glusterfs-cluster   ClusterIP   10.10.10.250   <none>        1/TCP            22d
httpd               ClusterIP   10.10.10.239   <none>        80/TCP           22d
kubernetes          ClusterIP   10.10.10.1     <none>        443/TCP          23d
nginx-service       NodePort    10.10.10.88    <none>        80:37445/TCP     22d
wordpress-mysql     ClusterIP   10.10.10.208   <none>        3306/TCP         1d
wordpress-nginx     NodePort    10.10.10.104   <none>        80:36206/TCP     22h
wordpress-php       ClusterIP   10.10.10.236   <none>        9000/TCP         23h
[root@master lnmp]#
```

图 13-20

单击“提交”按钮即可进入下一个页面，如图 13-21 所示，表示数据库设置成功。如果提示数据库连接失败，则需要检查数据库信息是否有误。

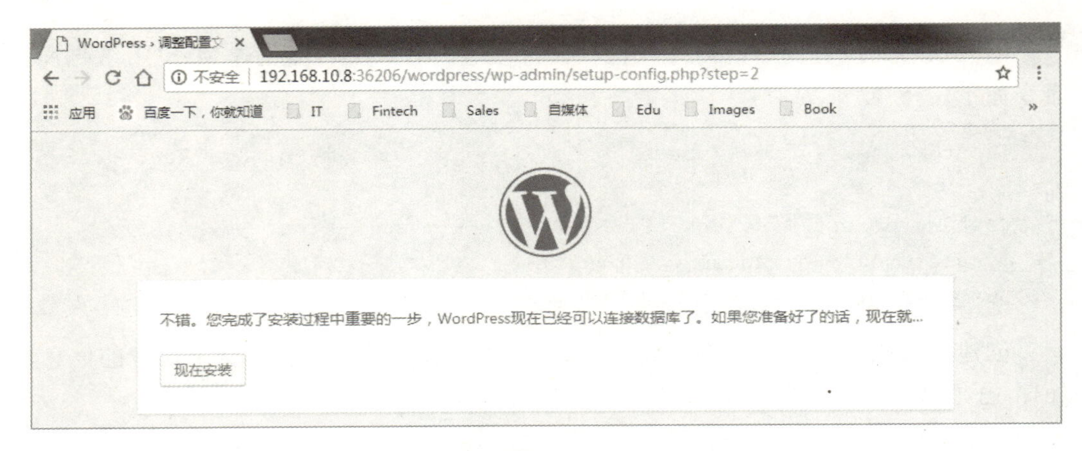

图 13-21

单击“现在安装”按钮即可安装 WordPress 博客程序，后面的操作不再进行演示，与安装其他开源网站程序的操作步骤一致。

安装完成后就可以访问 WordPress 博客程序的首页了，如图 13-22 所示。

图 13-22

至此，WordPress 开源博客程序就部署成功了。如果还想部署其他的开源程序，如 emlog、Discuz!、DedeCMS 和 PHPCMS 等，都可以按照上述步骤进行操作。

注意

需要注意的是，本小节所讲解的 WordPress 开源博客程序是存储在共享存储中的，而真实的生产环境一般都是打包到镜像中的。

13.3　Kubernetes高可用架构

13.3.1　高可用架构详解

Kubernetes 在之前做的演示和实践中采用的服务器是一台 Master 和五台 Node，而在实际的生产环境中需要考虑集群的高可用，集群中的任意一个节点宕掉后，都不会影响整个集群的使用。

Kubernetes 的 Node 服务器已经具备了高可用，因为 Pod 副本会分布在每个节点上，当任意节点宕掉后，其他节点就会启用，所以，此处我们无须去考虑。要重点考虑的是 Master 服务器，Master 节点宕掉后可能会导致集群不可管理或直接影响某些服务的正常使用，这在实际的生产环境中是不允许的。

下面就来看一下 Master 节点集群的高可用具体是如何操作的。如图 13-23 所示是 Kubernetes 官方所提供的高可用架构图。

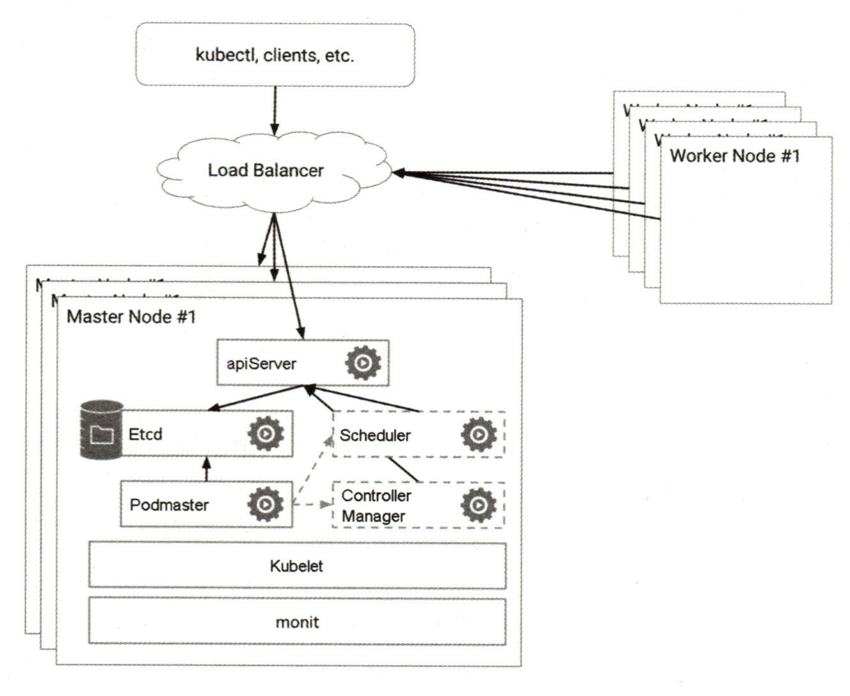

图 13-23

在图 13-23 中，最上面是 Kubectl 管理工具客户端连接 Load Balancer（LB，负载均衡），负载均衡下面是三个叠加的代理，实际上代理的就是 apiServer，由 Scheduler 和 Controller Manager 连接 apiServer。apiServer 还连接了 Etcd 的键值存储，用于存储集群的状态和一些配置信息，由此就组成了 Master 节点。Master 节点也会分配 Pod

任务，因为它运营 Kubelet。

综上所述，可以大致理解为从客户端连接 LB，LB 将请求转发到后端的 apiServer 中，右边则是 Kubernetes 的 Node，Node 连接 LB。从图 13-23 可以看出，高可用架构其实就是一个单独的 LB（负载均衡）代理的 apiServer，而 Node 连接 apiServer 实时地进行工作，然后获取 API 中的一些配置信息，并更新当前的状态。

笔者根据本小节要部署的 Kubernetes 高可用架构做了一张图，如图 13-24 所示。图中共有两个 Master 和五个 Node。Node 上部署了 Nginx 服务，Nginx 提供四层的负载均衡。Node-1 的 LB 代理的是两台 Master 节点的 apiServer，Node-2 也代理两台 Master 节点的 apiServer，Node-3 同样也是代理两台 Master 节点的 apiServer。而 apiServer 是连接 Etcd 键值存储的集群。

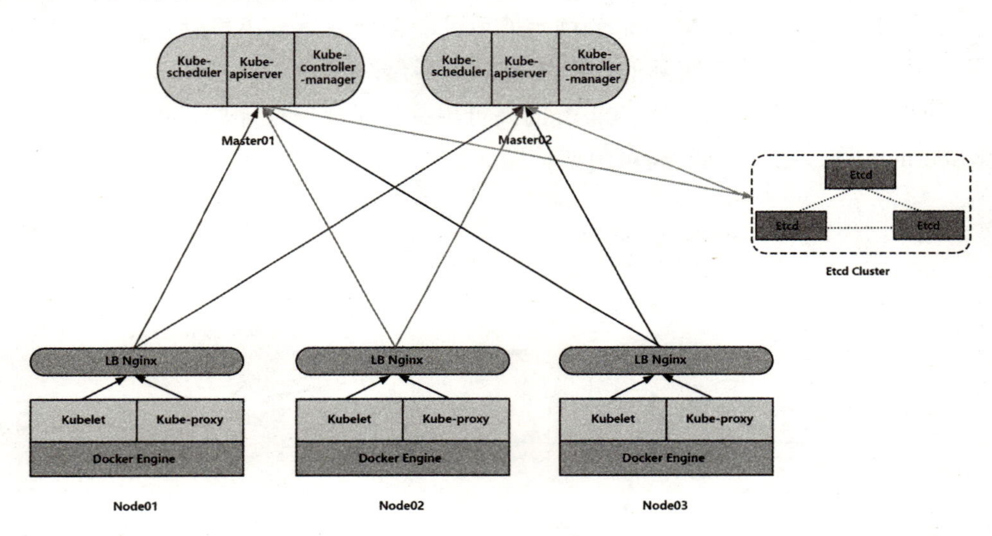

图 13-24

综上所述，部署高可用架构就是在 Node 上装一个 LB Nginx 去负载 Master 节点的 apiServer。

kube-scheduler 和 kube-controller-manager 这两个组件本身已经实现了高可用，通过 Etcd 做相应领袖（Leader）的选举，可以通过 ps aux|grep kube 命令查看 kube-controller-manager 和 kube-scheduler 启动了 Leader，如图 13-25 所示。当集群中有多个 kube-controller-manager 时，会做集群 Leader 的选举，保证集群的高可用，所以在此处无须考虑。需要考虑的是 apiServer，apiServer 在设计之初就是可扩展的，能够方便地对集群的架构进行扩展。

还有一种集群架构是一个单独的 LB，就像图 13-23 这种，是一个独立的 LB 做的 apiServer 负载均衡，Node 连接的是 LB，而本小节所采用的是图 13-24 这种方式的，这种方式的好处是在每个节点上部署的 Nginx 负载均衡都无须考虑它的单点，因为其中一个 Node 宕掉，其他 Node 不会受到任何影响，均可访问 apiServer，因为它自身就会管理一个负载均衡。单独的 LB 要考虑 LB 的高可用，因为是单点的，所以需要去维护它。

```
[root@master ~]# ps aux|grep kube-controller-manager
root      8880  0.2  0.3  94488 26024 ?        Ssl  21:37   0:00 /opt/kubernetes/bin/kube-controller-manage
 --logtostderr=true --v=4 --master=127.0.0.1:8080 --leader-elect=true --address=127.0.0.1 --service-cluster-i
p-range=10.10.10.0/24 --cluster-name=kubernetes --cluster-signing-cert-file=/opt/kubernetes/ssl/ca.pem --clus
ter-signing-key-file=/opt/kubernetes/ssl/ca-key.pem --service-account-private-key-file=/opt/kubernetes/ssl/ca
-key.pem --root-ca-file=/opt/kubernetes/ssl/ca.pem
root      8892  0.0  0.0 112660   972 pts/0    S+   21:39   0:00 grep --color=auto kube-controller-manager
[root@master ~]#
```

图 13-25

图 13-23 这种集群架构的好处是，比如在阿里云上部署，因为阿里云有 SLB，所以可以在 SLB 上做四层的内网负载均衡，这样也不会被收取费用。在企业架构中如果有 LB，就可以用现有的 LB 代理 apiServer。

13.3.2　Master 高可用部署

添加一台 Master 节点，主机名设置为 Master-2，命令如下。

```
[root@localhost ~]# hostnamectl set-hostname master-2
[root@localhost ~]# bash
[root@master-2 ~]#
```

在前面的章节中，使用 cfssl 命令生成了一系列的数字证书，但是由于本小节需要增加一台 Master 节点来设置高可用，所以需要将 Master-2 节点的 IP 地址添加到 Server.pem 数字证书中，添加 IP 地址后重新生成数字证书即可，命令如下。

```
[root@master ssl]# cat > server-csr.json <<EOF
{
    "CN": "kubernetes",
    "hosts": [
      "127.0.0.1",
      "192.168.10.5",
      "192.168.10.6",
      "192.168.10.7",
      "192.168.10.8",
      "192.168.10.9",
      "192.168.10.10",
      "192.168.10.11",
      "10.10.10.1",
      "kubernetes",
```

```
        "kubernetes.default",
        "kubernetes.default.svc",
        "kubernetes.default.svc.cluster",
        "kubernetes.default.svc.cluster.local"
      ],
      "key": {
         "algo": "rsa",
         "size": 2048
      },
      "names": [
         {
            "C": "CN",
            "L": "BeiJing",
            "ST": "BeiJing",
            "O": "k8s",
            "OU": "System"
         }
      ]
   }
   EOF
   [root@master ssl]# cfssl gencert -ca=ca.pem -ca-key=ca-key.pem
-config=ca-config.json -profile=kubernetes server-csr.json | cfssljson
-bare server
```

生成完毕后，重启 kube-apiserver 服务，使之应用新的数字证书，命令如下。

```
   [root@master ssl]# systemctl restart kube-apiserver.service
```

将 Master 节点上的/opt/kubernetes/目录下的所有数据文件复制到 Master-2 节点的/opt/目录下，如图 13-26 所示。在初期部署 Kubernetes 集群时，就要考虑后期的可维护性，所以将 Kubernetes 集群所用到的组件全部放到了/opt/kubernetes/目录下，再去用时会更加方便、快捷。

图 13-26

将 Master 节点上的 kube-apiserver、kube-scheduler.service 和 kube-controller-manager 等服务复制到 Master-2 节点上，如图 13-27 所示。

```
[root@master ssl]# scp -r /usr/lib/systemd/system/{kube-apiserver,kube-scheduler,kube-controller-manager}.service root@192.168.10.11:/usr/lib/systemd/system
kube-apiserver.service                                                                                        100%  282    98.2KB/s   00:00
kube-scheduler.service                                                                                        100%  281   132.6KB/s   00:00
kube-controller-manager.service                                                                               100%  317   121.3KB/s   00:00
```

图 13-27

Master-2 节点所需要的组件已经全部复制完毕，但是复制完是不可以直接使用的，需要更改 kube-apiserver 的 IP 地址，将监听地址--bind-address 和通告地址--advertise-address 的 IP 地址改为 192.168.10.11，如图 13-28 所示。

```
[root@master-2 cfg]# vim kube-apiserver


KUBE_APISERVER_OPTS="--logtostderr=true \
--v=4 \
--etcd-servers=https://192.168.10.5:2379,https://192.168.10.6:2379,https://192.168.10.7:2379,https://192.168.10.8:2379,https://192.168.10.9:2379,https://192.168.10.10:237
9 \
--insecure-bind-address=127.0.0.1 \
--bind-address=192.168.10.11 \
--insecure-port=8080 \
--secure-port=6443 \
--advertise-address=192.168.10.11 \
--allow-privileged=true \
--service-cluster-ip-range=10.10.10.0/24 \
--admission-control=NamespaceLifecycle,LimitRanger,SecurityContextDeny,ServiceAccount,ResourceQuota,NodeRestriction --authorization-mode=RBAC,Node \
--kubelet-https=true \
--enable-bootstrap-token-auth \
--token-auth-file=/opt/kubernetes/cfg/token.csv \
--service-node-port-range=30000-50000 \
--tls-cert-file=/opt/kubernetes/ssl/server.pem  \
--tls-private-key-file=/opt/kubernetes/ssl/server-key.pem \
--client-ca-file=/opt/kubernetes/ssl/ca.pem \
--service-account-key-file=/opt/kubernetes/ssl/ca-key.pem \
--etcd-cafile=/opt/kubernetes/ssl/ca.pem \
--etcd-certfile=/opt/kubernetes/ssl/server.pem \
--etcd-keyfile=/opt/kubernetes/ssl/server-key.pem"
```

图 13-28

修改完成后启动 Master-2 节点上的 kube-apiserver.service 服务，命令如下。

```
[root@master-2 cfg]# systemctl start kube-scheduler.service
```

查看 kube-apiserver 的进程是否启动，如图 13-29 所示，表示启动成功。

```
[root@master-2 cfg]# ps aux|grep kube
root      1294 79.0 28.7 426208 286444 ?      Ssl  21:44   0:21 /opt/kubernetes/bin/kube-apiserver --logtostderr=true --v=4 --etcd-servers=https://192.168.10.5:2379,htt
ps://192.168.10.6:2379,https://192.168.10.7:2379,https://192.168.10.8:2379,https://192.168.10.9:2379,https://192.168.10.10:2379 --insecure-bind-address=127.0.0.1 --bind-a
ddress=192.168.10.11 --insecure-port=8080 --secure-port=6443 --advertise-address=192.168.10.11 --allow-privileged=true --service-cluster-ip-range=10.10.10.0/24 --admissio
n-control=NamespaceLifecycle,LimitRanger,SecurityContextDeny,ServiceAccount,ResourceQuota,NodeRestriction --authorization-mode=RBAC,Node --kubelet-https=true --enable-boo
tstrap-token-auth --token-auth-file=/opt/kubernetes/cfg/token.csv --service-node-port-range=30000-50000 --tls-cert-file=/opt/kubernetes/ssl/server.pem --tls-private-key-f
ile=/opt/kubernetes/ssl/server-key.pem --client-ca-file=/opt/kubernetes/ssl/ca.pem --service-account-key-file=/opt/kubernetes/ssl/ca-key.pem --etcd-cafile=/opt/kubernetes
/ssl/ca.pem --etcd-certfile=/opt/kubernetes/ssl/server.pem --etcd-keyfile=/opt/kubernetes/ssl/server-key.pem
root      1323  2.1  1.9  42524 19336 ?       Ssl  21:44   0:00 /opt/kubernetes/bin/kube-scheduler --logtostderr=true --v=4 --master=127.0.0.1:8080 --leader-elect
root      1329  0.0  0.0 112704   968 pts/0   R+   21:44   0:00 grep --color=auto kube
```

图 13-29

此时，可以打开系统的日志信息查看 kube-apiserver 服务，发现已经从 Etcd 中获取了一些信息，如图 13-30 所示。

```
[root@master-2 cfg]# tail /var/log/messages
May 27 21:44:59 master-2 kube-apiserver: I0527 21:44:59.213025    1294 wrap.go:42] GET /api/v1/namespaces/default/services/kubernetes: (5.390085ms) 200 [[kube-apiserver/v
1.10.2 (linux/amd64) kubernetes/81753b1] 192.168.10.11:39050]
May 27 21:44:59 master-2 kube-apiserver: I0527 21:44:59.221408    1294 wrap.go:42] GET /api/v1/namespaces/default/endpoints/kubernetes: (7.337726ms) 200 [[kube-apiserver/
v1.10.2 (linux/amd64) kubernetes/81753b1] 192.168.10.11:39050]
May 27 21:44:59 master-2 kube-apiserver: W0527 21:44:59.222109    1294 mastercount.go:135] Resetting endpoints for master service "kubernetes" to &{{ } {kubernetes  defau
lt /api/v1/namespaces/default/endpoints/kubernetes 5ba3228e-61ad-11e8-a521-000c29551b6f 4062 0 2018-05-27 20:56:22 +0800 CST <nil> <nil> map[] map[] [] nil [] } [{[{192.1
68.10.11  <nil> <nil>}] [] [{https 6443 TCP}]}]}
May 27 21:44:59 master-2 kube-apiserver: I0527 21:44:59.235064    1294 wrap.go:42] PUT /api/v1/namespaces/default/endpoints/kubernetes: (12.128095ms) 200 [[kube-apiserver
/v1.10.2 (linux/amd64) kubernetes/81753b1] 192.168.10.11:39050]
May 27 21:45:00 master-2 kube-apiserver: I0527 21:45:00.778136    1294 wrap.go:42] GET /api/v1/namespaces/kube-system/endpoints/kube-scheduler: (8.350863ms) 200 [[kube-sc
heduler/v1.10.2 (linux/amd64) kubernetes/81753b1/leader-election] 127.0.0.1:33474]
```

图 13-30

kube-apiserver 服务监听的是 6443 端口，和 Master 节点上监听的端口一致，如图 13-31 所示。

启动 kube-scheduler.service 服务，如图 13-32 所示。

```
[root@master-2 cfg]# netstat -lntp
Active Internet connections (only servers)
Proto Recv-Q Send-Q Local Address           Foreign Address         State       PID/Program name
tcp        0      0 192.168.10.11:6443      0.0.0.0:*               LISTEN      1294/kube-apiserver
tcp        0      0 127.0.0.1:8080          0.0.0.0:*               LISTEN      1294/kube-apiserver
tcp        0      0 0.0.0.0:22              0.0.0.0:*               LISTEN      753/sshd
tcp        0      0 127.0.0.1:25            0.0.0.0:*               LISTEN      838/master
tcp6       0      0 :::10251                :::*                    LISTEN      1323/kube-scheduler
tcp6       0      0 :::22                   :::*                    LISTEN      753/sshd
tcp6       0      0 ::1:25                  :::*                    LISTEN      838/master
[root@master-2 cfg]#
```

图 13-31

```
[root@master-2 cfg]# systemctl start kube-scheduler.service
[root@master-2 cfg]# ps aux|grep kube
root      1294 79.0 28.7 426208 286444 ?      Ssl  21:44   0:21 /opt/kubernetes/bin/kube-apiserver --logtostderr=true --v=4 --etcd-servers=https://192.168.10.5:2379,htt
ps://192.168.10.6:2379,https://192.168.10.7:2379,https://192.168.10.8:2379,https://192.168.10.9:2379,https://192.168.10.10:2379 --insecure-bind-address=127.0.0.1 --bind-a
ddress=192.168.10.11 --insecure-port=8080 --secure-port=6443 --advertise-address=192.168.10.11 --allow-privileged=true --service-cluster-ip-range=10.10.10.0/24 --admissio
n-control=NamespaceLifecycle,LimitRanger,SecurityContextDeny,ServiceAccount,ResourceQuota,NodeRestriction --authorization-mode=RBAC,Node --kubelet-https=true --enable-boo
tstrap-token-auth --token-auth-file=/opt/kubernetes/cfg/token.csv --service-node-port-range=30000-50000 --tls-cert-file=/opt/kubernetes/ssl/server.pem --tls-private-key-f
ile=/opt/kubernetes/ssl/server-key.pem --client-ca-file=/opt/kubernetes/ssl/ca.pem --service-account-key-file=/opt/kubernetes/ssl/ca-key.pem --etcd-cafile=/opt/kubernetes
/ssl/ca.pem --etcd-certfile=/opt/kubernetes/ssl/server.pem --etcd-keyfile=/opt/kubernetes/ssl/server-key.pem
root      1323  2.1  1.9  42524 19336 ?       Ssl  21:44   0:00 /opt/kubernetes/bin/kube-scheduler --logtostderr=true --v=4 --master=127.0.0.1:8080 --leader-elect
root      1329  0.0  0.0 112704   968 pts/0   R+   21:44   0:00 grep --color=auto kube
```

图 13-32

启动 kube-controller-manager 服务，命令如下。

```
[root@master-2 cfg]# systemctl start kube-controller-manager.service
```

启动成功后，通过 Kubectl 客户端工具查看是否可以获取 Node，如图 13-33 所示。图中显示可以获取到，状态是 Ready，没有任何问题。

```
[root@master-2 cfg]# /opt/kubernetes/bin/kubectl get node
NAME            STATUS    ROLES     AGE       VERSION
192.168.10.10   Ready     <none>    55m       v1.10.2
192.168.10.6    Ready     <none>    1h        v1.10.2
192.168.10.7    Ready     <none>    50m       v1.10.2
192.168.10.8    Ready     <none>    52m       v1.10.2
192.168.10.9    Ready     <none>    53m       v1.10.2
[root@master-2 cfg]#
```

图 13-33

在前面的章节中，Node-1 节点使用的是 192.168.10.5 的 Kubectl 工具远程连接集群，现在将.kube/config 文件中的 IP 地址改为 192.168.10.11，如图 13-34 所示。使用 kubectl get node 命令查看是否可以获取 Node，如果可以，表示上述所有操作都没问题，如图 13-35 所示，显示没有任何问题。

```
[root@node-1 ~]# vim .kube/config

apiVersion: v1
clusters:
- cluster:
    certificate-authority: /root/ca.pem
    server: https://192.168.10.11:6443
  name: kubernetes
contexts:
- context:
    cluster: kubernetes
    user: cluster-admin
  name: default
current-context: default
kind: Config
preferences: {}
users:
- name: cluster-admin
  user:
    client-certificate: /root/admin.pem
    client-key: /root/admin-key.pem
```

图 13-34

```
[root@node-1 ~]# vim .kube/config
[root@node-1 ~]# kubectl get node
NAME            STATUS    ROLES     AGE       VERSION
192.168.10.10   Ready     <none>    23h       v1.10.2
192.168.10.6    Ready     <none>    23h       v1.10.2
192.168.10.7    Ready     <none>    23h       v1.10.2
192.168.10.8    Ready     <none>    23h       v1.10.2
192.168.10.9    Ready     <none>    23h       v1.10.2
[root@node-1 ~]#
```

图 13-35

在所有 Node 上安装 Nginx 服务，命令如下。安装 Nginx 是为了做负载均衡。现在有两个 Master，每个 Node 只能连接一个 Master 地址，Node 的 IP 地址要和 Master 的 IP 地址一致，后端是通过负载均衡实现的。

```
[root@node-1 ~]# yum install -y nginx
```

安装完 Nginx 服务后，编辑 Nginx 主配置文件/etc/nginx/nginx.conf。本实例主要做四层的负载均衡，在 Nginx1.9 版本后增加了四层的负载均衡，在之前的版本中，只是一个纯七层的负载均衡。四层和七层的区别在于，四层是通过 IP+Port 直接转发的，而七层需要先分区应用层的数据再进行转发。

将 nginx.conf 配置文件中 HTTP 下面的所有内容全部删除，添加一些内容到该文件中，如图 13-36 所示。

```
user nginx;
worker_processes auto;
error_log /var/log/nginx/error.log;
pid /run/nginx.pid;

# Load dynamic modules. See /usr/share/nginx/README.dynamic.
include /usr/share/nginx/modules/*.conf;

events {
    worker_connections 1024;
}

stream {

    log_format  main  '$remote_addr $upstream_addr - [$time_local] $status $upstream_bytes_sent';
    access_log  /var/log/nginx/k8s-access.log  main;
    upstream k8s-apiserver {
        server 192.168.10.5:6443;
        server 192.168.10.11:6443;
    }
    server {
        listen 127.0.0.1:6443;
        proxy_pass k8s-apiserver;
    }
}
```

图 13-36

将 Node 的三个.kubeconfig 文件中连接的 apiServer 地址改为本地的地址，命令如下。

```
[root@node-1 ~]# cd /opt/kubernetes/cfg/
[root@node-1 cfg]# ls *.kubeconfig
bootstrap.kubeconfig  kubelet.kubeconfig  kube-proxy.kubeconfig
[root@node-1 cfg]# ls *.kubeconfig | xargs -i sed -i
's/192.168.10.5/127.0.0.1/g' {}
```

修改完成后重启 Kubelet、kube-proxy 服务，命令如下。

```
[root@node-1 cfg]# systemctl restart kubelet.service && systemctl restart
kube-proxy.service
```

启动所有 Node 上的 Nginx 服务，命令如下。

```
[root@node-4 nginx]# systemctl start nginx && ps aux|grep nginx
```

连接好之后可以查看 Nginx 日志，看日志中是否会请求 Master 节点，如图 13-37 所示，请求了 Master 节点。

```
[root@node-5 cfg]# systemctl restart kubelet.service && systemctl restart kube-proxy.service
[root@node-5 cfg]# tail /var/log/nginx/k8s-access.log -f
127.0.0.1 192.168.10.11:6443 - [28/May/2018:22:46:46 +0800] 200 1115
127.0.0.1 192.168.10.5:6443 - [28/May/2018:22:46:46 +0800] 200 1115
127.0.0.1 192.168.10.11:6443 - [28/May/2018:22:51:09 +0800] 200 25337
127.0.0.1 192.168.10.5:6443 - [28/May/2018:22:51:09 +0800] 200 3737
127.0.0.1 192.168.10.11:6443 - [28/May/2018:22:51:10 +0800] 200 1115
127.0.0.1 192.168.10.11:6443 - [28/May/2018:22:51:10 +0800] 200 1115
^C
[root@node-5 cfg]#
```

图 13-37

如果其中一个 Master 节点宕掉，它会将请求分配给另外一个 Master 节点，保证 Kubernetes 集群的高可用。

13.4　Kubernetes集群监控

集群的监控在企业生产环境中是非常重要的，可以观察当前所有集群节点的运行状态和一些性能的指标信息；也可以在一些应用或节点出现问题时，通过监控去分析、定位问题；还可以查阅历史数据图去追溯一些问题。

官方给出了监控在 Kubernetes 集群中的解决方案，通过 cAdvisor、Heapster、InfluxDB 和 Grafana，这是一个开源组合的容器监控方案。

cAdvisor 和 Heapster 是 Google 公司自主研发的工具。cAdvisor 主要用于容器资源的收集。Heapster 主要用于 Kubernetes 集群节点之间的数据汇总，因为集群中存在 *N* 个 Node，所以要将它们汇总到一个地方。InfluxDB 主要用于存储 Heapster 汇总的数据。Grafana 是一个 Web 前端的仪表盘，可以通过不同的数据源展示不同的数据。

如图 13-38 所示，最左边的方框中是 Kubernetes 集群中的某个工作节点，在节点中会应用两个组件，分别是 Kubelet 和 kube-proxy。Kubelet 是管理容器和资源的，kube-proxy 用于管理网络规则。cAdvisor 集成到 Kubelet 中，默认已经安装，无须再次安装。

cAdvisor 收集的数据并不会进行持久化的存储，它是实时的。也就是说，要通过 Kubelet 从本机的 cAdvisor 中获取监控信息并汇报给 Heapster，Heapster 会主动向每个 Node 请求监控的资源，请求完之后将所有的资源汇总到 InfluxDB 中进行持久化存储，最后通过 Grafana 进行展示。

打开浏览器，在地址栏中输入 Node 的 IP 地址加 cAdvisor 默认的端口号 4194

就可以进行访问，如图 13-39 所示。图 13-39 中主要搜集当前节点的信息和容器的数据信息，主要收集的资源是 CPU、Memory、I/O 和 FileSystem。

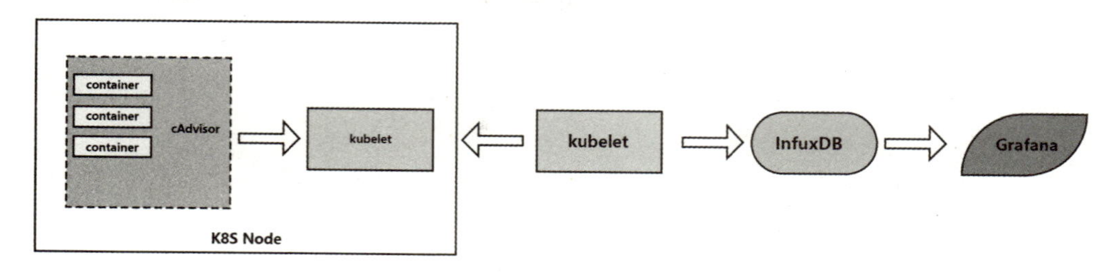

图 13-38

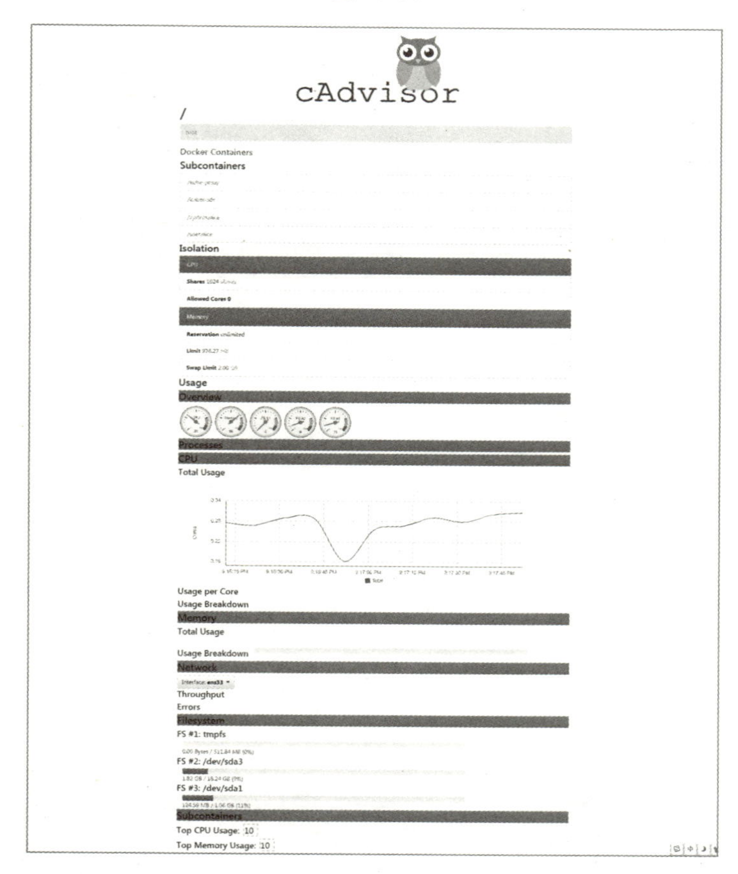

图 13-39

创建一个 monitor 目录用来部署 cAdvisor+Heapster+InfluxDB+Grafana，看看它主要监控哪些指标。

将事先准备好的 YAML 文件上传到 monitor 目录中，YAML 文件可在 GitHub 中进行下载，命令如下。

```
[root@master ~]# mkdir monitor
[root@master ~]# cd monitor/
[root@master monitor]# ls
grafana.yaml  heapster.yaml  influxdb.yaml
```

部署 InfluxDB 时序数据库，要先创建 InfluxDB 应用和服务，命令如下。

```
[root@master monitor]# kubectl create -f influxdb.yaml
deployment.extensions "monitoring-influxdb" created
service "monitoring-influxdb" created
```

创建完成后查看 InfluxDB 应用和服务是否启动成功，显示 Running 表示启动成功，如图 13-40 所示。

```
[root@master monitor]# kubectl get all -n kube-system
NAME                                        READY   STATUS    RESTARTS   AGE
pod/monitoring-influxdb-6cd98fd78f-rvt4w    1/1     Running   0          12s

NAME                          TYPE        CLUSTER-IP     EXTERNAL-IP   PORT(S)    AGE
service/monitoring-influxdb   ClusterIP   10.10.10.129   <none>        8086/TCP   12s

NAME                                  DESIRED   CURRENT   UP-TO-DATE   AVAILABLE   AGE
deployment.apps/monitoring-influxdb   1         1         1            1           12s

NAME                                             DESIRED   CURRENT   READY   AGE
replicaset.apps/monitoring-influxdb-6cd98fd78f   1         1         1       12s
```

图 13-40

部署完 InfluxDB 后再部署 Heapster，创建 ServerAccount、Deployment 和 Service，命令如下。

```
[root@master monitor]# kubectl create -f heapster.yaml
serviceaccount "heapster" created
clusterrolebinding.rbac.authorization.k8s.io "heapster" created
deployment.extensions "heapster" created
service "heapster" created
```

创建完成后查看 kube-system 命名空间中的 Heapster 的运行状态是否为 Running，如图 13-41 所示。

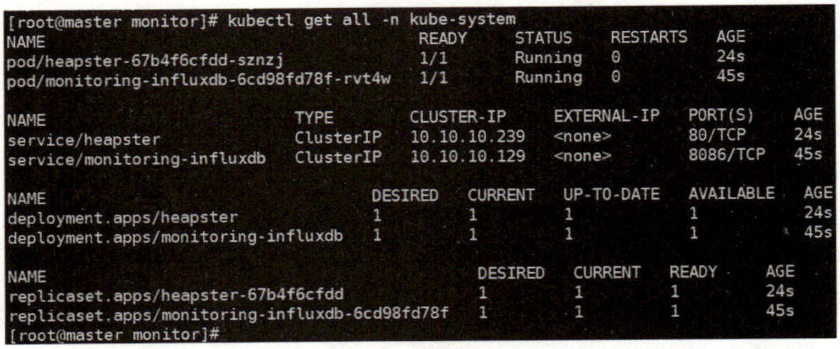

```
[root@master monitor]# kubectl get all -n kube-system
NAME                                        READY   STATUS    RESTARTS   AGE
pod/heapster-67b4f6cfdd-sznzj               1/1     Running   0          24s
pod/monitoring-influxdb-6cd98fd78f-rvt4w    1/1     Running   0          45s

NAME                          TYPE        CLUSTER-IP     EXTERNAL-IP   PORT(S)    AGE
service/heapster              ClusterIP   10.10.10.239   <none>        80/TCP     24s
service/monitoring-influxdb   ClusterIP   10.10.10.129   <none>        8086/TCP   45s

NAME                                  DESIRED   CURRENT   UP-TO-DATE   AVAILABLE   AGE
deployment.apps/heapster              1         1         1            1           24s
deployment.apps/monitoring-influxdb   1         1         1            1           45s

NAME                                             DESIRED   CURRENT   READY   AGE
replicaset.apps/heapster-67b4f6cfdd              1         1         1       24s
replicaset.apps/monitoring-influxdb-6cd98fd78f   1         1         1       45s
[root@master monitor]#
```

图 13-41

使用 kubectl get svc 命令指定 kube-system 命名空间，可查看 Heapster 和 InfluxDB 的 IP 和端口，如图 13-42 所示。

```
[root@master monitor]# kubectl get svc -n kube-system
NAME                  TYPE        CLUSTER-IP      EXTERNAL-IP   PORT(S)    AGE
heapster              ClusterIP   10.10.10.239    <none>        80/TCP     36s
monitoring-influxdb   ClusterIP   10.10.10.129    <none>        8086/TCP   57s
[root@master monitor]# kubectl get svc
NAME         TYPE        CLUSTER-IP   EXTERNAL-IP   PORT(S)   AGE
kubernetes   ClusterIP   10.10.10.1   <none>        443/TCP   12m
[root@master monitor]#
```

图 13-42

部署 Web 前端展示界面的 Grafana，命令如下。

```
[root@master monitor]# kubectl create -f grafana.yaml
deployment.extensions "monitoring-grafana" created
service "monitoring-grafana" created
```

查看 Grafana 的运行状态，如图 13-43 所示。

```
[root@master monitor]# kubectl get all -n kube-system
NAME                                       READY   STATUS    RESTARTS   AGE
pod/heapster-67b4f6cfdd-sznzj              1/1     Running   0          3m
pod/monitoring-grafana-646999ccb6-9jhxw    1/1     Running   0          40s
pod/monitoring-influxdb-6cd98fd78f-rvt4w   1/1     Running   0          3m

NAME                          TYPE        CLUSTER-IP     EXTERNAL-IP   PORT(S)        AGE
service/heapster              ClusterIP   10.10.10.239   <none>        80/TCP         3m
service/monitoring-grafana    NodePort    10.10.10.43    <none>        80:46845/TCP   39s
service/monitoring-influxdb   ClusterIP   10.10.10.129   <none>        8086/TCP       3m

NAME                                  DESIRED   CURRENT   UP-TO-DATE   AVAILABLE   AGE
deployment.apps/heapster              1         1         1            1           3m
deployment.apps/monitoring-grafana    1         1         1            1           40s
deployment.apps/monitoring-influxdb   1         1         1            1           3m

NAME                                             DESIRED   CURRENT   READY   AGE
replicaset.apps/heapster-67b4f6cfdd              1         1         1       3m
replicaset.apps/monitoring-grafana-646999ccb6    1         1         1       40s
replicaset.apps/monitoring-influxdb-6cd98fd78f   1         1         1       3m
[root@master monitor]#
```

图 13-43

想访问 Grafana，必须通过 apiServer 的非安全端口，非安全端口的 IP 地址最初设置的是 127.0.0.1，但是无法通过本地的浏览器进行访问。所以，要是有 apiServer 进行访问，需要让它监听一个可以访问的地址，再通过浏览器进行访问。允许临时设置可以访问的 IP 地址，如图 13-44 所示。

```
[root@master monitor]# vim /opt/kubernetes/cfg/kube-apiserver

KUBE_APISERVER_OPTS="--logtostderr=true \
--v=4 \
--etcd-servers=https://192.168.10.5:2379,https://192.168.10.6:2379,https://192.168.10.7:2379,https://192.168.10.8:2379,https://192.168.10.9:2379,https://192.168.10.10:237
9,https://192.168.10.11:2379 \
--insecure-bind-address=192.168.10.5 \
--bind-address=192.168.10.5 \
--insecure-port=8080 \
--secure-port=6443 \
--advertise-address=192.168.10.5 \
```

图 13-44

设置后重启 kube-apiserver 服务，命令如下。重启完成后监听的端口就是 192.168.10.5 了，通过 192.168.10.5:8080 加上 Grafana.yaml 文件中指定的/api/v1/namespaces/kube-system/services/monitoring-grafana/proxy/地址即可访问，如图 13-45 所示。

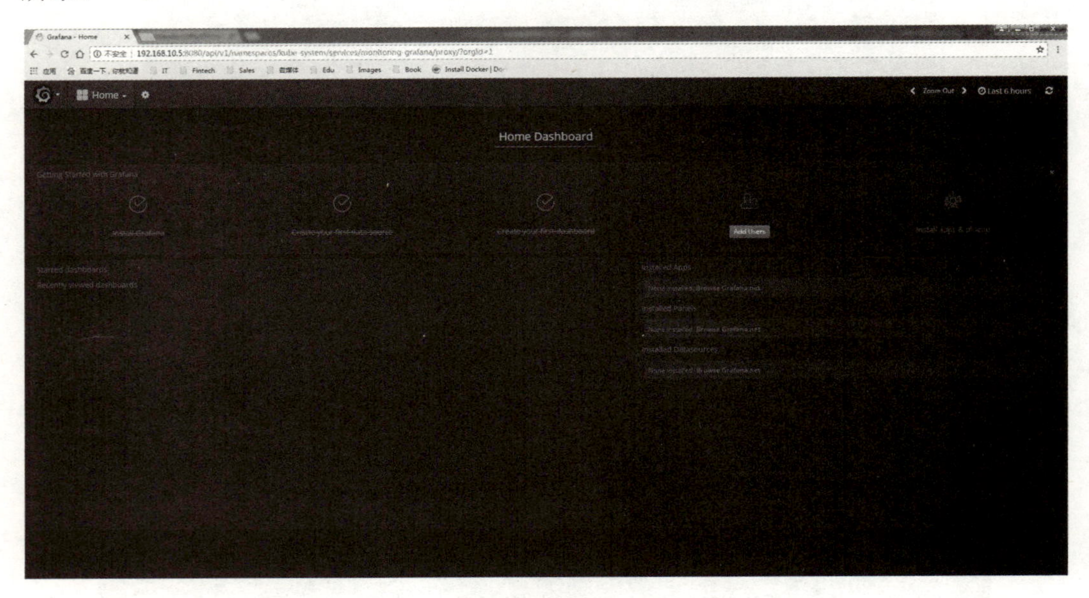

图 13-45

注意

在 Master 节点进行访问，需要在 Master 节点安装 Flanneld 网络。如未安装 Flanneld 网络，则无法进行访问。

还可以通过代理的方式进行访问，设置代理命令如下。

```
[root@master monitor]# kubectl proxy --address='192.168.10.5'
--port=8086 --accept-hosts="^*$"
Starting to serve on 192.168.10.5:8086
```

设置好 Proxy 后打开浏览器，输入 192.168.10.5:8086 和后缀也可以访问，如图 13-46 所示。

单击左上角的 Logo，在下拉菜单中选择 Data Sources，这时会发现 Influxdb-datacource 已经被添加进来了，说明数据源测试是成功的，如图 13-47 所示。如果连接不上，数据不会被展示出来，连接的网址是 http://monitoring- influxdb:8086。

系统默认创建两个仪表盘，分别监控集群节点的利用率和 Pod 的利用率，如图 13-48 所示。

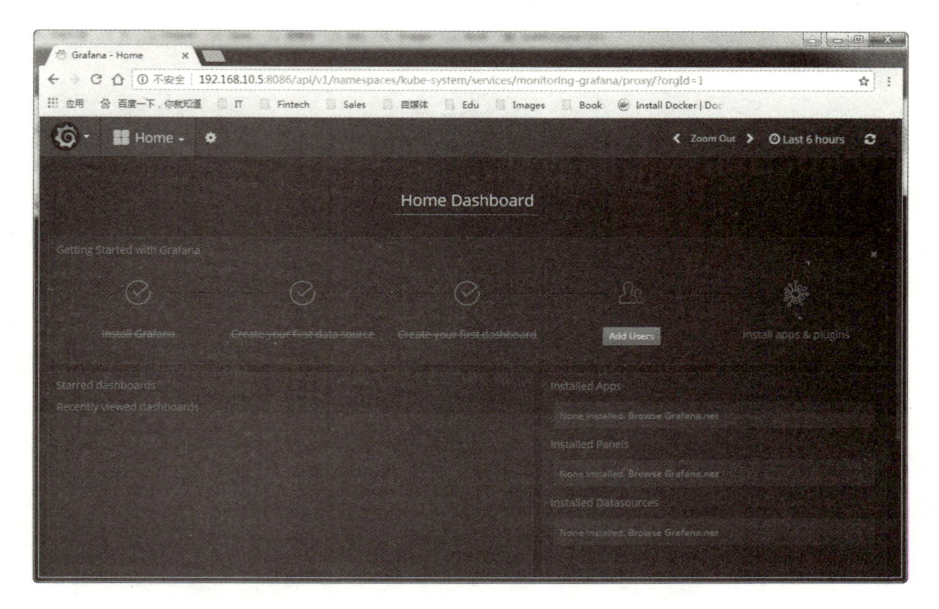

Grafana - Home
192.168.10.5:8086/api/v1/namespaces/kube-system/services/monitoring-grafana/proxy/?orgId=1
Home
Zoom Out
Last 6 hours
Home Dashboard
Getting Started with Grafana
Install Grafana
Create your first data source
Create your first dashboard
Add Users
Install apps & plugins
Starred dashboards
Recently viewed dashboards
Installed Apps
None installed. Browse Grafana.net
Installed Panels
None installed. Browse Grafana.net
Installed Datasources
None installed. Browse Grafana.net

图 13-46

Grafana - Home
192.168.10.5:8086/api/v1/namespaces/kube-system/services/monitoring-grafana/proxy/datasources/edit/1
Data Sources
Edit data source
Name
influxdb-datasource
Default
Type
InfluxDB
Http settings
Url
http://monitoring-influxdb:8086
Access
proxy
Http Auth
Basic Auth
With Credentials
TLS Client Auth
With CA Cert
InfluxDB Details
Database
k8s
User
root
Password
Default group by time
Success
Data source is working
Save & Test
Delete
Cancel

图 13-47

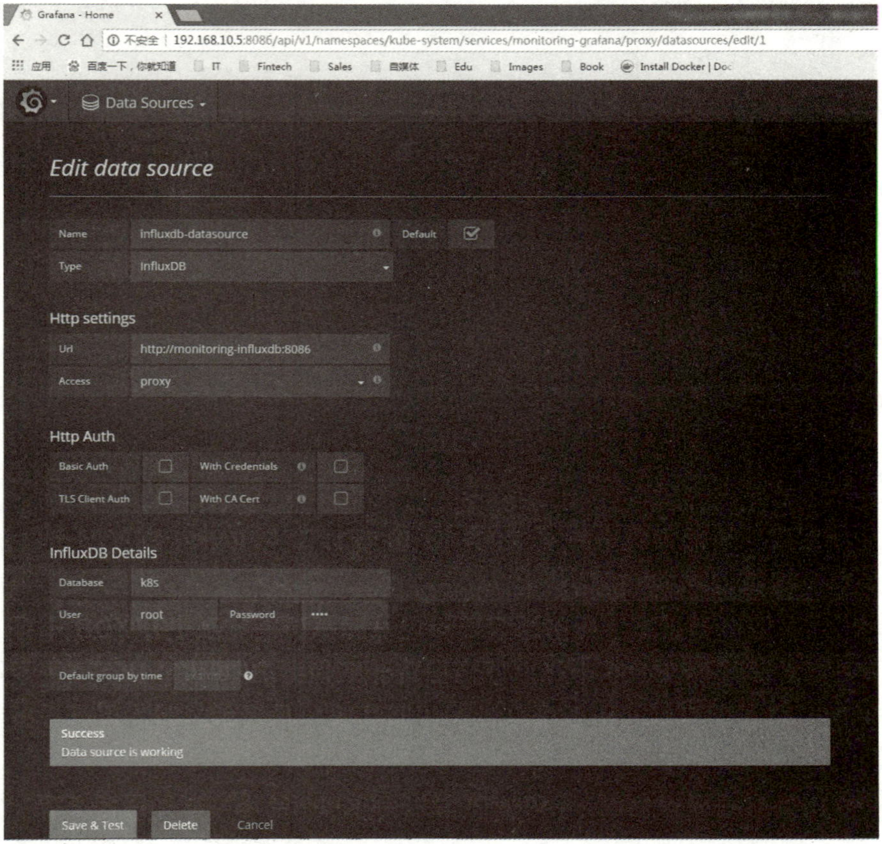

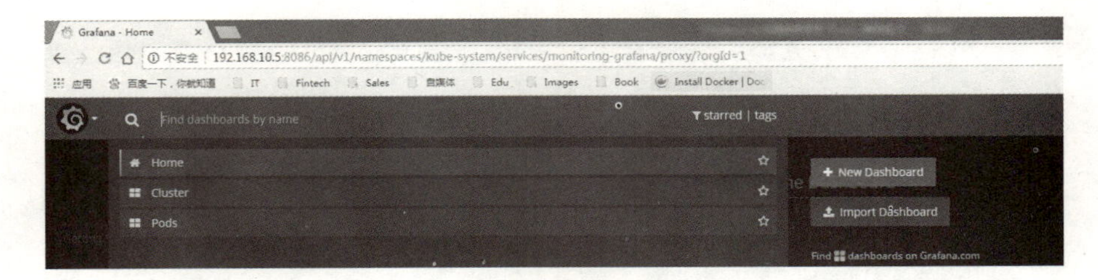

图 13-48

可以单击相应选项查看 Cluster 的利用率，如图 13-49 所示。

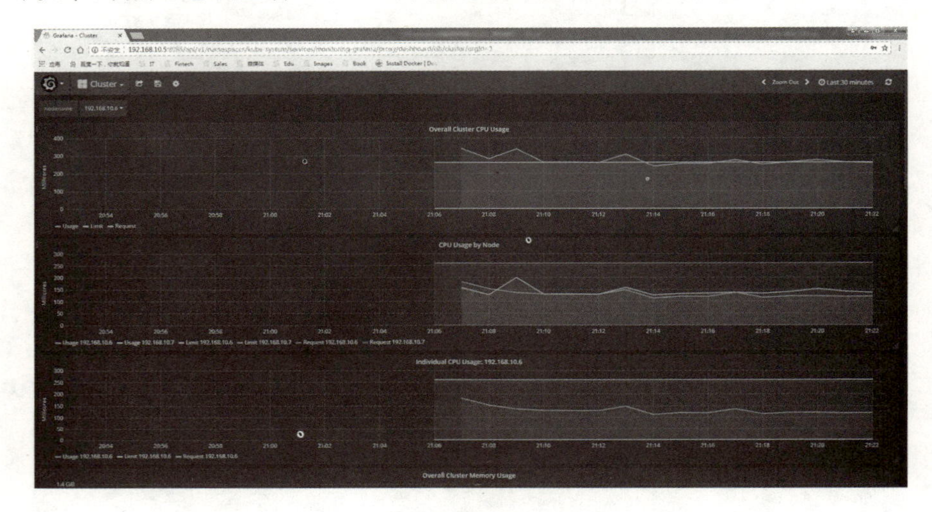

图 13-49

查看 Pod 的资源利用率，如图 13-50 所示。

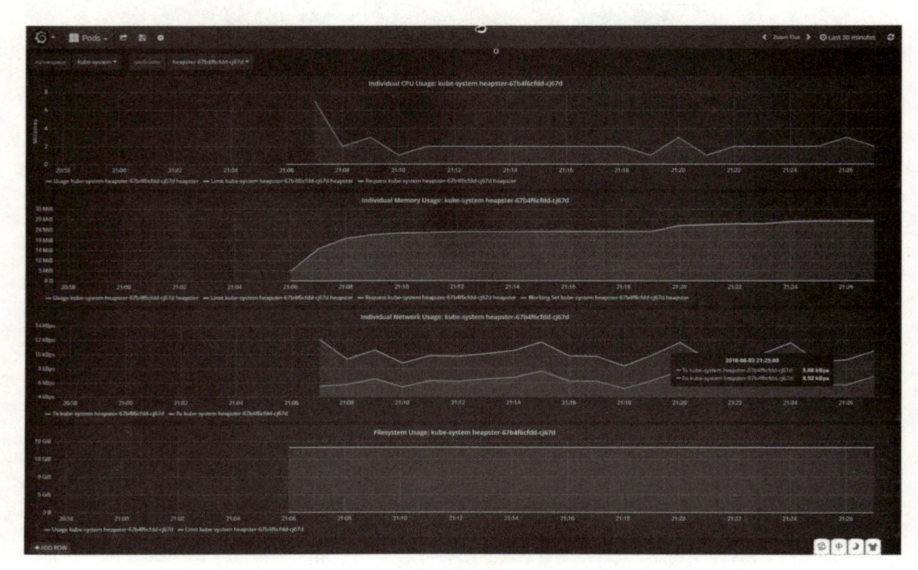

图 13-50

13.5　Kubernetes集群日志管理与应用

13.5.1　日志系统方案介绍

日志是 Kubernetes 集群中非常重要的一项，因为从日志中可以获取 Kubernetes 集群中很多非常有价值的信息，如运行一些程序时产生的日志、一些业务逻辑等，获取后可以进行收集、分析和展示。

Kubernetes 官方提供的日志收集解决方案是 fluentd-elasticsearch，GitHub 上的网址是 https://github.com/kubernetes/kubernetes/tree/master/cluster/addons/fluentd-elasticsearch。这种解决方案是非常好的，只不过只能默认收集每个 Pod 中控制台输出的日志。

在实际生产环境中，通常都是将日志落地到某一个具体的日志目录中，然后按天切割。所以，Kubernetes 官方所提供的日志收集解决方案是无法满足企业需求的。而且，在没有使用 Kubernetes 集群时，已经有了完善的日志分析系统。所以，可以直接使用现有的日志分析系统将应用日志进行收集，最后汇总到已有的日志系统中心进行展示和分析。

本小节所讲解的是一套 ELK 日志开源解决方案，如图 13-51 所示。最左边是 Kubernetes 的 Node，使用 Filebeat 进行日志收集，然后传输到 Logstash 中，最后在 Kibana 中展示。

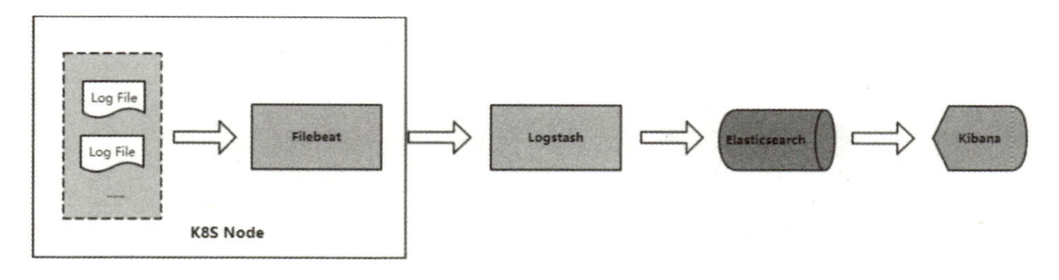

图 13-51

图 13-51 中共涉及了四个组件，其作用如下。

- Filebeat：日志采集工具，从指定的日志中进行增量式的采集，可以将日志传输到 Elasticsearch，也可以直接传输到 Logstash。
- Logstash：数据处理引擎，支持从各种数据源收集数据，并对数据进行过滤、分析等操作。
- Elasticsearch：分布式搜索引擎，用于全文检索，也可以形成一个集群，从而实现高可用。
- Kibana：可视化平台，能够搜索、展示存储在 ES 中的索引数据，可以很方便地以图表、表格、地图的形式进行展示。

上述组件是由 Elastic 公司进行开源的。打开该公司网站（https://www.elastic.colcnl），单击“产品”菜单项，就可以看到该公司所开发的产品，如图 13-52 所示。

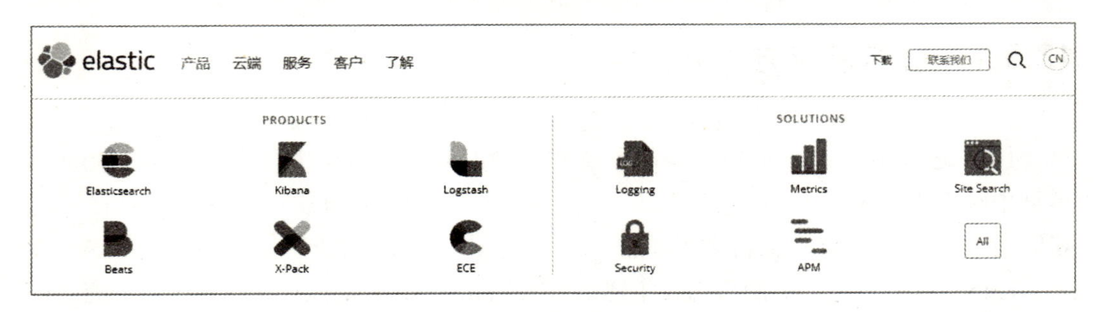

图 13-52

像 Elasticsearch、Kibana、Logstash 和 Beats 都会用到，而我们用的是 Beats 中的 Filebeat。Beats 是一系列产品的统称，单击 Beats 菜单项，可以查看都有哪些产品，如图 13-53 所示。

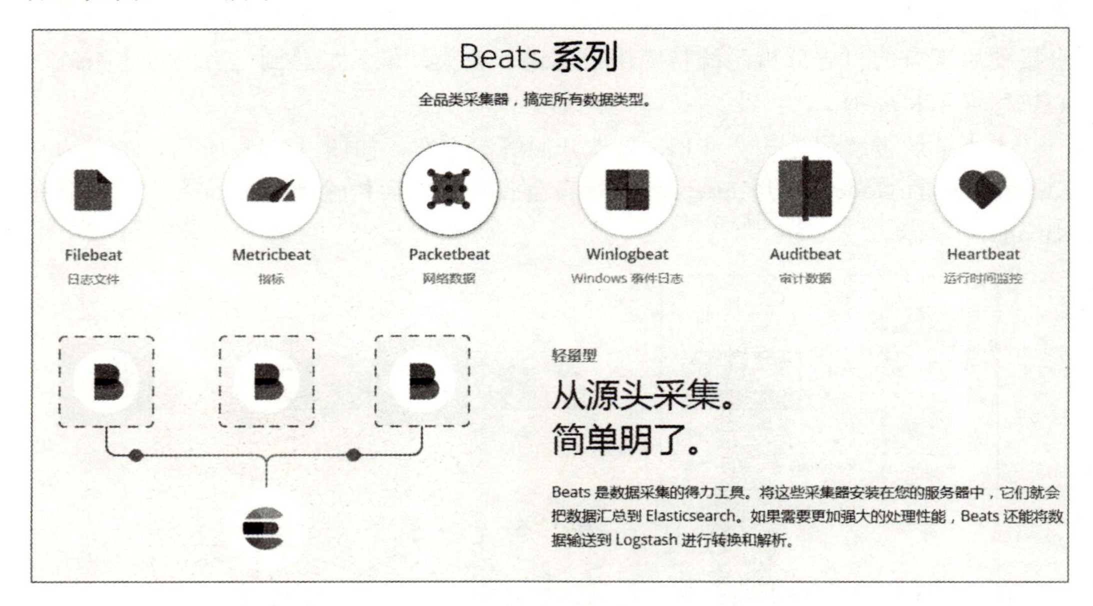

图 13-53

本小节所讲解的都是单点的工具。FileBeat+ELK 是一个可伸缩的架构。Elasticsearch 可以分布式部署多台，Logstash 也可以部署多台。Logstash 本身是过滤数据、分析数据的，当企业生产环境中节点非常多时，单个 Logstash 肯定处理不完数据，可以通过写多个 Logstash 进行分散，当然也可以增加缓存，如使用 Redis 和 Kafka 平缓网络的传输，这就是一个整体可伸缩的解决方案。

13.5.2 部署 ELK Stack

ELK 是基于 Java 环境开发的，所以需要先部署 Java 环境，可以使用二进制文件或 yum 命令安装 Openjdk。

新建一台虚拟机，将 Elasticsearch、Logstash 和 Kibana 二进制文件上传到该虚拟机上，命令如下。本小节全部采用最新的软件包版本进行演示。

```
[root@elastic ~]# cd /home/elk
[root@elastic elk]# ls
elasticsearch-6.2.4.tar.gz  kibana-6.2.4-linux-x86_64.tar.gz
logstash-6.2.4.tar.gz
```

上传完毕后分别解压上述三个软件包，命令如下。

```
[root@elastic elk]# tar xf elasticsearch-6.2.4.tar.gz
[root@elastic elk]# tar xf kibana-6.2.4-linux-x86_64.tar.gz
[root@elastic elk]# tar xf logstash-6.2.4.tar.gz
[root@elastic elk]# ls
elasticsearch-6.2.4  elasticsearch-6.2.4.tar.gz
kibana-6.2.4-linux-x86_64  kibana-6.2.4-linux-x86_64.tar.gz
logstash-6.2.4  logstash-6.2.4.tar.gz
```

全部解压完毕后，进行启动。启动顺序是：先启动 Elasticsearch，再启动 Kibana，最后启动 Logstash。

本地使用 Elasticsearch 无需进行任何配置，如果宿主机之外的其他节点进行访问，则需设置 config/elasticsearch.yml 监听地址，如图 13-54 所示。

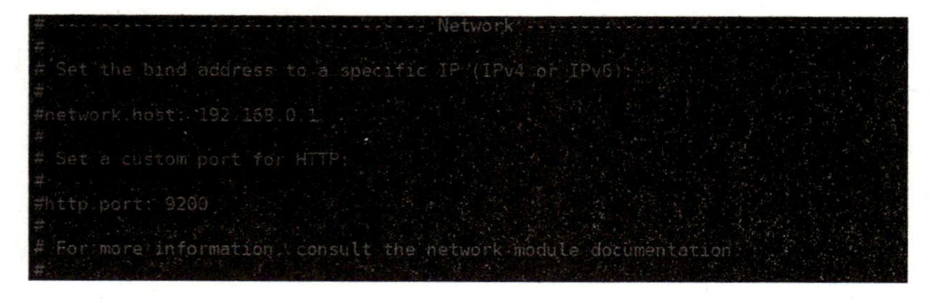

图 13-54

启动 Elasticsearch 服务，命令如下。

```
[root@elastic elasticsearch-6.2.4]# cd bin/
[root@elastic bin]# ./elasticsearch -d
```

启动后查看是否有日志输出，输出日志显示报错，信息是不能以 root 身份启动 Elasticsearch 服务的，如图 13-55 所示。所以需要创建一个普通用户 elk，命令如下。

```
[2018-06-05T22:32:26,955][ERROR][o.e.b.Bootstrap          ] Exception
java.lang.RuntimeException: can not run elasticsearch as root
        at org.elasticsearch.bootstrap.Bootstrap.initializeNatives(Bootstrap.java:105) ~[elasticsearch-6.2.4.jar:6.2.4]
        at org.elasticsearch.bootstrap.Bootstrap.setup(Bootstrap.java:172) ~[elasticsearch-6.2.4.jar:6.2.4]
        at org.elasticsearch.bootstrap.Bootstrap.init(Bootstrap.java:323) [elasticsearch-6.2.4.jar:6.2.4]
        at org.elasticsearch.bootstrap.Elasticsearch.init(Elasticsearch.java:121) [elasticsearch-6.2.4.jar:6.2.4]
        at org.elasticsearch.bootstrap.Elasticsearch.execute(Elasticsearch.java:112) [elasticsearch-6.2.4.jar:6.2.4]
        at org.elasticsearch.cli.EnvironmentAwareCommand.execute(EnvironmentAwareCommand.java:86) [elasticsearch-6.2.4.jar:6.2.4]
        at org.elasticsearch.cli.Command.mainWithoutErrorHandling(Command.java:124) [elasticsearch-cli-6.2.4.jar:6.2.4]
        at org.elasticsearch.cli.Command.main(Command.java:90) [elasticsearch-cli-6.2.4.jar:6.2.4]
```

图 13-55

```
[root@elastic bin]# useradd elk
[root@elastic bin]# id elk
uid=1000(elk) gid=1000(elk) groups=1000(elk)
```

elk 用户创建完毕后，将 ELK 目录下的所有文件和目录的所属的主和所属的组全部更改为 elk，命令如下。

```
[root@elastic bin]# chown -R elk.elk /root/ELK/
[root@elastic bin]# ls -l /root/ELK/
total 255504
drwxr-xr-x  8 elk elk       143 Apr 13 04:39 elasticsearch-6.2.4
-rw-r--r--  1 elk elk  29056810 Jun  5 22:07 elasticsearch-6.2.4.tar.gz
drwxrwxr-x 12 elk elk       232 Apr 13 04:57 kibana-6.2.4-linux-x86_64
-rw-r--r--  1 elk elk  85348919 Jun  5 22:08
kibana-6.2.4-linux-x86_64.tar.gz
drwxr-xr-x 11 elk elk       237 Jun  5 22:15 logstash-6.2.4
-rw-r--r--  1 elk elk 147224408 Jun  5 22:09 logstash-6.2.4.tar.gz
```

进入到普通用户 elk 下进行相关操作，命令如下。

```
[root@elastic bin]# su - elk
Last login: Tue Jun  5 22:40:50 CST 2018 on pts/1
[elk@elastic ~]$ ls
elasticsearch-6.2.4  kibana-6.2.4-linux-x86_64  logstash-6.2.4
```

进入 elasticsearch-6.2.4/bin 目录，启动 Elasticsearch，命令如下。

```
[elk@elastic ~]$ cd elasticsearch-6.2.4/bin/
[elk@elastic bin]$ ./elasticsearch -d
```

启动完成后查看日志是否报错，如图 13-56 所示，显示没有报错信息。

```
[elk@elastic bin]$ tail ../logs/elasticsearch.log -f
[2018-06-05T22:47:11,465][INFO ][o.e.n.Node               ] initialized
[2018-06-05T22:47:11,466][INFO ][o.e.n.Node               ] [V3CrHH-] starting ...
[2018-06-05T22:47:12,450][INFO ][o.e.t.TransportService   ] [V3CrHH-] publish_address {127.0.0.1:9300}, bound_addresses {[::1]:9300}, {127.0.0.1:9300}
[2018-06-05T22:47:12,517][WARN ][o.e.b.BootstrapChecks    ] [V3CrHH-] max file descriptors [4096] for elasticsearch process is too low, increase to at least [65536]
[2018-06-05T22:47:12,517][WARN ][o.e.b.BootstrapChecks    ] [V3CrHH-] max virtual memory areas vm.max_map_count [65530] is too low, increase to at least [262144]
[2018-06-05T22:47:15,971][INFO ][o.e.c.s.MasterService    ] [V3CrHH-] zen-disco-elected-as-master ([0] nodes joined), reason: new_master {V3CrHH-}{V3CrHH-rSLafifRPf5vXeA}{w8b-CY2FTnOQ2id_6XzXkg}{127.0.0.1}{127.0.0.1:9300}
[2018-06-05T22:47:15,988][INFO ][o.e.c.s.ClusterApplierService] [V3CrHH-] new_master {V3CrHH-}{V3CrHH-rSLafifRPf5vXeA}{w8b-CY2FTnOQ2id_6XzXkg}{127.0.0.1}{127.0.0.1:9300}, reason: apply cluster state (from master [master {V3CrHH-}{V3CrHH-rSLafifRPf5vXeA}{w8b-CY2FTnOQ2id_6XzXkg}{127.0.0.1}{127.0.0.1:9300} committed version [1] source [zen-disco-elected-as-master ([0] nodes joined)]])
[2018-06-05T22:47:16,107][INFO ][o.e.g.GatewayService     ] [V3CrHH-] recovered [0] indices into cluster_state
[2018-06-05T22:47:16,168][INFO ][o.e.h.n.Netty4HttpServerTransport] [V3CrHH-] publish_address {127.0.0.1:9200}, bound_addresses {[::1]:9200}, {127.0.0.1:9200}
[2018-06-05T22:47:16,168][INFO ][o.e.n.Node               ] [V3CrHH-] started
```

图 13-56

启动成功后监听的是本地的 9200 端口，如图 13-57 所示。

```
[elk@elastic bin]$ netstat -lantp | grep 9200
(Not all processes could be identified, non-owned process info
 will not be shown, you would have to be root to see it all.)
tcp6       0      0 127.0.0.1:9200          :::*                    LISTEN      18248/java
tcp6       0      0 ::1:9200                :::*                    LISTEN      18248/java
[elk@elastic bin]$
```

图 13-57

进入 Kibana 目录，修改 config 目录下的 kibana.yaml 文件，打开 Port、Host 和 URL，如图 13-58 所示。

```
# Kibana is served by a back end server. This setting specifies the port to use.
server.port: 5601

# Specifies the address to which the Kibana server will bind. IP addresses and host names are both valid values.
# The default is 'localhost', which usually means remote machines will not be able to connect.
# To allow connections from remote users, set this parameter to a non-loopback address.
server.host: "0.0.0.0"

# Enables you to specify a path to mount Kibana at if you are running behind a proxy. This only affects
# the URLs generated by Kibana, your proxy is expected to remove the basePath value before forwarding requests
# to Kibana. This setting cannot end in a slash.
#server.basePath: ""

# The maximum payload size in bytes for incoming server requests.
#server.maxPayloadBytes: 1048576

# The Kibana server's name.  This is used for display purposes.
#server.name: "your-hostname"

# The URL of the Elasticsearch instance to use for all your queries.
elasticsearch.url: "http://localhost:9200"
```

图 13-58

设置完成后启动 Kibana，命令如下。

```
[elk@elastic kibana-6.2.4-linux-x86_64]$ cd bin/
[elk@elastic bin]$ nohup ./kibana &>../kibana.log &
[1] 18332
```

查看日志信息是否报错，如图 13-59 所示。

```
[elk@elastic bin]$ tail ../kibana.log -f
nohup: ignoring input
{"type":"log","@timestamp":"2018-06-05T14:57:57Z","tags":["status","plugin:kibana@6.2.4","info"],"pid":18332,"state":"green","message":"Status changed from uninitialized
to green - Ready","prevState":"uninitialized","prevMsg":"uninitialized"}
{"type":"log","@timestamp":"2018-06-05T14:57:57Z","tags":["status","plugin:elasticsearch@6.2.4","info"],"pid":18332,"state":"yellow","message":"Status changed from uninit
ialized to yellow - Waiting for Elasticsearch","prevState":"uninitialized","prevMsg":"uninitialized"}
{"type":"log","@timestamp":"2018-06-05T14:57:58Z","tags":["status","plugin:timelion@6.2.4","info"],"pid":18332,"state":"green","message":"Status changed from uninitialize
d to green - Ready","prevState":"uninitialized","prevMsg":"uninitialized"}
{"type":"log","@timestamp":"2018-06-05T14:57:58Z","tags":["status","plugin:console@6.2.4","info"],"pid":18332,"state":"green","message":"Status changed from uninitialized
 to green - Ready","prevState":"uninitialized","prevMsg":"uninitialized"}
{"type":"log","@timestamp":"2018-06-05T14:57:58Z","tags":["status","plugin:metrics@6.2.4","info"],"pid":18332,"state":"green","message":"Status changed from uninitialized
 to green - Ready","prevState":"uninitialized","prevMsg":"uninitialized"}
{"type":"log","@timestamp":"2018-06-05T14:57:59Z","tags":["listening","info"],"pid":18332,"message":"Server running at http://0.0.0.0:5601"}
{"type":"log","@timestamp":"2018-06-05T14:58:02Z","tags":["status","plugin:elasticsearch@6.2.4","info"],"pid":18332,"state":"green","message":"Status changed from yellow
to green - Ready","prevState":"yellow","prevMsg":"Waiting for Elasticsearch"}
```

图 13-59

没有报错信息，查看 Kibana 的监听端口 5601 是否存在，如图 13-60 所示。

```
[elk@elastic bin]$ netstat -antlp|grep 5601
(Not all processes could be identified, non-owned process info
 will not be shown, you would have to be root to see it all.)
tcp        0      0 0.0.0.0:5601            0.0.0.0:*               LISTEN      18332/./../node/bin
[elk@elastic bin]$
```

图 13-60

上述检查都是正常的，此时可以在浏览器的地址栏中输入当前节点的 IP 地址和端口号进行访问，如图 13-61 所示，显示的是 Kibana 的图形界面。

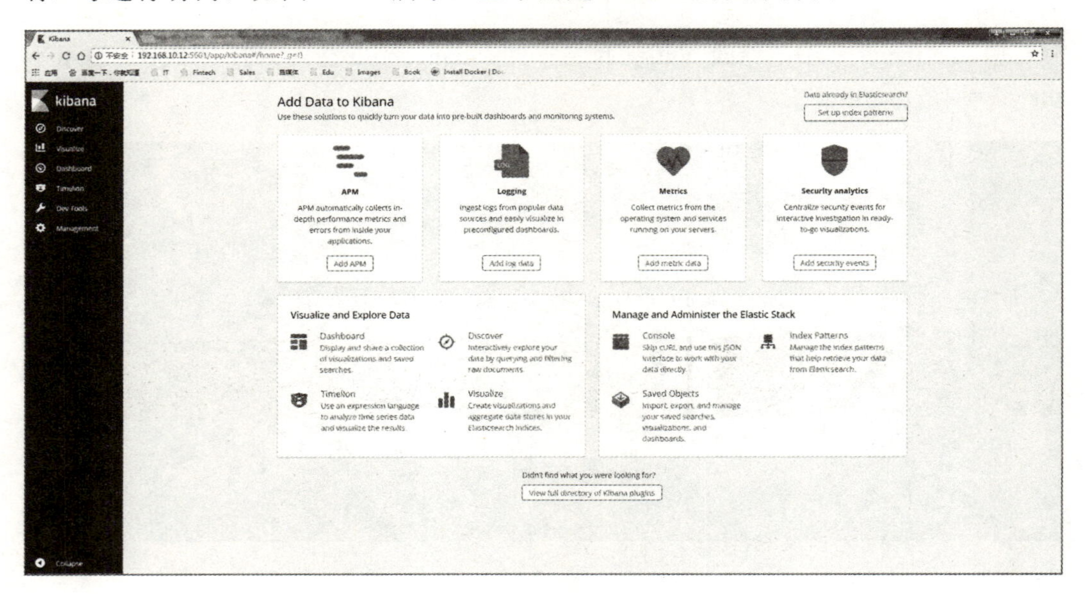

图 13-61

最后部署 Logstash，它用于接收从每个节点上收集的日志信息。Logstash 中过滤的规则需要进行自定义，只要在 config 目录下创建一个 logstash.conf 文件即可，命令如下。

```
[elk@elastic logstash-6.2.4]$ vim config/logstash.conf
input {
 beats {
   port => 5044
  }
}

output {
 if "nginx" in [tags] {
   elasticsearch {
     hosts => "localhost:9200"
     index => "nginx-access-%{+YYYY.MM.dd}"
   }
```

```
  }

  if "tomcat" in [tags] {
    elasticsearch {
      hosts => "localhost:9200"
      index => "tomcat-catalina-%{+YYYY.MM.dd}"
    }
  }
  stdout { codec => rubydebug }
}
```

进入 bin 目录，启动 Logstash，命令如下。

```
[elk@elastic logstash-6.2.4]$ cd bin/
[elk@elastic bin]$ ./logstash -f ../config/logstash.conf
```

Logstash 启动成功后，会显示光标信息，如图 13-62 所示。

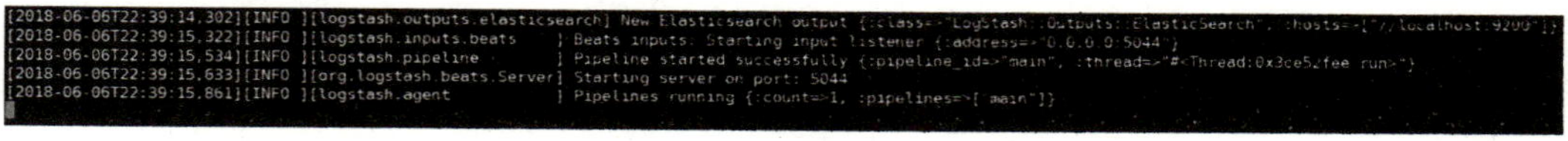

图 13-62

13.5.3 部署 Filebeat 日志收集工具

将 Filebeat 部署到 Kubernetes 集群中，将准备好的 filebeat-to-logstash.yaml 文件上传到 Master 节点中，命令如下。

```
[root@master ~]# mkdir elk
[root@master ~]# cd elk/
[root@master elk]# rz -E
rz waiting to receive.
[root@master elk]# ls
filebeat-to-logstash.yaml
```

创建 Filebeat 资源，命令如下。

```
[root@master elk]# kubectl create -f filebeat-to-logstash.yaml
configmap "filebeat-config" created
daemonset.extensions "filebeat" created
```

查看 filebeat-to-logstash 的运行状态，如图 13-63 所示，显示状态为 Running。

```
[root@master elk]# kubectl get all
NAME                                        READY   STATUS    RESTARTS   AGE
pod/busybox                                 1/1     Running   5          3d
pod/filebeat-9vmc9                          1/1     Running   0          4m
pod/filebeat-n92r6                          1/1     Running   0          4m
pod/nginx-65899c769f-g47bs                  1/1     Running   1          3d
pod/nginx-65899c769f-lt9cc                  1/1     Running   1          3d
pod/nginx-65899c769f-vbhmx                  1/1     Running   1          3d
pod/nginx-deployment-6b7b4d57b4-68qhg       1/1     Running   0          3d
pod/nginx-deployment-6b7b4d57b4-9cwfh       1/1     Running   0          3d
pod/nginx-deployment-6b7b4d57b4-lzlft       1/1     Running   0          3d

NAME                       TYPE        CLUSTER-IP     EXTERNAL-IP   PORT(S)           AGE
service/kubernetes         ClusterIP   10.10.10.1     <none>        443/TCP           4d
service/my-service         ClusterIP   10.10.10.171   <none>        80/TCP,443/TCP    3d
service/nginx-service      NodePort    10.10.10.183   <none>        80:35385/TCP      3d

NAME                      DESIRED   CURRENT   READY   UP-TO-DATE   AVAILABLE   NODE SELECTOR   AGE
daemonset.apps/filebeat   2         2         2       2            2           <none>          4m

NAME                                DESIRED   CURRENT   UP-TO-DATE   AVAILABLE   AGE
deployment.apps/nginx               3         3         3            3           3d
deployment.apps/nginx-deployment    3         3         3            3           3d

NAME                                          DESIRED   CURRENT   READY   AGE
replicaset.apps/nginx-65899c769f              3         3         3       3d
replicaset.apps/nginx-deployment-6b7b4d57b4   3         3         3       3d
[root@master elk]#
```

图 13-63